INTERNATIONAL UNION OF CRYSTALLOGRAPHY
BOOK SERIES

T0202220

IUCr Monographs on Crystallography

IUCr Texts on Crystallography

Polymorphism in Molecular Crystals

Second Edition

Joel Bernstein

Department of Chemistry
Ben-Gurion University of the Negev
New York University Abu Dhabi and Shanghai

OXFORD
UNIVERSITY PRESS

OXFORD

UNIVERSITY PRESS

Great Clarendon Street, Oxford, OX2 6DP,
United Kingdom

Oxford University Press is a department of the University of Oxford.
It furthers the University's objective of excellence in research, scholarship,
and education by publishing worldwide. Oxford is a registered trade mark of
Oxford University Press in the UK and in certain other countries

First Edition published in 2002
First published in paperback 2008
Second Edition published in 2020

Second Edition published in paperback 2023

Published in the United States of America by Oxford University Press
198 Madison Avenue, New York, NY 10016, United States of America

British Library Cataloguing in Publication Data
Data available

Library of Congress Cataloging in Publication Data
Data available

ISBN 978–0–19–965544–1 (Hbk.)
ISBN 978–0–19–887735–6 (Pbk.)

DOI: 10.1093/oso/ 9780199655441.001.0001

Printed and bound by
CPI Group (UK) Ltd, Croydon, CR0 4YY

To Tzipi

AT FIRST SIGHT

Once made, this stolid mauve
powder would seem forever;

but people intent on repro-
duction fire up pots next door

or across the sea, and out
of the odd one crystallizes

another, the same, but for
a tell-tale (to X-rays) part

that twists a tad; in a tango
of attractions and absences

molecules nestle in a variant
pattern. Neat, but from here on,

the first will not be made. So
it seems; the ur-makers once

patient hands grow limp—
has desire fled? In all flasks

the second precipitates. Who,
oh who, is to blame? Yes, lay it

to the other coming—as if
seed crystals flew the world.

But the first is the accident,
a small well in a chanced

landscape, a nicked knife edge,
the one parcel of phase space

never to be sampled again,
the vanishing polymorph...you.

                              Roald Hoffmann

Foreword

In January 2019 we were deeply saddened to suddenly lose Professor Joel Bernstein, a colleague, collaborator, parent, husband, grandfather, and a highly accomplished individual. Joel obtained his BA degree in chemistry from Cornell University in 1962 and an MSc in physical chemistry from Yale University in 1964. He obtained his PhD in 1967 from Yale, working with Basil G. Anex on spectroscopy of organic compounds. Following two postdoctoral positions, researching in X-ray crystallography with Kenneth Trueblood at UCLA and in organic solid-state chemistry with Gerhardt Schmidt at the Weizmann Institute of Science, he joined the Department of Chemistry of the newly established Ben-Gurion University of the Negev in Israel, where he later became the Barry and Carol Kaye Professor of Applied Science. Joel was instrumental in setting up the chemistry department at Ben-Gurion University. Later on, as a Distinguished Global Professor, he contributed to building the chemistry program and taught general chemistry at New York University's Global Network, particularly in their portal campuses in Abu Dhabi and Shanghai.

Joel was an incredible speaker, a dedicated mentor, a supportive colleague, a seasoned wine connoisseur, and a true friend to many of us in the solid-state research community. Those of us who had the privilege to be part of his life professionally will always cherish his immense contributions to the solid-state research, and will remember him the way he really was—a great, yet subtly humble person, a true gentle giant. Joel came with a combination of generosity and kindness that is uncommon with professionals of his caliber. He was able to talk to an undergraduate student or to a junior colleague with the same level of patience, appreciation and respect as he would with some of the most distinguished scientists or Nobel laureates. Joel taught his fellow faculty to approach problems with patience, and with a warm smile he was always ready to provide an endless support and encouragement to his colleagues and collaborators. He had an incredible and rare ability of narration with an extraordinary charisma that would make even a layperson want to hear and learn more about science. He had the ability to comfort others when they needed. He gained a respect from others by true interest, appreciation, sincerity and kindness. He had the ability to convey science with a great, sincere passion, and in addition to the polymorphism, a topic to which he dedicated most of his career, he will be remembered for his interest in popular science, particularly his research on the shroud of Turin, the fuzzy logic, and one his favorite subjects, the disappearing polymorphs. Joel's colleagues from his student days will always remember him for his generosity and readiness to help, his students will remember his inspirational and moving lectures, and those of us who had the privilege to collaborate with him will remember him for the rigor and depth of his scientific work. He believed in the power of knowledge as a universal human value that should be accessible to anyone, regardless of their religion, ethnicity, creed or color. He selflessly shared his knowledge with a great passion, and truly inspired so many of us.

Joel was inspired by the occurrence of *polymorphism*—the ability of solids to crystallize with different crystal structures and hence have different properties—and recognized its potential and the immense impact it could have for both science and practice very early on in his career, during one of his post-doctoral terms. Being inspired by the early works of Paul Heinrich von Groth, and particularly by the later works of Walter McCrone, he decided to pursue this subject as his life-long career. Over the years, he gave a number of lectures on the conceptual, terminological and scientific aspects or specific cases related to the topic, in what he used to refer to as his "adventures in polymorphland". Being the most comprehensive review on this immensely broad chemistry field to date, since the publishing of the first edition in 2002, Joel's book *"Polymoprhism in Molecular Crystals"* has quickly become the prime text on this important topic that has been appreciated as a precious source and comprehensive yet concise read equally by researchers and students. His book is very frequently cited as a reliable and relevant source of information on polymorphism in both academic and patent literature.

As with any active field of research, being a fundamental solid-state phenomenon, our understanding of polymorphism has continued to evolve and many new examples have been added to the related literature over the past ten years. This second edition of Bernstein's *"Polymorphism in Molecular Crystals"*, on which he started working tirelessly and passionately several years after the publication of the first edition to the day of his sudden passing, builds on the first edition by subliming the past 15 years of the author's research of the related literature. In this edition, as well as in the previous one, he shares his experience as one of the most eminent scholars in the solid-state chemistry, but also his expertise as an expert witness in multiple patent litigation cases in the court. The examples presented were carefully selected to illustrate a multitude of aspects of the polymorphism in molecular crystals, in communication with his colleagues, collaborators and professional contacts, his and others' experience with legal cases related to the topic, and from his own original research over the last decade. In selecting the examples, he carefully maintained the balance between examples from the older literature, and those from the most recent literature, particularly with respect to the computational techniques that have been developed in the meantime to explain or to predict the occurrence of polymorphism.

In over 600 pages of text and citing over 2060 references, the author delves in the basic aspects of polymorphism, such as the related thermodynamics, the nucleation processes, difference in structure between polymorphs, and the analytical methods that are commonly used to distinguish between polymorphs. With great passion and in considerable depth, he describes specific topics of this multifaceted subject, spanning a thorough historical overview of the subject, the different definitions and thermodynamics principles, and applications in pharmaceutical industry, dye and pigment technology, and electronics. In the book he is particularly elaborate on the topic of disappearing polymorphs, definitional issues related to polymorphism, and the conclusiveness of the presence of polymorphism by using various analytical techniques. This second edition of the book brings a much more elaborate reading on the computational aspects of predicting and detecting polymorphism, on crystallization as determining step in the evolution of polymorphism, and also provides an updated analytical techniques section. As some of the most important applications of polymorphs, the author has also included timely and thorough overview of the emerging directions and practical aspects in polymorph research—the polymorphism in dyes and pigments, high-energy materials that are becoming increasingly researched due to the potential global safety threats, and provides a set of selected examples of the relevance of polymorphism to the patent literature, particularly such that are related to the pharmaceutical industry. The narrative is comprehensive, with reference to the relevant research contributions from some of the leading research groups in the field, as well as to particular illustrative examples that are not commonly accessible to the wider readership. The writing style is accessible, while also being meticulous, thorough, and rigorous. Joel firmly believed in the value of a long-lasting, good, timeless science, such that has already passed or will most certainly pass the test of time, and which is based on facts, not on fiction—as he would often discuss in his lectures or articles on the topic—and he deeply appreciated quality over quantity, and rigor over hype in scientific research. The structure and contents of this second edition of the book reflects that approach. Together with Joel's own original publications and detailed reviews on polymorphism, the book will remain a testament of his valuable contribution to the chemistry research and his legacy to the solid-state chemistry community.

I would like to take this opportunity to thank all parties who kindly provided materials and information that was needed to complete the proofreading of this book posthumously.

Panče Naumov
Abu Dhabi, March 13, 2020

Preface to the second edition

Allow me to discourse on polymorphism
A subject oft greeted with cynicism
It's about multiple crystal forms
Whose behavior not always conforms
To every sacred scientific or linguistic formalism.

Nearly 20 years have elapsed since the publication of the first edition of this work. The exponential increase in the activity and interest in the subject far exceeded our expectations and our imaginations when the first edition was sent to the printer in July of 2001. The varieties of reasons we attribute to these developments are documented throughout the current volume.

We noted in the Preface to the first edition that the book was intended to provide a starting point for individuals encountering the phenomenon of polymorphism in molecular crystals for the first time. Therefore, considerable attention was paid to fundamentals. That intention has not changed, nor have most of the fundamentals (e.g., thermodynamics, structural principles, basic analytical techniques, some still classic systems, etc.) so those sections have very much been left intact. The change in the title of Chapter 3 from "Controlling the polymorphic form obtained" to "Exploring the crystal form landscape" reflects a change in the investigative nature of the search for solid forms and the increasing emphasis on discovering new forms.

In pursuit of that search there has been considerable progress in our understanding of the phenomenon of polymorphism, the experimental procedures for the exploration of the solid form landscape, the analytical techniques for identifying and characterizing polymorphs, the utilization of polymorphism to tune and improve the properties of materials, the application of polymorphism to the study of structure–property relations, and the impact of polymorphism in a wide variety of industrial/commercial applications, in particular those involving the development and protection of intellectual property. We attempt here to provide examples, indeed representative examples, of many of these very impressive developments, but the literature contains much more than can be contained in a single volume, and in preparing this work we were faced with a quandary of the rich. The choice of those examples to include is a difficult one, but the choice of what not to include is even more difficult. The first edition of this title contained approximately 1500 references, and this one contains at least 1000 more. Given the growing size of the community of practitioners in this field it is inevitable that some, indeed many, works that warrant inclusion or citation have not been included or cited due to lack of space. This is a subjective judgment; any slights to the increasing numbers of members of our very talented and very productive scientific community are unintended.

A number of the topics in the first edition have been the subjects of excellent review papers. As was our intent in the first edition to provide as useful a literature resource in one volume as possible to bring the reader up to date, we have tried to cite as many of those. In many cases, especially those for which the time lag between the publication date of the review and this edition was short, we briefly summarize the contents of the review and invite the reader to seek out the full review. As in the previous edition these reviews are meant to serve as literary milestones for future citations.

Any work of this sort cannot be completed with out the aid and support of many, and I wish to acknowledge them with deep-felt thanks. As before, the accomplished experts in specific areas of polymorphism responded with enthusiasm to my requests for information and key publications. But clearly first among them is Jan-Olav Henck from Bayer who contributed his vast experience and deep insight in reviewing and critiquing every chapter. Martin Schmidt from the University of Frankfurt lent his encyclopedic knowledge of pigments and dyes; Colin Pulham from the University of Edinburgh provided the latest developments in high pressure studies; Thomas Klapötke at the Ludwig Maximillian University of Munich contributed updated information on high energy materials.

Dario Braga, Fabrizia Grepioni, and Lucia Maini were consummate hosts during a three-month sojourn at the Institute for Advanced Study at the University of Bologna for much of the final writing, and the living quarters at the Collegio di Espagna in Bologna provided an ideal environment for contemplation and writing. The extraordinary library facilities of New York University were accessible from virtually anywhere in the world thanks to the efforts of Mike Ward in New York and his colleagues at other NYU locations; this in addition to Mike's constant sage scientific advice and support.

Others were crucial in our own work during the interim that is included in this volume. Among them special thanks and gratitude must go to Aurora Cruz-Cabeza from the University of Manchester and Ulrich Griesser from the University of Innsbruck. For education, guidance, and review on the preparation of Chapter 10 on the connection between polymorphs and patents I am deeply indebted to Howard Levine and Jill MacAlpine of Finnegan Henderson who combine scientific expertise with extensive experience in chemical and pharmaceutical patent issues.

There are always some specific visual inclusions that can be obtained only from unique sources. Two of those included here, and supplied with enthusiasm, are the Table on the solid forms of aripiprazole prepared by Anna Kowal, presently at Glaxo, and the excerpt on screening solvents from Walter McCrone's microscopy course manual provided by Gary Laughlin of the McCrone Research Institute.

Finally, my wife Tzipi has enthusiastically encouraged and devotedly supported this enterprise virtually from the day we met seventeen years ago; without that encouragement, support and wise counsel, which often left her in isolation, it could not have been completed. Dedicating this edition to her is only one symbolic way of expressing my deepest appreciation and gratitude.

Preface to the first edition

Sometime in the middle 1960s during an evening stint in the laboratory a fellow graduate student and I were struggling to determine the orientation of a known crystal on a quarter circle manual X-ray diffractometer. When things didn't turn out as expected he raised the possibility that it might be a polymorph (it wasn't). However, I recall being fascinated by the whole idea of a single molecule crystallizing in different structures, and the consequences of such a phenomenon. That fascination has not waned over the intervening years.

In the same period polymorphism has become a much more widely recognized and observed phenomenon, with both fundamental and commercial ramifications. The literature has grown enormously, albeit scattered in a variety of primary sources. In view of the growing interest in the subject there appeared to be the need for a monograph on the subject. Work in polymorphism and on polymorphs is quite interdisciplinary in nature, and as a result there is no single book that provides an introduction and overview of the subject.

The purpose of this book is to summarize and to bring up to date the current knowledge and understanding of polymorphism in molecular crystals, and to concentrate it in one source. It is meant to serve as a starting point and source book both for those encountering the phenomenon of polymorphism for the first time, and for more seasoned practitioners in any of the disciplines concerned with the organic solid state. It is intended to serve a readership from advanced undergraduate students through to experienced professionals. Much of the information in the book does appear in the open literature; however, because of the increasing commercial importance of the phenomenon, a significant portion of the information (for instance, on industrial applications, patents, or previously restricted distribution) is less accessible, and we have attempted to include both the information from those sources as well as full details of their citations. The intention is that even with the passage of time developments in many of the areas covered in the book can be followed by searching for the citations of the relevant papers cited here.

A work of this type cannot be completed without the help of many other people. This project was initiated during a sabbatical leave (in 1997–98) at the Cambridge Crystallographic Data Centre. My hosts there and in the contiguous Department of Chemistry at Cambridge University put all their resources at my disposal and simply let me go about my business of reading and writing. I am particularly grateful to Olga Kennard for encouraging me to spend that time there and to Frank Allen and his colleagues for making it so collegial and so congenial.

I have been particularly fortunate to have benefited from the assistance of a small army of bright, enthusiastic students who put up with my changing whims and wishes and managed the logistical aspects of organizing the reprints collection, obtaining reprint permissions, checking and completing the details of references, scanning, modifying, and preparing figures, etc.: Megan Fisher, Michal Stark, Avital Furlanger, Margalit Lerner, Noa Zamstein, Shai Allon, and Janice Rubin.

Over the years I have been in touch with countless colleagues—many of whom I have never met—who have willingly, indeed enthusiastically, provided me with preprints, obscure reprints, private documents, observations, and insights on a variety of polymorphic behavior and systems. To all of them I am grateful, and in the course of this work, a number of them provided exceptional assistance which made my task considerably easier and more enjoyable. They deserve special mention here. Peter Erk at BASF spent many hours helping put the connection between polymorphism and colorants into focus. His colleague Martin Schmidt at Clariant provided almost instantaneous responses and faxes to what must have seemed like an endless stream of questions and requests. The chapter on high energy materials probably could not have been written without the help of Charlotte Lowe-Ma of the Ford Motor Company. Following a brief conversation with her at a scientific meeting a courier showed up in my office with a box of historically important documents and personal notes and summaries on polymorphism of high energy materials that were invaluable. Richard Gilardi from the U.S. Naval Research Laboratories provided similar advice and assistance on many of the newer compounds and systems. Stephen Tarling from Birkbeck College availed himself of his time and experiences in a number of patent litigations involving crystal modifications to lead me to the appropriate cases. Michelle O'Brien of the firm of Finnegan, Henderson, Farabow, Garrett and Dunner, Washington, DC managed to get hold of every legal document I requested.

When it came to finding examples, systems, and references among the pharmaceuticals Jan-Olav Henck of Bayer and Ulrich Griesser of the University of Innsbruck were always ready and willing with immediate detailed answers, and faxed reprints if necessary.

Many graduate students and associates in my laboratory carried out the examples taken from our own work described here. I am grateful for their dedication and their contribution to the contents: Ilana Bar, Ehud Goldstein (Chosen), Leah Shahal, Liat Shimoni, Sharona Zamir, Arkady Ellern, Oshrit Navon, and the same Jan-Olav Henck mentioned above.

The exchanges with Roald Hoffmann of Cornell University on disappearing polymorphs in song and story were particularly memorable, and I am grateful for his permission to reprint his poem on the subject as the frontispiece of this tome. As he has been for a couple of generations of chemical crystallographers, Jack Dunitz was a constant inspiration and standard of excellence.

As a postdoctoral fellow myself I was very fortunate to have worked with two inspiring scientists whose scientific integrity and talent for precision in writing and expression have served as models throughout my career: K. N. Trueblood at UCLA and Gerhard Schmidt at the Weizmann Institute. Countless times in the course of preparing this work I found myself asking if they would have passed a sentence, a phrase or a scientific judgment or opinion that had just been written.

I hope they would, but as they also taught me, I alone am responsible for what follows.

My late wife Judy was a source of constant encouragement and support for nearly 35 wonderful years together, especially those during which this book developed and took shape. Dedicating it to her is but a minor recognition of her contribution and the life we shared.

Contents

Abbreviations

Abbreviations for specific high energy compounds are given in Chapter 9.

ADF	Atomic density functions
AFM	Atomic force microscopy
AIE	Aggregation-induced emission
ANDA	Abbreviated new drug application
API	Active pharmaceutical ingredient
BEDTT-TTF	See **ET**
BFDH	Bravais, Friedel, Donnay, Harker crystal morphology, a routine in the CSD Mercury software for determining theoretical crystal habits
CA	chloranil
CCDC	Cambridge Crystallographic Data Centre
COMPACK	A computer program for identifying crystal structure similarity using interatomic distances
Cp*	pentamethylcyclopentadienyl
CP/MAS	Cross polarization/magic angle spinning
CPU	Central processing unit
CSD	Cambridge Structural Database
CSP	Crystal structure prediction
CuPc	copper phthalocyanine
DFT	Density functional theory
DFT-D, DFT-D2, etc.	Density functional theory with various dispersion corrections
DOFlex	Degree of flexibility
DSC	Differential scanning calorimetry
EP	European Pharmacopoeia
esd	Estimated standard deviation
ESR	Electron spin resonance
ET	bisethylenedithio-tetrathiafulvalene
FDA	Food and Drug Administration
GBP	gabapentin
GSK	GlaxoSmithKline
HCB	hexabenzocoronene
HPMC	hydroxypropyl methylcellulose
ICDD	International Centre for Diffraction Data, the depository for powder diffraction data and a provider of software in support of that extensive data base
ILs	Ionic liquids
MfPc	Metal-free phthalocyanine
N-I	Neutral-ionic
NMR	Nuclear magnetic resonance

NQR	Nuclear quadrupole resonance
PC	Personal computer
Pc	phthalocyanine
PDF	Powder Diffraction File of the ICDD
POSA	Person of skill in the art
R-bonds	Rotatable bonds
RHCl	ranitidine hydrochloride
rmsd	Root mean square deviation
ROY	Iconic red–orange–yellow polymorphic system
SC	Supramolecular construct
SCDS	Semi-classical density sums
SIP	Substrate-induced phase
SMATCH	Simultaneous mass and temperature change
SSCI	Solid State Chemical Information, a service and consulting company in West Lafayette, Indiana
STM	Scanning tunneling microscopy
T_c	Critical temperature, below which a material becomes electrically superconducting
TCNQ	tetracyanoquinodimethane
TMAFM	Tapping mode atomic force microscopy
TORMAT	A computer program for the automated structural alignment of molecular conformations
TTF	tetrathiafulvalene
XRPD	X-ray powder diffraction
Z'	number of crystallographically independent molecules in the asymmetric unit
ZnPc	zinc phthalocyanine

1

Introduction and historical background

With the accumulation of data, there is developing a gradual realization of the generality of polymorphic behavior, but to many chemists polymorphism is still a strange and unusual phenomenon.

Buerger and Bloom (1937)

In spite of the fact that different polymorphs of a given compound are, in general, as different in structure and properties as the crystals of two different compounds, most chemists are almost completely unaware of the nature of polymorphism and the potential usefulness of knowledge of this phenomenon in research.

McCrone (1965)

1.1 Introduction

Notwithstanding chemists' occupation and fascination with structure and the connection between structure and properties, in McCrone's view in the nearly three decades following the observation of Buerger and Bloom there had not been any serious change in their awareness of polymorphism, its importance to chemistry, and its potential usefulness. In 2002 I wrote, "More than thirty-five additional years have passed, and that awareness is now increasing." As Figure 1.1 demonstrates, the number of publications, patents, and citations relating to polymorphism has increased exponentially. As analytical methods have become more sophisticated, more precise, and more rapidly carried out, the proliferation of data has revealed differences in structure and behavior that can be attributed to polymorphism. As I demonstrate in a number of instances in subsequent chapters, the increasingly rapid accumulation and archiving of structural data has allowed for the systematic search and retrieval of those data for the purpose of correlating both structural trends and structure with properties. In short, polymorphism in chemistry has moved from a "strange and unusual phenomenon" to one that is a legitimate and important area of research in and of itself that can also be utilized by chemists in unique and efficient ways for the study, understanding, development,

Polymorphism in Molecular Crystals. Second Edition. Joel Bernstein. © Joel Bernstein 2020. Published in 2020 by Oxford University Press. DOI: 10.1093/oso/ 9780199655441.001.0001

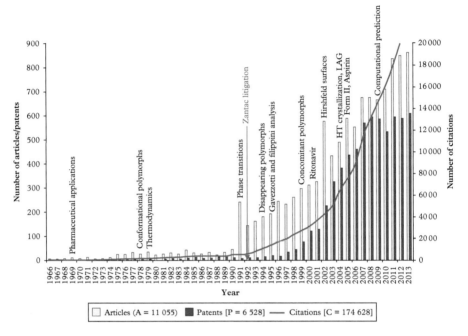

Figure 1.1 *Statistics for number of publications, citations to those, and patents related to polymorphism. Landmarks in the development of polymorphism are indicated and commented on further throughout the text. (Reproduced from Cruz-Cabeza, A., Reutzel-Edens, S., and Bernstein, J. (2015). Facts and fictions about polymorphism.* Chem. Soc. Rev., **44**, *8619–35, 10.1039/c5cs00227c with permission from The Royal Society of Chemistry.)*

and utilization of specific properties of solids and structure–property relations in those solids.

The vertical marker in Figure 1.1 at ca. 1991–1992 with the notation "Zantac litigation" demarks what I believe to be a seminal event in the increasing interest and activity in polymorphism. As detailed in Chapter 10, this patent litigation, which began in 1991, involved the world's largest selling drug (generically raniti-dine hydrochloride) and significant portions of the scientific aspects of that litiga-tion involved many of the aspects of polymorphism with which this volume deals. As I will also show, additional citation statistics testify to the importance of that litigation in the development of the field.

Structural diversity surfaces in almost every facet of nature. Chemistry in gen-eral is no exception, nor in particular is structural chemistry, and crystal polymorphism is one manifestation of that diversity. The emphasis in this treatise is on molecular crystals for a number of reasons. Inorganic compounds and minerals traditionally have been the purview of geologists and inorganic chemists and their innate inter-est in structure–property relationships led naturally to more organization and more awareness of polymorphism than in other pursuits. Monographs such as Wells'

(1984) *Structural Inorganic Chemistry* and Verma and Krishna's (1966) *Polytypism and Polymorphism in Crystals* are typical examples. On the other hand, organic solid-state chemistry is a relatively new discipline (or multidiscipline), founded (or re-founded) in the 1960s by the schools of G. M. J. Schmidt (1971) in Israel and I. C. Paul and D. Y. Curtin (1973, 1975) in Urbana, Illinois, so that information and knowledge of polymorphism in this area is scattered through a wider variety of literature. My aim is to provide within the framework of a single volume an introduction to the fundamental physical principles upon which this polymorphism is based, together with a variety of examples from the literature that demonstrate the importance of understanding polymorphism and, in McCrone's words (1965), the "potential usefulness of knowledge of this phenomenon in research." This work can then be used as a reference and source book for those encountering polymorphism for the first time, those embarking on polymorphism-related research, or those already involved in such endeavors who wish to find additional examples and an entrance to the related literature. The diversity of the field as well as its exponential development in the past few years makes a comprehensive survey prohibitive in terms of space and almost immediately out of date. As a necessary compromise I have attempted to choose examples which are meant to be representative of the phenomena they exhibit, as well as to provide leading references that can be updated with subsequent citations.

1.2 Definitions and nomenclature

1.2.1 Polymorphism

Polymorphism (Greek: *poly* = many, *morph* = form), specifying the diversity of nature, is a term used in many disciplines.[1] According to the Oxford English Dictionary the term first appears in 1656 in relation to the diversity of fashion. In the context of crystallography, the first use is generally credited to Mitscherlich (1823), who recognized different crystal structures of the same compound in a number of arsenate and phosphate salts. The historical development of polymorphism is discussed in Section 1.4.

As with many terms in chemistry, an all-encompassing definition of polymorphism is elusive. The problem has been discussed by McCrone (1965), whose working definition and accompanying caveats are as relevant today as when they were first enunciated. McCrone defines a polymorph as "a solid crystalline phase of a given

---

[1] In an internet search Threlfall (2000) found 1.5 million references to the term, of which 90% refer to video games in which creatures change shape. Ninety percent of the remainder refer to *genetic* polymorphism, which involves minor change of protein or DNA sequences that may lead, for instance, to particular sensitivity to drugs. Of references that refer to crystallographic polymorphism, approximately 90% are devoted to inorganic structures, which are not covered here. The remainder deal with molecular crystals.

compound resulting from the possibility of at least two different arrangements of the molecules of that compound in the solid state."

At first glance this definition seems straightforward. What are some of the complications? For flexible molecules McCrone would include *conformational polymorphs*, wherein the molecule can adopt different conformations in the different crystal structures (Corradini 1973; Panagiotopoulos et al. 1974; Bernstein and Hagler 1978; Bernstein 1987; Cruz-Cabeza and Bernstein 2014). But this is a matter of degree: dynamic isomerism or tautomerism would be excluded, because they involve the formation of different molecules. The "safe" criterion for classification of a system as polymorphic would be if the crystal structures are different but lead to identical liquid and vapor states. For dynamically converting isomers, this criterion invokes a time factor (Dunitz 1995). As with polymorphs, dynamic isomers will melt at different temperatures. However, the composition of the melt will differ. That composition can change with time until equilibrium is reached, however, and the equilibrium composition will be temperature dependent. Using these criteria, a system in which the isomers (or the limit conformers) were rapidly interconverting would be considered a polymorphic one, while a slowly interconverting system would not be characteristic of polymorphic solids. As Dunitz (1995) has pointed out, such a definition would lead to the situation in which a racemate and a conglomerate would be determined to be polymorphic when the rate of interconversion of enantiomers in the melt or in solution is fast, but would be classified as three different compounds when that rate of interconversion is slow. Since no time frame is defined for slow or fast, the borderline is indeed fuzzy. Dunitz has also noted that the distinction has important ramifications when considering the phase rule (see Section 2.2.1), since application of the phase rule requires definition of the number of components. In general, components are "chemically distinct constituents" whose concentrations may be varied independently at the temperature concerned. McCrone (1965) has attempted to clarify the distinction between polymorphism and dynamic isomerism. The latter involves chemically different molecules "more or less readily convertible in the melt state. The basic difference between two polymorphs can occur in the solid state, and the difference between any two polymorphs disappears in the melt state." A number of examples of these phenomena are described in Section 3.6.2. Chemists may certainly differ on precisely what comprises "chemically distinct molecules" and "more or less readily convertible" that can lead to the lack of precision in the definition of polymorphism.

Some additional aspects of the definition deserve mention here. Since polymorphism involves different states of matter with potentially different properties, debates about definitions of the phenomenon have centered alternatively on differences in thermodynamic, structural, or other physical properties. For instance, Buerger and Bloom (1937) cited Goldschmidt's use of "building blocks," "polarization properties," and "thermodynamic environment" to describe the state of the art and understanding of polymorphism at that time:

> …if a member of an isomorphous series is constructed of building blocks whose size and polarization properties lie near the limit which the structure of this series can

accommodate, changes in the thermodynamic environment may cause this limit to be exceeded and a new structure to be developed. This is polymorphism.

On the other hand McCrone's definition appears to have been simplified by Rosenstein and Lamy (1969) as "when a substance can exist in more than one crystalline state it is said to exhibit polymorphism." This simplified definition was apparently adopted by Burger (1983), "If these [solids composed of only one component] can exist in different crystal lattices, then we speak of polymorphism," which unfortunately confuses the concepts of *crystal lattice* and *crystal structure*. Some of these misconceptions have been carried through to more recent publications (Wood 1997).

There has also been an ongoing debate about the use and misuse of the terms allotropy and polymorphism (Jensen 1998). The former was originally introduced by Berzelius (1844) to describe the existence of different crystal structures of elements, as opposed to different structures of compounds. Findlay (1951) opposed the use of two terms for essentially the same phenomenon, and even proposed that polymorphism be abandoned in favor of allotropism as a description for the general phenomenon. The distinction between the two terms was debated by Sharma (1987) and Reinke et al. (1993). Sharma suggested that polymorphs be denoted as "different crystal forms, belonging to the same or different crystal systems, in which the identical units of the same element or the identical units of the same compound or the identical formulas or identical repeating units are packed differently." Reinke et al. invoked the modern language of supramolecular chemistry, by proposing "an extended and modified definition" for polymorphism as "the phenomenon where supermolecular structures with different, well defined physical properties can be formed by chemically uniform species both in the liquid and solid state." This line of thought has apparently come full circle, with Dunitz's (1991, 1996) description of the crystal as the "supermolecule par excellence" and on that basis, "If a crystal is a supermolecule, then polymorphic modifications are superisomers and polymorphism is a kind of superisomerism..."

As with many other concepts in chemistry, in a room full of chemists there is general agreement about the meaning, consequences, and relevance of polymorphism. Although the language of chemistry is constantly developing, McCrone's working definition noted at the beginning of this section appears to have stood the test of time, and is the one that would be recognized and used by most chemists today.

1.2.2 Some additional adjectival polymorphisms: pros and cons

The expanding research activity on polymorphism has spawned an increasing variety of nomenclature to describe presumably specific or unique phenomena. That new nomenclature is not always justified.

There is no reason to restrict polymorphism to a single compound. Why, for instance, can co-crystals not be polymorphic? Surely among molecular complexes (Pfeiffer 1922; Herbstein 2005), lately reincarnated as co-crystals (Almarsson and Zaworotko 2004), there are numerous examples of polymorphs that meet McCrone's criteria, as defined in Section 1.2.1. If the composition varies, then clearly they do not meet the criterion of having identical melts and they are not polymorphs; they are something else. Other hitchhikers on the nomenclature bandwagon should no longer be carried. For instance, *structural polymorphism* (e.g., Pravica et al. 2004; Singhal and Curatolo 2004; Piecha et al. 2008; Budzianowski et al. 2010) is simply redundant and belongs in the etymological trash heap. The same applies to *packing polymorphs* (Braun et al. 2008) and *synthon polymorphism* (Babu et al. 2010). And what criteria must a compound meet in order to be classified as a *pharmaceutical polymorph* (Nangia 2008)? Must it be a pharmaceutically active ingredient? Are excipients included? What if the compound is taken off the pharmaceutical market for some reason? Does it no longer qualify as a pharmaceutical polymorph? To quote Stahly (2007):[2]

> There is really no reason to classify organic compounds as "pharmaceutical" or "non-pharmaceutical" in discussing solid properties. Compounds used in the pharmaceutical industry are quite structurally varied; there is not any specific chemical attribute that renders them pharmaceutically active or warrants the term pharmaceutical polymorph.

On the other hand, there certainly are situations where a new term is helpful in recognizing and even describing a particular previously unobserved phenomenon. An example is *isotopomeric polymorphism* (Zhou et al. 2004), describing a change in crystal structure upon changing the isotopic identity of one or more of the atoms in a molecule. While seemingly an isolated incident when initially discovered, at least one other example has been reported (Crawford et al. 2009). There will undoubtedly be many others given the remarkable sensitivity of molecular crystal structure to the positions of hydrogen atoms (Price 2008a; Hughes and Harris 2009).

Perhaps somewhere in between these extremes of appropriate and inappropriate definitions is the case of *tautomeric polymorphism* (Bhatt and Desiraju 2007), particularly of omeprazole. In keeping with the spirit of the McCrone definition alluded to earlier, the appropriate questions to ask would be essentially: (1) Are the crystal structures different? (2) Do they give the same melt? The answers to both are somewhat ambiguous. Bhatt and Desiraju obtained five different forms with varying ratios of two tautomers. Three forms have been patented, distinguishable by their solid-state properties. Do they all give the same composition of tautomers in the melt? As the authors point out, this may be a matter of time, until equilibrium is attained; that also may be a complicating factor. The problem seems

---

[2] As noted on a monument in Beratzhausen, and a Memorial in Einsiedeln, Switzerland, Paracelsus, sometimes called the father of toxicology, wrote [German: Alle Ding' sind Gift, und nichts ohn' Gift; allein die Dosis macht, dass ein Ding kein Gift ist.] "All things are poison and nothing is without poison, only the dose permits something not to be poisonous." That is to say, substances considered toxic are harmless in small doses, and conversely an ordinarily harmless substance can be deadly if overconsumed. Even water can be deadly if overconsumed.

closely related to dynamic isomerism, also discussed by McCrone (1965). Given the difficulty in defining the limiting conditions, the legitimacy of *tautomeric polymorphism* remains questionable.

Three other adjectival polymorphisms—*conformational polymorphism, concomitant polymorphism,* and *disappearing polymorphs*—are discussed in subsequent sections of this volume.

The preceding discussion returns us to the question of *polymorphism in molecular crystals.* McCrone's definition first requires establishing the concept of molecularity, and in those cases the definition works very well. Even though McCrone's definition is still very useful, the last half century has led to a vastly expanded view of solids, which flaunts this concept wonderfully, so that even molecularity is an inherently fuzzy concept (Rouvray 1995, 1997). For instance, are molecular solids limited to neutral molecules? Are metal organic frameworks molecular solids? At what point is a solid no longer molecular?

What excites and motivates many of us about chemistry is the infinite variability that is possible and often observed. That variability defies precise definitions in many cases. In chemistry we use definitions to define essentially ideal cases in order to create a conceptual framework, and we then describe any particular situation as exhibiting or embodying features from more than one of those ideal situations. A classic example is that of the chemical bond—in many cases described as a covalent bond with a certain amount of ionic character. The two ideal states can be used to understand one that contains character of both. All the terms are clear and the meaning is clear. This is the language of chemistry and we can adopt a similar approach in the realm of multiple crystal forms. When a particular situation defies a precise description on the basis of our definitional framework, it does not necessarily warrant the creation of a new descriptive term. The perfectly acceptable alternative for special situations is to describe it as it is; it does not necessarily require an inclusive moniker. In the end, on the issue of nomenclature pragmatism should triumph over dogmatism.

1.2.3   Pseudopolymorphism, solvates, and hydrates

The literature on polymorphism and related phenomena has spawned a number of additional definitions and terms that potentially lead to confusion rather than clarification. The most outstanding of these is *pseudopolymorphism,* whose use was apparently proposed in the current context by Byrn (1982) in a rather limited (but now apparently mostly forgotten (or ignored) sense: "The classification scheme is based on the crystallographic behavior of solvates rather than the stability. Solvates that transform to another crystal form (different X-ray powder diffraction pattern) upon desolvation are *polymorphic solvates.* Solvates that remain in the same crystal form (similar X-ray powder diffraction pattern) are *pseudopolymorphic solvates*" (italics in original).

It is of interest that authors (McCrone 1965; Haleblian and McCrone 1969; Dunitz 1991; Threlfall 1995) who have given serious thought to the definition of polymorphism and its ramifications almost unanimously argued, strenuously in a number of cases, against the use of the term *pseudopolymorphism*. Typical is Seddon (2004) who argues against the use of the term "pseudopolymorph," since the scientific community gains no new understanding by its introduction, its use is pedagogically misleading, and a long-established and well-understood term "solvate" already exists. In support of Seddon I also pointed out the absurdity of "polymorphs of pseudopolymorphs" for polymorphic solvates (Bernstein 2005). The proponents did not desist, for example, arguing that its long term "wide acceptance" justified continued use (Desiraju 2004; Nangia 2006; Stahly 2007). A more detailed argument against the use of *pseudopolymorphism* was subsequently presented (Bernstein 2011). Even in arguing against its use, for the sake of completeness and to define some phenomena which are not to be considered as polymorphic behavior, it is unfortunately impossible to ignore the term and how it has been used and continues to be misused—*caveat emptor*![3] It is worthy of note that Byrn's scheme as defined previously has not generally been adopted in its original sense by most workers in the field.

McCrone (1965) and Haleblian and McCrone (1969) pointed out that pseudopolymorphism has been used to describe a number of phenomena that are related to polymorphism: among them are desolvation, second-order transitions (some of which may be considered examples of polymorphism), dynamic isomerism, mesomorphism, grain growth, boundary migration, recrystallization in the solid state, and lattice strain effects.

Probably the most common use, particularly prevalent in the pharmaceutical industry (David and Giron 1994; Henck et al. 1997), involves the confusion between solvates (including hydrates) and crystalline materials that do not contain solvent (anhydrates in the case of water). As noted by Byrn (1982), Byrn et al. (1999), Morris (1999), and Griesser (2006), crystal solvates exhibit a wide range of behavior. At one extreme, the solvent is tightly bound, and vigorous conditions are required for the desolvation process. In many of these cases the solvent is an integral part of the original crystal structure, and its elimination leads to the collapse of the structure and the generation of a new structure. At the other extreme are solvates in which the solvent is very loosely bound, and desolvation does not lead to the collapse of the original structure (Van der Sluis and Kroon 1989). Anything between the two extremes is also possible. Threlfall (1995) has noted that since a solvate and an unsolvated crystalline form are constitutionally distinct, they cannot be defined as polymorphs by any definition.

McCrone (1965) and Haleblian and McCrone (1969) proposed a simple experimental test to distinguish between a desolvation phenomenon and a true

---

[3] A 2016 *SciFinder* search on the term *pseudopolymorphism* showed an annual use of the term with ~285 hits in 2012, rising by about 15 in number per year for the decade preceding that. Specifics of the use were not further investigated.

polymorphic transformation, using the microscope hot stage (see Section 4.2). During heating of a crystalline sample, both a true polymorphic phase transition and a desolvation process will often lead to loss of transmission frequently accompanied by crystal darkening (due to formation of polycrystallites of the product phase). However, if the original sample is placed in a drop of solvent that is immiscible with the (suspected) solvent of crystallization then upon heating the liberated solvent will form an easily observable bubble in the surrounding droplet. No such observation can be made for a true polymorphic transformation. A more sophisticated technique, involving very much the same principle, is measurement by thermogravimetric analysis, which involves following the change in mass (in this case a loss in mass due to loss in solvent) corresponding to the heating process (Gruno et al. 1993; Perrenot and Widmann 1994) (see Section 4.3).

In spite of the objections to the use of pseudopolymorphism to describe solvated structures of a material, the term unfortunately has been used in this particular context, especially in the pharmaceutical industry, both in the characterization (Kitamura et al. 1994; Nguyen et al. 1994; Brittain et al. 1995; Kitaoka and Ohya 1995; Kitaoka et al. 1995; Caira et al. 1996; Gao 1996; Kalinkova and Hristov 1996; Kritl et al. 1996; De Ilarduya et al. 1997; Ito et al. 1997; De Matas et al. 1998) and production/processing aspects (Adyeeye et al. 1995; Hendrickson et al. 1995; Joachim et al. 1995).

McCrone (1965) also noted that second order phase transitions have been termed as pseudopolymorphic. Such transitions are difficult to detect by optical methods, because of the small structural changes that occur; hence, the origin of the prefix *pseudo* sometimes used to describe them. However, the birefringence of the crystals changes during such phase changes (see Section 4.2), so the use of crossed polarizers makes the phase change readily detectable.

A third phenomenon that has been described as pseudopolymorphism is dynamic isomerism (McCrone 1965). This takes us back to the problems defining polymorphism in general, where the questions of degree and time are raised. Dynamic isomers (including tautomers as well as geometric isomers) are generally considered to be chemically different. However, it is not always simple to make a distinction between geometric isomers and conformationally different species. Dynamic isomers exist in both the solid and the molten state, and are in equilibrium over a wide temperature range. Over that range, both isomers are stable in varying amounts depending on the temperature, and in solution, with solvent. Equilibrium between two polymorphs, on the other hand, can occur in the solid state, but upon melting the difference between the two polymorphs disappears. At any particular temperature only one polymorph is the thermodynamically stable one, except at a transition point, where two polymorphs are in equilibrium (see Section 2.2.2).

In principle, the distinction appears rather straightforward. However, a practical example will serve to demonstrate the difficulty. Matthews et al. (1991) described the crystal structures of three crystalline forms of 4-methyl-*N*-(4-nitro-α-phenylbenzylidene)aniline (**1-I**). In solution the material exists as a mixture of rapidly interconverting stereoisomers with *Z* and *E* configurations, hence dynamic isomers. In the solid state it is trimorphic. The so-called A crystal form has three molecules in the asymmetric unit, all exhibiting the *Z* configuration. The B form, which can be crystallized simultaneously with the A form at 0 °C from ethanol or hexane-ether, has two molecules in the unit cell, both exhibiting the *E* configuration. A third C form, obtained at room temperature from ethanol, also exhibits the *E* configuration. At ambient temperature the latter two forms are converted to A, with the appropriate molecular configurational change from *E* to *Z*.

1-I

While this system falls somewhere on the fuzzy line between polymorphism and dynamic isomerism we agree with McCrone (1965) and Threlfall (1995) that this phenomenon should not be described as pseudopolymorphism.

McCrone (1965) attempted to summarize the distinction using a number of important criteria, and again suggested some rather simple thermomicroscopic tests to determine it. They are worth noting here, since systems of this type have received little experimental attention, and the example cited demonstrates the problems well.

McCrone (1965) notes that polymorphs, existing only in the solid, can convert at least in one direction without going through the melt. On the other hand Curtin and Engelmann (1972) observed that the equilibrium in melt or in solution between the two configurational isomers may be shifted by crystallization or by chemical reaction to form a derivative of one of the isomers. In solution, the two isomers will have different solubilities, in the same way that different polymorphs can have different solubilities (see Sections 3.2 and 7.3.1). The solubility curves may cross, and with a change in temperature the solution can become saturated with one form. This is apparently what happens in the case of **1-I**, as the C form is obtained from the room-temperature crystallization, while at lower temperatures, a mixture of A and B is obtained. Dynamic isomers exist in both the melt and the solid state. Each isomer can exist in polymorphic forms, which is true for forms B and C. Details on the experimental techniques and observations are given in Chapter 4.

The thermomicroscopic differentiation between two phases that are known to be related either by polymorphism or dynamic isomerism is elegantly straightforward. The two phases should be melted side by side between a microscope slide and a cover slip, and then allowed to crystallize. Two possibilities exist for the crystallization events. In the case of polymorphism, the crystal fronts from the two melts will grow at a constant velocity until they come into contact, at which point one phase will grow through the other, due to a solid–solid transformation to the stable phase at that temperature. In the case of dynamic isomerism, the two crystal fronts would slow down as they approach each other, and in the so-called "zone of mixing" (McCrone 1965) a eutectic could appear.

Another suggestion for making the distinction between polymorphs and dynamic isomers is to melt each sample by the equilibrium melting procedure (McCrone 1957), and observe the melt as a function of time. For a polymorphic system, the melting point will not change unless a solid–solid transformation takes place. Such transformations are usually sudden, and the resulting melting point will not change. For the case of dynamic isomers, the melting point of each will decrease gradually with the attainment of equilibrium. The final melting point should be the same for each, since the same equilibrium composition will be attained for both. As the melting point is followed through the eutectic composition, one of the isomers should show an apparent phase transformation. In another test suggested by McCrone two crystals of the same compound suspected of being polymorphs are placed side by side on a microscope slide in a mutually suitable solvent. If they are polymorphs of different thermodynamic stability the more stable one will grow at the expense of the less stable one.

McCrone (1957, 1965) has also given detailed descriptions of the microscopic examinations and phenomena that can be used to distinguish polymorphism from other phenomena that sometimes have been mistakenly labelled as pseudopolymorphism: mesomorphism (i.e., liquid crystals), grain growth (boundary migration and recrystallization) and lattice strain.

1.2.4   Conventions for naming polymorphs

Part of the difficulty encountered in searching and interpreting the literature on polymorphic behavior of materials is due to the inconsistent labelling of polymorphs. In many cases, the inconsistency arises from lack of an accepted standard notation. However, often, and perhaps more important, it is due to the lack of various authors' awareness of previous work or lack of attempts to reconcile their own work with earlier studies (see, for instance, Bar and Bernstein 1985). While many polymorphic minerals and inorganic compounds actually have different names (e.g., calcite, aragonite, and vaterite for calcium carbonate or rutile, brookite, and anatase for titanium dioxide) this has not been the practice for molecular crystals, which have been labelled with Arabic (1, 2, 3…) or Roman (I, II, III…)

numerals, lower or upper case Latin (a, b, c...or A, B, C...), or lower case Greek (α, β, γ...) letters, or by names descriptive of properties (red form, low temperature polymorph, metastable modification, etc.).

As Threlfall (1995) and Whitaker (1995) have commented, arbitrary systems for naming polymorphs should be discouraged to avoid confusion surrounding the number and identity of polymorphs for any compound. Relative stability and/or order of melting point, as well as a specification of the monotropic or enantiotropic nature of the polymorphic form (see Section 2.2.4) have also been suggested as a basis for labelling (Herbstein 2001) but these do not allow for the discovery of forms with intermediate values, in addition to the fact that small differences in stability or melting point might lead to different order and different labelling by different workers. McCrone (1965) proposed using Roman numerals for the polymorphs in the order of their discovery, with the numeral I specifying the most stable form at room temperature. By Ostwald's rule (Ostwald 1897) (Section 2.3), the order of discovery should in general follow the order of stability. McCrone also supported the suggestion by the Koflers (Kofler and Kofler, 1954) that the Roman numeral be followed by the melting point in parentheses. In fact, the successors of the Koflers at the Innsbruck school have very much followed this practice (Kuhnert-Brandstätter 1971), although in general it has not been adopted by others. The use of melting points is complicated by the fact that while this datum has a clear thermodynamic definition, a number of techniques are employed to determine the melting point (or melting point range, in many cases) so that real or apparent inconsistencies may arise from such a designation (see Sections 4.2 and 4.3).

In view of the body of literature already existing and the questions surrounding the definition of a polymorph it does not appear to be practical to define hard and fast rules for labelling polymorphs. The Kofler method has clear advantages, since the melting point designation may eliminate some questions of identity. But the downside of adopting such an approach is the number of techniques that may be employed to measure a melting point. In addition, in practice many "melting points" are recorded as a range of temperatures, further confusing the issue, as for instance when two ranges overlap. For those studying (and naming) polymorphic systems it is important to be fully aware of previous work, to try to identify the correspondence between their own polymorphic discoveries and those of earlier workers, and to avoid flippancy in the use of nomenclature in the naming of truly new polymorphs.

The problem appears to be particularly egregious in the naming of crystal forms in the patent literature. A perhaps extreme, but nevertheless representative, example is presented in Table 1.1. The fact that virtually all of the named forms were granted patents implies that at least the patent examiners were convinced that the applicant(s) had prepared new and different forms from those in the prior art. Clearly, sorting out any possible identities would be a formidable task, but in a very practical sense it is one that is increasingly faced in pharmaceutical patent litigations (see Chapter 10).

Table 1.1 *Collection of all the names for the various crystal forms in patents and publications of aripiprazole*

Publication	Title	Crystalline forms
Proceedings of the Fourth Japanese– Korean Symposium on Separation Technology (October 6–8, 1996)		Type-I Type-II
Nanubolu et al. (2012)	Sixth polymorph of aripiprazole—an antipsychotic drug	The authors report the existence of the sixth polymorph of aripiprazole (APPZ) as characterized by single-crystal X-ray diffraction and present its structural and lattice energy comparison with five other polymorphs of APPZ in the Cambridge Structural Database (CSD). Incidentally, APPZ with six well-characterized polymorphs happens to be the second most polymorphic system in the CSD after the classic ROY molecule which has a record number of seven characterized polymorphs. The extensive polymorphism in the title compound is attributed to a very high degree of conformational freedom, significant differences in the hydrogen bonding, and the influence of crystal packing effects.
Zeidan et al. (2016)	An unprecedented case of dodecamorphism: the twelfth polymorph of aripiprazole formed by seeding with its active metabolite	A new polymorph of APPZ has been discovered, making it the most polymorphic drug to date with twelve reported anhydrous forms, and a record-breaking ninth olved crystal structure.
Braun et al. (2009b)	Conformational polymorphism in aripiprazole: preparation, stability and structure of five modifications	

continued

Table 1.1 Continued

Publication	Title	Crystalline forms
Braun et al. (2009a)	Stability of solvates and packing systematics of nine crystal forms of the antipsychotic drug aripiprazole	
Morissette et al. (2004)	High-throughput crystallization: polymorphs, salts, co-crystals and solvates of pharmaceutical solids	

Patent no./company	Title	Crystalline forms/amorphous
WO2003026659 Otsuka Pharmaceutical	Low hygroscopic aripiprazole drug substance and process for the preparation thereof	Conventional hydrate Hydrate A Conventional anhydrate Anhydrate B Anhydrate C Anhydrate D Anhydrate E Anhydrate F Anhydrate G
WO2004106322 Cadila	Polymorphs of aripiprazole	Polymorph I Polymorph II Polymorph III Polymorph IV
US8008490 Sandoz	Polymorphic forms of aripiprazole and method	Form X Ethanol hemisolvate Methanol solvate
WO2007004061 Medichem	Syntheses and preparations of polymorphs of crystalline aripiprazole	Form J Form L
WO2004083183 Hetero Drugs Limited	Novel crystalline forms of aripiprazole	Form I Form II
WO2005058835 Teva Pharma	Methods of preparing aripiprazole crystalline forms	Form I Form II Form VI Form VIII

Patent no./company	Title	Crystalline forms/amorphous
		Form X
		Form XI
		Form XII
		Form XIV
		Form XIX
		Form XX
WO2005009990 Hetero Drugs Limited	Aripiprazole crystalline forms	Form III Form IV Form VI
EP2082735 Helm AG	Amorphous aripiprazole and process for the preparation thereof	Amorphous
WO2006053780 Synthon	Crystalline aripiprazole solvates	Form B (methanolate and hemiethanolate)
WO2006012237 Shanghai Institute of Pharmaceutical Industry	Aripiprazole crystalline forms and associated methods	Anhydrous form
WO2008020453 Unichem Laboratories Limited	A process for the preparation of a novel crystalline polymorph of aripiprazole	Form U
US20160083381 Raqualia Pharma Inc.	Polymorph forms	Form I (L-tartrate salt but inventor calls this salt a polymorph)
WO2006077584 Chemagis	New crystalline forms of aripiprazole	Form AETl Form AETH Form AM2 Form AMI

1.3   Is this material polymorphic?

1.3.1   Occurrence of polymorphism

Perhaps the most well-known statement about the occurrence of polymorphism is that of McCrone (1965): "It is at least this author's opinion that every compound has different polymorphic forms and that, in general, the number of forms known for a given compound is proportional to the time and money spent in research on that compound." As a corollary to this rather sweeping, even provocative, statement, McCrone noted that "all the common compounds (and elements) show polymorphism," and he cited many common organic and inorganic examples.

These echo similar statements by Findlay (1951, p. 35), "[polymorphism] is now recognized as a very frequent occurrence indeed," Buerger and Bloom (1937), "polymorphism is an inherent property of the solid state and that it fails to appear only under special conditions," and Sirota (1982),

> [polymorphism] is now believed to be characteristic of all substances, its actual non-occurrence arising from the fact that a polymorphic transition lies above the melting point of the substance or in the area of yet unattainable values of external equilibrium factors or other conditions providing for the transition.

Such statements tend to give the impression that polymorphism is the rule rather than the exception. The body of literature in fact indicates that considerable caution should be exercised in making them. It appears to be true that instances of polymorphism are not uncommon in those industries where the preparation and characterization of solid materials are integral parts of the development and manufacturing of products (i.e., those on which a great deal of time and money is spent): silica, iron, calcium silicate, sulfur, soap, pharmaceutical products, dyes, and explosives. Such materials, unlike the vast majority of compounds that are isolated, are prepared not just once, but repeatedly, under conditions that may vary slightly (even unintentionally) from time to time. Similarly, in the attempt to grow crystals of biomolecular compounds, much time and effort is invested in attempts to crystallize proteins under carefully controlled and slightly varying conditions, and polymorphism is frequently observed (Bernstein et al. 1977; McPherson 1982; McPherson and Gavira 2014). Even with the growing awareness and economic importance of polymorphism, many documented cases have been discovered by serendipity rather than through systematic searches. Some very common materials, such as sucrose and naphthalene, which certainly have been crystallized innumerable times at ambient conditions, have not been reported to be polymorphic.[4] The *possibility* of polymorphism may exist for any particular compound, but the conditions required to prepare as yet unknown polymorphs are by no means obvious. Even with the accumulated experience of the past twenty-five years there are as yet no comprehensive systematic methods for feasibly determining those conditions. Moreover, we are almost totally ignorant about the properties to be expected from any new polymorphs that might be obtained.

There have been a number of efforts to provide a statistical basis for the expectation of multiple crystal forms of any particular molecular entity. The true

---

[4] After nearly a decade of experiments carried out at high pressure, Katrusiak and colleagues succeeded in preparing and determining the crystal structure of a high pressure form of sucrose (Patyk et al. 2012).

occurrence of polymorphism is very difficult to determine and depends to a large extent on the choice of the data sample. This is demonstrated for three attempts summarized in Table 1.2. During the period 1948–1961 McCrone regularly reported the results of crystal growing experiments with approximately 25% of the organic compounds exhibiting polymorphism. A more recent survey of the pharmaceutical compounds in the European Pharmacopoeia yielded 42% polymorphism (Braun 2008). A summary of 248 compounds studied by the commercial analytical consulting firm SSCI with the specific goal of screening for crystal forms yielded 48% exhibiting polymorphism (Stahly 2007).

The difficulty in compiling statistics on polymorphism is evident from two sets of statistics on polymorphism that were recently compiled: (i) from the CSD and (ii) from 157 solid form screens performed at Lilly Company over more than fifteen years of polymorph screenings (Table 1.3) (Cruz-Cabeza et al. 2015). On the one hand, the CSD dataset contains a very large amount of information— much larger than any single group could ever compile—but the degree of form screening for the reported compounds may vary enormously. On the other hand, the Lilly set of compounds is much smaller but all of them have been intensively screened for multiple crystal forms. The statistics on polymorphism together with their 95% confidence intervals are presented in Table 1.3.

From these data 34% of unique compositions in the CSD are polymorphic compared to 25% in the Lilly dataset. When the statistics are broken into different sub-groups of compositions, the results remain quite homogeneous for the CSD dataset but are extremely heterogeneous for the Lilly dataset. This is partly because the Lilly dataset is derived from solid form screenings targeted specifically to identifying commercially viable drug crystal forms, whilst the CSD represents more of a homogeneous representation of all types of compounds.

Table 1.2 *Some early statistics on the occurrence of polymorphism*

Source	Data type	Compounds	Polymorphism occurrence
Microscopy studies by McCrone (1948–1961)[a]	Organic compounds	140	25%
From European Pharmacopoeia (1964–2004)[b]	Single component organic compounds	598	42%
From SSCI polymorph screens of organic compounds (1991–1997)[c]	Organic compounds	245	48%

[a] See text;
[b] Griesser (2011);
[c] Stahly (2007).

Table 1.3 *Occurrence of polymorphism in the CSD and Eli Lilly datasets*

Data type	CSD			Lilly		
	Unique compositions	Pol. Occ. (%)	95% C.I.* (%)	Unique compositions	Pol. Occ. (%)	95% C.I.* (%)
All compositions	8 035	34	(33, 35)	564	25	(21, 29)
Single-component	**5 941**	**37**	**(36, 38)**	**68**	**66**	**(55, 77)**
Neutral multicomp.	1 108	27	(24, 30)	138	13	(7, 19)
Solvates + Co-crystals	721	31	(28, 34)	88	11	(4, 18)
Hydrates	387	20	(16, 24)	50	16	(6, 26)
Salts	986	35	(32, 38)	110	28	(20, 36)
Unhydrated salts	**820**	**37**	**(34, 40)**	**56**	**54**	**(41, 67)**
Hydrated salts	166	28	(21, 35)	54	2	(0, 6)

*Margin of error is calculated within a 95% confidence interval.

The occurrence of polymorphism in single component compounds was found to be 37±1% in the CSD dataset compared to 66±11% in the Lilly dataset. The difference in polymorph occurrence is due in part to the inherent nature of the data. Whilst the CSD dataset contains compounds crystallized at least twice as single crystals suitable for study by X-ray diffraction, the Lilly dataset contains data from extensive polymorph screenings where many crystallization conditions were explored. Moreover, characterization of the forms comprising the Lilly data-set extended beyond single crystal X-ray diffraction, in some cases, increasing the probability of identifying polymorphs.

Polymorphism occurrence for neutral multicomponent systems is 27±3% for the CSD, but only 13±6% for the Lilly dataset. Potential causes for this difference may be i) the inherently different nature of the data and ii) the difficulty to accurately determine stoichiometries in hydrates and solvates. With regard to the different nature of the data, the Lilly dataset of neutral multicomponent systems uniquely contains hydrates and solvates, whilst the CSD dataset contains a broader range of compositions. A further breakdown of the CSD dataset of multicomponent systems into co-crystals (and solvates) and hydrates reveals further differences between these groups. Hydrates appear to be slightly less prone to being poly-morphic (20±4%) than co-crystals and solvates (31±3%). Non-solvated salts similarly display a high propensity to polymorphism, that being 37±3% and 54±13% from the CSD and Lilly datasets, respectively.

Many of the attempts to understand the appearance of polymorphism have been based on an assumed (very often hindsight) connection between some molecular property and the propensity for polymorphism, often with contradictory conclusions. For instance, polymorphism has been attributed to conformational freedom in a molecule (Aitipamula et al. 2014), or it has been attributed to conformation flexibility *and* the potential for hydrogen bonding (Ahn et al. 2006). It is claimed that several conformers are available in the crystallization milieu (solution or melt phase) to form different hydrogen bond synthons and close-packing motifs (Nangia 2008). This approach has become textbook dogma, polymorphism being promoted by the fact that drug molecules tend to contain functional groups that are flexible and capable of forming hydrogen bonds (Desiraju et al. 2011). To add to the confusion the *absence* of hydrogen bonding and conformational flexibility have also been attributed to the propensity for polymorphism (Dey and Desiraju 2006).

One of the most enigmatic aspects of polymorphism is how the molecular structure of a compound might relate to its ability to exhibit polymorphism. We recently compiled the statistics to investigate such a relationship (Cruz-Cabeza et al. 2015), as presented in Table 1.4.

It is seen that rigid molecules were found to be as likely to be polymorphic as flexible molecules. As for molecular size, again, small molecules (number (N) of heavy atoms $\leq 18/M_r \leq 245$) were found to be as prone to polymorphism as large molecules, in agreement with previous findings by Sarma and Desiraju (1999).

Table 1.4 *Molecular structure and polymorphism occurrence (adapted from Cruz-Cabeza et al. 2015)*

Property type	Property	$N[+]$	Compounds contain property [+]			Compounds do not contain property [−]		
			Pol. occ. [+] (%)	95% C.I. [+] (%)		$N[-]$ Pol. occ. [−] (%)	95% C.I. [−] (%)	
Flexibility	R-bonds	4 536	38	(37, 39)	1 405	36	(35, 37)
	DOFlex	5 471	37	(36, 38)	470	40	(38, 42)
Size	$M_r \leq 245$	2 300	38	(37, 39)	3 641	37	(36, 38)
	Heavy atoms ≤ 18	2 823	38	(37, 39)	3 118	37	(36, 38)
Drug-likeness	Lipinski RO5	4 471	36	(36, 37)	1 470	41	(40, 42)
HB capacity	HB groups	2 991	40	(39, 41)	2 950	34	(33, 35)
Chirality	Chiral centre	2 083	26	(25, 27)	3 858	43	(42, 44)

Most of the molecules in the CSD dataset (4 471 out of 5 941) could, based on Lipinski's rule of five (RO5) (Lipinski 2004), be classified as drug-like compounds. The difference in polymorphism occurrence found for drug-like and non-drug-like molecules is very small (5%), with non-drug like compounds being found to have a slightly higher occurrence (41±1%). Molecules able to form hydrogen bonds were found to be just 6% more likely to be polymorphic than those that are not able to hydrogen bond.

Finally, chirality was the only molecular property found to be significant in determining the proclivity of a molecule to display polymorphism. The polymorphism occurrence of chiral molecules was found to be only 26±1% compared to a 43±1% of achiral molecules. These statistics for chiral molecules do not change significantly whether or not both enantiomers are present in the same crystal structure. We contrasted this observation with the Lilly dataset and found a similar trend: 60% of chiral compounds were found to be polymorphic compared to 73% of non-chiral compounds.[5]

The original paper should be consulted for a detailed discussion of these statistics. With the growing awareness among chemists of the phenomenon of polymorphism its actual occurrence in any particular system may not be as great a surprise as a generation or two ago. The predicted existence of any particular polymorphic structure for a single compound, the conditions and methods required to obtain it, and the properties it will exhibit are still problems that will challenge researchers for many years to come.

1.3.2   Literature sources of polymorphic compounds

As noted above, the phenomenon of polymorphism is not new to chemistry. Nineteenth century chemists were very much aware of the properties of solids, and in the decades preceding the development of spectroscopic and X-ray crystallographic methods, the characterization of solids was a crucial aspect of the identification of materials. Chemists grew crystals carefully in order to obtain characteristic morphologies and then determined physical properties such as color, interfacial angle, indices of refraction, melting point, and even taste (Schorlemmer 1874; Orndorff 1893; Senechal 1990; Kahr and McBride 1992).

---

[5] While perhaps anecdotal, the following appears to be a good measure of the state of our knowledge about the "pervasiveness" of polymorphism. In assigning a research project to a graduate student, a research advisor assumes a certain risk that the project will not succeed. One could imagine as a perfectly reasonable project the assignment to prepare and characterize the polymorphic forms of a *single* compound of interest, which is, of course, the practical manifestation of all the quotations on the expectations for polymorphism. Unexpected results can constitute the basis for a PhD thesis, but the absence of results, that is, the inability to obtain *any* polymorphs, would constitute a total failure of the project. This author has yet to encounter an academic research advisor who would be prepared to take the responsibility of assigning such a research project to a PhD student. That is, in spite of the hyperbole of McCrone's statement and the notoriety it has received, and the increasing importance of polymorphism in the market place, there is not sufficient confidence in its veracity to risk the career of a PhD student on any single particular compound.

Being critically observant was essential, for there was little other information to rely on.

A great deal of information on crystalline properties, including polymorphism, was summarized in the five-volume compendium covering over 10 000 compounds by P.H.R. von Groth, published between 1906 and 1919. The first two of these tomes (Groth 1906b, 1908) deal with elements and inorganic compounds, while the last three (Groth 1910, 1917, 1919) are concerned with organic materials. The genesis of this opus is vividly described:

> Groth's most stupendous work was the *Chemische Kristallographie,* five volumes which appeared between 1906 and 1919, comprising in toto 4208 pages and 3342 drawings and diagrams of crystals. The manuscript was written entirely by Groth in his fine hand and corrected over and over again by him until there was hardly a white spot left on the manuscript and again on the galley proofs. Oh for the admirable compositors in the Leipzig printing centers in the days before the general use of typewriters! The volumes contain a review of all crystallographic measurements…Each section is preceded by a survey of the crystal-chemical relations and includes many hints of gaps which should be filled by further work. In many instances Groth doubted the correctness of the work reported in the literature, and wherever possible, he got his pupils, assistants or visiting colleagues to prepare the same substances again, and to recrystallize and re-measure them…Altogether measurements on between 9000 and 10,000 are critically discussed in *Chemische Kristallographie,* an astounding feat considering the small number of the team….

The work thus contains a thorough, checked survey of the physical properties of many of the crystals that had been studied up to its publication. Typical pages of the "crystal-chemical relations" for dimorphic diphenyl malonic anhydride are shown in Figure 1.2, in which the methods for obtaining both structures are described.

das **Diphenylmaleïnsäureanhydrid** $= C_6H_5 \cdot C \!=\!\!=\! C \cdot C_6H_5$ und
$$CO.O.\dot{C}O$$

in der Tat zeigen nun beide Körper eine ähnliche Verwandtschaft in krystallographischer Beziehung wie jene, indem sie die gleiche Symmetrie und sehr nahe übereinstimmende Werte zweier Axen (*a* und *b*) besitzen; die ungesättigte Verbindung existiert aber außerdem noch in einer metastabilen Modification, deren monokline Krystalle sich neben der stabilen in wässeriger Acetonlösung bilden, aber sich sehr bald umwandeln, wenn sie mit Krystallen der stabilen Form in Berührung sind; durch Erwärmung kann die Umwandlung bei jeder Temperatur bewirkt werden, niemals die umgekehrte (Monotropie); die metastabile Modification entsteht außerdem durch Unterkühlen der Schmelze, wenn keine Spur der stabilen vorhanden ist.

Figure 1.2 *A typical entry from Groth's* Chemische Kristallographie. *(a) Textual description of the dimorphic diphenyl maleic anhydride; (b) physical data for the stable modification melting at 155 °C; (c) physical data for the metastable modification melting at 146 °C.*

Diphenylmaleïnsäureanhydrid $= C_6H_5 . C\!\!=\!\!\!=\!\!\!=\!\!C . C_6H_5 .$
$$CO . O . CO$$

Stabile Modification.

Schmelzpunkt $155°$.

Spec. Gew. $1,340$ Drugman[44].

Rhombisch bipyramidal.

$$a : b : c = 0,5176 : 1 : 0,7024 \text{ Drugman}[44].$$

Aus Aceton entsteht die Combination (Fig. 2575): $m\{110\}$, $o\{111\}$, $q\{011\}$, ebenso aus Benzol, Chloroform, Äther und Alkohol, aus letzterem nach der c-Axe dünn nadelförmig. Einmal wurden

Fig. 2575.   Fig. 2576.

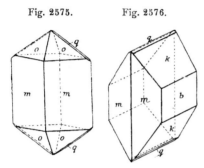

aus Aceton kleine oktaëderähnliche Krystalle erhalten, die nur $m\{110\}$ und $q\{011\}$ zeigten; aus etwas harzhaltiger Lösung wurden Combinationen mit untergeordneten Flächen von $x\{112\}$, $y\{122\}$, $c\{001\}$ und $b\{010\}$ beobachtet. Die Krystalle aus Toluol zeigen die Formen (Fig. 2576): $m\{110\}$, $k\{021\}$, $b\{010\}$, untergeordnet: $q\{011\}$, $c\{001\}$, seltener $o\{111\}$; die hier vorhandene Verlängerung nach der a-Axe tritt noch mehr hervor an den Krystallen aus Xylol, welche dieselben Formen, aber mit besser ausgebildetem $o\{111\}$, zeigen. Eine solche nach der a-Axe prismatische Combination mit untergeordnetem $a\{100\}$ hatte früher bereits Jenssen[43] (l. c. 64) an den von Anschütz und Bendix aus Äther erhaltenen Krystallen beobachtet, ihr aber eine andere Aufstellung gegeben.

		Berechnet:		Beobachtet:		
				Drugman:		Jenssen:
$m : m =$	$(110) : (1\bar{1}0) =$	—		$*54° 44'$		$54° 42'$
$q : q =$	$(011) : (0\bar{1}1) =$	—		$*70$	10	—
$o : o =$	$(111) : (1\bar{1}1) =$	$45°$	$14'$	45	13	—
$o : q =$	$(111) : (011) =$	48	0	48	1	—
$o : m =$	$(111) : (110) =$	33	12	33	15	—
$o : m =$	$(111) : (1\bar{1}0) =$	61	$6\frac{1}{2}$	61	7	—
$q : m =$	$(011) : (110) =$	74	41	74	42	—
$k : b =$	$(021) : (010) =$	35	27	35	5 ca.	34 43
$k : m =$	$(021) : (110) =$	68	$0\frac{1}{2}$	67	55 »	67 49
$x : o =$	$(112) : (111) =$	19	25	19	13	—
$x : q =$	$(112) : (011) =$	37	23	37	24	—
$y : q =$	$(122) : (011) =$	29	$2\frac{1}{2}$	29	$2\frac{1}{2}$	—

Keine deutliche Spaltbarkeit.

Doppelbrechung positiv; Axenebene $a\{100\}$, 1. Mittellinie Axe c; Axenwinkel klein.

$\alpha = 1,505$ *Li*, $1,511$ *Na*, $1,517$ *Tl* (alle optischen Angaben von
$\beta = 1,505$ ca. » $1,5115$ » $1,518$ ca. »   Drugman).
$\gamma = 1,811$ » $1,836$ » $1,865$ »

44) Drugman, Zeitschr. f. Krystall. 1912, **50**, 576.

Figure 1.2 (Continued)

Metastabile Modification.

Schmelzpunkt 146°.

Spec. Gew. 1,345 Drugman[44]).

Monoklin prismatisch.

$$a : b : c = 2,5615 : 1 : 2,3275; \quad \beta = 101°33' \quad \text{Drugman}[44].$$

Diese Modification erhielt Drugman neben der
stabilen aus wässeriger Acetonlösung mit den Formen: $c\{001\}$ (oft sehr stark vorherrschend), $a\{100\}$,
$m\{110\}$, $o\{111\}$; aus Xylol bilden sich kleine Krystalle der gleichen Form mit $n\{210\}$ (Fig. 2577).
Einmal wurde ein Zwilling nach $a\{100\}$ beobachtet.

Fig. 2577.

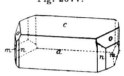

			Berechnet:	Beobachtet:
$m : a =$	$(110) : (100) =$		—	*68°15'
$a : c =$	$(100) : (001) =$		—	*78 27
$o : c =$	$(111) : (001) =$		—	*64 13
$m : c =$	$(110) : (001) =$	85°45'		85 45
$n : c =$	$(210) : (001) =$	82 50		—
$n : a =$	$(210) : (100) =$	51 25		51 21
$o : n =$	$(111) : (210) =$	24 28		24 17
$o : o, =$	$(111) : (\bar{1}1\bar{1}) =$	65 58		65 57
$o : m' =$	$(111) : (\bar{1}10) =$	51 5		51 7
$o, : n =$	$(\bar{1}1\bar{1}) : (210) =$	66 21		66 21

Spaltbarkeit nach $a\{100\}$ und $c\{001\}$ vollkommen.

Ätzfiguren auf $c\{001\}$ nach $b\{010\}$ symmetrisch.

Ebene der optischen Axen $b\{010\}$, durch $a\{100\}$ und $c\{001\}$ je ein Axen-
bild, durch eine Schliffläche $\| \{101\}$ beide sichtbar; starke Dispersion.

Figure 1.2 (Continued)

A few pages on appear the entries for the description of crystal habit, melting
point, solvent, appropriate reference(s), interfacial angles, and indices of refraction, if reported in the literature. Many of the substances had been reported to be
polymorphic, and Groth recorded those facts, along with methods for preparing
the polymorphs and the original literature references. It is a remarkable work, and
one which should be consulted to check for the existence of polymorphism in a
specific material, as well as for the source of physical phenomena, once observed,
but since forgotten.

A second rich collection of references on the polymorphic behavior of organic
materials is the compilation by Deffet (1942). This contains information and references to primary sources on 1188 substances that exhibit polymorphism at
atmospheric pressure and another 32 that exhibit polymorphic behavior at elevated pressures. A typical entry contains the number of reported polymorphic
forms, their melting points, temperature(s) of transition, crystal system, some
physical properties, and literature references, of which there are nearly 1000.
Substances are organized by empirical formula with an index organized by compound name (in French).

A third compilation intended to be devoted to polymorphic materials is that of Kuhnert-Brandstätter (1971). The body of this book is an identification table for hot stage studies of pharmaceutical materials (see Sections 4.2 and 7.2), in which entries are arranged by increasing melting point, with eutectic data for mixtures with azobenzene and benzil. There is considerable descriptive detail on the melting behavior and identification and description of polymorphic forms, albeit only microscopic determinations, for approximately 1 000 pharmaceutically important compounds. There is no formula index, and the subject index contains only a partial listing of the compounds included. Nevertheless, the book contains some very useful information about the existence of polymorphism and the characterization of its behavior in many of these commercially important materials. In this context, it is perhaps noteworthy that the Merck Index (2016) describes polymorphic behavior for only fifty-five of over 10 000 entries, many of which appear in the Kuhnert-Brandstätter compilation.

There are a number of additional sources for consultation on information on polymorphism of particular compounds. As noted in the previous section (Table 1.1), from 1948 to 1961, McCrone edited a regular column in *Analytical Chemistry* entitled "Crystallographic Data," in which were published the details on crystal growth, physical properties, and polymorphic behavior of approximately 200 compounds. The series was undertaken at the time "because optical crystallography is neglected as an analytical tool because too few compounds have been described," and with the desire to "…initiate a process which [would] enable a group of crystallographers to complete the tabulation of crystal data for most of the common everyday compounds" (Grabar and McCrone 1950). About 140 of these were organic compounds, and 25% of these exhibited polymorphism. Even in the cases where there is no evidence of polymorphism, these reports contain detailed descriptions of conditions for growth of crystals with well-defined faces, and the characterization of crystal habit very much in the tradition of Groth. It is information that future investigators will be able to utilize for a variety of studies. The need for recording the detailed description of crystal growth, crystal habit, and crystal properties was later echoed in an appeal by Dunitz (1995) to authors of crystallographic structure analyses:

> …please give the color (easy to observe) and melting point of crystals studied (easy to measure); if possible, also the heat of fusion and of any observed phase transitions (only slightly more difficult to measure): report also any "unusual" behavior, any observed change of physical properties or of the diffraction pattern.

The short reports solicited and edited by McCrone are models of the kind of data that should be required and included in descriptions of crystals and crystal structure reports, even if only in deposited form (Section 1.3.3).

Some additional literature sources should also be consulted to check for earlier reports of polymorphism. The Barker index (Porter and Spiller 1951, 1956; Porter and Codd 1963) made use of the characteristic interfacial angles for purposes of identification of crystals. The index is based on Groth's earlier compilation (which is organized by chemical composition) and is arranged by increasing interfacial angle within a crystal system. There are some additional compounds, with totals of 2 991 in tetragonal, trigonal, and orthorhombic space groups (Volume I) (Porter and Spiller 1951), 3 572 in monoclinic (Volume II) (Porter and Spiller 1956), and 871 in triclinic (Volume III) (Porter and Codd 1963) space groups. However, the method of arrangement means that polymorphs of a compound crystallizing, say, in monoclinic and orthorhombic space groups requires that the compound be checked in all three volumes.

Another approach was taken by Winchell (1943, 1987), who prepared a compilation of "all organic compounds whose optical properties are sufficiently well known to permit identification by optical methods." The compilation is arranged in the same fashion as the fourth edition of **Beilstein**'s *Handbuch der Organischen Chemie* (Beilstein 1978), and at the time of its publication was meant to include all organic compounds whose indices of refraction had been measured. Since indices of refraction differ among them, polymorphs could be easily recognized by different optical properties. The book does contain references to primary sources and drawings of crystals, as illustrated in a typical entry Figure 1.3.

p-Methylbenzophenone or phenyl *p*-tolyl ketone $[C_6H_5 \cdot CO \cdot C_6H_4(CH_3)]$ has two phases. The stable phase is monoclinic with $a:b:c = 1.012:1:0.412$, $\beta = 95°7'$. Crystals {010} tablets or equant with {110}, {210}, {100}, {011}, {001}, etc. Figs. 60, 61. No distinct cleavage. M.P. 60°. The optic plane is 010 for red to green and normal thereto for blue and violet. X $\wedge$ $c = +37°$. (-)2E = 49°11' Li, 35°15' Na, 6°55' Tl, 49°32' blue. The matastable phase is ditrigonal pyramidal with $c/a = 1.225$. Crystals show both trigonal prisms, {10$\bar{1}$0} and {$\bar{1}$100}, etc. Fig. 62. M.P. 55°. Uniaxial negative with $N_O = 1.7067$ Li, 1.7170 Na, 1.7250 Tl; $N_E = 1.5564$ Li, 1.5629 Na, 1.5685 Tl; $N_O - N_E = 0.1541$ Na.

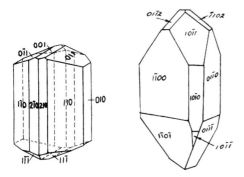

Figs. 60, 61. Phenyl-β-tolyl ketone.

Figure 1.3 *Typical entry from Winchell's* Optical Properties of Organic Crystals *for dimorphic* p-*methylbenzophenone (reproduced, with permission).*

Another useful compilation of crystallographic data as a source of examples of polymorphic systems is *NIST Crystal Data* (NIST 2001), which contains the principal crystallographic data on over 237 000 organic and organometallic entries. Each entry contains cell constants, space group, and other crystallographic information and bibliographic citations. In some cases the fact that a crystalline compound is one of a polymorphic system is specifically noted. In other cases the polymorphism may be recognized by the fact that a compound has more than one entry either in the formula index or the compound name index.

In addition to these compilations of crystal data in which instances of polymorphism may be recorded, a number of texts on the subject of the solid-state properties of organic compounds contain many examples of polymorphism. Since these books are based in part, at least, on work by the authors not published elsewhere, they may be considered as primary literature sources. Particularly noteworthy in this regard are the books by McCrone (1957), Kofler and Kofler (1954), and Pfeiffer (1922).

The usual search strategies for information on the preparation and properties, such as use of Chemical Abstracts and Beilstein, can also be useful for determining if a particular compound has been reported to be polymorphic. However, reference to the primary sources on the preparation and the characterization of the compound may reveal unusual behavior (e.g., melting points or colors which differed from one crystallization to the next) which testifies to the possible existence of polymorphic forms, behavior that is not specifically noted in the abstracted material.

1.3.3   Polymorphic compounds in the Cambridge Structural Database

The Cambridge Structural Database (CSD) is the repository for the results obtained from the X-ray crystal structure analysis of organic and organometallic compounds (Allen et al. 1991, 1994; Allen and Kennard 1993; Kennard 1993). As of the May 2016 release, the database contains over 800 000 entries, and as of this date approximately 50 000 structures are added annually. It is now also a depository for crystallographic data that may not be published elsewhere. In the past five decades the database has increasingly influenced the way structural chemists carry out their trade. An enormous amount of geometric and structural information is available in a very short time for searches, correlations, model compounds, packing arrangements, reaction coordinates, hydrogen bonding patterns, and a variety of studies. The rapid increase in the data availability that has been accompanied by increasingly sophisticated software has opened opportunities that could not have been imagined even a quarter of a century ago. Formerly accessible only on mainframes or work stations it has recently become available online.

As the repository for all organic and organometallic crystal structures, the CSD naturally contains entries for polymorphic materials. Each entry in the CSD contains one- (1D), two- (2D), and three-dimensional (3D) information. The 2D information is used to generate the structural formula and chemical connectivity, which clearly will be the same for polymorphs. The 3D information contains the results of the X-ray structure determination: cell constants, space group, atomic coordinates, and atomic attributes needed to generate the three-dimensional molecular and crystal structures. The 1D data contain bibliographical and chemical information (name and empirical formula), including qualifying phrase(s) such as "neutron study," "absolute configuration," etc. It is here that the CSD notes that the material is polymorphic with a qualifying phrase such as "red phase," "meta-stable polymorph," or "Form II" *if the author of the primary publication noted this feature or if the abstractors recognized that the structure was one of a polymorphic system.* In many cases note is taken of the fact that this is some special crystal form only when a second (or third, etc.) structure of a polymorphic series is being reported. The first report may not contain such a notation, since the author may not have been aware that the material is polymorphic. (This may be the case for subsequent structure determinations as well. In the early days of the CSD some polymorphic structures were archived with different REFCODEs—the unique identifier for each chemical species. The more sophisticated archiving software used now prevents such duplication and has eliminated many of the older "orphans," but some may still exist.) Once one member of a polymorphic set of structures has been identified care should be taken to extract all entries of that compound. Many, if not most, of these potential pitfalls and problems in the search for true polymorphs in the CSD have been addressed and solved by van de Streek and Motherwell (2005) but the generation and identification of new polymorphs may still be fraught with uncertainty. The absence of a descriptor indicating that a material belongs to a polymorphic system is not a foolproof indication that the material is not polymorphic. Other literature sources should be consulted to make that determination.

An early example of the caution that must be exercised in performing such searches and the numbers obtained was given by Gavezzotti and Filippini (1995). The search was defined for organic compounds (containing only C, H, N, O, F, Cl, or S) and for which the crystal structure of more than one polymorphic form had been determined. A total of 163 "clusters" were obtained, where a cluster is a group of polymorphic crystal structures of the same compound. Of the 163 clusters, 147 contained two structures, thirteen had three, and three had four structures. The authors note that these numbers are "first evidence of the high frequency of polymorphism in organic crystals," although the number of clusters is a relatively small percentage of the entries in the database. The number of these clusters is probably more a measure of certain authors' interest in the particular polymorphic system in question. In a more recent study (Cruz-Cabeza and Bernstein 2014), 1297 polymorphic systems were identified, 89.2% of which have two polymorphs, 8.8% have three polymorphs, and only twenty-six molecules have four polymorphs or more. A more realistic measure (although certainly not precise because of the caveats mentioned above) of the frequency of polymorphism in these compounds would be the fraction of compounds in the database known to be polymorphic, whether multiple structure determinations have been carried out or not.

1.3.4 Powder Diffraction File

The second crystallographic database that can serve as a source of examples of polymorphic structures is the Powder Diffraction File (PDF; Jenkins and Snyder 1996; ICDD 2016). This is the depository for over 500 000 powder diffraction patterns of solids (2015 release) of which more than 250 000 have atomic coordinates, roughly divided into organic, inorganic, and metallic compounds, of which organics are about 98%. Bibliographic searches may be run on compound name or formula, and again, the existence of polymorphism for a particular compound may be recognized by the presence of more than one entry for a compound. An example of identifying polymorphism from the bibliographic entries (formula index and compound name index) of the PDF is shown in Figure 1.4.

1.3.5 Patent literature

As polymorphism has become an increasingly important factor in the commercial aspects of many solid materials, the number of patents relating to the discovery and use of particular polymorphic forms has increased. This is particularly important for pharmaceuticals, pigments and dyes, and explosive materials, which

	Sulphamethylthiazole	$C_{10}H_{11}N_3O_2S_2$	7.80_x	4.34_x	6.80_6	8–521
i	Sulphamidochrisoidine	$C_{12}H_{13}N_5O_2S$	3.86_x	5.13_9	3.27_8	39–1610
o	Sulphamidochrysoidine	$C_{12}H_{13}N_5O_2S$	4.51_x	3.96_8	13.9_5	39–1611
*	β-Sulphanilamide	$C_5H_8N_2O_2S$	6.12_x	3.90_8	4.91_6	41–1909
*	Sulphanilamide	$C_6H_8N_2O_2S$	4.49_8	3.78_8	6.57_8	38–1710
*	Sulphanilamide	$C_5H_8N_2O_2S$	4.47_8	3.70_8	7.82_8	38–1709
o	α-Sulphanilamide	$C_5H_8N_2O_2S$	4.23_x	3.36_x	3.57_7	30–1944
	2-Sulphanilamidopyrimidine Sodium	$C_{10}H_9N_4NaO_2S$	9.16_x	5.17_7	4.06_7	5–112
*	Sulphanilic Acid	$C_6H_7NO_3S$	4.91_x	6.96_5	3.48_2	30–1945
o	Sulphaphenazole	$C_{15}H_{14}N_4O_2S$	4.37_x	7.29_5	3.90_2	30–1946
	Sulphapyrazine	$C_{10}H_{10}N_4O_2S$	5.59_x	7.21_7	4.79_7	5–213
*	Sulphapyridine	$C_{11}H_{11}N_3O_2S$	5.48_x	3.57_5	4.01_5	37–1695
*	Sulphapyridine	$C_{11}H_{11}N_3O_2S$	4.77_x	4.13_5	3.81_8	37–1698
i	Sulphapyridine	$C_{11}H_{11}N_3O_2S$	3.81_x	4.76_7	6.49_5	37–1700
	Sulphasalazine	$C_{18}H_{14}N_4O_5S$	3.77_x	5.73_8	4.28_7	29–1928
	Sulphathiazole	$C_9H_9N_3O_2S_2$	5.81_x	4.12_x	4.02_x	5–206
	Sulphathiazole	$C_9H_9N_3O_2S_2$	5.77_x	4.03_x	4.33_x	29–1930
	Sulphathiazole	$C_9H_9N_3O_2S_2$	5.59_x	5.06_x	4.75_x	29–1931
	Sulphathiazole Sodium Hydrate	$C_9H_8N_3NaO_2S_2 \cdot 1.50H_2O$	6.85_x	4.50_x	3.77_x	8–684
	Sulphathiazole Sodium Hydrate	$C_9H_8N_3NaO_2S_2 \cdot 1.5H_2O$	6.80_x	12.3_3	3.96_8	8–802

Figure 1.4 *Example of the bibliographic entries in the PDF for substances listed by compound name. Each name is followed by the formula and the* d*-spacings of the three strongest diffraction lines, with the relative intensity as a subscript. The last column on the right is the card number in the PDF. Multiple entries with different principle lines are indications of polymorphic systems, for instance the three entries for sulfapyridine, but additional bibliographic information should be obtained from the entries themselves.*

are discussed in Chapters 7–9. Some examples of the role of polymorphism in legal litigation are described in detail in Chapter 10. The patent literature on the U.S. Patents and Trademarks Office site is readily searchable using terms such as "crystal form," "polymorph," etc., and since polymorphic behavior often forms the basis of a patent (as opposed to many journal publications, where it may be peripheral to the main point of the paper) instances of polymorphism are relatively straightforward to locate.

1.3.6 Polymorphism of elements and inorganic compounds

Berzelius (1844) introduced the term "allotropy" as the phenomenon of polymorphism in elements. There has been some debate about the necessity of a special term to designate the polymorphism of elements, as opposed to compounds (Sharma 1987; Reinke et al. 1993), but the term is still introduced in first year chemistry texts, so it has become part of the chemical language. Sharma (1987) has given some examples of allotropism, and Sirota (1982) has noted that "54–55 elements" exhibit the property (Samsonov 1976; Smithells 1976). More complete descriptions can be found in the texts by Wells (1984) and Donohue (1974).

The inorganic equivalent of the CSD is the Inorganic Crystal Structure Database (ICSD) (Bergerhoff et al. 1983; FIZ 2001). This currently contains over 185 000 entries (as of May 2016) with two updates per year, and may be searched in a manner similar to that used for the CSD. Another useful source is the inorganic section of the PDF (Jenkins and Snyder 1996; ICDD 2016). For older references, the first two volumes of Groth (1906b, 1908) are particularly valuable.

1.3.7 Polymorphism in macromolecular crystals

Protein crystal structures are archived in the Protein Data Bank (PDB) (Bernstein et al. 1977; Berman et al. 2003). About 5% of the approximately 124 000 (July 2016) entries (~12 500 proteins, peptides, and viruses, ~900 nucleic acids, ~600 protein/nucleic acid complexes, ~20 carbohydrates) contain the qualifier "form" in the compound name/descriptor field, and most of those refer to polymorphic varieties. In biomolecular crystallography, great efforts are expended varying crystallization conditions in the attempts to obtain single crystals suitable for structural investigations (McPherson 1982, 1989, 1999; McPherson and Gavira 2014). These myriad attempts and the variety of conditions have led to the acquisition of many polymorphic forms, especially for those compounds on which a great deal of work has been done. For instance, the extensively studied lysozyme has entries in the PDB for triclinic, monoclinic, orthorhombic, trigonal, tetragonal, and hexagonal modifications; human hemoglobin has been studied in monoclinic, orthorhombic, and tetragonal modifications. The amount of effort expended in a typical protein crystal structure analysis means that the isolation of crystals and the determination of cell constant and space group is an accomplishment worthy of publication in and of itself. Thus much of the information on polymorphism in macromolecular structures can be found in the primary literature (King et al. 1956, 1962; Kim et al. 1973; Cramer et al. 1974; McClure and Craven 1974; Falini et al. 1996). One secondary source, which should be of increasing importance as the number of proteins studied increases, is the Biological Macromolecule Crystallization Database and the NASA Archive for Protein Crystal Growth Data (Tung and Gallagher 2009). In 2016 this database contained nearly 43 406 crystal entries from about 2300 biological macromolecules. McPherson (1982) summarized the crystallization procedures for 331 proteins. Of these, twenty-three (or about 7%) were listed as being polymorphic. Another primary source is the citations of the McPherson book (1982); of the nearly 700 citations by early 1998, twenty were for polymorphic systems. For smaller proteins, at least some of the incidents of polymorphism have been included in the abovementioned NIST Crystal Data Compilation.

1.4   Historical perspective

Following the historical development of a particular scientific concept or discipline helps to recall the way certain modes of thinking developed, were debated and accepted as new facts came to light and perhaps were abandoned. Tracing that development serves as a reminder that the field is dynamic, with new techniques and new findings changing our ideas and the problems we are seeking to solve. As in any human activity, knowing where we have come from and where we are helps to define where we have to go, and it is certainly true for the field of polymorphism. An early account may be found in Hartley (1902) and a later one in Verma and Krishna (1966).

Mitscherlich is generally credited with the first recognition of the phenomenon of polymorphism (e.g., Tutton 1911a). Early in his career in 1818 he discovered that crystals of certain phosphates and arsenates were very similar. He termed this phenomenon *isomorphism*, and pursued further investigations with Berzelius in Stockholm on the pairs of salts $NaH_2PO_4 \cdot H_2O–NaH_2AsO_4 \cdot H_2O$ and $Na_2HPO_4 \cdot H_2O–Na_2HAsO_4 \cdot H_2O$ and the corresponding ammonium and potassium salts. Among the measurements he carried out were the interfacial angles of the crystals, then a standard technique for characterizing solids (Romé de I'sle 1783; Lima-de-Faria 1990). Mitscherlich (1822) found that the members of the first pair of compounds usually have different crystals, but that the phosphate sometimes crystallizes in the same form as the arsenate. Typical of so many other subsequent discoveries of polymorphism, this one also appears to have been serendipitous:

> Whilst I was still seeking a difference in chemical composition [in the different crystals of the phosphate] I succeeded several times, in the recrystallization of the phosphate, in obtaining crystals having the same form as the acid arsenate. Since I knew

definitely that there was no difference between the two salts I proceeded with the investigation of this phenomenon, and the whole solution of the acid phosphate crystallized several times in the form of the arsenate.

Hence it is established that one and the same body, composed of the same substances in the same proportions, can assume two different forms. This is easily understood from the atomic theory: different forms can result according as the position of the atoms with respect to one another is changed, but the number of different forms remains quite restricted.

Mitscherlich's mentor, Berzelius, considered the discoveries of isomorphism and *dimorphism*, as it was initially called, "the most important made since the doctrine of chemical proportions, which depends on them of necessity for its further development."

Mitscherlich followed this paper shortly thereafter with another one on the dimorphism of sulfur (Mitscherlich 1823). Actually, others had earlier identified more than one crystal form for a number of materials. Klaproth (1798) had recognized that calcite and aragonite have the same chemical composition and Davy had recognized that diamond was a form of carbon (Encyclopaedia Britannica 1798). This prompted Thenard and Biot (1809) to reach nearly the same conclusion as Mitscherlich, in stating that:

> the same chemical elements combined in the same proportions can form compounds differing in their physical properties either because the molecules of these elements have the intrinsic faculty of combining in different ways or because they acquire this faculty through the temporary influence of a foreign agent which afterwards disappears without destroying itself (Webb and Andersen 1978).[6]

Monoclinic sulfur (in addition to the more common orthorhombic form) had also been recognized and documented by a number of other people (see, e.g., Partington 1952, which also contains many early references to polymorphism and polymorphic materials).

The microscope played a crucial role in research on polymorphism, and as this analytical tool became of wider and more sophisticated use, so polymorphism became the subject of increasing interest and study (Lima-de-Faria 1990; Authier 2013). Frankenheim's (1839) early investigation into the polymorphism of potassium nitrate is one of the classic studies of that period. He demonstrated that phase changes could be brought about by solvent moderation and by physical perturbations of a crystal, such as scratching or physical contact with another polymorph. With a detailed study of the mercuric iodide septum he also established many of the principles still recognized today regarding the nature of polymorphism. Some of these are as follows:

---

[6] The controversy that arose about the nature of these discoveries and who should get credit for them prompted correspondence, among others, between Berzelius and the pioneering French crystallographer Haüy. Detailed accounts have been given by Amorós (1959, 1978) and Authier (2013).

- Polymorphs have different melting and boiling points and their vapors have different densities.
- The transition from a low temperature form (A) to a high temperature form (B) is distinguished by a specific temperature of transition.
- The low temperature form (A) cannot exist at a temperature above the transition point to form B, but B can exist below the transition point; below the transition point it is a metastable form.
- At temperatures below the transition point, B will transform to A upon contact with A, the transition proceeding in all directions, but with differing velocities.
- In some cases, B can be converted without contact with A by mechanical shock or by scratching.
- Heat is absorbed upon the transition from A to B.

As early as 1835, Frankenheim was particularly concerned with cohesive forces in different states of aggregation, and suggested that in the various solid states of a material the attractions which lead to the aggregation in different solids are different, and are characterized by different special symmetry relations (Frankenheim, 1835).

The first *polarizing* microscope, an instrument that was destined to play such an important role in the development of chemical crystallography in general and polymorphism in particular, was invented by Amici (1844). It was also at about this point that Berzelius (1844), Mitscherlich's early mentor, suggested that the pyrite-marcasite polymorphism of FeS_2 was due to the polymorphism of the sulfur in the two solids, while the iron was the same in the two, although the concept of structure, per se, had not yet really crept into the lexicon of chemical crystallography. As Hartley (1902) pointed out, in spite of the investigation of many polymorphic modifications, the middle decades of the nineteenth century were not noted for any new generalizations in terms of the characterization and understanding of the phenomenon itself.

In the 1870s things started to change rapidly. Mallard (1876, 1879) had been concerned with geometrical crystallography and had considered the structural basis for polymorphism in an 1876 paper. He considered crystals as being built up of minute elementary crystallites that can pack in a number of ways giving rise to different crystal forms. The ideal form is that with the closest packing thereby being the most dense, and different forms have different packing which results in different physical properties such as optical properties and density. He attributed the differences in physical properties to differences in the arrangement of these elementary crystallites. In general, though, he still saw a great deal of similarity in the structures of two forms of the same substance:

It has been known for a long time that when the same substance displays two fundamentally incompatible forms, often belonging to two different chemical systems, these two forms are always only slightly different and the symmetry of the less symmetrical is very similar to that of the other.

As an early pioneer of chemical crystallography, (particularly of organic compounds) Lehmann's PhD thesis, much of which was published in the first issues of *Zeitschrift für Kristallographie* (founded by Groth; Lehmann 1877a, 1877b), already contained some new concepts for polymorphic systems (Lehmann 1891). He characterized two different types of polymorphism. The first, which he termed *monotropic*, involves two forms in which one undergoes an irreversible phase change to the second form. The second form is termed *enantiotropic*, in which the two phases can undergo a reversible phase transition (see Chapter 2). An increase in temperature tends to lead to the transformation to the more stable form.[7] Lehman also showed that many organic compounds crystallize from the melt as monotropic forms, and that these tend to be the less stable form with a lower melting point.[8]

Lehmann further reduced Mallard's "structural crystallites" to be aggregates of "physical molecules." Then the structural crystallites could differ in the number or in the arrangement of the physical molecules of which they were composed, thereby constituting the difference between two polymorphs. These distinctions were then related to the transformation phenomena: an enantiotropic transformation was characterized by Lehmann as a reversible polymerization, that is, with an increase in temperature, elementary particles of a large size were transformed into elementary particles of a smaller size. In a monotropic transition, according to Lehmann, there is no such relationship between temperature and the mode of rearrangement.

The problem of distinguishing between molecular isomerism and polymorphism arose in this period as well. For instance, in a manner similar to Berzelius' arguments about the pyrite-marcasite system, Geuther (1883) postulated that the calcite-aragonite polymorphism arose from the existence of two carbonic acids. Wyrouboff (1890) differed in his view, claiming that polymorphs differ only in their physical properties. Crystals with different molecular isomers would give different products upon reaction, whereas true polymorphs would give the same reaction products. Polymorphic products, according to Wyrouboff, are distinguishable only by

---

[7] It is remarkable how particular systems attain the status of "classics." Hartley (1902) noted α and β sulfur (transition temperature 95.6 °C), red and yellow mercuric iodide (transition temperature 126 °C), and the four modifications of ammonium nitrate as examples of enantiotropic behavior. These three systems are given as archetypical experiments in Chamot and Mason's (1973) book on chemical microscopy.

[8] It is of interest to note Tutton's optimistic assessment of Lehmann's definition of monotropism and enantiotropism, published just prior to the dawn of the age of structural crystallography: "It thus appears that any general acceptance of Lehmann's ideas will only tend to amplify and further explain the nature of polymorphism on the lines here laid down, the temperature conversion of one form into another being merely that at which either a different homogeneous packing is possible, or that at which the stereometric relations of the atoms in the molecule are so altered as to produce a new form of point-system without forming a new chemical compound" (Tutton 1911b).

their physical properties.[9] He also differed with Lehmann's classification of polymorphs based on monotropic and enantiotropic phase transformations, choosing a scheme based essentially on the physical manifestations of the phase changes. For most materials, labelled heteroaxial by Wyrouboff, the starting crystal loses homogeneity upon transformation, becoming optically clouded and the transformation results in the breaking up of the crystal into many smaller crystallites. The heteroaxial designation results from the lack of any correspondence between axes of the initial and product phases. In the second class, labelled isoaxial, the phase transformation takes place without the crystal losing its optical transparency. If it does break up into smaller crystals they remain parallel to each other and to the axes of the parent crystal.

Following the elaboration of many of the principles of thermodynamics in the latter three decades of the nineteenth century, a major development in polymorphism came with the work by Ostwald (1897) on the relative stability of different polymorphs, and the reason for the mere existence of less stable forms. Among the findings was the fact that unstable polymorphic forms have a greater solubility than the more stable forms in a particular solvent, and that monotropic forms have a lower melting point than enantiotropic forms. Ostwald related these findings to the phenomena of supersaturation and supercooling. The result is Ostwald's so-called "Rule of Steps" or "Law of Successive Reactions," although as Findlay (1951) has pointed out, the designation "law" is not justified since many exceptions are known, but as a guideline or rule of thumb, it is still a useful concept. In Ostwald's words (1897), "...that on leaving any state, and passing into a more stable one, that which is selected is not the most stable one under the existing conditions, but the nearest" (i.e., that which can be reached with the minimum loss of free energy). Groth (1906a) provided an explanation for the phenomenon, which is discussed in detail in Chapters 2 and 3. The phenomenon described by Ostwald is in fact often (unknowingly) observed by synthetic chemists. The first synthesis of a new material with a melting point above room temperature may result in a metastable form, which eventually (either spontaneously or through an intentional recrystallization) will yield a more stable form. The metastable form may not always be recognized or the stable form may not appear immediately—it may take years until the appropriate constellation of conditions exists (Davey et al. 2013). However, once seeds of the more stable form exist in a particular environment, it may be difficult to obtain the metastable form (Dunitz and Bernstein 1995) (see Section 3.5). An example of the stable form crystallizing out of the metastable one over a period of days is shown in Figure 1.5.

Ostwald (1897) was aware of the fact that his "rule" was tenuous, since it was not based on a very large set of observations. In addition, if the metastable region

---

[9] On first glance this seems consistent with our definition above. However, the topochemical principles, first enunciated by Cohen and Schmidt (1964) were actually developed from the fact that different polymorphs of a substance (*trans*-cinnamic acid) undergo different photochemical reactions, leading to different products (see Section 6.4).

Figure 1.5 *Example of Ostwald's rule of successive reaction. 2,4-Dibromoacetanilide initially crystallizes from alcoholic solutions as small needle-shaped crystals, forming the voluminous mass in (1). Successive photos (2, 3, and 4) of the same crystallization vessel, taken at two day intervals show the transformation to the more stable chunky rhombic crystals (from Findlay 1951, with permission).*

were to shrink to a vanishingly small value, then sufficient time would not be allowed for crystallization of Form I to appear, and the "rule" would be invalidated. In fact, this does happen in many cases. Nevertheless, Ostwald's rule has remained in the lexicon of crystal chemists, probably because it is generally observed that if a succession of polymorphic forms is obtained, those which appear later are generally more stable than those which appear earlier.

The turn of the twentieth century brought to play a convergence of many experimental techniques and theoretical developments in the investigation and understanding of polymorphism. Experimentally, hot stage microscopy (Lehmann 1891), dilatometry, precise vapor pressure and solubility measurements, heat capacity and transition point determinations all served to provide data by which models and theories could be tested. Theoretically, thermodynamic relations, in particular the Gibbs phase rule and the Clapeyron equation applied to the solid state, established the equilibrium relationship that exists among polymorphic forms. There appears to have been a real symbiotic relationship between theory and experiment here, as improved theories required more precise measurements and data, which in turn provided the impetus generated for theoretical refinements. Most of Frankenheim's conclusions were shown to be correct, and Lehmann's assertions about monotropism and enantiotropism also were validated (Tutton 1911b).

The study of organic crystals had gained considerable momentum during this period, particularly in Germany. This was a period of intense activity in organic chemistry. As noted earlier, in those days preceding the invention and application of today's armory of spectroscopic methods for the characterization of physical properties of a solid material, often involving their own sensual perception for such characteristics, chemists employed somewhat simpler techniques. By the end of the 19th century, one of the principal tools of the organic chemist for characterizing and identifying materials was the polarizing microscope, which by 1879 had essentially developed to the state we recognize today (Lima-de-Faria 1990). Many of the observations on organic crystals are summarized in the last three volumes of Groth's five-volume compendium (Groth 1910, 1917, 1919). By the time the last volume was published X-ray crystallography was well on the way to overtaking polarized light microscopy as the principal method for examining and characterizing solids. The view of polymorphism turned from phenomenological to structural.

Gustav Tammann from Göttingen was one who bridged this period, publishing a book on crystallization and melting prior to the advent of X-ray crystallography (Tammann 1903) and one on the structural aspects of crystals well into the age of X-ray crystallography (Tammann 1926). Tammann considered two polymorphic molecular crystals as being identical molecular species being arranged on different lattices, and he made a point of distinguishing between the outer form of the crystal and the inner structure of the crystal (habit vs. form in modern terms; see Section 2.4.1). He revisited Lehmann's division of enantiotropic and monotropic forms, noting that this distinction in principle is really not sufficient, but in practice works quite well, since most investigators do not operate outside of the domain of one atmosphere. For a universal criterion for distinction among polymorphs, he proposed the relative thermodynamic stability, namely the thermodynamic potential per unit of mass. This is the ς-surface of Gibbs for which the less stable forms have higher values than the stable form.

Tammann demonstrated the use of this thermodynamic measure with three-dimensional $P/T/\varsigma$ plots. If the surfaces for two polymorphs intersect then each is partly stable and partly unstable, depending on the P-T domain. If they don't intersect, then one form may be considered "totally stable" while the other form is "totally unstable." Tammann (1926) reviews the thermodynamic details of a number of possible cases of relative stability, but he summarizes a still-used rule of thumb in stating that the "relative magnitudes of the volumes and of the heats of fusion of two forms differing in stability are to be regarded as indications of total or partial stability." In fact, these generalizations are echoed in the so-called "density rule" (Section 2.2.10) and "heat-of-fusion rule" (Section 2.2.7) enunciated more than half a century later on the basis of a large body of experimental data (Burger and Ramberger 1979a). Tammann also proposed models for nucleation and the relationship between molecular structure and the possibility of polymorphism.

By 1925, Niggli (1924) had proposed a model for enantiotropism that was based on structural *changes*. If two forms are similar in structure, then the structural change upon a phase transformation is not large, and the two forms were considered to be enantiotropically related. A second type of relationship exists for polymorphs in which the structural differences are large (e.g., calcite–aragonite or graphite–diamond) and the phase transformation perforce results in much greater structural change.

The emerging detailed knowledge of crystal structures in the 1920s changed the direction from thermodynamic to structural, and stimulated interest in the relationship between the structure and properties of materials, as enunciated by Goldschmidt (1929): "The task of crystal chemistry is to find systematic relationships between chemical composition and physical properties of crystalline substances..." But there still was an abiding interest "...in relating crystal structure to chemical composition especially to find how crystal structure, the arrangement of atoms in crystals, depends on chemical composition." After noting "the very extensive wealth of observations [of an earlier] epoch, which has been treasured by v. Groth...," Goldschmidt considered polymorphism to be the result of "thermodynamic alteration. The substance, under different conditions, may no longer be isomorphous with itself...The amount of thermal energy, involved in polymorphic changes are mostly rather small, compared for instance with the heats of chemical reactions" (Goldschmidt 1929).

By the 1930s, though, interest in polymorphism had very much waned, as the chemical and physical aspects of crystallography were relegated to a relatively minor role in the shadows of the rapidly developing discipline of structural crystallography, with its capability of revealing molecular structure at the atomic level. For about thirty years interest and activity in polymorphism of organic materials was limited to a few devoted (and probably isolated) practitioners, notably L. and A. Kofler and their successors in Innsbruck, and McCrone (1957)[10] in the US. In

---

[10] In 1956 McCrone founded McCrone Associates, a private analytical laboratory in which the principal analytical technique employed was polarized light spectroscopy. Over the years he and his staff learned to visually identify over 30 000 particles (McLafferty 1990). McCrone Associates specialized in the identification of polymorphs, asbestos samples, and airborne impurities, among others. McCrone eventually endowed a chair of chemical microscopy to Cornell University, his Alma Mater.

both of these centers of activity the optical microscope, often equipped with a hot stage, was a principal tool of investigation (McCrone 1957; Kuhnert-Brandstätter 1971), which may have played an important role in saving it from extinction in the search for and investigation of organic polymorphic materials.

McCrone's (1965) comprehensive chapter on polymorphism and his later review on its growing pharmaceutical importance (Haleblian and McCrone 1969) are indeed landmark publications on this subject. However, as can be seen from Figure 1.6, they were both virtually ignored (there were even six years in which the McCrone chapter was not cited at all), again until the 1991–1992 time frame, when the number of citations started to rise dramatically, in concert with the number of publications, patents, and citations, as shown in Figure 1.1. This is further evidence that the seminal event for these developments was Zantac litigation, and the central role of polymorphism in that litigation.

These two publications have set the stage for subsequent developments. In summarizing their 1937 account of the historical development of polymorphism, Buerger and Bloom (1937) raised two "major questions that demand answer:"

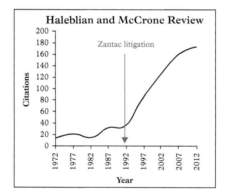

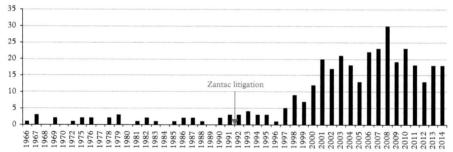

Figure 1.6 *Plots of the citations of two seminal papers on the polymorphism of molecular crystals. The arrows in the 1991–1992 time frame note the beginning of the first of a number of patent litigations involving the antiulcer drug the antac®, at the time world's largest selling drug. Statistics for upper: Haleblian and McCrone (1969); lower: McCrone (1965).*

1. What fundamental property of crystalline matter calls for different forms of the same chemical substance in compounds of all types and causes the appearance of discontinuous jumps in physical properties at definite transformation points?

2. What are the factors that determine what particular crystalline form will be generated from vapor liquid or solution when the thermodynamic phase region of the solid state is entered?

In the interim a great deal has been learned about the nature and consequences of polymorphism, but the fact that over eighty years following that publication, and nearly fifty years following the two McCrone publications both of these questions are still largely unanswered is a challenge to chemists, physicists and crystallographers. What has been learned is the subject of the remainder of this book and is meant to provide a jumping-off point for the investigation of these questions.

1.5 Commercial/industrial importance of polymorphism—some additional comments

The discovery, whether accidental or intended, of polymorphs is unlikely to be greeted with enthusiasm by senior management, and the situation is better treated as an opportunity rather than as a problem. Opportunities are likely to exist for increasing patent cover, for retaining a competitive edge through unpublished knowledge and in formulating pharmaceutical products. A metastable polymorph can be used in capsules for tableting, and the thermodynamically stable one for suspensions.

Bavin (1989)

As I document further in a number of the later chapters of this book, polymorphism can and does play an important role in a number of industrial and commercial applications. Again, the overriding reasons for this are the differences in properties that often accompany differences in structure. Obviously if two polymorphs exhibited the same properties required for a certain product specification, there would be no concern about which polymorph (or what mixture of polymorphs) was actually present. A number of examples will serve to demonstrate the scope of this role.

As Bavin (1989) has succinctly noted, in the real world of chemical processing "few compounds reach development and fewer still are marketed." Since in many ways a new polymorph is a new compound, the characterization and control of the polymorphic behavior is an integral part of that development and marketing.

In the photographic industry, differences in solubility of polymorphs can pose problems during manufacture, and can lead to impaired performance (Nass 1991). The transformation to undesired structures may take place through a solvent-mediated phase transformation (see Section 3.3), during a manufacturing process.

On the other hand, recognition of such a process may lead to its utilization to obtain a desired polymorph. In an example from the food additives industry, for L-glutamic acid, later converted to the monosodium salt (MSG) used for taste enhancement, it is crucial to obtain the α polymorph rather than the β form. The latter can lead to a situation in which the crystallizing slurry coagulates into a gel and can no longer be processed (Sugita 1988). A study by Garti and Zour (1997) indicates that the addition of selected surface-active agents can lead to the preferential crystallization of the α polymorph (see Section 3.4).

Polymorphism plays an important role in the huge industries of fat-based food products, for instance, ice cream, chocolate, and margarine (Weissberger 1956; O'Connor 1960; Garti and Sato 1988; Loisel et al. 1998; Roth 2005), as well as in the chemistry involved with long alkyl chains (Robles et al. 1998). The melting point and melting behavior are clearly important physical properties of such materials, but the appearance and the properties perceived by the consumer are to a great extent determined by the structure of the solid fat phase. The solid fats provide structure to the marketed product, with other additives supplying taste, stability, and color. One of the key ingredients of these fats is the triacylglycerols, and the different polymorphic forms play an important role in both the designing of properties and the processing of the product (Hagemann and Rothfus 1993; Johansson et al. 1995; Jovanovic et al. 1995; Herrera and Rocha 1996). Hence considerable effort is often expended in characterizing the polymorphic forms of many of these materials, their properties and the methods required to control which polymorph is obtained (Sato et al. 1985; Suzuki and Ogaki 1986; Ng 1990; Kellens et al. 1992; Sato 1999).

Similarly, the chain-like structures of polymers also result in a proliferation of polymorphic structures (e.g., Keller and Cheng 1998; Lotz 2000; Rastogi and Kurelec 2000). The differences in structure lead to a variety of properties (e.g., Calleja et al. 1993; Chunwachirasiri et al. 2000), which of course is one of the driving forces for the development of new polymeric materials. Although not referenced specifically many of the principles and examples presented in subsequent chapters apply equally well to many of the polymorphic molecular systems.

2

Fundamentals

Of course, one cannot ignore the physics, and, in particular, one has to pay due respect to thermodynamic considerations...

Dunitz (1991)

...a major deficiency in our current knowledge and understanding concerns the relationship among the members of a polymorph cluster—what is their relative mutual stability, how do they transform, one into another, what are the thermodynamic factors governing their mutual stability, what are the kinetics of the transitions. Answers to the last of these questions are very important to users of, and sufferers from, polymorphism.

Herbstein (2001)

2.1   Introduction

Polymorphic structures of molecular crystals are different phases of a particular molecular entity. In order to understand the formation of those phases and relationships between them we make use of the classic tools of the phase rule, and of thermodynamics and kinetics. In this chapter I will review the thermodynamics in the context of their relevance to polymorphism and explore a number of areas in which they have proved useful in understanding the relationship between polymorphs and polymorphic behavior. This will be followed by a summary of the role of kinetic factors in detecting the growth of polymorphic forms. I will then provide some guidelines for presenting and comparing the structural aspects of different polymorphic structures, with particular emphasis on those that are dominated by hydrogen bonds.

2.2   The thermodynamics of molecular crystals

It is beyond the scope of this book to provide a comprehensive review of the thermodynamics of molecular crystals. The field was very adequately covered in the classic chapter by Westrum and McCullough (1963), a work that has stood the test of time remarkably well, and can serve as an excellent resource on this general

Polymorphism in Molecular Crystals. Second Edition. Joel Bernstein. © Joel Bernstein 2020. Published in 2020 by Oxford University Press. DOI: 10.1093/oso/ 9780199655441.001.0001

subject. An earlier useful reference is the chapter on polymorphism in the classic book by Tammann (1926).

2.2.1   The phase rule

The phase rule was first formulated by Gibbs (1876) on the basis of thermodynamic principles and then applied to physical chemistry by Roozeboom (1911). As with so much of chemistry the apparently absolute physical principles stated in the Phase Rule must be tempered by real chemical situations (Dunitz 1991; Brittain 1999c). However, in order to establish a working language, it is necessary to define terms and then indicate the difficulties that may arise in the practical use of those definitions.

The phase rule is simply stated as follows:

$$F = C - P + 2,$$

where F is the number of degrees of freedom of the system, C the number of components, and P the number of phases.

A phase is defined (Glasstone 1940; Findlay 1951) as *any homogeneous and physically distinct part of a system which is separated from other parts of the system by definite bounding surfaces*. By this definition for any substance there is one gaseous phase and one liquid phase, since these must be physically and chemically homogeneous. Each crystalline form constitutes an individual phase, for example, the different forms of ice. A mixture of two polymorphs contains two solid phases, but a homogeneous solid solution or an alloy of two totally miscible metals is only one phase. Problems arise when one must consider the dimensions of the structural realm for the definition of "homogeneous"—uniform throughout. As both Findlay (1951) and Dunitz (1991) have pointed out, at the molecular level such a definition certainly breaks down, and even some molecular substances that exhibit, say, the X-ray diffraction patterns expected from single crystals (i.e., a single phase), have been shown on closer inspection to be inhomogeneous mixed crystals (Weissbuch et al. 1995) or inhomogeneous racemic mixtures (Green and Knossow 1981; Ramdas et al. 1981). In other cases, it has been claimed that a single crystal was actually a hybrid in which two polymorphs coexist (Freer and Kraut 1965; Fryer 1997; Coppens et al. 1998; Fomitchev et al. 2000; Bond et al. 2007; Vogt et al. 2009). Similar phenomena have been described as "composite crystals" by Coppens et al. (1990).

The number of *components* is the minimum number of independent species required to define the composition of all of the phases in the system. The simplest example usually cited to demonstrate the concept of components is that of water, which can exist in various equilibria involving the solid, liquid, and gas. In such a system there is one component. Likewise for acetic acid, even though it associates into dimers in the solid, liquid, and gaseous state, the composition of each phase can be expressed in terms of the acetic acid molecule and this is the only component.

The important point for such a system is that the monomer–dimer equilibrium is established very rapidly, that is, faster than the time required to determine, say, the vapor pressure. In the cases in which the equilibrium between molecular species is established more slowly than the time required for a physical measurement, the vapor pressure, for example, will no longer be a function only of temperature, but of the composition of the mixture, and the definition of a component acquires a kinetic aspect.

The *number of degrees of freedom* (sometimes also referred to as the *variance*) is the number of variable factors, such as temperature, pressure, and concentration, that must be fixed in order to define the condition of a system at equilibrium. Thus, a one-component system in one phase, say a gas, would have two degrees of freedom; a one-component system in two phases (liquid and gas) would have one degree of freedom. A system of one component and three phases would have no degrees of freedom. For the water system this would correspond to the familiar triple point. These relationships are often described, respectively, as bivariant, monovariant, and invariant (Trevor 1902).

For polymorphic systems of a particular material we are interested in the relationship between polymorphs of one component. A maximum of three polymorphs can coexist in equilibrium in an invariant system, since the system cannot have a negative number of degrees of freedom. This will also correspond to a triple point. For the more usual case of interest of two polymorphs the system is monovariant, which means that the two can coexist in equilibrium with either the vapor or the liquid phases, but not both. In either of these instances there will be another invariant triple point for the two solid phases and the vapor on the one hand, or for the two solid phases and the liquid on the other hand. These are best understood in terms of phase diagrams, which are discussed below, following a review of some fundamental thermodynamic relationships that are important in the treatment of polymorphic systems.

2.2.2 Thermodynamic relations in polymorphs

In terms of thermodynamics one of the key questions regarding polymorphic systems is the relative stability of the various crystal modifications and the changes in thermodynamic relationships accompanying phase changes and different domains of temperature, pressure, and other conditions. Buerger's (1951) treatment of these questions provides the fundamentals upon which to base further discussion.

For simplicity, and to demonstrate the principles of these considerations, we will generally limit ourselves to a discussion of two polymorphic solids, although the extension to more complex systems is based on precisely the same principles. The relative stability of the two polymorphs depends on their free energies, the most stable form having the lowest free energy. Because of this energy relationship those forms that are less stable will be energetically driven to transform into the most stable form, although kinetic factors may prevent the transformation (such is the case, for instance, for diamond and graphite). Since we are dealing with solids the differences in volume between polymorphs are small fractions of the volumes of the solids themselves; then we can neglect volume and pressure changes with energy. Under these conditions (of essentially constant temperature and pressure) the free energy of a solid phase may be represented by the Helmholtz relationship,

$$A = E - TS,$$

in which E is the internal energy, T the absolute temperature, and S the entropy.

At absolute zero TS vanishes and the Helmholtz free energy equals the internal energy. As a consequence, at absolute zero the most stable polymorphic modification should have the lowest internal energy.[1] Above absolute zero the entropy term will play a role which may differ for the two polymorphs. Hence the behavior of the free energy as a function of temperature can differ for the two polymorphs, as presented by the curves A_I and A_{II} in Figure 2.1. Form I is more stable at absolute zero and the two curves behave differently, crossing at the transition temperature $T_{p,I/II}$. Above the transition temperature Form II is more stable. At the transition temperature the free energy of the two forms is identical, but since the internal energy of Form I is less than that of Form II a quantity of energy ΔE is required to be input for the phase transition, which must be endothermic. Buerger (1951) has also demonstrated the endothermic nature of any transformation that takes place upon raising the temperature.

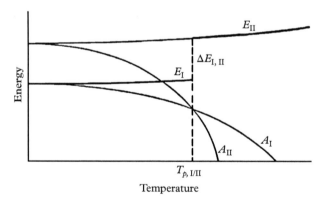

Figure 2.1 *Energy versus temperature curves for two polymorphs I and II. A is the Helmholtz free energy and E is the internal energy. Consistent with the labeling scheme proposed by McCrone (see Chapter 1), Form I is assumed to be the stable form at room temperature. (From Buerger 1951, with permission.)*

[1] Buerger (1951) defines the internal energy as being composed of the "structural energy" and the zero-point energy. He objects to non-crystallographers equating the term "lattice energy" with "structure energy." In the nearly seventy years since his chapter was written, "lattice energy" appears to have become accepted usage and will be used further here as well.

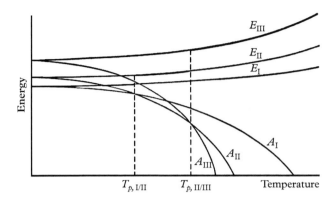

Figure 2.2 *Energy versus temperature curves for three polymorphs I, II, and III. A is the Helmholtz free energy and* E *is the internal energy. (Adapted from Buerger 1951, with permission.)*

Moving from low temperature to high temperature for a trimorphic system the positive heat of transformation at each step (ΔE) corresponding to a crossing point of the corresponding free energy curves leads to the behavior in Figure 2.2. If the entropy (which is the slope of the free energy curve) increases at a uniform rate, then each pair of curves crosses only once. To put these diagrams into perspective it is perhaps useful at this point to get some idea of the magnitudes of some of the differences in lattice energy and vibrational entropy between polymorphic forms. Values of ΔE for real polymorphs are often in the range 0–10 kJ mol^{-1} (Kitaigorodskii 1970; Kuhnert-Brandstätter and Solinger 1989; Chickos et al. 1991) and probably do not exceed 25 kJ mol^{-1}, while ΔS values are less than 15 kJ mol^{-1} (Gavezzotti and Filippini 1995).

2.2.3   Energy versus temperature diagrams—the Gibbs free energy

Diagrams qualitatively very similar to Figures 2.1 and 2.2 may be prepared on the basis of the Gibbs free energy,

$$G = H - TS,$$

rather than the Helmholtz free energy. In such a case E is replaced by the enthalpy H, and A is replaced by G, the Gibbs free energy. Since the data required to produce these are experimentally more readily accessible, the diagrams based on the Gibbs free energy are more commonly in use, and will be described in some detail here (Grunenberg et al. 1996; Herbstein 2004; Kawakami 2007; Upadhyay et al. 2012). The utility of these diagrams is that they contain a great deal of information in a compact form, and provide a one page visual and readily interpretable summary of what can be complex interrelationships among polymorphic modifications.

As their utility is recognized these diagrams should become more widely used for the characterization of polymorphic systems and the phase relationships among various polymorphs (Henck and Kuhnert-Brandstätter 1999).

A typical energy versus temperature diagram using the Gibbs relationship is given in Figure 2.3. Compared to the analogous Figure 2.1 there are two additional isobars: the H_{liq} curve (above the two (H_I and H_{II}) solid curves), and the G_{liq} curve. The H versus temperature curves may be constructed experimentally by determination of the heat capacity C_p, from

$$\left(\frac{\delta H}{\delta T}\right) = C_p,$$

the fundamental relationship between the enthalpy and the heat capacity, as demonstrated in Figure 2.4. A number of points concerning this plot are worthy of note. It can be shown from the third law of thermodynamics that the heat capacity of an ideal crystal is zero at 0 K. Therefore the slope of the curve in Figure 2.4 (and those of H in Figure 2.3) at 0 K must also be zero.

From the expression for the Gibbs free energy the partial derivative with respect to temperature is

$$\left(\frac{\delta G}{\delta T}\right)_p = -S.$$

Since S is always positive, G is a constantly decreasing function, as seen in Figure 2.3. The G isobars can follow different paths, and their intersections represent

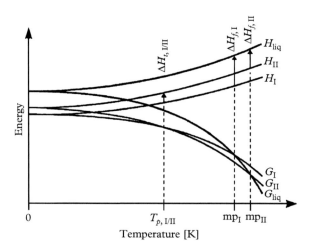

Figure 2.3 *Energy versus temperature (E/T) diagram of a dimorphic system. G is the Gibbs free energy and H is the enthalpy. This diagram represents the situation for an enantiotropic system, in which Form I is the stable form below the transition point, and presumably at room temperature, consistent with the labeling scheme for polymorphs proposed by McCrone (see Chapter 1). (Adapted from Grunenberg et al. 1996, with permission.)*

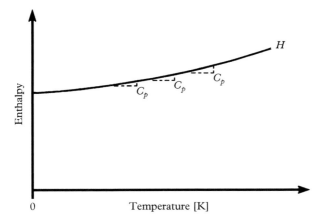

Figure 2.4 *Plot of enthalpy* H *versus temperature indicating the relationship with the heat capacity,* C_p. *(Adapted from Grunenberg et al. 1996, with permission.)*

transition points between phases. Buerger (1951) suggested the geometric possibility for two of the *G* isobars to cross twice, but Burger and Ramberger (1979a) have shown by statistical-mechanical arguments that only one crossing is physically possible.

2.2.4 Enantiotropism and monotropism

Enantiotropism and monotropism were referred to in their historical context in Chapter 1. I now provide a thermodynamic basis for these two important descriptors of polymorphic behavior.

In Figure 2.3 it is seen that the thermodynamic transition point $T_{p,\mathrm{I/II}}$, defined by the point at which G_{I} and G_{II} cross, falls at a temperature below the melting point of the lower melting form, mp_{II}. This is the thermodynamic definition of an enantiotropic polymorphic system. The melting point itself is defined by the crossing of the *G* curves for Form II and the liquid (and similarly for the melting point of Form I). The enthalpies of transition (ΔH_t) or fusion (ΔH_f) appear at the corresponding temperatures as the vertical differences between the appropriate *H* curves.

The energy versus temperature diagram for a monotropic relationship between two polymorphs is shown in Figure 2.5. In this case, however, the free energy curves do not cross at a temperature below the two melting points.

2.2.5 Phase diagrams in terms of pressure and temperature

Because pressure and temperature are two readily measured experimental quantities (e.g., Griesser et al. 1999), the relationships among the vapor, liquid, and polymorphs of a substance are often represented on diagrams of pressure versus

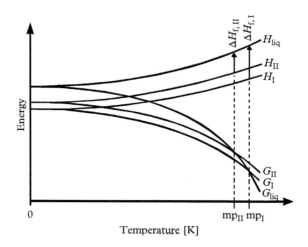

Figure 2.5 *Energy versus temperature (E/T) diagram for a monotropic dimorphic system. The symbols have the same meaning as in Figure 2.3. Form I is more stable at all temperatures; the crossing of the G_I and G_{II} curves (not shown) will be above the melting point for Form I and Form II. (From Grunenberg et al. 1996, with permission.)*

temperature. These are also very useful in summarizing the polymorphic behavior of a system.

Figure 2.6 shows the prototypical plots of pressure versus temperature for the enantiotropic and monotropic cases. These are best understood by proceeding along various curves, which represent equilibrium situations between two phases. The l./v. line in the high-temperature region of Figure 2.6(a) is the boiling point curve. Moving to lower temperatures along that line one encounters the II/v. line, which is the sublimation curve for Form II. The intersection of the two curves is the melting point of Form II. Under thermodynamic conditions Form II would crystallize out at this point and the solid part of the II/v. line would govern the behavior. However, if kinetic conditions prevail (e.g., if the temperature is lowered rapidly) the system may proceed along the broken l./v. line to the intersection with the I/v. line, at which point Form I would crystallize. Continuing downward along the solid part of the II/v. sublimation curve, the crossing point with the II/v. sublimation curve is the transition point between the two polymorphic phases. Once again, if thermodynamic conditions prevail Form II will be transformed to Form I. Under kinetic conditions Form II may continue to exist (even indefinitely in some cases) along the II/v. sublimation curve. Figure 2.6(a) represents the enantiotropic case because the transition point between the two solid phases is at a temperature below the melting point of Form II and Figure 2.6(b) is the diagram for the monotropic case.

As McCrone (1965) has pointed out, the complete pressure–temperature diagram must also contain the curves representing the phase boundaries between the two solid forms (transition temperature curve) and between each form and

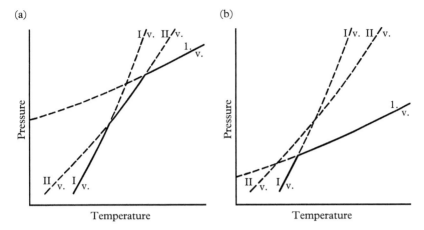

Figure 2.6 *Pressure versus temperature plots. I/v. and II/v. represent sublimation curves; l./v. is the boiling point curve. Broken lines represent regions that are thermodynamically unstable or inaccessible. (a) Enantiotropic system; (b) monotropic case. The labeling corresponds to earlier figures to indicate that Form I is stable at room temperature which is below the transition point in the enantiotropic case. (From McCrone 1965, with permission.)*

the liquid (melting point curves). These are shown included in Figure 2.7 for the same system as in Figure 2.6. The transition temperature curve rises from the intersection of the two solid–vapor curves and goes through the intersection of the two solid–liquid curves. For the enantiotropic case the curve is a solid line, indicating the existence of true thermodynamic equilibrium between the two phases. The II ↔ l. line is also solid, since it originates from a thermodynamically accessible point. The I ↔ l. line is a broken one, however, since it originates from a thermodynamically inaccessible, but kinetically accessible, point. Similar arguments describe the three additional lines for the monotropic diagram (Figure 2.7(b)).

A review of some of the features of these diagrams in terms of the phase rule is enlightening (Findlay 1951). A system composed of two different solid forms of a substance will have one component and two solid phases. In the absence of a further definition of the system there will be one degree of freedom. In Figure 2.7 this is either the temperature or the pressure along the I↔II line. Choosing either variable fixes a point and defines the system. However, suppose that we are interested in the situation when the two phases are in equilibrium with the liquid or the vapor. Each one of those is an additional phase, making three in total and, by virtue of the phase rule, rendering the system invariant. Invariance results in a triple point for each case, defined by the intersection of the I ↔ II curve with the I ↔ v. and II ↔ v. curves on the one hand and the intersection of the I ↔ II curve with the I ↔ l. and II ↔ l. curves on the other hand.

Recalling the classic definition of the triple point, say, for water as the intersection of the solid–vapor and the liquid–vapor curves, the analogy in Figure 2.7 is the intersection of the II ↔ v. and I ↔ v. curves. Below the triple point only one of

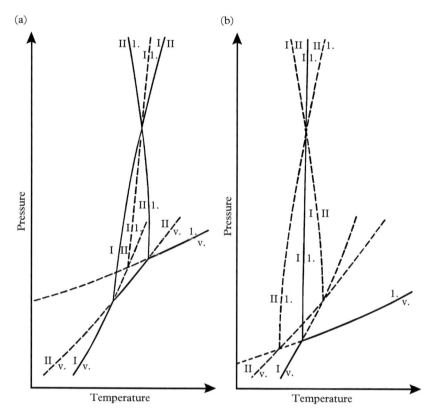

Figure 2.7 *Transition temperature (I ↔ II) and meting point (I ↔ melt, II ↔ melt) curves added to Figure 2.6: (a) enantiotropic; (b) monotropic. (From McCrone 1965, with permission.)*

the solid phases (I) can exist in stable equilibrium with the vapor; above the triple point only II can exist in equilibrium with the vapor. The triple point I–II–v. may be looked upon as a point at which there is a change in the relative stability of the two phases. In a general way, then, phase changes may be viewed as what can transpire at the triple point.

Which of the two representations (energy/temperature or pressure/temperature) is preferred? Westrum and McCullough (1963) have pointed out that most textbooks use the pressure/temperature representation because of the availability of such data for inorganic systems and examples that can be given using those data. On the other hand for organic systems, energy data are more abundant and more convenient to use (Chickos 1987; Chickos et al. 1991). The sources of those data and their use in generating (semi-quantitative) energy–temperature diagrams are given in Chapters 4 and 5. Examples of enantiotropic and monotropic energy–temperature diagrams for two real systems are given in Figure 2.8.

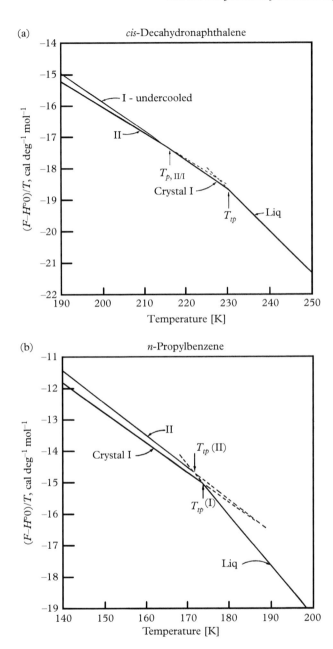

Figure 2.8 *Diagrams illustrating (a) enantiotropic and (b) monotropic phase relationships for two organic compounds,* cis-*decahydronaphthalene and* n-*propylbenzene, respectively. Note that the y scale is actually given in units of entropy calculated from the energy terms.* $T_{p,\,II/I}$ *represents the transition point between phases I and II, being above the melting point in (a) and (by extrapolation) below the melting point in (b). (From Westrum and McCullough 1963, with permission.)*

Consideration of the details of these phase diagrams leads to a number of "rules" which are helpful in characterizing, understanding, and predicting the behavior of polymorphic systems, including monotropism and enantiotropism as well as generating the energy–temperature diagrams from experimental measurements and observations. A number of these rules were originally developed by Tammann (1926) and then expanded by Burger and Ramberger (1979a; 1979b) and by Grunenberg et al. (1996).

2.2.6   Heat-of-transition rule

As indicated in Figures 2.3–2.5, ΔH and ΔS are usually positive, and it is assumed that the H curves do not intersect. Also as discussed above, the G curves intersect only once. This set of circumstances leads to a statement of the heat-of-transition rule (Burger and Ramberger 1979a; Grunenberg et al. 1996):

> If an endothermic phase change is observed at a particular temperature, the transition point lies below that temperature, and the two polymorphs are enantiotropically related. If an exothermic transition is observed, then there is no thermodynamic transition point below that transition temperature. This can occur when the two modifications are monotropically related or when they are enantiotropically related and the thermodynamic transition point is higher than the measured transition temperature.

Burger and Ramberger (1979b) claim that this rule is observed in at least 99% of the cases examined.

2.2.7   Heat-of-fusion rule

The heat-of-fusion rule states that in an enantiotropic system the higher melting polymorph will have the lower heat of fusion. If the higher melting polymorph has a higher heat of fusion the two are related monotropically. This is a direct consequence of the relationship between the H curves, and the rule will be valid so long as the thermodynamic behavior for the two cases can be represented by Figures 2.3 and 2.5. Deviations will arise when the H curves diverge significantly, or the melting points are not close together ($\Delta mp \approx 30$ K), or both. In such cases it may be preferred to use the entropy-of-fusion rule or the heat-capacity rule as guidelines, rather than the heat-of-fusion rule.[2] Nevertheless the success rate of this rule is essentially as high as that of the heat-of-fusion rule (Burger and Ramberger 1979b).

---

[2] Burger and Ramberger (1979b) discuss the error in calculation of the difference of heats of fusion of two polymorphs in detail.

2.2.8 Entropy-of-fusion rule

The melting point is thermodynamically defined as the (single) temperature at which the liquid is in equilibrium with the solid so that the difference in Gibbs free energy between the two phases is zero. The entropy of fusion can then be expressed as

$$\Delta S = \frac{\Delta H_f}{T}$$

According to the rule, if the polymorph with the higher melting point has the lower entropy of fusion, the two modifications are enantiotropically related. If the lower melting form has the lower entropy of fusion then the two forms are monotropically related (Burger 1982a, 1982b).

2.2.9 Heat-capacity rule

For a pair of polymorphs, if the modification with the higher melting point also has a higher heat capacity at a given temperature than the second polymorph, then there exists an enantiotropic relationship between them. Otherwise, the system is monotropic.

2.2.10 Density rule

Kitaigorodskii (1961) enunciated the principle of closest packing for molecular crystals. Briefly this principle states that the mutual "orientation of molecules in a crystal is conditioned by the shortest distances between the atoms of adjacent molecules," and because the periphery of molecules is often dominated by hydrogen atoms that these distances will usually "be determined by the interactions between hydrogen atoms or the interaction of hydrogen atoms with other atoms of other elements." What determines the existence or non-existence of a crystal structure is the free energy, and the energetic manifestation of these distance arguments is that the most stable structure energetically should be expected to correspond to the one that has the most efficient packing. In other words, on the multidimensional energy surface the lowest lying among all the deep minima at zero degrees should also correspond to the structure with the highest density. Additional polymorphic structures, each located at a minimum with higher free energy than the minimum energy structure at zero degrees, will be expected to have less efficient packing and correspondingly lower density.

This rule is quite general for ordered molecular solids that are dominated by van der Waals interactions. Exceptions are not unexpected when other interactions, such as hydrogen bonds, dominate the packing, since some energetically favorable hydrogen-bond-dominated packing arrangements can lead to large voids in the crystal structure with correspondingly lower density.

While the density is arguably the easiest to obtain of the physical properties noted for deriving these rules (see Section 4.10), an increasing portion of crystallographic investigations no longer include the *experimental* determination of density. The *calculated* crystal density is routinely obtained from the unit cell dimensions and contents, but at the very minimum this requires an indexed X-ray powder diffraction pattern. While the density of polymorphs can be very useful in ranking the relative stability of polymorphs for which no X-ray data are available, the determination of experimental densities is becoming a lost art (Tutton 1922; Reilly and Rae 1954; Richards and Lindley 1999). Although accuracy as good as 0.02% can be obtained for the experimental determinations, Stout and Jensen (1989) have estimated that the agreement between calculated and experimental values of density is normally about 1–1.5%. Since densities of polymorphs often differ by very close to that amount, a *caveat* must be associated with the use of this rule, which Burger and Ramberger (1979b) indicate is correct 90% of the time, excluding cases of density differences of less than 1%.

2.2.11 Infrared rule

Burger and Ramberger (1979a) have also proposed an "infrared rule" for the highest frequency infrared absorption band in polymorphic structures containing strong hydrogen bonds. The formation of strong hydrogen bonds is associated with a reduction in entropy and an increase in the frequency of the vibrational modes of those same hydrogen bonds. The intramolecular N–H or O–H bond is correspondingly weaker (if the hydrogen participates in a hydrogen bond) with a reduction in the frequency of the associated bond stretching modes. The assumption is that these vibrations are only weakly coupled to the rest of the molecule, which is usually true for the O–H and N–H stretching vibrations and the NH_2 symmetric stretch. Under these conditions, the infrared rule says that the hydrogen-bonded polymorphic structure with the higher frequency in the bond stretching modes may be assumed to have the larger entropy.

Many cases could not be tested by Burger and Ramberger using this rule, since there was no difference in the highest frequency infrared absorption on going from the solid to the melt. Also, compounds containing the strong hydrogen bonding group –CO–NH were prominent among the exceptions. However, after eliminating these approximately 10% exceptional cases all the remaining 113 cases examined behaved according to the rule (Burger and Ramberger 1979b).

None of these rules is foolproof. However, they are useful guidelines, and the combination of relatively simple techniques can often be used to obtain a good estimate of the relative stability of polymorphs under a variety of conditions, information which is useful in understanding polymorphic systems, the properties of different polymorphs and the methods to be used to selectively obtain any particular polymorph (see Section 3.2). As noted, much of that information can be included in the energy–temperature diagram, and the actual preparation of that diagram from experimentally determined quantities is described in

Sections 4.2 and 4.3 following the description of the techniques used to obtain those physical data.

2.3 Kinetic factors determining the formation of polymorphic modifications

The starting point for a discussion of the kinetic factors is the traditional energy–reaction coordinate diagram, Figure 2.9. This shows G_0, the free energy per mole of a solute in a supersaturated fluid which transforms by crystallization into one of two crystalline products, I or II, in which I is the more stable ($G_{II} > G_I$). Associated with each reaction pathway are a transition state and an activation free energy which is implicated in the relative rates of formation of the two structures. Unlike a chemical reaction, crystallization is complicated by the nature of the activated state since it is not a simple bi- or trimolecular complex as would be expected for a process in which a covalent bond is formed; rather the activated state relates to a collection of self-assembled molecules having not only a precise packing arrangement but also existing as a new separate solid phase.

The existence of the phase boundary between the solid and liquid phase complicates matters since a phase boundary is associated with an increase in free energy of the system that must be offset by the overall loss of free energy. For this reason the magnitudes of the activated barriers are dependent on the size (i.e., the surface-to-volume ratio of the new phase) of the supramolecular assembly (crystal nucleus). This was recognized in 1939 by Volmer in his development of the kinetic theory of nucleation from homogeneous solutions and remains our best

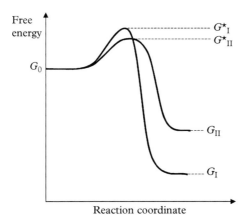

Figure 2.9 *Schematic of the reaction coordinates for crystallization in a dimorphic system, showing the activation barriers for the formation of polymorphs I and II. (Adapted from Bernstein et al. 1999, with permission.)*

model today (Volmer 1939), but see also the discussion on nucleation in Chapter 3.

One of the key outcomes of this theory is the concept of critical size that must be achieved by an assembly of molecules in order to be stabilized by further growth. The higher the operating level of supersaturation the smaller is this size (typically a few tens of molecules). Now, in Figure 2.10 the supersaturation with respect to II is simply $G_0 - G_{II}$ and is lower than $G_0 - G_I$ for structure I. However, it can now be seen that if for a particular solution composition the critical size is lower for II than for I then the activation free energy for nucleation is lower and kinetics will favor form II. Ultimately Form II will have to transform to Form I, a process that I discuss later. Overall we can say that the probability that a particular Form II will appear is given by

$$P(i) = f(\Delta G, R), \tag{2.1}$$

in which ΔG is the free energy for forming the ith polymorph and R is the rate of some kinetic process associated with the formation of a crystal by molecular aggregation. Thus, for example, if we follow the above reasoning we could equate the rate process with $\mathcal{J}$, the rate of nucleation of the form. If all polymorphs had the same rates of nucleation then their appearance probability would be dominated by the relative free energies of the possible crystal structures.

The rates of nucleation as expressed by the classical expression of Volmer are related to various thermodynamic and physical properties of the system, such as surface free energy (g), temperature (T), degree of supersaturation (σ), and solubility (hidden in the pre-exponential factor A_n) which will not be the same for each structure but will correctly reflect the balance between changes in bulk and surface free energies during nucleation. This is seen in the following equation which relates the rate of nucleation to the above parameters (v is the molecular volume):

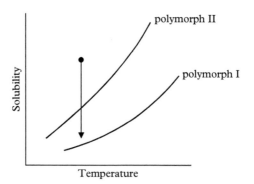

Figure 2.10 *Schematic solubility diagram for a dimorphic system (polymorphs I and II) showing a hypothetical crystallization pathway (vertical arrow) at constant temperature. (Adapted from Bernstein et al. 1999, with permission.)*

$$\mathcal{J} = A_{n}\exp(-16\pi\gamma^{3}v^{2}/3k^{3}T^{3}\sigma^{2}).$$ (2.2)

From this analysis it is clear that the trade-off between kinetics and thermodynamics is not at all obvious. For simplicity consider a monotropic, dimorphic system whose solubility diagram is shown schematically in Figure 2.10. It is quite clear that for the occurrence domain given by solution compositions and temperatures that lie between the form II and I solubility curves only polymorph I can crystallize. However, the outcome of an isothermal crystallization that follows the crystallization pathway indicated by the vector in Figure 2.10 is not so obvious since the initial solution is now supersaturated with respect to both polymorphic structures, with thermodynamics favoring form I and kinetics (i.e., supersaturation) form II.

Experimentally, the reality of this overall scenario of kinetic versus thermodynamic control was known long before the development of nucleation theory and is encompassed by Ostwald (1897) in his Rule of Stages (Cardew and Davey 1982; Davey 1993). The German scientific literature between 1870 and 1914 contains many organic and inorganic examples in which crystallization from melts and solutions initially yield a metastable form that was ultimately replaced by a stable structure. On the basis of such a phenomenon Ostwald was led to conclude that "when leaving a metastable state, a given chemical system does not seek out the most stable state, rather the nearest metastable one that can be reached without loss of free energy" (Ciechanowicz et al. 1976). Figure 1.5 shows an example of Ostwald's rule.

There are significant flaws in Ostwald's conclusion that led to his rule. When a crystallization experiment yields only a single form there is the question of whether it contradicts the Rule or whether the material is simply not polymorphic. Moreover, there is no way of answering this question. However, a sufficient number of cases of successively crystallizing polymorphic forms have been observed to warrant considering the principles behind Ostwald's rule as guidelines for understanding the phenomenon of the successive crystallization of different polymorphic phases.

By making use of Volmer's equations some attempts have been made by Stranski and Totomanov (1933), Becker and Döring (1935), and Davey (1993) to explain the rule in kinetic terms. In doing this it becomes apparent that the situation is by no means as clear cut as Ostwald might have us believe. Figure 2.11 shows the three possible simultaneous solutions of the nucleation equations that indicate that by careful control of the occurrence domain there may be conditions in which the nucleation rates of two polymorphic forms are equal and hence their appearance probabilities are nearly equal. Under such conditions we might expect the polymorphs to crystallize concomitantly (see Section 3.3). In other cases there is a clearer distinction between kinetic and thermodynamic crystallization conditions, and that distinction may be utilized to selectively obtain or prevent the crystallization of a particular polymorph.

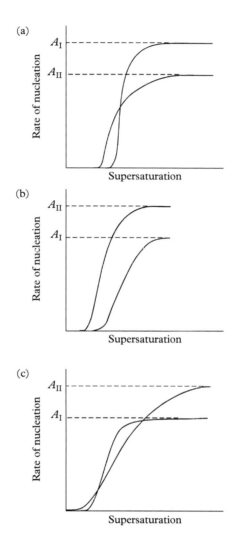

Figure 2.11 *The rates of nucleation as functions of supersaturation for the dimorphic system defined in Figure 2.10. The three diagrams (a), (b), and (c) represent the three possible solutions for the simultaneous nucleation of two polymorphs each of which follows a rate equation of the form of equation (2.2). Note that solutions (a) and (c) both allow for simultaneous nucleation of the forms at supersaturations corresponding to the crossover of the curves. (From Bernstein et al. 1999, with permission.)*

2.4   Structural fundamentals

Together with thermodynamics, structure is a fundamental and usually distinguishing property of polymorphs. In this section I will deal with the definition of that structure, how it may be best viewed and understood, and some of the ways in which polymorphic structures may be compared; others are dealt with in Chapter 5. The determination of crystal structure, by X-ray methods is the subject of many fine texts (e.g., Dunitz 1979; Stout and Jensen 1989; Giacovazzo et al. 1992; Glusker et al. 1994), which should be referred to, if necessary, as background for this section.

2.4.1 Form versus habit

In the description of crystals and crystal structures the two terms *form* and *habit* have very specific and very different meanings. *Form* refers to the internal *crystal structure* and etymologically is the descendant of the Greek *morph*. Hence *polymorph* refers to a number of different crystal modifications or different crystal structures, and the naming of different structures as "Form I" or "α Form" follows directly from this definition and usage. As we have seen, the difference in crystal structure is very much, although not exclusively, a function of thermodynamics. Certainly, only the structures that are thermodynamically accessible can ever exist, but there often is a question of thermodynamic versus kinetic control over which particular structure may be obtained under any particular set of crystal growth conditions.

Habit, on the other hand, derives from the Latin and Old French word for mode of growth, and describes the shape of a particular crystal.[3] That shape is influenced greatly by the environment. It is essentially a manifestation of kinetic factors determining the relative rate of growth along various directions of the crystal, and hence the preferential growth or inhibition of the development of the different crystal faces that ultimately define the shape of the crystal. Examples of definitions of different habits and the variation in habit resulting from changes in crystal growth conditions are given in Figure 2.12.

Unfortunately, the distinction between *form* and *habit* is often blurred, and consequently some confusion has crept into the literature. Especially in the older crystallographic literature, *form* was used to describe a set of crystal faces that are alike or symmetry related, and the *habits* then described as the collection of the forms that are exhibited (Chamot and Mason 1973). Ideally, the external shape of the crystal reflects internal symmetry of the crystal. Crystallographers refer to this symmetry as the *crystal class*, while chemists traditionally refer to it as the point group (Hahn and Klapper 1992). The study of the external shape and symmetry of crystals is called crystal morphology. An excellent collection of crystal habits according to crystal class, their descriptors, and drawings of both mineral and organic examples, can be found in Buerger's (1956) text.

The important point is that differences in external crystal shape, *habit*, or crystal *morphology* may not necessarily indicate a change in the polymorphic *form*

---

[3] According to the Oxford English Dictionary (OED), the initial use for *habit* was in zoology and botany, indicating the characteristic mode of growth and general appearance of an animal part. The usage was transferred to crystallography to indicate the characteristic mode of formation of a crystal. For instance, the OED cites an 1895 usage by M. H. N. Story-Maskelyne: "Such differences, then, may generally be held to indicate a mero-symmetrical habit."

(a)

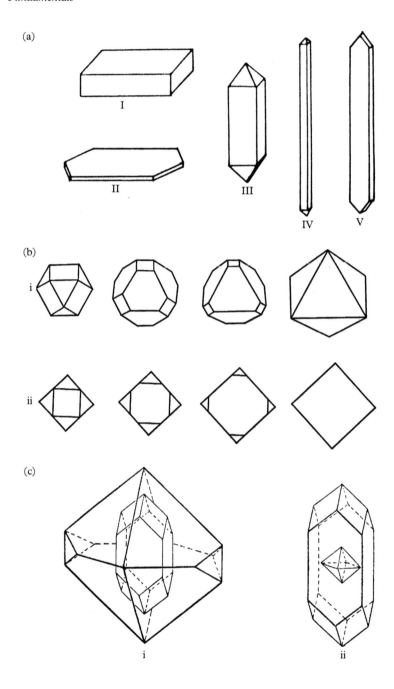

(d)

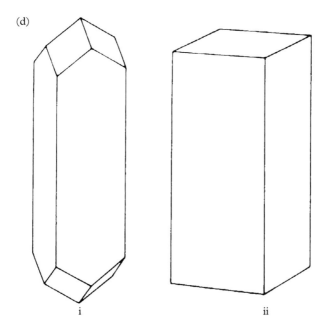

i   ii

Figure 2.12 *(a) Some different crystal habits of crystals and their descriptions: I, tabular; II, platy; III, prismatic; IV, acicular; V, bladed (from Hartshorne and Stuart 1964, with permission). (b) Demonstration of the difference of rate of growth of cubic or octahedral faces of a crystal as governed by the rate of deposition on different crystal faces. Starting on left-hand side, both crystals have equal development of the cubic and octahedral faces. Deposition is faster on the cubic faces in i and the octahedral faces in ii, leading eventually to the crystal habit on the right (from Chamot and Mason 1973, with permission). (c) Demonstration of the effect of solvent on the crystal habit of anthranilic acid: i, crystal initially grown in water, then in ethanol; ii, crystal form i transferred to acetic acid and allowed to grow (from Wells 1946, with permission). (d) Demonstration of the use of* form *to describe the family of faces bounding a crystal. In the commonly used terminology both orthorhombic crystal exhibit the same prismatic habit, but different combinations of* forms. *i is bounded by (110), (101), and (011) faces, while ii is bounded by (100), (010), and (001).*

or polymorphic crystal structure. An example of this distinction is given in Figure 2.13.

Control of crystal habit and habit modification is an important aspect of the preparation and ultimate use of solids (Davey et al. 1994; Wood 1997), and a great deal of work has been done in this area. Some early leading references are those of Buckley (1951) and Tipson (1956), and there have been important theoretical (Clydesdale et al. 1994a, 1994b, 1996) and experimental developments, for example, in the more recent work of Bennema and Hartman (1980), Addadi et al. (1985), Black et al. (1990), Clydesdale et al. (1997), and Lahav and Leiserowitz (1993).

Figure 2.13 *Demonstration of the difference between crystal habit and polymorphic form for the three polymorphs of naphthazarin. Upper left, polymorph A; upper right, polymorph B; lower, polymorph C in two different habits. (Crystals prepared and photos taken by R.B. Kress and M.C. Etter, personal communication.)*

The renaissance of crystal control and modification and its application to molecular crystals has also flourished since the pioneering publications of Lahav and Leiserowitz in the 1980s and 1990s. The earlier work focused primarily on the *macroscopic* manifestations of changing conditions, while the more recent work of the same group and others deals with interactions at the *molecular level*. There have been many developments in this area, and the research strategies employed can, and often do, influence the polymorphic result of an experiment, even if unintentionally, which warrants some familiarity with the increasingly molecular approach. A variety of theoretical and experimental techniques have been developed and applied to this issue. I present but a small, and necessarily selective, sampling here to give some flavor of the kinds of approaches followed.

The Lahav–Leiserowitz approach is based on a detailed analysis of the crystal structure of the growing crystal, the prominent crystal faces and direction of growth and the interactions of an incoming molecule—either of the crystallizing substance or a "tailor-made" additive or solvent molecule—with the molecules projecting on those faces of the crystal. The same authors have reviewed how these principles apply to the growth of polar crystals (Lahav and Leiserowitz 2006). The habit changes in the lysozyme polymorphic system have also been the subject of an interpretation based on these principles (Heijna et al. 2008). The effects of impurities, solvent, and additives were demonstrated to influence the crystal habit and polymorphic occurrence of the antiviral/HIV drug stavudine (Mirmehrabi et al. 2006). Solvent (Algra et al. 2005), impurities (Poornachary et al. 2008), and minor components and additives (Smith et al. 2011) have been shown to be influential in the development of specific crystal habits.

On the theoretical level, a number of approaches have been reported. For instance, Gervais and Hulliger (2007) studied the influence of surface symmetries on growth-induced physical properties. Deij et al. (2007) used Monte Carlo simulations to examine the influence of all possible (non-flat) growth site configurations, taking the crystalline surface, steps, and other configurations into account. In another Monte Carlo simulation Van Enckevort and Los (2008) studied the details of the influence of tailor-made additives on the growth and etching of crystals. These studies, together with the sophisticated experimental investigations are providing details on the mechanism of crystal growth and dissolution.

Even with the significant developments in learning to control both crystal habit and crystal form, a single modification may crystallize in a number of habits, one of which has more desirable processing and packaging characteristics, other chemical and physical properties being equal, for example, aspartame (see Section 10.4.7). Hence these two intimately related characteristics of internal structure and external shape must often be considered together, which has been the approach of most of the modern researchers in this area.

2.4.2 Structural characterization and comparison of polymorphic systems

As Herbstein (2001) has noted, consideration of polymorphism in terms of classical thermodynamics ignores the structural aspects of the phenomenon, an approach he termed as purely physical. At the other extreme of his proposed physics–chemistry scale for polymorphism, molecules of a substance would exhibit "appreciable chemical differences between the chemical entities in the two polymorphs." At the chemical end of the scale one is dealing with structural characterization and comparison, and a number of typical examples along the physics–chemistry scale are discussed in detail by Herbstein.

In dealing with a polymorphic system perhaps the fundamental question is how similar or how different are the various crystal structures. Gavezzotti (2007) has discussed many aspects of this question in some detail twice with the hiatus of a decade (Gavezzotti and Filippini (1995)) with no new conclusions, even with the doubling of the size of their database. That discussion is based on applying quantitative measures to this comparison, and I review those developments in detail in Chapter 5. However, I believe that any investigation of the question of similarities and differences in polymorphs, or structures claimed to be polymorphs, should include a close visual examination of the crystal structures as outlined in the next section.

2.4.2.1 *Presentation of polymorphic structures for comparison*

Even for the trained and practiced eye, a single crystal structure of a molecular solid is rarely understood with ease, so that *comparison* of two or more crystal structures, even involving the same molecule in polymorphic structures can be an exercise in frustration. Very often the best means for examining and comparing polymorphic crystal structures is to plot them *on the same molecular reference plane*, and orient that plane in the same way for all of the structures. Such a strategy allows a ready comparison of the immediate environment about the reference molecule and the intermolecular interactions that dominate the structures. When the molecule in question can exhibit conformational flexibility then the molecular reference plane chosen for projection of the structure should be a rigid part of the molecule (e.g., a phenyl ring) or an appropriate group of three atoms. If the object of the figure is to demonstrate the differences in conformation as well as the differences in crystal structure, then the reference plane should be chosen such that the torsion or dihedral angles that lead to the largest differences in molecular conformation are immediately adjacent to the reference plane. However, such figures can be confusing in terms of understanding the differences in *packing*, simply because the projection may contain many overlapped atoms, in which case an alternative is to choose a reference plane which is distant by a number of bonds from the ones which are the major source of the conformational difference. If these views are given in stereo (a mode of presentation which unfortunately has been rapidly going out of style), then views for the different structures to be compared should appear above and below each other to facilitate comparison. Exemplary samples of this presentation strategy are shown in Figure 2.14.

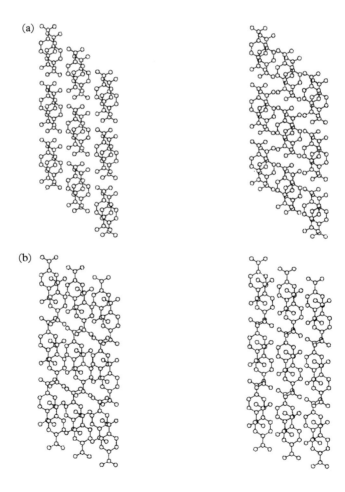

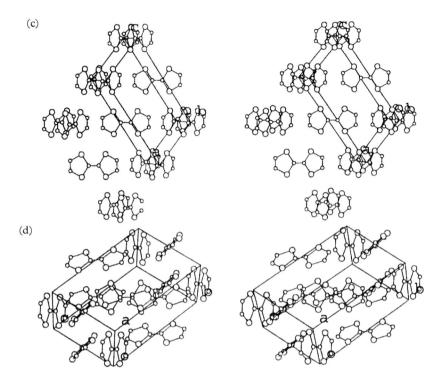

Figure 2.14 *Examples of comparative stereoplots of the packing of polymorphic structures, with the view chosen to convey similarities and differences in packing. (a, b) Terephthalic acid (Bailey and Brown 1967, 1984; Berkovich-Yellin and Leiserowitz 1982), plotted on the plane of the phenyl ring. Both polymorphs are triclinic, $P\overline{1}$, and are composed of very similar layers (which is essentially the plane of the paper) composed of linear chains of hydrogen-bonded molecules. The lateral offset of chains within a layer is also very similar for the two structures. The structures do differ in the manner in which subsequent layers are offset from each other (after Davey et al. 1994). (c, d) Tetrathiafulvalene (Cooper et al. 1974; Ellern et al. 1994). (c) Form 1 is monoclinic, $P2_1/c$; (d) Form 2 is triclinic $P\overline{1}$. In (c) and (d) the two structures are plotted on the best plane of the central molecule and oriented so that the long axis of that molecule is horizontal. Especially in mono view (c) suggests a layered structure for Form 1; however, a rotation of slightly less than 90° about the horizontal axis (to better enable viewing of all of the atoms of the molecule) would indicate that the packing is better described as a herringbone motif. Note that the views chosen here, especially that for Form 2, may not be the best one for interpreting or discussing that structure individually. The views chosen do, however, greatly facilitate comparison of the polymorphic structures.*

This strategy unfortunately is contrary to the usual practice in preparing packing diagrams, most of which are given as the view down a particular crystallographic axis or on a particular crystallographic plane. Such views carry very little information with regard to similarities or differences in polymorphic structures. However, in some cases, polymorphs have some similar cell constants or symmetry elements in common, and the similarity may be manifested in the crystal packing.

In such a case, plotting the structures in such a way as to view and compare those ostensibly common features is essential for the analysis. An example is the α- and β-modifications of L-glutamic acid (Lehmann et al. 1972). The two structures crystallize in the same orthorhombic space group, $P2_12_12_1$. Comparison of the cell constants in Table 2.1 suggests a possible axial relationship as follows:

Table 2.1 *Comparison of cell constants for the two polymorphs of L-glutamic acid*

	α form	β form
a (Å)	7.068	5.519
b (Å)	10.277	17.30
c (Å)	8.755	6.948

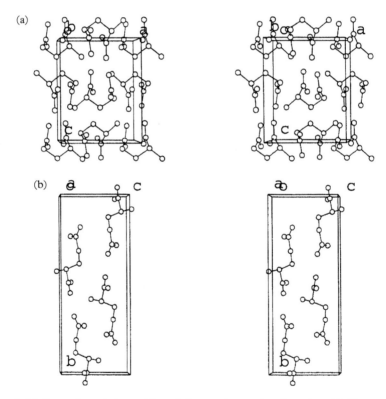

Figure 2.15 *Stereoplots of the packing of the two forms of L-glutamic acid. The figures are oriented to facilitate comparison of the axes of the two forms, which are apparently either similar or related by whole number ratios (Table 2.1). (a) α-Form; (b) β-Form (adapted from Bernstein 1991a, with permission).*

$$a_\alpha \cong c_\beta,$$

$$b_\alpha \cong 2a_\beta,$$

$$2c_\alpha \cong b_\beta.$$

Therefore, plotting the structures so that the possibly common $a_\alpha \cong c_\beta$ axis is horizontal and the possibly doubled $2c_\alpha \cong b_\beta$ axis is vertical greatly aids in making this comparison (Figure 2.15). In fact, it is seen in Figure 2.15 that in spite of the identity in space group and the apparent possible relationship between some cell constants of the two polymorphs, there is no similarity in the crystal structures.

An additional example will serve to demonstrate the similarities in structures that can be revealed by taking care in the way they are plotted. The cell constants for the two forms of 5-methyl-1-thia-5-azacyclooctane 1-oxide perchlorate (Table 2.2; Paul and Go 1969) suggested the following axial relationships:

$$2a_\alpha \cong a_\beta,$$

$$b_\alpha \cong b_\beta,$$

$$c_\alpha \cong 2c_\beta.$$

Plotting both structures with a view down the *b* axis (Figure 2.16) reveals the close similarity between them, as discussed in detail in the original paper.

The two examples given in Figures 2.15 and 2.16 demonstrate another point that deserves emphasis, especially to non-crystallographers. The identity or non-identity of space groups for polymorphic structures is purely coincidental. No structural similarity can be associated with polymorphic structures on the basis of space group identity. Also, the lack of structural similarity between polymorphs cannot be assumed simply on the basis of a comparison of the space groups. Space group symbols provide information on the symmetry elements present in the crystal structure. They do not provide details on the intermolecular relationships or on the environment of any particular molecule in a crystal structure. For the

Table 2.2 *Comparison of cell constants for the two polymorphs of 5-methyl-1-thia-5-azacyclooctane 1-oxide perchlorate*

	α form	**β form**
a (Å)	9.87	20.10
b (Å)	8.78	8.89
c (Å)	13.26	6.77
β (°)	97.90	97.80
Sp. Gr.	$P2_1/c$	$P2_1/a$

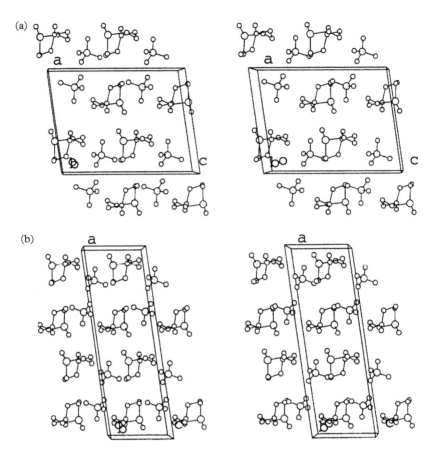

Figure 2.16 *Stereoplot of the two structures of 5-methyl-1-thia-5-azacyclooctane 1-oxide perchlorate. In both cases the view is along the* b *crystallographic axis. Note the similarity in the arrangement of the molecules, although a casual glance at the unit cell dimensions might suggest that the structures are very different. (a) α-Form; (b) β-Form. (Adapted from Paul and Go 1969, with permission.)*

non-crystallographer it should be noted that $P2_1/c$, $P2_1/a$, and $P2_1/n$ are equivalent space groups (No. 14 in Hahn 1987).

2.4.2.2 *Characterization of hydrogen-bonded structures*

Because hydrogen bonds are the strongest and most directional of intermolecular interactions, they have been the subject of study since the 1920s (Latimer and Rodebush 1920; Pimentel and McClellan 1960). The strength and directionality of these bonds might suggest a high incidence of polymorphism for those compounds containing a number of potential hydrogen bond donors and hydrogen bond acceptors, since they might be combined in a number of different ways. However, Etter (1990, 1991) pointed out that in the formation of hydrogen bonds not all combinations of donor and acceptor are equally likely, since strong hydrogen donors (strongly acidic hydrogens) will tend to form hydrogen bonds preferentially with strong hydrogen acceptors (atoms with electron pairs, etc.). Hence, the preference for certain hydrogen bonds to form might tend to lead to the preservation of hydrogen bond patterns among polymorphs rather than the proliferation of polymorphs for molecules with many potential hydrogen bonds. Indeed, as recently demonstrated (Cruz-Cabeza et al. 2015) in the statistics presented in Table 1.3 there is no greater propensity for molecules with hydrogen bonding potential to exhibit polymorphism than for those lacking hydrogen bond functionality. Nevertheless, for those compounds that do exhibit polymorphism, comparing the hydrogen bonding patterns can provide considerable insight into the structural nature of that polymorphism.

The traditional approach for comparing the hydrogen bonding among polymorphs is to list the geometry (bond lengths and angles) of the individual hydrogen bonds for each structure, sometimes including the crystallographic symmetry elements which generated each of the hydrogen bonds (e.g., Bernstein 1979). This approach carries very little information that can serve as a basis for *comparing* the *patterns* of hydrogen bonds among the polymorphs. A much more useful method involves the utilization of graph sets (Hamilton and Ibers 1968; Kuleshova and Zorky 1980; Zorky and Kuleshova 1980; Wells 1984), also originally proposed and demonstrated in its current context by Etter (1985, 1990, 1991) and Etter et al. (1990a), with subsequent modifications and refinements (Bernstein et al. 1995; Grell et al. 1999). This method can be quite useful and informative in comparing polymorphs dominated by hydrogen bonds; hence I review the fundamental principles of the graph set method in the next section and follow that with a section describing a number of typical examples. Although the Etter approach was developed and applied to hydrogen-bonded networks in molecular crystals, Wells' (1984) earlier application of graphs in inorganic systems and subsequent developments (Navon et al. 1997; Grell et al. 1999) suggest the potential widespread use of graphs for characterizing crystal structures in general (e.g., Moers et al. 2000), and comparing polymorphic structures in particular.

2.4.2.2.1 Some basics of graph set notation for the description of hydrogen bonds

2.4.2.2.1.1 The designator A remarkable feature of the graph set approach to analysis of hydrogen-bond patterns is the fact that most complicated networks can be reduced to combinations of four simple patterns, each specified by a designator: chains (**C**), rings (**R**), intramolecular hydrogen-bonded patterns (**S**), and other finite patterns (**D**). Specification of a pattern is augmented by a subscript **d** designating the number of hydrogen-bond donors (most commonly covalently bonded hydrogens, but certainly not limited to them), and a superscript **a** indicating the number of hydrogen bond acceptors. When no subscript and superscript are given, one donor and one acceptor are implied. In addition, the number of bonds

n in the pattern is called the degree of the pattern and is specified in parentheses. The general graph set descriptor is then given as $G_d^a(n)$, where **G** is one of the four possible designators.

These four patterns and their descriptors are best illustrated by examples. A chain whose "link" is composed of four atoms as in **2-I** is specified as **C(4)**. Similarly, the intramolecular hydrogen bond in **2-II** would be specified as **S(6)**, for the six atoms comprising the intramolecular pattern. When the donor and acceptor are from two (or more) *discrete* entities (molecules or ions), as in **2-III**, the designation of the hydrogen bond is **D**. The entities may differ on chemical grounds (different ions or molecules) or on crystallographic grounds (chemically identical but not related by a crystallographic symmetry operation). In **2-III** there is only one donor and one acceptor, and the pattern involves only one hydrogen bond. The fourth possible pattern is the ring **2-IV**. In the example shown, the two hydrogen bonds in the ring could be different but in this case they are related by a crystallographic inversion center. The pattern contains a total of eight atoms, two of them donors and two acceptors, and hence is designated $R_2^2(8)$.

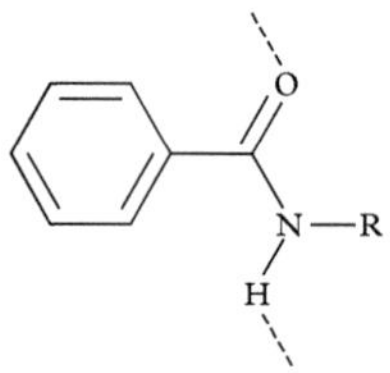

2-I C(4)

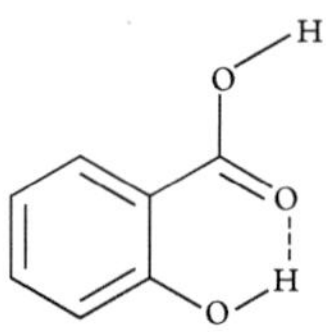

2-II S(6)

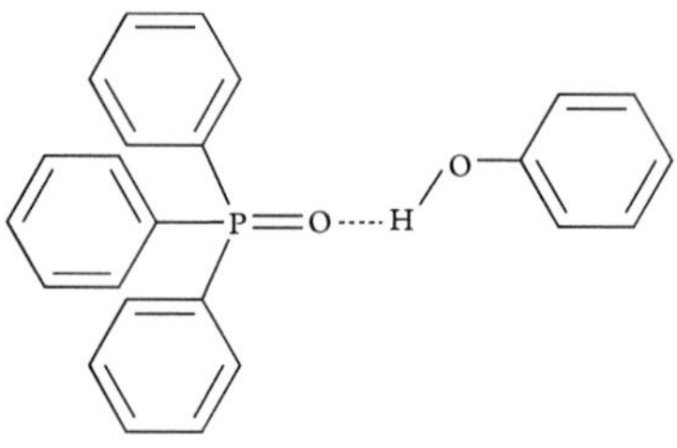

2-III D

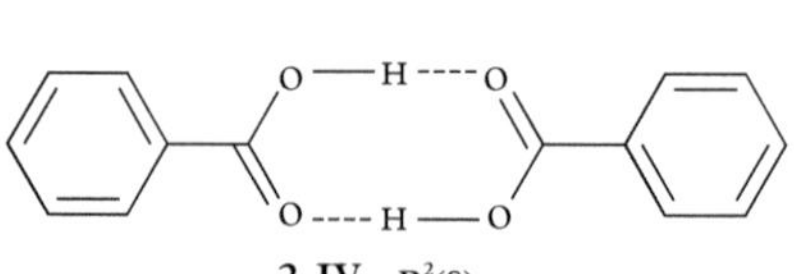

2-IV $R_2^2(8)$

2.4.2.2.1.2 Motifs and levels A motif is a pattern containing only one type of hydrogen bond (Etter et al. 1990a). Specifying the motif for each different hydrogen bond in a network according to one of the four pattern descriptors mentioned leads to a description of the network in the form of a list of the motifs. This is the *unitary*, or *first level*, graph set, noted as $\mathbf{N}_1$.

The chemically interesting or topologically characteristic patterns of a system often appear when more than one type of hydrogen bond is included in the description, that is, in higher level graph sets (Etter et al. 1990a; Bernstein et al. 1990, 1995; Bernstein 1991b). This will be true for **S**, **D**, and **C** patterns. Suppose a structure contains three distinct hydrogen bonds, designated *a*, *b*, and *c*. There will now be several possible *binary* (or *second level*) graph sets, each one describing a pattern formed by two of these hydrogen bonds—that is, $\mathbf{N}_2(\boldsymbol{ab})$, $\mathbf{N}_2(\boldsymbol{ac})$, and $\mathbf{N}_2(\boldsymbol{bc})$. These concepts are illustrated in Figure 2.17. The *ternary* or *third level* graph sets are those that involve three hydrogen bonds (Bernstein et al. 1995).

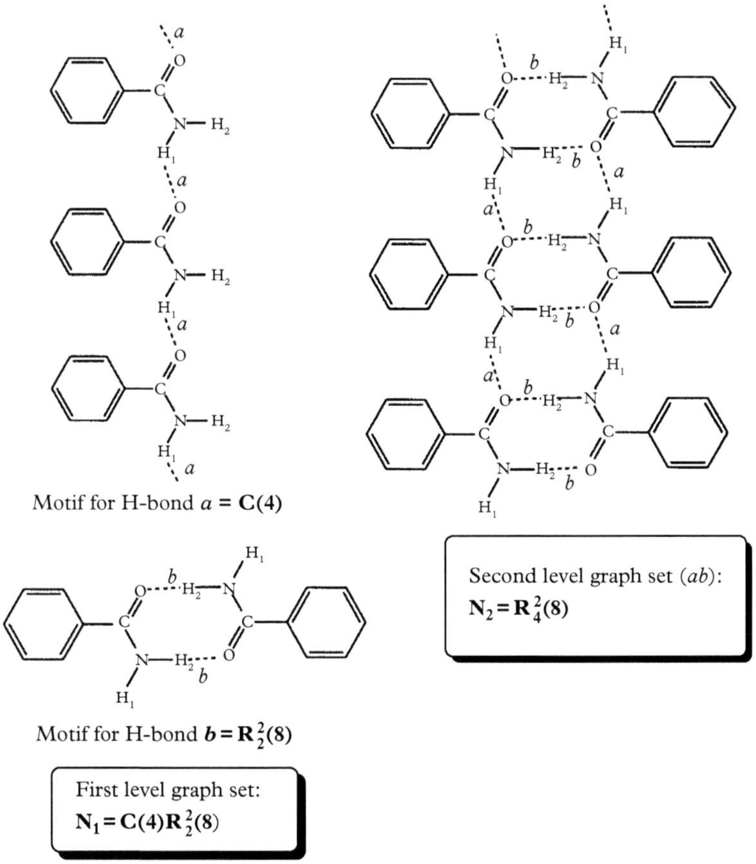

Motif for H-bond a = **C(4)**

Motif for H-bond b = $\mathbf{R}_2^2(8)$

Second level graph set (*ab*):
$\mathbf{N}_2 = \mathbf{R}_4^2(8)$

First level graph set:
$\mathbf{N}_1 = \mathbf{C(4)}\mathbf{R}_2^2(8)$

Figure 2.17 *Examples of the use of graph set descriptors to define motifs and first and second level graph sets for schematic representations of the hydrogen-bond patterns in the crystal structure of benzamide. (From Bernstein and Davis 1999, with permission.)*

Different pathways might also be found that include the same set of hydrogen bonds, but with different degrees (**n**). To describe this situation, the term *basic* was suggested to describe the graph set of the *lowest degree* and the term *complex* to describe ones of *higher degree*. Consideration of some of the choices for the binary graph set for α-glycine (Power et al. 1976) illustrates this point (Figure 2.18). The shortest path involving hydrogen bonds *a* and *b* gives the binary graph set $C_2^2(6)$; this is the *basic* binary graph set for *a* and *b*. A longer chain, $C_2^2(10)$, represents a *complex* binary graph set for the two hydrogen bonds. Neither of these, however, describes the most obvious feature of this array, the ring structure, which is denoted by another complex binary graph set $R_4^4(16)$. In addition, in this structure it is possible to define an infinite number of increasingly larger ring systems. In such a case it remains for the chemist or crystallographer to choose a ring or those rings that best characterize the particular structure in question (Bernstein et al. 1995).

Many additional details on the assignment of graph sets with examples of applications to polymorphic systems are given in Bernstein et al. (1995).

The early work on graph sets required a fairly tedious process of plotting the structures, identifying the patterns and counting the number of atoms, donors, and acceptors in order to determine the descriptor. It was also a process fraught with possibilities for errors and inconsistencies. A significant advance was achieved with the automatic identification and visualization of hydrogen bonding patterns and the assignment of the graph set in the standard software of the CSD (Grell et al. 1999; Motherwell et al. 1999). These developments have greatly facilitated the use of graph sets in the study and comparison of polymorphic systems.

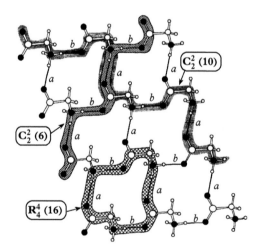

Figure 2.18 *Graph set assignments for the binary level of α-glycine. As in Figure 2.17, different types of hydrogen bonds (solid lines) are distinguished by labeling with lower case bold letters; carbon and hydrogen atoms are shown as open circles, oxygen atoms as solid circles, and nitrogen atoms as shaded circles. (From Bernstein and Davis 1999, with permission.)*

2.4.2.2.1.3 Does hydrogen bonding necessarily differ among polymorphs? There are many cases where the hydrogen bonding is essentially identical among the polymorphs of a system. A simple, and simply understood, case is that of terephthalic acid **2-V** (Davey et al. 1994) (Figure 2.14 a,b). The presence of two identical carboxyl groups in the *para* position of the benzene ring essentially dictates a chain motif, and the two polymorphs differ slightly (but measurably) in the relationship of neighboring chains to each other (to form layers) and the subsequent relationship of layers to each other in the two polymorphs (see Section 2.4.2.1). A second example may be found in the two polymorphs of 2,2-aziridinecarboxamide **2-VI** (Brückner 1982) (Figure 2.19), again in which identical chains of $R_2^2(8)$ rings are formed in the two structures. The combination of space groups and molecules in the asymmetric unit ($P4_12_12$ ($Z = 4$) and P_1 ($Z = 16$), respectively) in this pair of polymorphs is quite unusual for organic materials. The difference in the relationship among neighboring layers is shown in Figure 2.19b.

2-V

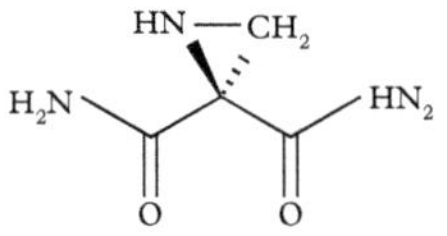

2-VI

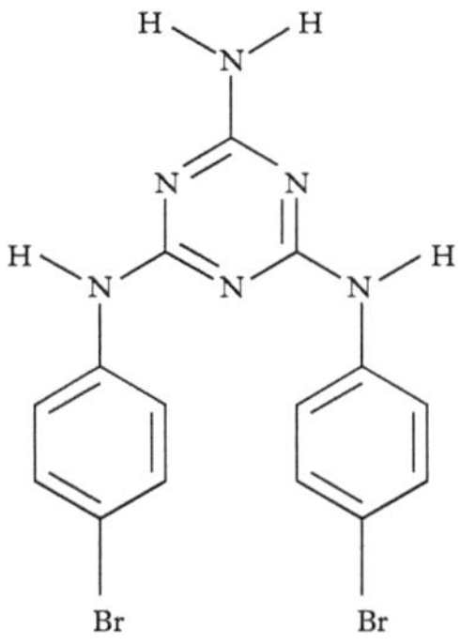

2-VII

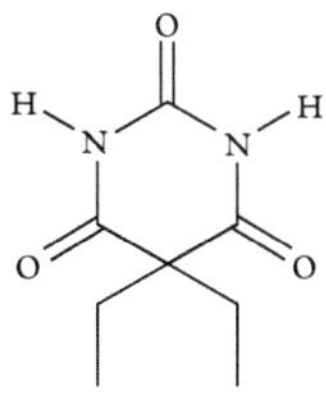

2-VIII

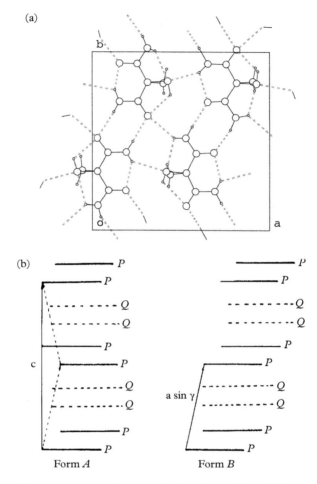

Figure 2.19 *(a) The chain structure (running horizontally) of 2,2-aziridinecarboxamide* **2-VI** *common to the two polymorphic structures. (b) The schematic relationship between stacking of the layers composed of chains in the two forms. The views are along the* b *axis in the tetragonal Form A and along* c × a* *in the triclinic Form B. (adopted from Brückner 1982, with permission).*

An additional example is for a *co-crystal* of N,N'-bis(p-bromophenyl)melamine *(2-VII)* and barbital *(2-VIII)* (Zerkowski et al. 1997). There are clearly a number of potentially different hydrogen bonding donor and acceptor sites on each molecule (in fact that was part of the rationale for studying these co-crystalline materials). The two polymorphic structures have identical tape hydrogen bonding patterns, but the tertiary structure, as defined by the arrangement of the tapes within the crystal is quite different (Figure 2.20).

(a)   Form I

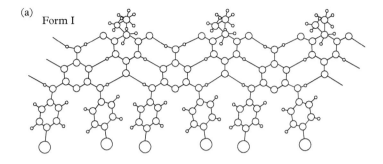

(b)   Form II

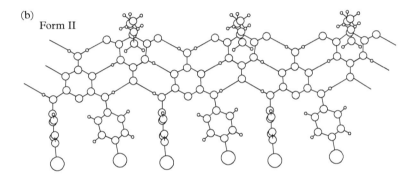

(c)

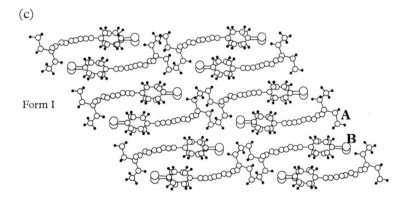

Form I

A

B

Figure 2.20 *Comparison of the crystal structures in the two polymorphs of the co-crystal of 2-VII and 2-VIII. (a, b) The hydrogen-bonded tapes. Note the difference in the orientation of the bromophenyl substituents along the chain. (c, d) End-on view of the packing of the tapes in the two polymorphs. (From Zerkowski et al. 1997, with permission.)*

(d)

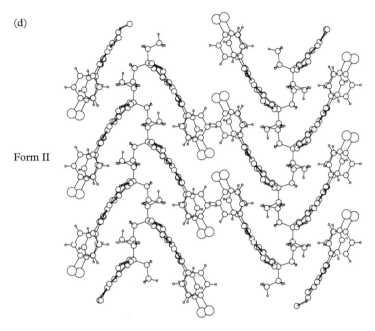

Form II

Figure 2.20 (Continued)

2.5   Thermodynamics and kinetics of crystallization—signs of a changing paradigm

The discussion in this chapter is based on classical thermodynamics as developed by Gibbs in the latter decades of the nineteenth century, and classical physical chemistry as pioneered by Ostwald slightly later. Those concepts are based on and experimentally verified by macroscopic systems: the smallest crystal visible to the human eye contains $\sim 10^{16}$ particles. That was the world of Gibbs and Ostwald, and traditionally the laws of thermodynamics are considered immutable.

For chemists, at least, the opposite end of the size scale is atoms and electrons, whose behavior is understood today with the aid of the quantum mechanics developed during the first three decades of the twentieth century. That revolution forced chemists to consider probability along with thermodynamic immutability. They are not mutually exclusive; after all Gibbs' development of thermodynamics was based on the statistical mechanics of Maxwell and Boltzmann.

The size regime about which we know and understand very little is that associated with the nucleation of solids, in particular crystals. If that is still a quantum world it is extremely difficult to treat with current theoretical and computational tools. If "macroscopic" world principles apply there are many basic questions about the numbers of particles in an ensemble, the structure(s) and dynamics of those particles, critical sizes and numbers, etc. Yet, these are crucial questions if we are to understand the fundamental nature of the formation of crystals. This is the world of nanoparticles, and after approximately two decades of increasingly intensive study it is clear that the nano world is different from the macro world and the quantum world—not totally different, but different in some ways that require an alteration to how we think about these systems.

There are some signs that these differences are being revealed by some significant experimental and theoretical studies, with a *caveat* regarding the latter, that *small* simulation scales (i.e., those often defined by computing limitations) are inherently inappropriate for applications of bulk thermodynamics (Anwar and Zahn 2011). In particular, in contrast to the inviolability of macro thermodynamics it has been shown that the order of relative thermodynamic stabilities of two polymorphs can be altered upon going from the macro to the nano size regime (McHale et al. 1997; Navrotsky et al. 2008; Belenguer et al. 2016).

3

Exploring the crystal form landscape

Crystal growth is a science and an art. The scientist's role in the crystal growth process is that of an assistant who helps molecules to crystallize. Most molecules, after all, are very good at growing crystals. The scientific challenge is to learn how to intervene in the process in order to improve the final product.

Etter (1991)

I am very sorry, that to the many…difficulties which you meet with, and must therefore surmount, in the serious and effectual prosecution of Experimental Philosophy, I must add one discouragement more, which will perhaps as much surprise you as dishearten you; and it is, that besides that you will find…many of the experiments published by Authors, or related to you by the persons you converse with, false or unsuccessful,…you will meet with several Observations and Experiments, which though communicated for true by Candid Authors or undistrusted Eye-witnesses, or perhaps recommended to you by your own experience, may upon further tried disappoint your expectation, either not at all succeeding constantly, or at least varying much from what you expected.

This Advertisement may seem of so discouraging a nature that I should much scruple the giving it to you, but that I suppose the trouble at that unsuccessfulness which you may meet with Experiments, may be somewhat lessened, by your being forewarned of such contingencies. And that you should have the luck to make an Experiment once, without being able to make the same thing again, you might opt to look upon such disappointments as the effect of and unfriendliness in Nature or Fortune to your particular attempts, as proceeding from a secret contingency incident to some Experiments, by whomever they be tried.

Boyle (1661)

Control nevertheless is important in science—tremendously so—not as an end but as a component of proof. The ability to control is the strongest possible demonstration of true understanding. Many doubted whether Becquerel, Curie, Bohr, Oppenheimer, and the rest really understood what causes what inside the atom. But after July 16, 1945, when the day dawned prematurely to the northwest of Alamagordo at White Sands, New Mexico, no one could possibly doubt any

Polymorphism in Molecular Crystals. Second Edition. Joel Bernstein. © Joel Bernstein 2020. Published in 2020 by Oxford University Press. DOI: 10.1093/oso/ 9780199655441.001.0001

more, for the atom bomb was plainer than the sun. With a demonstrated ability to control, the good scientist may sign off like the mathematician at the end of a proof: Quod erat demonstrandum.

Huber (1991)

3.1 General considerations

Crystallization is a process that has fascinated both scientists and casual observers throughout the ages. It is indeed remarkable that upwards of 10^{20} molecules or ions, distributed essentially randomly throughout some fluid medium (gas, liquid, or solution) coalesce, very often spontaneously, to form a regular solid with a well-defined structure, or in the case of polymorphs, with a limited number of well-defined structures. Those structures are invariant across a wide variety of conditions, in some cases almost under any conditions for which crystals form. Two of the principal questions to be asked for such a process are how it begins and how it proceeds, especially in the context of polymorphic systems. A great deal of work has been devoted in attempts to answer these questions, and in spite of considerable progress especially on experimental and empirical fronts, there is still much to be learned in developing current models. Historical treatments of the classic notions of crystallization and recrystallization, including many important references, have been given by Tipson (1956) and van Hook (1961). A more recent thorough account may be found in Mullin's (2001) book.

For any substance it is possible in principle to define experimentally the solvents, temperature range, rate of evaporation or cooling, and many of the other conditions under which it will crystallize. This collection of conditions has been called the *occurrence domain* (Sato and Boistelle 1984). That domain exists for any substance, but rarely, if ever, are its contents completely known. The contents of the occurrence domain for any material—in the present context, any polymorph—are not necessarily unique. In regions in which there is an intersection of domains, one may expect that two or more polymorphs would crystallize under essentially identical conditions. On the other hand, determining which regions of the domain are unique to a particular polymorph can be advantageous in determining crystallization strategy. This chapter deals with a number of the factors which should be considered in making such a determination, along with examples of the phenomena associated with competitive polymorphic crystallizations.

3.2 Aggregation and nucleation

The thermodynamics and kinetics outlined in Chapter 2 treat the question of crystallization on the macroscopic scale. As noted there, these ideas developed in the late nineteenth century from the thermodynamics of Gibbs and the physical chemistry of Ostwald. On the microscopic (nano) scale we would like to be able to answer questions about the critical size, structure and energetics of a collection of molecules that will grow into the eventual crystal. In particular for the present discussion, how and when will polymorphs be obtained or be prevented from forming? The first stage of crystallization is viewed as *nucleation,* the spontaneous formation or introduction of a nucleus, or center of crystallization, in the crystallization medium from which crystals may grow, although the process is viewed as dynamic and nuclei may also be destroyed before growing into larger crystals. The size of such nuclei has been a matter of considerable discussion. For instance, on the one hand, Ostwald (1902) initially claimed that particles containing between 10^8 and 10^{12} molecules are not sufficiently large to induce crystallization from supersaturated solutions, but later work indicated that much more modest numbers (e.g., 10–10^5) may be considered a critical size to generate crystals (McIntosh 1919; Tammann et al. 1931; Erdemir et al. 2009; Harano et al. 2012). This amounts to a cube of approximately 100 Å on an edge and a "crystal" nucleus weighing as little as 10^{-18} g. Additional aspects of the question of the size of a crystal nucleus are discussed by Mullin (2001).

The last few years have witnessed rapidly increasing interest in the nucleation phenomenon. As increasingly powerful and sophisticated experimental and computational techniques have recently been developed for treating nanoscale systems, those techniques have been applied with the aim of understanding— and eventually controlling—the nucleation of crystals. Kashchiev's (2000) book supplements Mullin's (2001) chapter on the subject with considerably more detail, thus providing a benchmark of the development and understanding of the phenomenon to that date. The major portion of that work involved what is now called *classical nucleation theory* (CNT) generally attributed to Volmer (1939); the more recently developed model is known as a *two-step nucleation* (Erdemir et al. 2009; Davey et al. 2013) and there have been many recent developments in the examination, characterization, and application of the two models (e.g., *Faraday Discussions,* 2015). A detailed discussion of these two models is beyond the scope of this work, but in view of the crucial role of nucleation in the formation of crystals—and thus of polymorphs—I present here some of the basic concepts and terms in use with both of these models, as outlined in the following diagram:

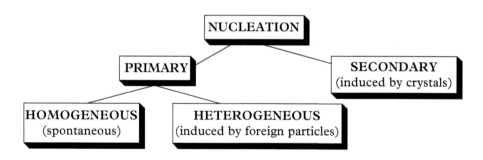

Primary nucleation refers to those systems that do not contain crystalline matter. When no foreign bodies are present, that is, the crystallization results from the spontaneous formation of nuclei of the crystallizing material, the process is referred to as homogeneous. The presence (intentional or unintentional) of foreign particles can also induce nucleation, which is then termed heterogeneous.

Secondary nucleation deals with the situation in which nuclei are generated in the vicinity of crystals of the solute already present in a supersaturated solution. The solute crystals may have resulted from primary nucleation or may be deliberately added. This subject has also been covered by Mullin, as well as in a number of reviews (Strickland-Constable 1968; Botsaris 1976; DeJong 1979; Garside and Davey 1980; Garside 1985; Nyvlt et al. 1985; Agrawal and Paterson 2015) and two *Faraday Discussions* (2007, 2015).

Mullin argued that the minimum number of molecules in a stable crystal nucleus can vary from about ten to several thousand. However, a model based on the simultaneous collision of this number of molecules with the degree of order required for it to be recognized by additional molecules as a crystal is highly unlikely. According to CNT a more likely scenario is that the nucleus would be generated by a sequence of bimolecular additions in which a so-called critical cluster—critical in terms of size and thermodynamic stability—would be built up stepwise:

$$A + A \rightleftharpoons 2A$$
$$A + 2A \rightleftharpoons 3A$$
$$A + (n-1)A \rightleftharpoons nA(\text{critical cluster } A_1).$$

In Mullin's model, further molecular additions to the critical cluster results in nucleation.

A solution or melt can contain a variety of clusters, $A_1, ..., A_n$, in a system of competing equilibria. Each cluster in turn is a potential critical cluster for the nucleus of one or more polymorphic crystal modifications. In the context of polymorphic structures, in particular those which crystallize under similar conditions, there must be a number of processes of this type, all involved in competing equilibria. This is the idea behind Etter's (1991) and Weissbuch et al.'s (2003) extension of this model, describing the clusters as aggregates, which must contain the

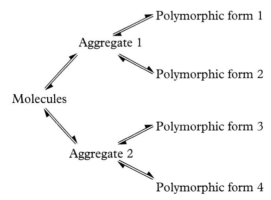

structural essence of the eventual crystal structure(s), and therefore are likely to be dominated by the same intermolecular interactions. Because such a system involves multiple equilibria, once nucleation occurs for one of the polymorphic forms, the equilibrium will be displaced in favor of that form at the expense of other forms. On a qualitative basis this demonstrates the competition between kinetic and thermodynamic factors. For instance, even if polymorphic form 1 were the thermodynamically most stable one, polymorph 3 might be the only one obtained if aggregate 2 nucleated crystal growth faster than aggregate 1. When these factors are equal, or very nearly so, then two or more modifications may result from the same aggregate or from different aggregates, leading to *concomitant polymorphs* (see Section 3.6). An important assumption for this model, a component of CNT, is that the resulting crystal structure reflects the structure of an aggregate.

Some experimental evidence for the presence of different aggregates in solution leading to polymorphic structures was presented by Näther et al. (1996), and there were early attempts to relate nucleation rates with proposed structures in solution based on molecular modeling (Petit et al. 1994).

Twenty years before Etter's model for competing aggregate structures in the formation of polymorphs, Powers (1971) clearly stated the fundamental question and the outstanding problem regarding the challenge of understanding the nucleation process:

> It would appear almost certain that the development of these at least transient aggregates is the precursor to the development of phase transitions to the ordered solid nucleus, in harmony with the local thermodynamic conditions. Yet though a vast amount of study—conferences, books, etc.—have been devoted to nucleation, it is still uncertain how this last transition takes place.

That is, despite its simplicity there was still a great deal that was not understood about nucleation on the basis of this model. For instance, Erdemir et al. (2009) also pointed out the route to crystallization has been found to be more complex than portrayed by the classical theory. Among the "long-standing puzzles of crystal nucleation in solution" noted by Vekilov (2010) are nucleation rates which are many orders of magnitude lower than theoretical predictions, and in the case of proteins, the significance of the dense protein liquid prior to crystallization.

An alternative model for nucleation, originally prompted by observations of protein crystallization is a so-called "two-step model" in which the first step is the formation of a metastable cluster of solute molecules, and upon reaching a sufficient size in the second step the cluster undergoes reorganization into an ordered structure. Vekilov (2010) has suggested that the first-stage metastable clusters consist of dense liquid and are suspended in the solution and he has estimated their size to be several hundred nanometers. Gebauer and Cölfen (2011) have shown that these pre-nucleation clusters, which have been detected experimentally, themselves behave very much like molecules. They also subsequently reviewed in detail the issue of pre-nucleation clusters (Gebauer et al. 2014), although with only passing reference to molecular crystals. Vekilov (2010) has shown how the presence of these dense clusters accounts for the previously noted large discrepancy in the crystallization kinetics predicted by CNT.

In reviewing the situation, Davey et al. (2013) presented a schematic diagram to distinguish between the two models (Figure 3.1) and noted that little progress had been made regarding the nature of nucleation of small organic molecules:

> The biggest challenge is to identify the structure of low-concentration, nanosized dynamic clusters of molecules...no relevant results have yet been reported for small organic molecules...Ultimately, however, we need new approaches to the detection and analysis of molecular clusters at low concentrations in solution. These approaches will most likely develop from current state-of-the-art X-ray scattering and spectroscopic techniques in combination with sophisticated modeling of the properties of potential clusters.

Indeed, in the relatively short period preceding this writing from that publication, there has been a great deal of activity in exploring and validating these competing models, including a number of contributions from the 2015 *Faraday Discussions*.

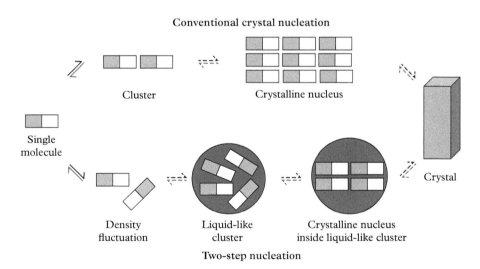

Conventional crystal nucleation

Cluster   Crystalline nucleus

Single molecule

Crystal

Density fluctuation   Liquid-like cluster   Crystalline nucleus inside liquid-like cluster

Two-step nucleation

Figure 3.1 *The two alternative structural models currently used for the dynamics of cluster formation during crystal nucleation from supersaturated solutions. (a) Density fluctuations are concomitant with fluctuations in the order parameter, so that molecular packing within clusters reflects the possible polymorphs of the solute. (b) Fluctuations in density are disconnected from those in order, so that the initial clusters that form are liquid-like, and crystalline order appears later on. (From Davey, R. J., Schroeder, S. L. M., and ter Horst, J. H. (2013). Nucleation of organic crystals—a molecular perspective.* Angew. Chem. Int. Ed., **52**, *2166–79. With permission of John Wiley and Sons.)*

One of those (Davey et al. 2015) notes the importance of "an obscure publication by Dunning and Shipman" (Dunning and Shipman 1954) on the nucleation of sucrose. Recalling the classic connection between kinetics and mechanism, Davey et al. invoke the "reaction coordinate" and transition state theory, originally proposed by Eyring et al. (1935) and employed by Dunning and Shipman to combine structural and kinetic aspects of nucleation. In any kinetic study the rate determining step plays a crucial role in the overall mechanism and Davey et al. demonstrate that the desolvation of specific functional groups is the principal contributor to the rate determining step in the nucleation process.

The following is demonstrative of the nature of that activity, but is necessarily selective. There are many recent developments in this area of research on understanding and controlling nucleation. I note here that the Davey et al. (2015) approach contains a consideration of the nature of the molecule that is crystallizing, which potentially can have important implications for polymorph screening and discovery. As I have demonstrated in Chapter 1, there are no molecular properties that can serve as predictors of polymorphic behavior. But chemistry is certainly involved in the crystallization process and may be invoked to aid in understanding and controlling the crystallization. Therefore, any molecular property that can serve to aid crystallization in general, and the search for polymorphs in particular is of particular significance and value in designing strategies for crystallizations and the search for solid forms.

In concert with the two-step mechanism, Shahar et al. (2016) have demonstrated experimentally that the crystallization of a perylene diimide originates with an amorphous phase. In spite of the fact that this first stage is amorphous, subsequent crystallization depends critically on two factors: the structure of that amorphous "embryo" and the nature of the solvation. Varying either can lead to different crystallization outcomes. Control of the process can be gained by varying the hydrophobicity of the environment at various stages of the crystallization, even to the extent of preventing crystallization.

On the other hand, Lu et al. (2015) investigated the primary nucleation behavior of aspirin using an innovative experimental platform to study gas-segmented flow produced by a microfluidic T-junction device. Assuming nucleation follows the homogeneous Poisson rate process, the authors concluded that the experimental data followed the fitted sigmoidal curve compatible with CNT.

In a third example Du et al. (2015) attempted to apply the combination of solution-phase experiments and computational studies suggested earlier by Davey et al. (2013) to a specific case of tolfenamic acid. They found that in solution the conformationally flexible molecules undergo rapid fluctuation between solvated monomers or dimers, depending on the solvent used. In spite of the fact that crystallization of tolfenamic acid from ethanol results in conformational polymorphs (Cruz-Cabeza and Bernstein 2014) this study did not reveal any connection between the solution chemistry and crystallization outcomes. They point out the importance, in performing these calculations, of selecting a suitable level of theory to achieve the most reliable results, and the potential preference for measurement of nucleation rates over the more common practice (noted above) of associating the crystal structure resulting from crystallization with nucleation phenomena.

Finally, as noted in Chapter 2, an important study, adhering to the Davey et al. prescription of "…spectroscopic techniques in combination with sophisticated modeling of the properties of potential clusters" by Belenguer et al. (2016) elegantly demonstrates the inversion of relative thermodynamic stability for clusters in the nanoscale size regime. Similar studies will no doubt lead to much greater understanding and control of the crucial process of nucleation and thus of the crystal form obtained (see Section 3.5), but as with many other aspects of polymorphism of molecular materials it is clear that while we may develop increasingly powerful experimental and computational tools for investigating the phenomenon of nucleation, every system will be unique, and will provide unique challenges to the crystal chemist.

From a practical point of view, control over nucleation, and in cases of polymorphic systems, control over the polymorph obtained as a result of nucleation, has been the concern of those industries for which crystallization is a crucial or final step in the production process, for example, sugars, amino acids, pharmaceuticals, oils, and fatty acids. A number of examples of studies regarding polymorphic variation and preferences for nucleation may be found in the literature from those disciplines.

The initiation of crystal growth has been a problem for the sugar industry since its infancy. Aqueous solutions of sugars often tend to form syrups—indeed that has become one form of marketing, although clearly generally not the preferred one. As noted previously, Powers (1971) has reviewed the role of nucleation in the sugar industry, including much of the accumulated experience involving sucrose. As in many industries successful techniques were developed over decades or centuries and were considered trade secrets or even commonplace practice without being scientifically recognized or understood. Thus, traditionally, sucrose crystallizations, carried out from the huge copper vats in which solutions were concentrated, were initiated (i.e., nucleated) by the mechanical shock of hammering on the vat (Figure 3.2).

Yet it was only in 1912 that Young described mechanical shock as a factor in nucleating supersaturated solutions (Powers 1971). It is noteworthy that in spite of the fact that sucrose was successfully crystallized innumerable times over centuries there was no evidence of any polymorphism. In fact it was often cited, along with naphthalene as a common compound that does not exhibit polymorphism. Following nearly a decade of experimentation at high pressure Patyk et al. (2012) succeeded in obtaining a second polymorph of sucrose (which is not stable at ambient pressure).

Three other curiosities related to nucleation and polymorphism of sugars demonstrate some of the difficulties encountered. Turanose was long considered to be a liquid at room temperature (Powers 1971), until it spontaneously crystallized; following that event fresh batches of the material always crystallized. In another case more closely related to polymorphism, α-D-mannose had been prepared routinely until the appearance of β-D-mannose, following which the α-form could not

CRYSTALLIZING PANS

Figure 3.2 *Detail from an 1850 drawing of the London sugar refineries of Messrs. Fairrie and Co., showing a copper crystallizing pan for sugar. The worker to the right of the pan is holding a mallet which was used to bang on the pan to induce nucleation of the crystallization process. (Reproduced from Fairrie 1925, with permission.)*

be induced to crystallize in the same laboratory (Levene 1935; Dunitz and Bernstein 1995; Bučar et al. 2015). Xylitol, the sweetener in sugarless chewing gum, was first prepared as a syrup in 1891, but did not crystallize until 50 years later (Wolfrom and Kohn 1942) as a metastable hygroscopic form melting at 61 °C. Shortly thereafter there appeared another form melting at 94 °C (in concert with Ostwald's Rule of Stages) resulting in the disappearance of the lower melting form (Carson et al. 1943). As Powers noted, these three cases can be attributed to unintentional seeding, a topic treated in more detail in Section 3.7.

Glutamic acid has been an iconic subject of crystallization studies, including many on nucleation. For instance, Black and Davey (1988) described a number of the interrelationships and practical aspects of the control of nucleation, crystal growth and polymorphic transformation and elucidated those relationships for the primary nucleation of L-glutamic acid, including temperature, critical nucleus, relationship of interfacial tension to solubility, thermal history, induction time, agitation, and effect of additive.

Kitamura (1989) studied many of these nucleation factors in the competitive crystallization of the α and β forms of L-glutamic acid. He found that at 25 °C only the α modification nucleates and grows. In this system, at least, the effect of temperature on the relative nucleation rates of the two polymorphs is more "remarkable" than the effect of the supersaturation ratio: as the temperature is increased with a constant supersaturation ratio, the amount of α decreases. He also reported that the β form tends to nucleate in stagnant solutions, while at 25 °C essentially only α nucleated homogeneously.

More recently Schöll et al. (2006) have determined the nucleation kinetics during reaction precipitation upon mixing aqueous monosodium glutamate with hydrochloric acid. Srinivasan and Dhamasekaran (2011) claim to have selectively nucleated the α and β polymorphs by swift cooling combined with variable degrees of supersaturation. The continuing interest in this system demonstrates many of the challenges to be met in understanding and controlling nucleation in order to obtain a specific polymorph robustly and consistently. Sala et al. (2010) have demonstrated the ability to selectively nucleate the monoclinic E form by using a kinetically controlled technique called depressurization of an expanded liquid organic solution (DELOS) which is particularly suitable for scaling up this industrially important organic solid material.

Finally, it is worthy of note that the mere existence of polymorphic structures can be used as a probe of the nucleation process. For instance in considering the aggregation process in supersaturated solutions of 2,6-dihydroxybenzoic acid, Davey et al. (2000) found a direct link between the relative occurrence of two polymorphic forms (from toluene and chloroform solutions) and the solvent-reduced self-assembly (aggregation) of the molecule.

3.3 Thermodynamic versus kinetic crystallization conditions

Physical organic chemists have long been accustomed to making the distinction between "thermodynamic" and "kinetic" conditions when referring to reactions and reaction mechanisms. In chemical parlance thermodynamic conditions essentially means those conditions under which thermodynamic equilibrium is maintained or very nearly maintained. Kinetic conditions refer to situations that are far from equilibrium (Van Hook 1961).

In terms of crystallization (Ward 1997), thermodynamic conditions might refer to a slow evaporation, a very slow cooling, a slow crystallization from the melt at a constant temperature only slightly below the melting point, a slow sublimation for which there is only a small difference between the temperature of the solid and that of the cold finger on which the sublimate is crystallizing, etc. Another thermo-dynamic condition is the so-called stirring experiment, where one crystal form or a mixture of two or more crystal forms is stirred in a solvent or solvent system of suitable solubility at a given temperature for a certain period of time (usually several days). Under these conditions the thermodynamically stable crystal form will grow or remain (McCrone 1965; Cardew and Davey 1985). On the other hand, kinetic conditions might refer to a high degree of supersaturation, rapid cooling of a solution or melt, rapid evaporation of solvent, large temperature dif-ference between the sample and cold finger in a sublimation, precipitation by rapid addition of an anti-solvent, etc. A number of examples serve to demon-strate how these principles have been applied to the crystallization of different modifications in polymorphic systems. *p*-chlorobenzylidene-*N*-*p*-chloroaniline **3-I** is dimorphic (Bernstein and Schmidt 1972; Bernstein and Izak 1976). The thermodynamically more stable orthorhombic form may be obtained by slow evaporation of a solution in which ethanol or methylcyclohexane is the solvent. Prismatic crystals usually grow in hours to days, depending on the initial con-centration of the solution. The metastable triclinic form is obtained by dissolv-ing the maximum amount of substance in a boiling ethanolic solution, which is then immediately placed in a desiccator freshly charged with calcium chloride. Needle-like crystals appear within minutes, and the desiccant aids in accelerating the rate of evaporation. The crystals are metastable and may begin to spontan-eously transform to the orthorhombic modification in periods ranging from hours to days.

3-I

Berman et al. (1968) noted that "mannitol is unusual among carbohydrates in that it exists in several polymorphic forms," indicating that a number of these are often obtained simultaneously. They describe the preparation of a number of these modifications. The α form is obtained by slow crystallization from 96% ethanol, the α' form by evaporation from 100% ethanol and the β form from aqueous ethanolic solutions, all apparently under thermodynamic conditions. On the other hand the γ form is obtained kinetically by rapid cooling of a 1:1 water–ethanol solution. An additional κ form was obtained (unexpectedly) upon evaporation of a boric acid/methanol solution (Kim et al. 1968).

Bock has studied a number of systems in which different polymorphs were obtained under thermodynamic and kinetic conditions. (2-Pyridyl)(2-pyrimidyl)-amine **3-II** is dimorphic. Modification I is readily crystallized thermodynamically "from any solvent" (toluene was actually used) while modification II is obtained kinetically by fast evaporation of an etheral solution or by resolidification of the melt (Bock et al. 1997).

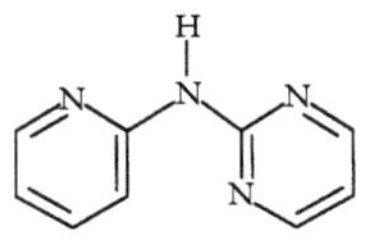

H

N — N — N

3-II

H₃CO, S, OCH₃ / H₃CO, S, OCH₃

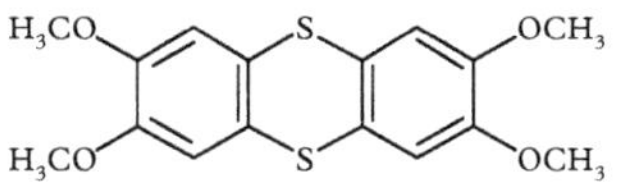

3-III

In 2,3,7,8-tetramethoxythianthrene **3-III**, the less stable (lower in both density and absolute value of lattice energy) monoclinic modification is obtained under kinetic conditions: rapid crystallization from polar diisopropyl ether, whereas the more stable (higher density and lattice energy) orthorhombic modification is thermodynamically obtained from a non-polar hydrocarbon solvent.

In pharmaceutical applications the choice of polymorphic modification for formulation depends very much on the robustness of the crystallization process as well as the properties and characteristics of the preferred modification. For instance, an antiarrhythmic under development (McCauley et al. 1993) was shown to exist in two anhydrous polymorphs, two dihydrated enantiotropic polymorphs, a monohydrate, and the solvates of several organic solvents. Following characterization of all of these modifications it was desired to selectively obtain one of the dihydrates, termed modification A, which is thermodynamically less stable at room temperature than another dihydrate, D, in contact with aqueous solutions, but A is more stable over a wider range of relative humidities. The enantiotropic transition point is 37 °C. Procedures were developed for obtaining A preferentially. Above the transition point a thermodynamic crystallization is carried out at 50 °C, using type A seeds as an added precaution to force the crystallization to type A. The desired type A can also be obtained under kinetic conditions by spontaneous crystallization below the transition point followed by rapid filtration and removal of excess water. The latter procedure prevents a transformation from the A state (metastable below the transition temperature) to the D form in the crystallization medium. Similar considerations were applied to develop procedures for the selective crystallization of the α and β modifications of glutamic acid (Kitamura 1989).

3.4 Monotropism, enantiotropism, and crystallization strategy

The examples cited in the previous section involved distinguishing between thermodynamic and kinetic conditions for crystallization. For a practicing chemist that distinction is often made instinctively rather than consciously. However, it can be related directly to the monotropic or enantiotropic relationship between two polymorphic forms, as expressed in the energy–temperature diagrams (see Sections 2.2.2 and 2.2.3). Practical means for determining the monotropic/enantiotropic relationship of two phases are given in Chapter 4. This information in turn can be used to design strategies for attempts to obtain desired crystalline forms at the expense of the less desired ones. Here I summarize the ramifications of the particular monotropic/enantiotropic relationship on the crystallization strategy once that relationship has been determined, preferably by generating the energy–temperature diagram. For a dimorphic system there are four possibilities:

- Obtaining the thermodynamically stable form in a monotropic system: no transformation can take place to another form, and no precautions need be taken to preserve the stable form or to prevent a transformation.

- Obtaining the thermodynamically stable form in an enantiotropic system: precautions must be taken to maintain the thermodynamic conditions (temperature, pressure, relative humidity, etc.) at which the G curve for the desired polymorph is below that for the undesired one.

- Obtaining the thermodynamically metastable form in a monotropic system: a kinetically controlled transformation may take place to the undesired thermodynamically stable form. To prevent such a transformation it may be necessary to employ drastic conditions to reduce kinetic effects (e.g., very low temperatures, very dry conditions, storage in the dark, etc.).

- Obtaining the thermodynamically metastable form in an enantiotropic system: the information for obtaining and maintaining this form is essentially found in the energy–temperature diagram.

3.5 The polymorph (solid form) screen

3.5.1 General comments

As noted elsewhere in this volume since the publication of the earlier edition there has been an enormous increase in the awareness of polymorphism and the importance of exploring and understanding the solid form landscape for any material with intended or potential use in its solid form. That awareness has led to the development of a variety of techniques for exploring the crystal form landscape. There is a vast variety of conditions that can affect crystallization, and for any polymorphic system it is indeed difficult to single out a particular factor that might dominate, in particular at the onset of the crystal form screen. However, it is certain that seeding can and does play an important role in determining the fate of a crystallization, especially one in which competitive processes can lead to individual polymorphs, or a mixture of them.

The important point is that the determination of the crystallization conditions for various polymorphic forms need not be a completely random process. Thoughtful observation combined with experience and consideration of all the available chemical structures and thermodynamic information can provide extremely useful guidelines, if not for success, then at the very least for further experiments. Crystallization is almost never a surefire procedure, especially when one is trying to selectively produce a particular polymorph, and one that has proven consistently or suddenly elusive.

What are the techniques for carrying out a so-called "polymorph (or solid form) screen" to survey and understand the solid form landscape: how many solid forms (including amorphous) comprise that landscape? How can we make them consistently unique and pure? How do we characterize and identify and distinguish among them? What are their relevant properties for any ultimate use?

A solid form screen is often referred to as "routine." In all aspects of solid-state research solid-state screening is routinely conducted to understand the material and for ultimate commercial utilization, to identify commercially viable solid forms, along with the forms that crystallization processes must be designed around and upon which control strategies for processing and storage of drug products are based. The fact that polymorph screening is a component of the "routine" of research and development of a solid active pharmaceutical ingredient (API) by no means implies that a particular polymorph screen is a routine procedure. There are no "cookbook" recipes for carrying out such screens. As noted below just for solution crystallizations every screen must be individually designed based on solvent, solubility, solvent mixtures, heating and cooling program, compound stability, amount of compound available, and general familiarity with the compound and its physical and chemical properties. This says nothing about all the other techniques available for exploring the solid form landscape. The fiction that carrying out a crystal form search is a routine procedure has been propagated by some, notably in intellectual property circles (Amundsen 2013). The complexities of any individual crystal form screen have recently been detailed by Lee et al. (2011a). Thus, while the crystal form screen is a component of the routine of drug product development, the crystal form screen itself for any particular compound is by no means routine.

With no way of predicting if a molecule will crystallize, let alone in what forms or under what conditions, solid form screens are designed broadly to survey solid form space with the goal of promoting different crystal nucleation and growth pathways. In principle, an agreeably comprehensive screen could be devised to encompass diverse crystallization conditions and applied to all compounds. However, in practice no two solid form screens are alike. Why must screens be adapted, sometimes significantly and on the fly, to find forms? More importantly, how, with all that is at stake in, say, drug development to find the right form that will reliably deliver the drug in the commercial presentation, can forms, especially more stable ones, be missed?

The reasons why polymorph screens fail to find forms are many and varied, ranging from limitations on how materials are generated and characterized to how fully the screen is resourced. The inability to design a single comprehensive form screen that will find all forms stems from the fact that each compound has unique solubility properties and different crystallization tendencies, either of which can severely limit the experiments that can be meaningfully conducted in search of crystal forms. Compounds with very high or low solubility, for example, make achieving solvent diversity in solution-based crystallization screens difficult. When the solubility is high, oftentimes films, oils or solutions are produced. On the other hand, when the solubility in a given solvent is low, the starting material may be incompletely dissolved and if sufficient care is not taken, residual seed crystals could be carried through that would inevitably bias the experimental outcome. Even experiments that are tailored to the solubility properties of the molecule may fail when the compound is either reluctant to crystallize, that is, slow to nucleate, or in the opposite extreme, willingly crystallizes in a single form regardless of the conditions. It stands to reason that if a kinetic form is the first to crystallize, there is simply no assurance that it will eventually convert to a more stable form. For drug compounds high temperatures are usually exploited to promote the crystallization of neat forms directly from solution or in the alternative, to produce them by desolvation.

Throughout the course of a solid form screen, many materials will be generated and characterized in order to unequivocally establish that new forms have been found. This is not a trivial exercise. Form identification not only relies on using the right combination of techniques to chemically and physically characterize the solid products, but it also hinges on the ability to discern forms amongst materials that are frequently mixtures of phases, poorly crystalline or isolated in poor yield, potentially limiting the analyses that can be done. Complicating matters is the fact that some forms are only ever isolated as a minor phase impurity, while others are seen early on at small scale only never to be reproduced or to disappear once a new, more stable form crystallizes (see Section 3.7). In cases for which the characterization data are incomplete or ambiguous and the phases cannot be reproduced, bona fide forms may have been generated, but not identified as such. Finally, especially in an industrial environment there are often restraints of time, resources, and even material that can limit the extent of the exploration undertaken.

At the time of the first edition there was essentially only one review on the subject generally classified as solid form screens (Guillory 1999). The number since then testifies to the importance, the complexity, and the rapid developments of this activity; a representative, but not comprehensive sampling of those reviews include those by Hilfiker et al. (2006), Stahly (2007), Llinàs and Goodman (2007), Cains (2009), Florence (2009), Aaltonen et al. (2009), Lee et al. (2011a), and Newman (2013). All of these contain extremely useful, even if occasionally differing, advice on the pursuit of a crystal form screen, and should be consulted both for general considerations and for solutions to specific problems and situations. I present here an attempt at an overview of the entire process, with some of the considerations to be weighed in the course of a screen, and over the course of the "lifetime" of a compound. It is to be emphasized that this is a rapidly developing and rapidly changing activity with new techniques, particularly the preparation and/or the detection and characterization of new forms.

3.5.2 When to carry out a solid form screen?

Chemistry in general is all about control: control of processes, control of structure, and control of properties. The degree of control that is obtained depends on the level of knowledge and understanding of the particular system in question. Regarding the solid form landscape of a molecular compound, as demonstrated in Figure 3.3, obtaining that control essentially requires constant experimentation, testing, awareness of conditions, and exploration for new forms over the entire lifetime of its use. There can never be an assurance that the solid form landscape is completely known.

But like any research activity the extent of the effort is limited by resources and time. In the case of a potential API at the early stage of development there is often a very limited amount of material. Nevertheless, for initial clinical and toxicology testing some solid form must be chosen which requires some screening at the earliest stages of development. As that development proceeds accompanied by process scale up, dissolution profiles, stability testing, etc., crystallization procedures (e.g., equipment, heating/cooling rates, sources and purities of reagents, impurity profiles) change, often revealing new solid forms or possibly requiring new forms with specific processing characteristics found lacking in the initially

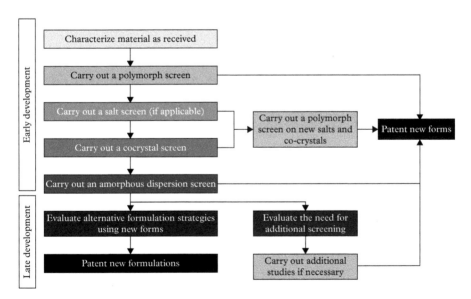

Figure 3.3 *Timeline for solid form screening over the course of the lifetime of a molecular entity. (Reproduced from Newman, A. (2011). An overview of solid form screening drug development, PPXRD-10, May 18, 2011, www.seventhstreetdev.com. Courtesy of Ann Newman.)*

chosen form. As the project matures and the product approaches launch intellectual property issues enter with increasing significance. In a manufacturing environment the solid form ultimately chosen for product development must be protected by patents. It behooves any manufacturer to be as familiar as possible with the solid form landscape, and to protect that landscape with appropriate patents, so that competitors cannot launch a product with a different—and unpatented by the innovator—solid form. Hence, the exploration of the solid form landscape essentially commonly continues over the entire lifetime of the molecule.

3.5.3   What comprises a solid form screen?

The goal of a solid form screen is to determine and understand as much of the solid form landscape as possible. Since that landscape is defined by thermodynamic variables with various crystal forms occupying different regions, the surveying of the landscape requires the preparation, identification, and characterization of as many solid forms as possible. In some cases, as for naphthalene, this might turn out in retrospect to be almost trivial, since only one form is known. At the other extreme, for instance in the case of atorvastatin calcium (the API in Lipitor®) with 78 reported solid forms (Jin 2012), the task is indeed daunting. But since nothing is known about the solid form landscape at the start of a screen it is totally unexplored and unknown territory.

Stahly (2007) has provided a convenient flow sheet for the solid form screening and selection (Figure 3.4). The first stage "characterize starting material" is again most often limited by the quantity available for such a task—sometimes just tens of milligrams. In such a case "characterization" would very often be limited to the melting point and perhaps some spectral data. However, it is important to stress the amount of information that may be obtained about the behavior of the material with a few hours investigation with a very small quantity of material on the hot/cold stage polarizing microscope: any tendencies for thermal phase transitions (above or below room temperature), for example, melting, recrystallization, and remelting, the presence of solvent/water, tendency for sublimation, decomposition, etc. (see Section 4.2 for details). These initial observations can be augmented with hot-stage microscopy, differential scanning calorimetry, and thermogravimetric analysis measurements, which again require small quantities of (generally unrecoverable) material, to further serve as a preliminary exploration of the phase space and to provide some guidelines for the design of subsequent crystallization experiments (see Section 4.2 for details). For instance the optical or calorimetric detection of a phase change above a certain temperature suggests attempting subsequent crystallizations both below and above that temperature.

The next step in the solid form screen generally involves attempts to crystallize the material from solution. One of the first decisions in performing those experiments is the choice of solvent. Generally the material has been obtained either directly from the synthetic procedure or via recrystallization from some solvent or mixture of solvents. This is a starting point, but the object of the solid form landscape

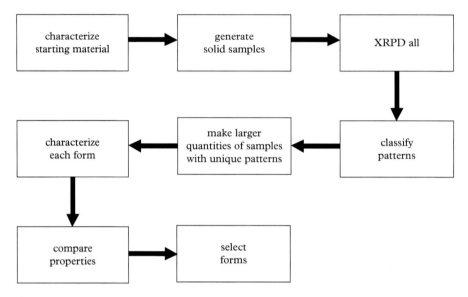

Figure 3.4 *Process of solid form screening and selection. (From Stahly, G. P. (2007). Diversity in single- and multiple-component crystals. The search for and prevalence of polymorphs and co-crystals.* Cryst. Growth Des., *7, 1007–26. With permission of American Chemical Society.)*

search is to find as many forms as possible so a wide choice of solvents and mixtures of solvents is recommended. [Note that different polymorphs may crystallize from the same solvent under the same conditions (*concomitant polymorphs*, see Section 3.6) *vide infra* which can often be revealed by careful optical examination of any crystallized product, again on the polarizing microscope.]

Mullin (2001) presents general criteria for choice of solvent, while a number of authors have suggested potential solvents for such crystallizations (Guillory 1999; Cains 2009; Newman 2013). Guillory and Cains suggest fifteen and twenty-two, respectively, ranking them by decreasing boiling point, while Newman (citing Miller et al. 2005 and Marcus 1998) suggests twenty-four, ranking them by decreasing dielectric constant. Many of these are indeed commonly used solvents, but there is no reason why crystallization experiments should be limited to them; again, the object is to explore as much of the landscape as possible, although in the pharmaceutical industry there may be a bias towards limiting the solvent to those generally recognized as safe (GRAS).

In the end the choice of solvent can be determined only by experiment. But solution crystallizations can be time-and substance-consuming. For instance, a one hundred experiment crystallization screen on the angiotensin II antagonist agent MK-996 required 10 g of substance and required 32 weeks to observe nine crystalline forms (Jahansouz et al. 1999). On the other hand, McCrone proposed a still little-known and rarely used microscope technique for screening many potential

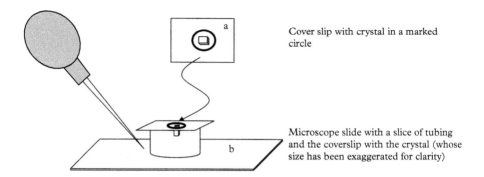

a

Cover slip with crystal in a marked circle

b

Microscope slide with a slice of tubing and the coverslip with the crystal (whose size has been exaggerated for clarity)

Figure 3.5 *Excerpt from McCrone (1983) describing the (McCrone 1982 repeated) experiments carried out to screen for appropriate solvents on a nanogram single crystal. The experimental setup has been redrawn to add clarity and include additional details. (McCrone, W. C. (1983). Particle characterization by PLM. Part III. Crossed polars.* The Microscope, *Vol.* **31:2**, *187–206. Courtesy of the McCrone Research Institute.)*

solvents that uses a minimum of material. The method is best described in McCrone's own inimitable words (Figure 3.5; McCrone 1983):

> There was a time when I might not have believed it is possible, microscopically, to determine the solubility of a single subnanogram particle in any number of different solvents at a rate of one to two minutes per solvent and still have all of the material left when you finish. This is, however, quite possible, and easily carried out.

Following the choice of solvent the order of experiments for solvent-based crystallizations has been summarized schematically by Florence (2009) in Figure 3.6. Such flowcharts can be deceivingly simplistic (which is the primary reason for their use). Note that the first step calls for "vary[ing the] amount to alter concentration and supersaturation" and "us[ing] a range of different solvent or solvent mixtures," hence suggesting an unspecified and unlimited variety of conditions. The second step of dissolving the material also indicates that "temperature/stirring" are unspecified variables, also indicating an unspecified variety and number of conditions. The precise conditions for any single crystallization must be determined by experimentation.

The choice of solvent is but the first step in the solid form screen. Aaltonen et al. (2009) and Florence (2009) have pointed out the advantages of the Design of Experiment (DoE) approach to carrying out the exploration. Stahly (2007) noted the common belief that solvent alone controls the solid form produced is misplaced. Clearly a solvate cannot be obtained in the absence of that solvent in the crystallizing solution (although it may actually be an unintended minor component; Rafilovich et al. 2005), but in general it is the degree of supersaturation at which nucleation occurs that controls which solid form is obtained. Stahly (2007) has also listed some of the variations in solution crystallization that should be attempted in the search: solvents with a variety of polarities and structures, crystallization by

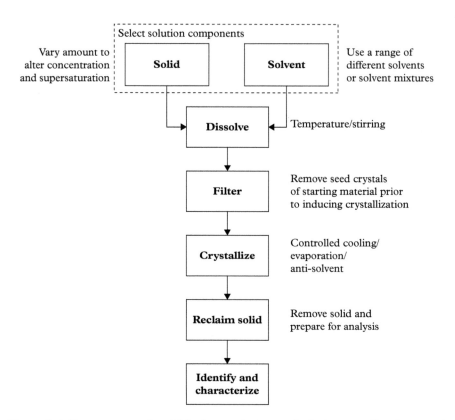

Figure 3.6 *Process steps and variables in a typical manual solution crystallization experiment. In the context of an experimental physical form search the aim is to explore factors that may influence the thermodynamic and/or kinetic control of nucleation and crystal growth. (From Florence, A. J. (2009). Approaches to high-throughput physical form screening and discovery. In* Polymorphism in Pharmaceutical Solids, *2nd edn (ed. H. G. Brittain), pp. 139–84. Informa Healthcare, New York. With permission of Taylor and Francis Group LLC Books.)*

evaporation, cooling of solutions, addition of anti-solvent to solutions, extended agitation of slurries. Thermodynamic (i.e., near equilibrium) crystallization conditions should be accompanied by kinetic (i.e., far from equilibrium) conditions such as rapid cooling or rapid addition of anti-solvent. Aaltonen et al. (2009) have given a graphical demonstration (Figure 3.7) of the distinction between thermodynamic and kinetic crystallizations for a dimorphic system, based on the relationship with the free energies of the two forms.

Stahly (2007) points out that screening is a "hunt for seeds" and that "without knowing what forms are available, there is no way to plan the appropriate concentration at which to bring about nucleation in order to crystallize one form or another. Rather, the attempt should be to provide a variety of conditions under which nucleation may occur."

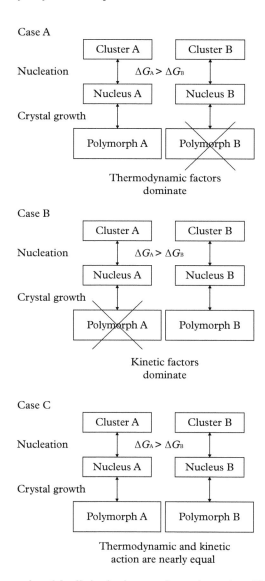

Figure 3.7 *Demonstration of the distinction between thermodynamic and kinetic crystallization for a dimorphic system (Forms A and B) for which Polymorph A is the more stable form with a higher free energy barrier (ΔG) to nucleation. (From Aaltonen, J., Alleso, M., Mirza, S., Koradia, V., Gordon, K., and Rantanen, J. (2009). Solid form screening—a review.* Eur. J. Pharm. Biopharm., *71, 23–37. With permission of Elsevier.)*

Table 3.1 *Methods to generate various solid forms (from Hilfiker, R., De Paul, S. M., and Szelagiewicz, M. (2006). Approaches to polymorphism screening. In* Polymorphism in the Pharmaceutical Industry *(ed. R. Hilfiker), pp. 287–308. Wiley-VCH Verlag GmbH, Weinheim. With permission of John Wiley and Sons.)*

Method	Degrees of freedom
Crystallization by cooling a solution	Solvent, cooling profile, concentration, mixing
Solvent evaporation	Solvent, initial concentration, evaporation rate, temperature, pressure, ambient relative humidity
Precipitation	Solvent, anti-solvent, rate of anti-solvent addition, mixing, temperature
Vapor diffusion	Solvents, temperature, concentration
Suspension equilibration	Solvent, temperature, solubility, temperature programs, mixing, equilibration time
Crystallization from the melt	Temperature changes (min, max, gradients)
Quench cooling the melt	Cooling rate
Heat-induced transformations	Temperature changes
Sublimation	Temperature gradient, pressure, surface type
Desolvation of solvates	Temperature, pressure
pH change	Temperature, rate of change, acid/conjugate base ratio
Mechanical treatment (i.e. milling, cryo-grinding)	Milling time, mill type
Freeze-drying	Solvent, concentration, temperature programs
Spray drying	Solvent, concentration, drying temperature

But solution crystallization (in all of its variations) is only one of an increasing number of techniques available to search for (those seeds of) crystal forms. Some additional "traditional" techniques, with the associated degrees of freedom (i.e., experimental variables), have been given by Hilfiker et al. (2006) in Table 3.1. Stahly has noted that a number of these do not involve the use of solvent.

The search for new solid forms has stimulated the development of a number of non-traditional crystallization techniques.

Capillary crystallization (Llinàs and Goodman 2007; Stahly 2007) confines a solution to a small capillary volume, thus considerably reducing the surface to volume ratio and altering the kinetics of crystallization and providing a wide range of supersaturation levels using small volumes. This is particularly relevant to control of nucleation, since it has been demonstrated that the rate of nucleation of metastable forms becomes competitive with the stable forms (Childs et al. 2004).

The scale of this approach has been further reduced to the nano level with *nanoscopic confinement* of the crystallizing medium (Ha et al. 2004; Hamilton et al. 2012). It was pointed out that in the nanoscale environment melting points and enthalpies of fusion can differ drastically from macroscale values. In an elegant combined computational and experimental study (Belenguer et al. 2016) it has been demonstrated that such a confinement can actually lead to a reordering of the thermodynamic stability of two polymorphic forms of a single compound compared to the order on the macroscopic scale. Thus, this technique may have particularly important ramifications in the search for solid forms.

Recently, in the intermediate *micro* size regime between macro and nano, microfluidic emulsion-based crystallization has been used to facilitate polymorph selection (Leon et al, 2015). The chemical principle behind this approach is that polymorph selection is controlled with evaporating emulsion drops containing API–excipient mixtures via the kinetics of two simultaneously solvent evaporation process occurring processes: liquid–liquid phase separation and supersaturation generation.

Providing a nucleating medium by *templating* has been demonstrated in a number of different experimental manifestations. Designing or utilizing a template takes advantage of the fact that the templating surface may be compatible with the molecule to be crystallized, and thus provides a structural and energetic medium for the attachment of molecules that can lead to nucleation and subsequent growth. A traditional one, of course, is *seeding*, but since the object of a solid form search is obtaining new forms, the intentional seeding per force must involve a form that is not known, and generally will involve a different substance, with presumably a crystal structure or a planar section of a crystal structure that is similar to that desired for the material being studied (Beckmann 2000; Mangin et al. 2009).

Creating suitable surfaces for *templating* has been actively pursued in a variety of creative ways, leading to new solid forms and to new ways to control the crystallization of solid forms. For instance, early work on ledge-directed epitaxy (Bonafede and Ward 1995) was developed further by the same group (Hooks et al. 2001) and has been followed by demonstration of the applicability of polymer-induced heteronucleation to a variety of substances (Lang et al. 2002; Price et al. 2005; Diao et al. 2012), among them extended solids (Grzesiak et al. 2006), pharmaceutical compounds (Grzesiak and Matzger 2007; Lu and Rohani 2009b), and co-crystals (Porter et al. 2008).

The surface of a crystal or of a polymer is not necessarily a required substrate for inducing crystallization; self-assembled monolayers (Love et al. 2005) have been shown to be effective in this regard as well (Hiremath et al. 2005). With the same experimental philosophy, but at the other end of the structural regime, *microporous membranes* have been employed in the search for new crystal forms (Curcio et al. 2003). The principle in this case is that the membrane separates two isothermal solutions and the movement of solvent between them alters the concentration, leading to changes in supersaturation and the relative influence of kinetic and thermodynamic factors. The technique was originally developed for protein crystallization, but was applied to the selective crystallization of γ-glycine in competition with the α and β forms (Di Profio et al. 2007), and to pharmaceutically important molecules and co-crystals (Caridi et al. 2012).

In a related technique *soluble additives* may inhibit the nucleation and/or growth of a known form, thus facilitating the nucleation and growth of a different form (Weissbuch et al. 1995; Davey et al. 1997; Torbeev et al. 2005; Kwon et al. 2006). While this approach has been increasingly applied for solid form selection among known forms, the intentional inhibition of a known form can lead to the appearance of new forms. Such an approach is not only valuable in the discovery of new forms; it constitutes part of the strategy for pursuing the development of the new form (Liu et al. 2014).

Clearly, a change in the medium for crystallization can also lead to additional solid forms. One of those that has been known for a long time, but only occasionally employed is that of crystallization in gels (Henisch 1996; Choquesillo-Lazarte and Garcia-Ruiz 2011; Moreno and Mendoza 2015). The method is particularly useful for compounds of low solubility. Since the gel environment has a higher viscosity than most solutions the crystallization is essentially diffusion-controlled and the influence of sedimentation and convection are decreased; these conditions thus approximate those obtained in microgravity crystal growth experiments in space. Different gels may be prepared for organic or ionic (i.e., salts) materials (Choquesillo-Lazarte and Garcia-Ruiz 2011; Menéndez-Taboada et al. 2012).

Another alternative crystallization medium that has shown promise of yielding new solid forms is that of supercritical carbon dioxide as a solvent (Bouchard et al. 2007; Pasquali et al., 2008; Padrela et al. 2015; Pando et al. 2016; Rodriguez et al. 2016). One increasingly important aspect of this method is that it involves green chemistry and eliminates the need for the disposal or the recovery of solvent. Practically, it is also very much based on chemical engineering principles, which facilitates a great deal of control in developing and maintaining conditions for the discovery, selection, and control of the crystal form obtained. Some other advantages include control of particle size (narrow size distribution), the elimination of the need for (physical) micronization (e.g., drying, milling), the possibility of processing thermolabile molecules, possible addition of other solvents (e.g., EtOH, acetone, H_2O) in the process, and the possibility of forming solid complexes, composites, or co-crystals in the process.

While *electrochemical* crystallizations have been employed for many years in the preparation of charge transfer compounds and salts (Wang et al. 1989) (see Section 3.7) they are also being used for crystallization of single component systems. (e.g., Miyahara et al. 2013).

Physical and chemical perturbations of solutions have also been employed to induce the formation of new solid forms. *Sonocrystallization* (Luque de Castro and Priego-Capote 2007; Gracin et al. 2015) and *non-photochemical laser-induced nucleation* (Zaccaro et al. 2000; Li et al. 2016) have been used to successfully produce solid forms. Metastable forms have been obtained at the edge of an evaporating solution—a so-called *contact line crystallization*—and it has been shown that under such conditions the metastable form may not undergo a solvent-mediated transformation to a stable form (Capes and Cameron 2007; Poornachary et al. 2013).

All of the previously mentioned strategies in the search for crystal forms involve solutions. The importance of non-solution crystallization techniques in the exploration of the solid form landscape should not be overlooked. I have earlier noted the utility of hot/cold stage microscopy. But physical manipulation such as grinding and compression, heat stressing of both crystalline and amorphous forms, dehydration of hydrates and desolvation of solvates can lead to new forms that may otherwise appear in the downstream development stage of a material, where many of these physical stresses are inherent in the production process.

Also, the role of creative chemistry should be encouraged. For example, the almost intuitively obvious co-crystal of caffeine with benzoic acid resisted crystallization for a number of years. A co-crystal of pentafluorobenzoic acid with caffeine was prepared on the assumption that it would likely have a structure similar to the non-fluorinated derivative. That co-crystal was used to seed the benzoic acid–caffeine co-crystal crystallization which readily yielded the co-crystal in four different laboratories which had earlier failed to produce the desired co-crystal. (Bučar et al. 2013; Dubey and Desiraju 2015; Corpinot et al. 2016).

This survey is intended to provide but a general introduction to the variety of techniques employed in the search for solid forms. Each one of these techniques involves many experimental variables, all of which can influence the crystal form outcome. Combinations of these techniques are in use, and increasingly imaginative and sophisticated experimental approaches in the search for solid forms are constantly under development. The overriding fact that one can never know if all the solid forms of a compound have been obtained is both an incentive and a catalyst for pursuing the search for new ones.

While this volume deals with molecular crystals, the significant developments related to crystal growth in the closely related discipline of protein crystallography can provide guidelines for new developments as well. In general the ultimate goal of those practitioners is to obtain X-ray quality single crystals, but they have developed sophisticated methods for high throughput experiments with a large variety of conditions on samples containing small amounts of potentially crystallizing material (McPherson 2004, 2016). Although, as discussed below, the application of high throughput methods in general have not fulfilled their earlier promise, these methods may be applicable in some specific situations.

3.5.4   What is the time frame for a solid form screen?

Having surveyed a number of the possible techniques available in exploration of the solid form landscape one can inquire as to the time scales for the various types of experimental techniques. How long does a solid screen form take? In general, thermodynamic (i.e., close to equilibrium) techniques inherently favoring stable polymorphs involve longer time scales than kinetic approaches favoring metastable polymorphs. Anderton (2004) has graphically summarized the time demands for some of the more commonly applied experimental approaches (Figure 3.8). The summary clearly gives a relative time scale for various procedures. As noted numerous times herein, every substance is an individual case for which

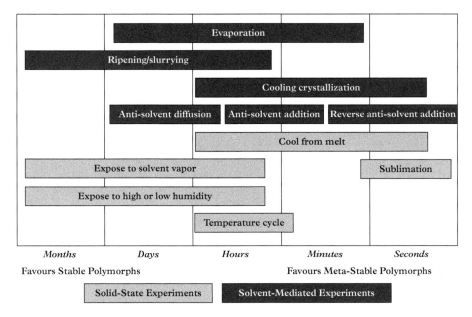

Figure 3.8 *Timescales for various crystallization experiments in a solid form screen (from Anderton 2004, with permission).*

these general time frames may or may not apply. For instance, as noted also in Section 3.2, xylitol, the common artificial sweetener in sugarless chewing gum, was first prepared in 1891 but did not crystallize for over 50 years (Wolfrom and Kohn 1942). A second polymorph was reported within one year (Carson et al. 1943).

For a potentially industrially important material, at what stage during the lifetime of the material should a solid form screen be pursued, and on what scale? The answer to the first part of the question is that the search for crystal forms continues for the lifetime of the compound (Newman and Wenslow 2016). Scales of experiments change; demands on the physical properties of the product change; new forms are discovered serendipitously, and difficulty may be encountered in preparing the original (or previously preferred) form; hence, as noted earlier the solid form screen continues at least throughout the commercial lifetime of the compound. Some of the general considerations in terms of purpose, amount of material available (or required), the nature of the investigations, and the appropriate stage in the product lifetime are given in two graphics in Figure 3.9, prepared by two experienced practitioners in this discipline.

Newman (2011) has noted that while the initial investigation may involve less than 500 mg of material and tens of experiments, the solid form selection stage often involves 2–5 g of material and hundreds of experiments, while subsequent and continuing investigations may involve thousands of investigations requiring significantly larger amounts of material.

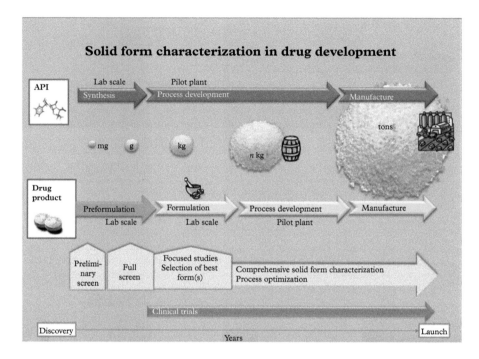

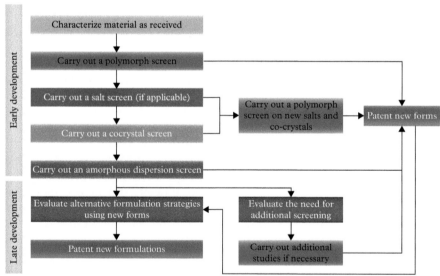

Figure 3.9 *Two schematic diagrams of the nature of the continuing study of the solid form landscape during the lifetime of a commercial compound. Due to the variety of approaches taken the two are not meant to be comprehensive, but rather to give two complementary views of the scope of such an endeavor. (Upper, from Griesser 2011, with permission; lower, from Newman, A. (2013). Specialized solid form screening techniques.* Org. Process Res. Dev., *17, 457–71. With permission of American Chemical Society.)*

Some compounds have acquired iconic status, and hence are often the subjects of basic research intended to demonstrate the principles and success of a new experimental technique, as well as proof-of concept examples for particular, say, pharmaceutical applications. An example of the first is red, orange, and yellow (ROY) crystals (Yu 2010; Hamilton et al. 2012) while that of the latter is carbamazepine (Aaltonen et al. 2009; Chieng et al. 2011).

3.5.5   What are the variables in a crystallization?

Some of the variables in crystallization techniques have been given by Morissette et al. (2004), as shown in Table 3.2.

These have been condensed into graphical form by Florence (2009), as shown in Figure 3.10. The overriding message in Table 3.2 and Figure 3.10 is that the variety of conditions for crystallization is essentially infinite. Thus the execution of a solid form screen requires careful design combined with observation and evaluation of results as they are obtained to modify and tune the screen, in order to maximize results both in terms of numbers of identified forms and to develop the facility to constantly and robustly obtain every identified form. This variety also indicates why a solid form screen can never be considered complete: even a small change in one parameter can be sufficient to generate a new form.

3.5.6   High throughput (HT) screening for solid forms

While this volume deals with polymorphism of molecular crystals in a general way, this chapter bears particular relevance to the pharmaceutical industry. The importance of identifying the thermodynamically stable solid form early in drug product development cannot be overstated. That does not mean that the most stable form is always the natural choice for the API. Many examples of marketed APIs of less stable forms have been given by Byrn et al. (1999), while examples of marketed amorphous forms have been given by Guillory (1999). More recent examples include Accolate® (zafirlukast), Ceftin® (cefuroxime axetil), Accupril® (quinapril hydrochloride), and Viracept® (nelfinavir mesylate) (Newman 2010). In addition to potentially delaying regulatory submission and marketing approval, form changes prompted by the late discovery of more stable forms will inevitably incur significant costs associated with redeveloping new crystallization and formulation processes, repeating toxicology and stability studies, establishing bioequivalence, and potentially adjusting dose strengths. The consequences of missing a more stable crystal form only to have it appear after the product is on the market can be catastrophic, minimally threatening market supply and in the worst case, forcing product withdrawal, cf. ritonavir (Morissette et al. 2003). Hence, the need for continuing the exploration of the solid form landscape through the lifetime of the compound, as described previously.

Clearly understanding the importance of finding all forms with the potential to be intentionally or unintentionally introduced in a given drug product, the pharmaceutical industry has invested heavily in solid-state form screening tools and technologies to

Table 3.2 *Experimental variables for a variety of crystallization attempts with differing compositions (from Morissette, S. L., Almarsson, Ö., Peterson, M. L., Remenar, J. F., Read, M. J., Lemmo, A. V., Ellis, S., Cima, M. J., and Gardner, C. R. (2004). High-throughput crystallization: polymorphs, salts, co-crystals and solvates of pharmaceutical solids. Adv. Drug. Deliv. Rev.,* **56**, *275–300. With permission of Elsevier.)*

Composition type		Process variables[a]				
Polymorph/solvates	Salts/co-crystals	Thermal	Anti-solvent	Evaporation	Slurry conversion	Other variables
■ Solvent/solvent combinations	■ Counter-ion type	■ Heating rate	■ Anti-solvent type	■ Rate of evaporation	■ Solvent type	■ Mixing rate
■ Degree of supersaturation	■ Acid/base ratio	■ Cooling rate	■ Rate of anti-solvent addition	■ Evaporation time	■ Incubation temperature	■ Impeller design
■ Additive type	■ Solvent/solvent combinations	■ Maximum temperature	■ Temperature of anti-solvent addition	■ Carrier gas	■ Incubation time	■ Crystallization vessel design (including capillaries, etc.)
■ Additive	■ Degree of supersaturation	■ Incubation temperature(s)	■ Time of anti-solvent addition	■ Surface–volume ratio	■ Thermal cycling and gradients	
	■ Additive type and concentration	■ Incubation time				
	■ pH					
	■ Ionic strength					

[a] Applicable to all types of screens.

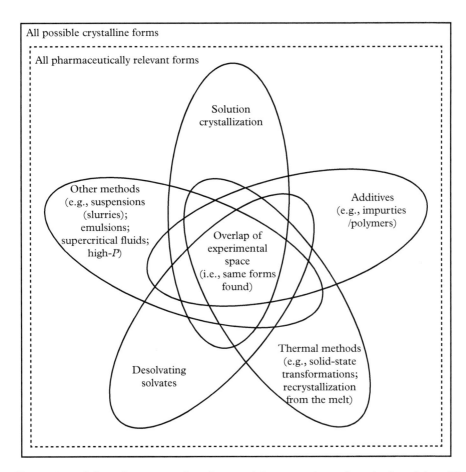

All possible crystalline forms

All pharmaceutically relevant forms

Solution
crystallization

Other methods
(e.g., suspensions
(slurries);
emulsions;
supercritical fluids;
high-*P*)

Additives
(e.g., impurties
/polymers)

Overlap of
experimental
space
(i.e., same forms
found)

Desolvating
solvates

Thermal methods
(e.g., solid-state
transformations;
recrystallization
from the melt)

Figure 3.10 *Schematic representation of many of the approaches to investigation of the solid form landscape of a molecular compound. (From Florence, A. J. (2009). Approaches to high-throughput physical form screening and discovery. In* Polymorphism in Pharmaceutical Solids, *2nd edn (ed. H. G. Brittain), pp. 139–84. Informa Healthcare, New York. With permission of Taylor and Francis Group LLC Books.)*

explore novel approaches to find forms. Most notable was the flurry of activity in the development and use of high throughput (HT) screening techniques in various crystallization applications, for example, salt screening, crystal form screening, and co-crystal screening. These developments at the end of the 1990s and the beginning of this century followed on the heels of the development of automated techniques (robotics and informatics) applied to combinatorial chemistry and the human genome project. A number of innovative companies provided access to HT crystal form screening, either as a stand-alone equipment package (e.g., Symyx) or as a service provider (e.g., Transform Pharmaceuticals), with the expectation of automating the search for crystal forms and the definition of the crystal form landscape.

The principles and practice of HT solid form searches have been described in some detail by practitioners in a number of publications (Morissette et al. 2004; Hilfiker et al. 2006; Florence 2009) so only a brief description will be given here. The fundamental idea behind this approach is to cover as many as possible of the potential conditions for crystallization in a minimum amount of time as automatically as possible, and thus in principle to generate quickly the solid form landscape. An example of the difference between scales of time and quantities in manual and HT methods has been given by Florence (Table 3.3). An example of the application of HT methods to a specific system has been given by Peterson et al. 2002.

In spite of initial optimism and enthusiasm for HT methods, practical aspects resulted in a number of problems. As recounted elsewhere (Cruz-Cabeza et al. 2015) in one company's experience the automation was repeatedly tested for polymorphs. Although the screens were rapidly conducted, the vast majority resulted in failed crystallization attempts (for a variety of reasons). In the end, there were too many instances, in which compounds that crystallized with relative ease by manual methods, had failed to do so with the HT automation. Forms were missed entirely and there was the very common (and initially unanticipated) problem of not being able to scale up the microgram "hits" from the automation. This experience, considered to be typical of that across the industry, showed the extent to which the development of robotic instrumentation, analytics, and informatics has outpaced the understanding of crystal nucleation/growth, the development of computational methodologies to predict/model crystal structures, and strategies/means to target forms of interest. The unique nature of the crystal chemistry of each compound, coupled with the occasional (but crucial!) failure of HT methods to find crystal forms obtained by other means, has led to HT automation becoming but one of the many tools in the armory for exploring the crystal form landscape, especially polymorphism. While HT automation is still in use at many sites, in a number of them it has been relegated almost entirely for salt screening where reaction chemistry can help to drive salt crystallization, and the number of entries in the literature has

Table 3.3 *Comparison of performance criteria for manual and high throughput crystallization studies on an individual compound (from Florence, A. J. (2009). Approaches to high-throughput physical form screening and discovery. In* Polymorphism in Pharmaceutical Solids, *2nd edn (ed. H. G. Brittain), pp. 139–84. Informa Healthcare, New York. With permission of Taylor and Francis Group LLC Books.)*

	Manual methods	High-throughput methods
Number of crystallizations implemented	100	1 500
Time scale of study	Approximately 32 weeks	4 weeks
Total amount of compound used	More than 10 grams	Less than 5 grams
Number of crystalline forms observed	9	18

decreased significantly. For a recent assessment of the applications of HT methods, see Bhardwaj (2016). In that light it is worthy of note that a number of the initial commercial entries into this field (e.g., Symyx and Transform) no longer exist.

3.5.7   A specific example of a solid form screen—axitinib

As in any exploratory effort, the possible investigative methods and techniques are familiar to the explorer, but the outcome is the unknown for which the search is undertaken. In the case of a solid form screen of a molecular compound, when the compound is first isolated absolutely nothing is known about that landscape, although an experienced and observant chemist may already recognize some properties even from the first batch of solid material. I have outlined in a general way how a solid form screen might proceed. As noted elsewhere, every compound presents different challenges and will produce different results. Virtually no information can be taken from the landscape of one compound to another. For instance, the difference between the solid form landscape between what appears to be two quite similar molecules **3-IV** and **3-V** is shown in Figure 3.11. Hence navigating the exploration depends on the skills, experience, and intuition (and even serendipitous luck) of the explorer(s).

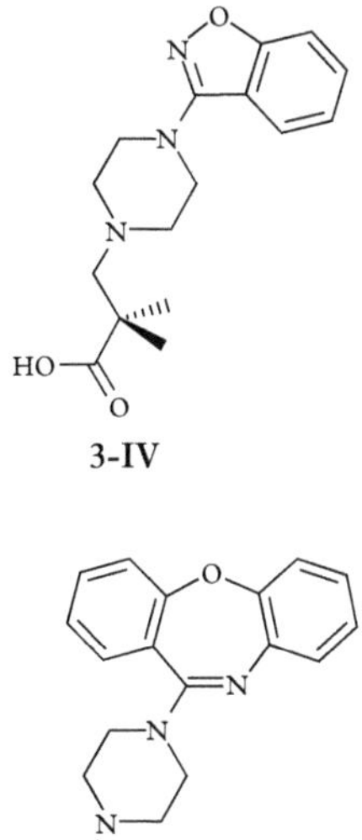

3-IV

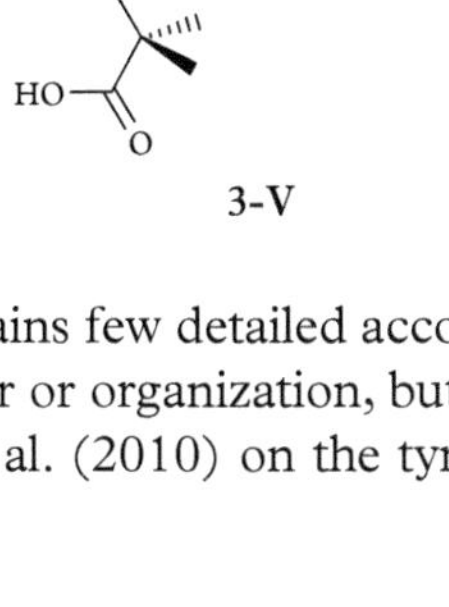

3-V

The literature actually contains few detailed accounts of a solid form screen on the part of a single investigator or organization, but two recent reports by Chekal et al. (2009) and Campeta et al. (2010) on the tyrosine kinase inhibitor axitinib

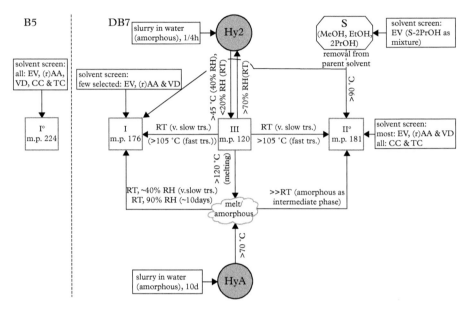

Figure 3.11 *Demonstration of the different solid form landscapes for two similar compounds* **3-IV** *and* **3-V**. *The original caption provides definitions of symbols used in the figure. (From Braun, D. E., McMahon, J. A., Koztecki, L. H., Price, S. L., and Reutzel-Edens, S. M. (2014). Contrasting polymorphism of related small molecule drugs correlated and guided by the computed crystal energy landscape.* Cryst. Growth Des., **14**, *2056–72.With permission of American Chemical Society.)*

(Inlyta®) (**3-VI**), a potential treatment for tumors, document such a screen. This compound turned out to be particularly rich in solvates as well as number of polymorphs, but again, none of that can be known at the start of a screen. The overall effort is best described in the authors' own words:

> The presented case illustrates an extensively polymorphic compound with an additional propensity for forming solvates. In all, five anhydrous forms and 66 solvated forms have been discovered. After early polymorph screening using common techniques yielded mostly solvates and failed to uncover several key anhydrous forms, it became necessary to devise new approaches based on an advanced understanding of crystal structure and conformational relationships between forms. With the aid of this analysis, two screening approaches were devised which targeted high temperature desolvation as a means to increase conformational populations and enhance overall probability of anhydrous form production. Application of these targeted approaches, comprising over 100 experiments, produced only the known anhydrous forms, without any appearance of any new forms.

The overall effort, as recounted in the above two papers has been graphically summarized by Newman (2013) (Figure 3.12). The resulting forms are summarized in Table 3.4.

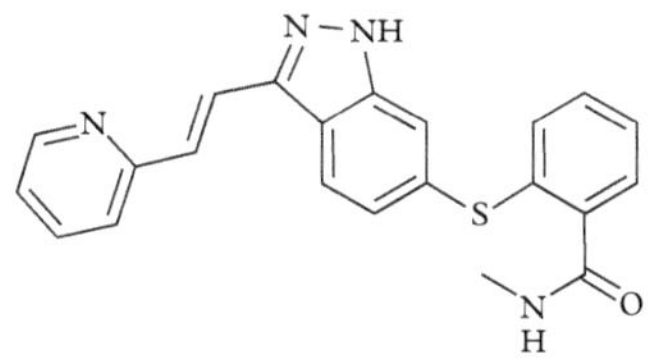

3-VI

Screen 1 resulted in mostly solvated forms
-Forms I to VIII
-Form IV found as most stable form during screen and used in early clinical studies

Two additional screens (Screens 2 and 3) conducted with Form IV material
-Forms IX to XXIV found

Screen 4 conducted using Form IX (hydrate) as starting material
-Mostly slurries
-No new patterns found

Form XXV produced during process development studies and determined to be more stable than Form IV

Screen 5 performed with XXV material to confirm stability
-Forms I to VIII
-Form IV found as most stable form during screen and used in early clinical studies

During large scale manufacturing campaign Form XII is produced
-Found to be most stable form, total of 5 anhydrous forms and 66 solvated forms identified to date
-Most stable form was not found during any screen

Devised new approaches targeting high temperature desolvation to ensure Form XLI was stable form
-Vacuum drying at 100 °C
15% of solvates did not desolvate
57% desolvated to Form IV
5% produced Form XLI
-Solvent-based desolvation at high temperatures
Slurry in ethanol at 90 °C: *97% produced Form XLI*
Higher solubility allows transformation to more stable form
Slurry in heptane at 105 °C: *33% produced Form XLI, 45% desolvated to Form IV*
Slurry in *p*-cymene at 150 °C: *83% produced Form XLI*

Figure 3.12 *Graphical summary of the solid form screen of axitinib. See Table 3.4 for the identity of each of the solid forms noted here. (From Newman, A. (2013). Specialized solid form screening techniques.* Org. Process Res. Dev., ***17****, 457–71. With permission of American Chemical Society.)*

Table 3.4 *Summary of axitinib solid forms*

Name	Form	Solvent
Form I (1)	Anhydrate	–
Form II (2)	Hydrate	Water
Form III (3)	Solvate	Ethyl acetate (EtOAc)
Form IV (4)	Anhydrate	–
Form V (5)		No solid form designation
Form VI (6)	Anhydrate	–
Form VII (7)	Solvate	Isopropyl alcohol (IPA), IPA/water
Form VIII (8)	Solvate	Dioxane, tetrahydrofuran (THF)
Form IX (9)	Hydrate	Water
Form X (10)	Solvate	Dimethylformamide (DMF), DMF/water
Form XI (11)	Solvate	THF/water, THF
Form XII (12)	Solvate	Dichloromethane (DCM), ethanol (EtOH)
Form XIII (13)	Solvate	Acetonitrile (ACN)
Form XIV (14)	Solvate	Acetic acid
Form XV (15)	Solvate	EtOH
Form XVI (16)	Solvate	IPA
Form XVII (17)	Solvate	Acetone
Form XVIII (18)	Solvate	Methylisobutyl ketone (MIBK)
Form XIX (19)	Solvate	Methylethyl ketone (MEK)
Form XX (20)	Solvate	Methyl benzoate
Form XXI (21)	Solvate	$2,2,2-CF_3CH_2OH$/ether/hexane
Form XXII (22)	Solvate	1-Pentanol
Form XXIII (23)	Solvate	Pyridine
Form XXIV (24)	Solvate	Chloroform
Form XXV (25)	Anhydrate	–
Form XXVI (26)	Solvate	THF/water, THF
Form XXVII (27)	Solvate	Dimethylsulfoxide (DMSO)
Form XXVIII (28)	Solvate	Benzyl alcohol
Form XXIX (29)	Solvate	Trichloroethylene
Form XXX (30)	Solvate	Dimethylformamide (DMF)/octanol (1:1)
Form XXXI (31)	Solvate	Octanol
Form XXXII (32)	Solvate	Methanol
Form XXXIII (33)	Solvate	1-Butanol
Form XXXIV (34)	Solvate	3-Methyl-1-butanol
Form XXXV (35)	Solvate	MEK
Form XXXVI (36)	Solvate	Pyrrole/1-pentanol pyrrole/*p*-cymene

Name	Form	Solvent
Form XXXVII (37)	Solvate	Allyl alcohol
Form XXXVIII (38)	Solvate	Pyrrole allyl alcohol
Form XXXIX (39)	Solvate	Acetic acid
Form XL (40)	Solvate	EtOH
Form XLI (41)	Anhydrate	–
Form XLII (42)	Solvate	2-Butanol
Form XLIII (43)	Solvate	2-Methyl THF
Form XLIV (44)	Solvate	2-Methyl THF
Form XLV (45)	Solvate	Toluene
Form XLVI (46)	Solvate	N-Methylpyrrolidone
Form XLVII (47)	Solvate	Isoamyl acetate
Form XLVIII (48)	Solvate	Methylcyclohexane
Form XLIX (49)	Solvate	Cyclohexanone
Form L (50)	Solvate	Cyclohexanone
Form LI (51)	Solvate	1,2-Dichloroethane
Form LII (52)	Solvate	Propionic acid
Form LIII (53)	Solvate	*Tert*-butanol
Form LIV (54)	Solvate	Dimethoxymethane
Form LV (55)	Solvate	2-Pentanone
Form LVI (56)	Solvate	Dimethyl acetate (DMA)
Form LVII (57)	Solvate	Nitromethane
Form LVIII (58)	Solvate	1,2,3,4-tetrahydronaphthalene
Form LIX (59)	Solvate	Tetramethylene sulfone
Form LX (60)	Solvate	Methyl acetate
Form LXI (61)	Solvate	p-Xylene
Form LXII (62)	Solvate	Trichloroethylene
Form LXIII (63)	Solvate	n-Butyl acetate
Form LXIV (64)	Solvate	Isobutyl alcohol
Form LXV (65)	Solvate	Cyclohexanol
Form LXVI (66)	Solvate	Isopropyl acetate
Form LXVIII (67)	Solvate	p-Cymene/pyrrole (1:1)
Form LXIX (68)	Solvate	t-Amyl alcohol
Form LXX (69)	Solvate	4-Methyl-2-pentanone
Form LXXI (70)	Solvate	Cyclohexane
Form LXXII (71)	Solvate	1,2-Dichlorobenzene
Form LXXIII (72)	Solvate	p-Cymene/acetone (1:1)

Although it is clear from the solvate obtained that a wide variety of solvents and solvent combinations were employed in this screen it is worthy of note that many additional ones were also tried. As noted in Figure 3.12, the unsolvated anhydrate Form XLI was found to be the most stable and was chosen for further development and formulation. The actual detailed preparation of Form I may be found in Example 1(d) of the U.S. Patent application US20100179329 A1.

3.6 Concomitant polymorphs

In situations where there is overlap between the occurrence domains of two or more polymorphs the modifications may appear simultaneously or in overlapping stages so that a particular procedure or process yields more than one form under identical conditions. This phenomenon is termed *concomitant polymorphism* and has been reviewed in considerable detail (Bernstein et al. 1999).

The situations in which polymorphs concomitantly crystallize are determined by the experimental conditions in relation to both the free energy–temperature relationships and the relative kinetic factors. These situations may arise because either specific thermodynamic conditions prevail or the competing kinetic processes have equivalent rates. In thermodynamic terms we have seen that polymorphs can coexist in true equilibrium only at the thermodynamic transition temperature (where the free energy G curves cross). The chance of carrying out a crystallization precisely at such a temperature must be small with the inevitable conclusion if concomitant polymorphs are produced that kinetics play at least some role in the overall process of crystallization. The final consequence of this situation is that a system of concomitantly crystallizing polymorphs will be subject to change in the direction favoring the formation of the most stable structure. If the crystals have grown from and remain in contact with solution then the most likely route for this transformation is via dissolution and recrystallization. This situation is commonly exploited in production processes as a slurrying procedure to produce the most stable polymorph (McCrone 1965; Cardew and Davey 1985). If the crystals have formed from the melt or vapor phase or have been isolated from their mother phase, a solid-state transformation is possible (Cardew et al. 1984).

Concomitant crystallization offers the investigator the opportunity to maximize the data from a single crystallization experiment. Since concomitant polymorphs are energetically equivalent structures, or very nearly so, they provide excellent and demanding benchmarks for theoretical and computational models for predicting crystal growth and crystal structure (Reed et al. 2000). The programs and force fields employed in these programs must reproduce the near equivalency of the crystal energetics, even if the absolute energy is not reproduced or even not known (Price et al. 2016; Price and Reutzel-Edens 2016). These issues are treated in more detail in Chapter 5.

The phenomenon of concomitant polymorphs has been recognized since Wöhler and Liebig (1832) discovered the first polymorphic organic substance, benzamide. Although Von Groth (1906b) collected a number of examples of the phenomenon in both organic and inorganic systems in his book published nearly 75 years later, he still characterized the phenomenon as a "peculiarity," which is "analogous to the...indifference towards direct transformation between polymorphic forms."

In spite of von Groth's guarded skepticism there are many examples of concomitant polymorphism crossing a wide range of crystallization techniques and diverse chemical systems. I cite some of them here, since they demonstrate that diversity as well as the interplay and often delicate balance of thermodynamic and kinetic factors governing the competition between different crystal modifications. Additional examples and more details can be found in the above referenced review and in the original papers. The phenomenon is likely considerably more widespread than is generally appreciated, but is difficult to recognize or search for in the literature since it is rarely specifically abstracted or noted, and is usually buried in the experimental section of a paper. With increased awareness of the phenomenon and the information it can provide on polymorphic systems it is worthy of note that an increasing number of publications now make specific reference to the phenomenon (Lee et al. 2008; Jiang and Ward 2014; Varughese 2014; Yu 2010). For the purpose of this chapter I have chosen to concentrate on examples of concomitant polymorphism that demonstrate how the fine tuning of crystallization conditions may be used to *control* the polymorphic form obtained. The more common *screening* techniques of varying conditions (solvent, temperature, rate of evaporation, etc.), *vide infra,* are used to identify the occurrence domain of each of the crystal modifications. By definition, regions in which occurrence domains do not overlap are those in which control over the polymorphic result is maintained. Concomitant polymorphs occur in the regions of overlap between unique occurrence domains, and controlling the polymorph attained essentially involves defining the borders between the overlap regions and the unique ones.

3.6.1   Concomitant polymorphs—crystallization methods and conditions

The crystallization behavior of the copper complex **3-VII** demonstrates the effect of concentration on the simultaneous appearance of polymorphs and the role of concentration in altering the preference for one or the other. Kelly et al. (1997, 2001) prepared blue crystals of **3-VII**, which precipitates from "reasonably concentrated [acetonitrile] solutions" with square planar coordination geometry around the copper. Filtering off the precipitate and treating the remaining mother liquor with diethyl ether, followed by cooling in a freezer, leads to concomitant crystallization of the blue crystals with green crystals, in which the molecular structure exhibits pseudotetrahedral coordination geometry around the copper center.

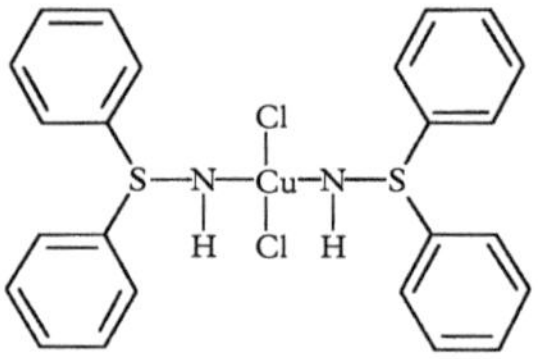

3-VII

There is ample evidence that this also is a genuine case of polymorphism. The blue, square planar form can be dissolved to obtain the green tetrahedral one; indeed the authors state that the two form intergrowths, which apparently grow coincidentally. The green form is obtained only from dilute solutions, and once it is obtained the blue form cannot be regenerated. The sequence of events and the experimental conditions are compatible with Ostwald's Rule of Stages (Section 2.3), the slight thermodynamic preference for the green form over the blue one and the phenomenon of disappearing polymorphs (Dunitz and Bernstein 1995; Bučar et al. 2015) (Section 3.7). Once seeds of the (even slightly) preferred green form are present in the immediate environment the blue form cannot be obtained.

The effect of temperature is demonstrated in the polymorphic behavior of the "diphenylcarbamide" **3-VIII**, one of the systems originally cited by Von Groth (1906b), and more recently studied by Etter et al. (1990b), Huang et al. (1995), and Rafilovich et al. (2005).

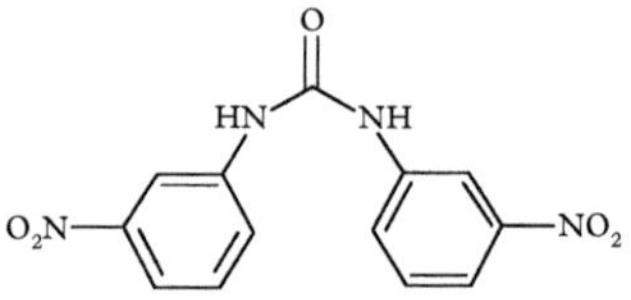

3-VIII

The three originally reported forms were: α, yellow prismatic needles; β, white needles; and γ, yellow tablets. Upon crystallization from 95% ethanol in the range 30–75 °C, the α and β forms always crystallized concomitantly, even in the presence of seeds of one of them. The relative amounts can be regulated by varying the temperature, α being preferred at higher temperatures. Upon evaporation of the mother liquid at "the ordinary temperature (13 °C)" (*sic*) γ crystallizes out alone; warming the same mother liquor leads also to some crystals of β; at 40 °C γ no longer appeared, but small amounts of α did appear. The authors concluded that on the basis of the experiments with 95% alcoholic solutions γ is the stable form at room temperature (considerably lower than room temperature today), and with increasing temperature β and then α are the stable forms. However, this behavior may be modified by solvent. A subsequent detailed investigation using a variety of analytical techniques, revealed a new polymorphic form, δ, and a monohydrate, in addition to the previously characterized α and β forms. The monohydrate is apparently the earlier reported γ form (Rafilovich et al. 2005). This is clearly a rich system with a very delicate balance between the relative stability of the three forms and the monohydrate, as well as the kinetic factors that govern their appearance.

Concomitant crystallization is by no means limited to crystallization from solution, or to preservation of constant molecular conformation. As noted in Section 2.2.5, the classic pressure versus temperature phase diagram for two solid phases (Figure 2.6) of one material exhibits two lines corresponding to the solid/vapor equilibrium for each of two polymorphs. At any one temperature one would expect the two polymorphs to have different vapor pressures. This, in fact, is the basis for purification of solids by sublimation. Nevertheless there are examples where the two have nearly equal vapor pressures at a particular temperature and thus co-sublime. This could be near the transition temperature or simply because the two curves are similar over a large range of temperatures or in close proximity at the temperature at which the sublimation is carried out. For instance, the compounds **3-IX** and **3-X** both yield two phases upon vacuum sublimation (Cordes et al. 1992a), while **3-XI** yields three differently colored modifications (Griffiths and Monahan 1976).

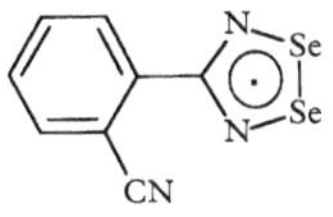

3-IX

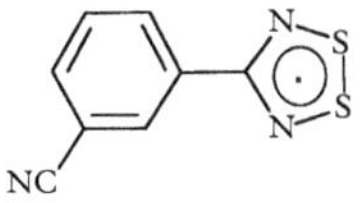

3-X

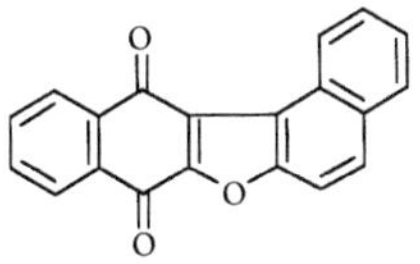

3-XI

For **3-IX** slow sublimation at 140 °C and 0.1 Torr led to a mixture of a "few feathery needles" (mp 192–3 °C) among the main product of "lustrous coppery blocks" (mp 220–3 °C), which could be separated manually for further characterization. **3-X** was also purified/crystallized by vacuum sublimation (120 °C/10^{-2}

Torr) yielding manually separable deep red needles (α phase, mp 157–60 °C) and blocks (β phase, mp 165–8 °C). The crystal structure determinations of the two phases of **3-XI** indicated significantly different molecular geometry, a cofacial dimer in the α phase, and a *trans* antarafacial dimer in the β phase.

Another example of co-sublimed phases is **3-XII** for which it appears that the sublimation conditions were systematically varied in an attempt to obtain a second polymorph (Cordes 1992b). The α form was initially prepared as a single phase of golden needles by slow sublimation over several weeks at 10^{-6} Torr during which the sample was heated to 180 °C and the cold finger was maintained at 100 °C. Raising the pressure to 10^{-1}–10^{-2} Torr and increasing the sample temperature to 220 °C, with a cold finger in the range 120–40 °C also led to the same α-phase needles, but accompanied by blocks of an additional β phase. Clearly, the authors moved along the solid/vapor line of the α phase towards the intersection with the solid/vapor line of the β phase (see Figure 2.6).

3-XII

Electrocrystallization, in particular of complexes of two components, is a technique that affords additional degrees of freedom of crystallization parameters compared to "conventional" crystallization: voltage, current density, counter ions, supporting electrolyte, electrode materials, etc. For the current and voltage one can also obtain a variable range of conditions, from thermodynamic (with a minimum perturbation from equilibrium) to kinetic (with a large or even systematically varying perturbation). Electrocrystallizations are employed particularly in the field of organic conductors and magnets for the preparation of complexes and salts. Considerable effort is generally expended in growing crystals and fine tuning conditions for obtaining crystals of sufficient quality for further structural and physical characterization. Thus a wide variety of crystallizing conditions are attempted and many individual crystals from a single crystallization might be subjected to structural and physical characterization (Wang et al. 1989). These are favorable circumstances for the preparation and recognition of polymorphism and concomitant polymorphs.

Perhaps most prominent of these materials are the organic conducting and superconducting salts based on so-called "ET" compounds, in which BEDT-TTF **3-XIII** is the donor (cation in the salt), generally in a 2:1 ratio with the acceptor (anion in the salt) (Williams et al. 1991). One of the most widely studied of these salts is $(ET)_2I_3$, for which at least fourteen different phases have been reported (Carlson et al. 1990; Williams et al. 1991), although the α and β phases tend to dominate (Carlson et al. 1990; Shibaeva et al. 1990). It has been shown that α is the kinetically favored product (>90%) under conditions of high current density and a small amount of water or oxidant added to the crystallization medium. Under more nearly equilibrium conditions (i.e., much lower current density) and dry solvent (tetrahydrofuran), pure β phase can be obtained, suggesting that it is the thermodynamically preferred form. Intermediate conditions apparently lead to concomitant crystallization of the two forms (Carlson et al. 1990; Williams et al. 1991).

3-XIII

When the crystallization is carried out as an oxidation of ET in tetrahydrofuran with a mixture of $(n\text{-}C_4H_9)_4NI_3$ and $(n\text{-}C_4H_9)_4NAuI_2$ (16:1 w/w) with an intermediate current to that described above, the α form is the main product, concomitantly crystallized with small amounts of θ and κ polymorphs (Kato et al. 1987). The authors note the difficulty in identifying all three forms on the basis of crystal shape alone, and the three were characterized by a combination of X-ray diffraction and other physical measurements.

3-XIV

Montgomery et al. (1988) showed that in systems in which attempts were made to prepare alloys of $\beta\text{-}(ET)_2I_3$ with $\beta\text{-}(PT)_2I_3$ (PT = **3-XIV**) they apparently obtained "single crystals" of $(ET)_2I_3$ which were in fact mixtures of α and β forms. The phenomenon in this system was initially detected and then confirmed by ESR measurements, which were subsequently used to develop a quantitative procedure for the determination of the polymorphic composition of such "mixed single crystals." There are other scattered reports in the literature of the phenomenon (e.g., Freer and Kraut 1965), which has been termed "composite crystals" and has been discussed in detail by Coppens et al. (1990), Fryer (1997), and Bond et al. (2007).

Interdiffusion of saturated solutions (as opposed to electrocrystallization) is another method for obtaining crystals of the potentially conducting salts. In most of the preparations of TTF[Pd(dmit)$_2$]$_2$ (dmit: **3-XV**, X = Y = Z = S) by diffusion of $(TTF)_3(BF_4)_2$ and $(n\text{-}Bu_4N)[Pd(dmit)_2]$ mainly black shiny needles of the α phase were obtained. However, some experiments yielded, in addition, a so-called α' phase (due to its structural similarity to the α phase, but different electrical behavior), and occasionally a third δ phase of plate-shaped crystals could also be physically separated from the batch (Legros and Valade 1988; Cassoux et al. 1991).[1]

---

[1] Cassoux et al. (1991) noted that "The occurrence of several phases for this compound complicates its study." While the separation of concomitantly crystallizing phases may have been an experimental complication in their particular study, I believe that the existence of polymorphs greatly *facilitates* the study of structure–property relations, since all chemical parameters are constant among polymorphs of a particular substance and differences in properties can be related directly to differences in structure. Chapter 6 is devoted to this subject.

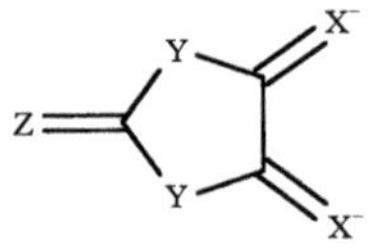

X,Y, Z = S, Se

3-XV

The technique of solvent diffusion has also led to concomitant polymorphs, which implies that the different modifications have very similar solubility in the same *mixture* of solvents. Such is the case for 1,5,9,13-tetrathiacyclohexadecane **3-XVI** (Blake et al. 1993).

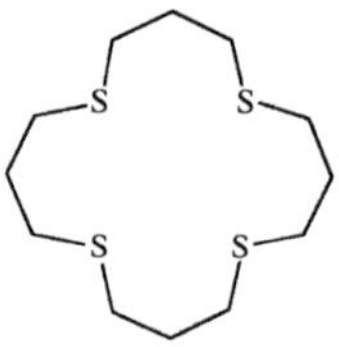

3-XVI

The material is reported to be trimorphic, all three structures crystallizing in polar space groups [needles, *Pbc2*$_1$ (α); plates, *P2*$_1$ (β); and twinned, apparently *Fdd2* (γ)]. The α and β forms are obtained concomitantly at ambient temperature by diffusion of hexane into methylene chloride solutions of **3-XVI**. Lowering the temperature to 130 °C for the same diffusion process leads to the γ form exclusively. The forms that crystallize concomitantly have similar but unusual molecular conformations. The melting points for the three modifications (α = 59.5–60.2 °C, β = 57.8–59.0 °C, γ = 60.0–60.9 °C) are very similar, which is not surprising for concomitant polymorphs, and the authors used the technique of mixed melting points to verify the existence of the three polymorphs.

3.6.2 Concomitant polymorphs—examples of different classes of compounds

In the previous section the emphasis was on the variety of different crystallization techniques that have led to concomitant crystallization of polymorphs. In this section I survey the additional diversity of chemical entities which have exhibited this phenomenon, since such diversity can also provide guidelines to developing methods for controlling the polymorph obtained.

Among the sugars, mannitol was mentioned earlier. It was one of the first concomitant polymorphs noted by von Groth (1906b), citing work by Zepharovich (1888). In fact because of the pharmaceutical uses of mannitol as a solid excipient, its polymorphic behavior has continued to be of considerable interest (Burger et al. 2000). In another rather intensively studied example from the pharmaceutical industry, sulfathiazole has been shown to crystallize in at least four polymorphs and numerous solvates (Threlfall and Hursthouse 2000), often appearing together or sequentially, depending on the solvent and crystallizing conditions

(Blagden et al. 1998a, 1998b). The similarities and differences of the polymorphs have been summarized in some detail (Gelbrich et al. 2008).

m-Nitrophenol has had a long history (Bernstein et al. 1999), although more recent interest was aroused because of its significant second harmonic associated with a non-centrosymmetric crystal, in spite of the fact that it appeared to crystallize in a centrosymmetric space group (Pandarese et al. 1975). The dilemma was resolved when it was discovered that approximately 20% of the crystals from a benzene-grown batch belonged to the non-centrosymmetric space group $P2_12_12_1$, while the remainder crystallized in $P2_1/c$ thus explaining the source of the second harmonic signal. A method for purifying *m*-nitrophenol and growing single crystals was developed by Wojcik and Marqueton (1989) and the structure of the orthorhombic form was published by Hamzaoui et al. in 1996.

Two-component (i.e., two molecules, more recently described as co-crystals) systems also exhibit concomitant polymorphism, implying a balance for the equilibrium situations governing the formation of the isomeric complexes as well as the kinetic and thermodynamic factors associated with the crystallization processes. The often serendipitous nature of the discovery of concomitant polymorphs is also illustrated by an example of a hydrogen-bonded two-component system, pyromellitic acid **3-XVII** and 2,4,6-trimethylpyridine **3-XVIII** (Biradha and Zaworotko 1998). The first polymorph (A) was obtained by reacting **3-XVII** with four equivalents of **3-XVIII** in a methanolic solution.

3-XVII

3-XVIII

Using **3-XVIII** as the solvent yielded a second polymorph (B) in fifteen minutes. Modification A was found to have crystallized as well in the same reaction vessel after about 24 hours. These two polymorphs are not readily distinguishable by their morphology. However, the authors point out that the experimental evidence indicates that Form B is the kinetically controlled one, while Form A is the thermodynamically preferred one.

An example of a π-bonded complex is the remarkable cyanine:oxonol system, **3-XIX**:**3-XX**, for which at least fourteen different polymorphs or solvates have been identified (Etter et al. 1984). Two of these, a gold and a red form (each containing a molecule of $CHCl_3$ solvent per 1:1 complex, and hence true polymorphs) crystallize concomitantly and have been structurally characterized (Etter et al. 1984).

Three of these polymorphs are shown in Figure 3.13. Despite the fact that both of these dyes are known to be individually *self*-aggregating (Cash 1981) the two structures exhibit mixed stacks, with very significant differences in the relative orientation of neighboring molecules along the stack (Figure 3.14) in the case of the gold structure the molecular long axes are oriented nearly perpendicular to each other, while in the red polymorph the molecular long axes are very nearly parallel.

3-XIX

3-XX

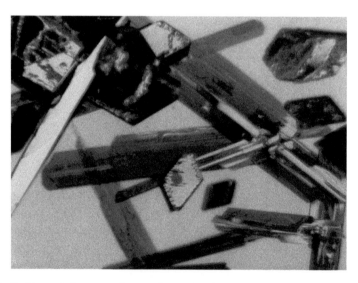

Figure 3.13 *Three of the concomitant polymorphs of the cyanine:oxonol dyes **3-XIX:3-XX**. Gold (by reflection; otherwise red by transmission) and red forms mentioned in the text are easily distinguishable. The third form is a purple one, normally diamond shaped as on the middle right, but many of these crystals are undergoing conversion, as indicated by varying degrees of mottled surfaces (from Bernstein et al. 1999, with permission).*

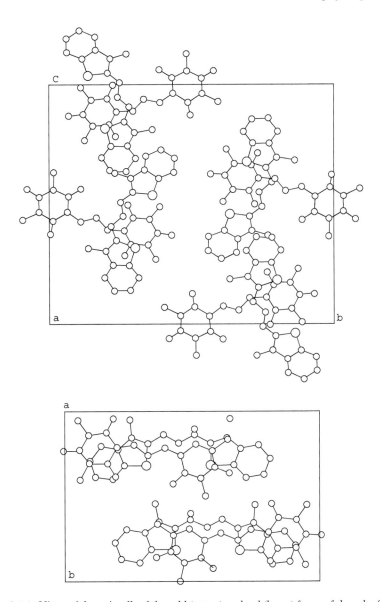

Figure 3.14 *Views of the unit cells of the gold (upper) and red (lower) forms of the salt of* **3-XIX** *and* **3-XX**. *In both cases the view is on the plane of the three central atoms of the oxonol molecule, with the bisector of the angle formed by the three oriented vertically. Chloroform molecules of solvation have been eliminated for clarity. Both figures indicate the relative orientation of cation and anion which is one feature of the overall packing (from Etter et al. 1984, with permission).*

Another π-complex system demonstrates concomitant polymorphism being manifested in differences in color and in habit of different polymorphic modifications. **3-XXI** is a relatively strong donor that was an early subject of study as a component of potential electrically conducting complexes (Kronick and Labes 1961; Kronick et al. 1964), and quite a few complexes of this donor were prepared (Matsunaga 1965, 1966; Koizumi and Matsunaga 1972). The complex of **3-XXI** with **3-XXII** was investigated in detail (Goto et al. 1996) with results that indicate both the utility and some of the limitations of studies of polymorphic systems. The 1:1 complex is polymorphic, with a dark brown α form and a green β form. The two forms may be obtained simultaneously from benzene solution by both slow cooling and "prolonged" slow evaporation. Similar results from these two equilibrium methods of crystallization indicate a similarity of the G versus T curves over the range of temperatures in the slow cooling process. A number of cases of concomitant polymorphs in which the factor of time or the equilibrium composition of the solution might be factors demonstrate the principles discussed in Section 1.2.

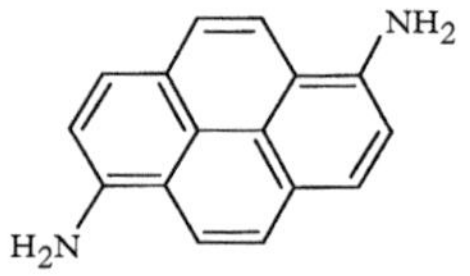

3-XXI

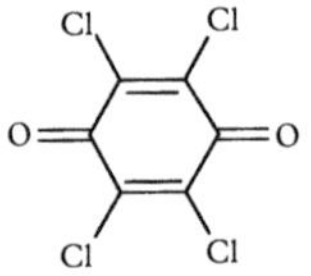

3-XXII

Ojala et al. (1998) reported on the crystallization of acetone tosylhydrazone **3-XXIII**. A triclinic form and a monoclinic form are both obtained from anhydrous ethanol—sometimes together. If the crystallizing solution is allowed to evaporate completely, only the monoclinic form is obtained, suggesting that it is the thermodynamically preferred form at room temperature. This is consistent with Ostwald's Rule of Stages and McCrone's (1965) test for relative stability of polymorphs according to which the more stable polymorph will grow at the expense of the less stable one. The crystal structure determinations indicate that the conformations differ by ~15° about the S–C exocyclic bond. In this case then the solution has an equilibrium mixture of (at least) these two molecular conformations. Lattice energy calculations (Cerius2) are consistent with this observation, indicating that the triclinic polymorph is more stable than the monoclinic by *ca.* 1 kcal/mol, which suggests that the composition of conformers can vary with temperature.

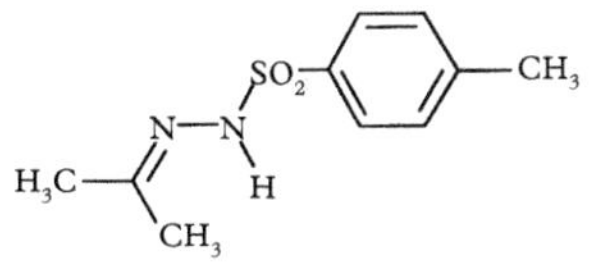

3-XXIII

As noted in Section 1.2, another issue in the definition of polymorphs is that of *tautomerism*. An example of the crystallizing tautomeric structures of 2-amino-3-hydroxy-6-phenylazopyridine **3-XXIV** has been reported by Desiraju (1983), **3-XXIVa** being the "low temperature" form as lustrous blue needles and **3-XXIVb** being the "high temperature" form, both melting at 181–2 °C. They were obtained simultaneously from a recrystallization of the crude synthetic product from ethanol, but the relative amounts varied from batch to batch in subsequent crystallizations in which concentration and temperature of the crystallization were varied. The high temperature form was always obtained; the amount of low temperature form varied with conditions. The tautomeric separation clearly takes place upon crystallization. To complete the picture and to make this arguably a case of polymorphism it must be shown that the blue and red crystals dissolve to yield the same equilibrium mixture. An increasing number of cases of tautomeric polymorphism has been recognized, for example, 2-thiobarbituric acid (Chierotti et al. 2010) and omeprazole (Bhatt and Desiraju 2007; Mishra et al. 2015). A case of concomitant crystallization which involves *configurational isomerism* of the benzophenone anil **3-XXV** has been reported by Matthews et al. (1991), as was discussed in Section 1.2.3.

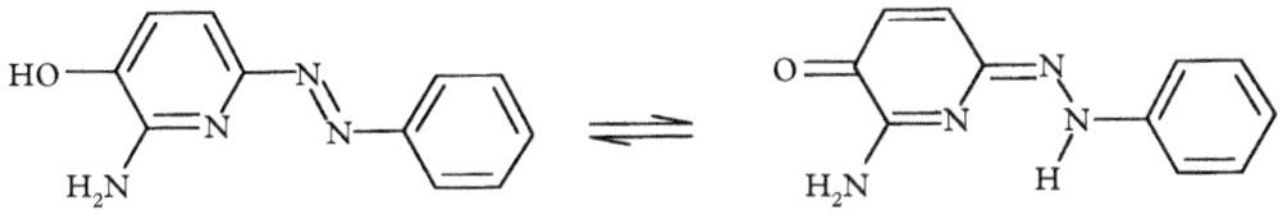

3-XXIV

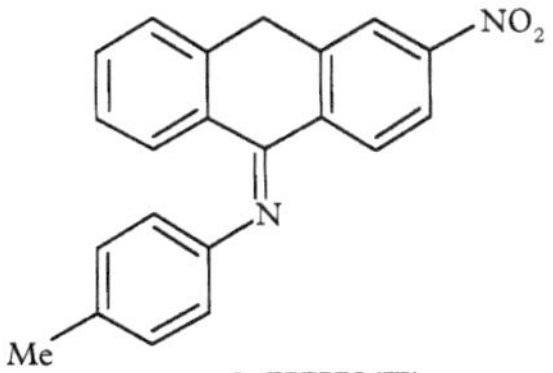

3-XXV(E)

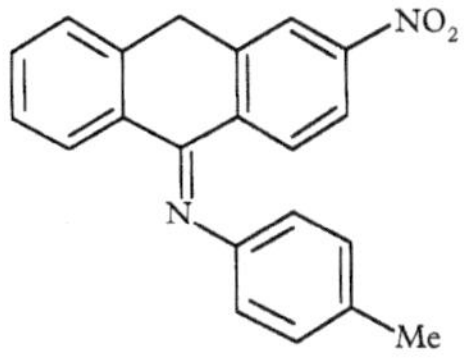

3-XXV(Z)

One of the issues raised in the discussion definition of polymorphs (Section 1.2.1) deals with racemic mixtures versus enantiomerically pure crystals or conglomerates (McCrone 1965; Dunitz 1995; Threlfall 1995). In principle, enantiomerically pure crystals are different from racemic ones, but if the enantiomers racemize quickly upon dissolution and/or racemates in solution spontaneously resolve upon crystallization it is still debatable whether the respective crystals are to be considered polymorphic substances. Masciocchi et al. (1997) characterized a concomitantly crystallized system that incorporates and illustrates many of these features, and in addition provides an example of a substance in which the synthetic approach to the material apparently plays a role in determining which polymorph is initially obtained.

The molecule under study was $Pd[(dmpz)_2(Hdmpz)_2]_2$, where Hdmpz = 3,5-dimethylpyrazole, **3-XXVI**. The material is trimorphic. The reaction mixture yields mostly (90%) the monoclinic (*C2/c*, racemic) α phase, the remainder being the triclinic ($P\bar{1}$, racemic) γ phase. The latter can be removed by recrystallization from 1,2-dichloroethane, which suggests that it is the more soluble and hence less stable polymorph in that solvent. Masciocchi et al. (1997) found that mixtures of various amounts of α and γ polymorphs could be obtained by varying the solvent and precipitation temperature (–70 °C to +50 °C), with α preferred at higher temperatures, consistent with the earlier observation of relative stability. Pure polymorph γ may be obtained by a different synthetic route which, when employing an excess of 3,5-dimethylpyrazole, leads to an approximately 50:50 mixture of polymorphs α and γ. This system thus also represents a case in which the polymorph obtained, or the polymorphic mixture obtained, depends on the synthetic route to the desired material. It is probably more correct to state that as usual, the polymorph or polymorphic mixture depends on the crystallization conditions, and these will clearly differ in the solvent/reagent/product compositions resulting from different synthetic conditions and routes.

3-XXVI

The tetragonal (*I422*, chiral) β "polymorph" is obtained quantitatively by a solid–liquid synthesis. The product is a conglomerate of enantiomeric crystals, which the authors claim does not transform into the α phase because of the impossibility of a solid/solid transformation. Dissolution of the β phase in 1,2-dichloroethane and subsequent evaporation quantitatively restores a mixture of α and γ forms. Despite different space groups the gross features of packing modes are very similar, the molecules being arranged about a pseudo (α, γ) or real (β) fourfold axis. Such a view of the crystal structure is consistent with observations of Gavezzotti and Desiraju (1988) and Braga and Grepioni (1991) on the general similarities of packing features and coordination numbers of organic and organometallic compounds.

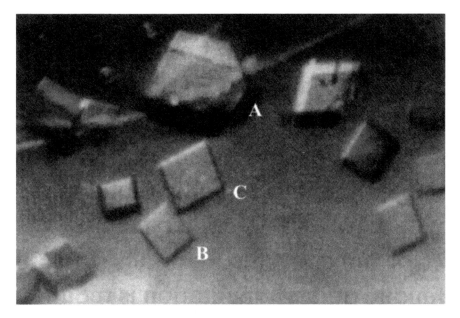

Figure 3.15 *Photograph of concomitantly crystallizing forms of rat liver glutathione S-transferase. The three forms are labeled. (From Fu et al. 1994, with permission.)*

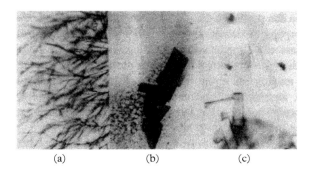

(a)   (b)   (c)

Figure 3.16 *Example of the stage-like growth in cytochrome c from* Valida membranaefaciens. *(a) Tree-like arrays of small triclinic crystals. (b) Triclinic crystals, which are frequently seen to dissolve at the expense of the orthorhombic prism, shown in (c). (From Day and McPherson 1991, with permission.)*

In view of the many experiments carried out in achieving the crystallization of proteins and subsequently refining those conditions to maximize crystal size and quality for X-ray structure determination, it is not surprising that examples of concomitant crystallization are found among proteins. I cite two here. Fu et al. (1994) reported the simultaneous crystallization of three polymorphs of a μ-class glutathione S-transferase from rat liver (Figure 3.15). Day and McPherson (1991) reported crystallization of two crystalline forms in stages in accord with Ostwald's Rule for cytochrome c from *Valida membranaefaciens* (Figure 3.16). In every case of crystallization they obtained arrays of thin triclinic plates (Figure 3.16a), some of which grew up to 0.5 mm in the largest physical dimension (Figure 3.16b). In some cases, some of these dissolved (in accord with Ostwald's Rule of Stages (Section 2.3)) to give rectangular prisms (Figure 3.16c), which turned out to be orthorhombic.

3.6.3 Concomitant polymorphs—the structural approach

To this point this account of instances of concomitant polymorphs has been phenomenological. I have discussed the thermodynamic and kinetic crystallization of polymorphs. There is still the question if any insight concerning controlling the polymorph obtained can be gained from the study of the crystal structure of concomitantly crystallized polymorphs. A qualitative attempt was made to see if details of the crystal structures may provide clues to the reasons for concomitant crystallization, and the near energetic equivalence of the two forms which can be assumed from that concomitant crystallization. The squarylium dye **3-XXVII** crystallizes from methylene chloride in a triclinic violet form and a monoclinic green form (Bernstein and Goldstein 1988). The cell constants (Table 3.5) do not suggest any similarity of the structures.

3-XXVII

Table 3.5 *Crystallographic cell constants for the two polymorphic forms of 3-XXVII*

	Triclinic form	Monoclinic form
a (Å)	11.911	15.72
b	7.401	7.283
c	6.501	9.591
α (°)	92.78	90
β	111.9	106.11
γ	98.08	90
Space group	$P\bar{1}$	$P2_1/c$

However, a projection on the molecular plane which includes two neighboring molecules (Figure 3.17) indicates that the stacking of the planar molecules is virtually identical in the two structures. In both cases the two molecules are related by a lattice translation: in the triclinic structure it is along the c axis and in the monoclinic structure along the b axis. Although the views appear the same to the eye, the vertical separation between planes differs (3.40 Å *versus.* 3.86 Å), which is a manifestation of the different axial lengths involved. The similarity of these diagrams strongly suggests that stacking is the dominant influence in the crystal growth process for both—hence the concomitant crystallization. The

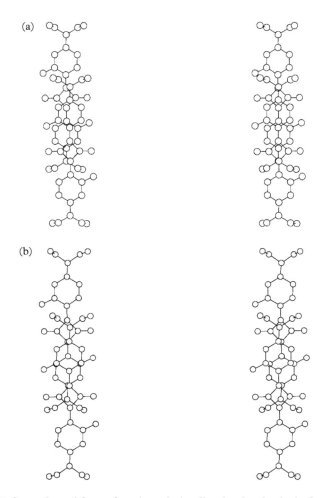

Figure 3.17 *Stereoviews of the overlap of translationally related molecules in the two structures of 3-XXVII. In both cases the view is on the plane of the reference molecule. (a) Triclinic structure, c-axis translation; (b) monoclinic structure, b-axis translation. (Adapted from Bernstein and Goldstein 1988, with permission.)*

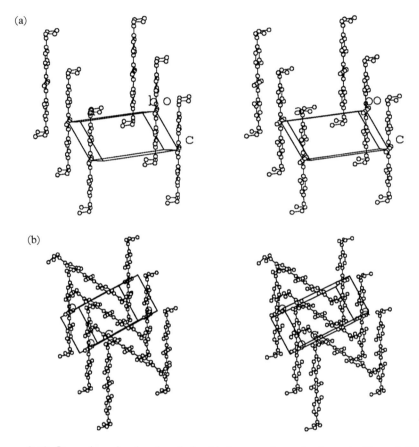

Figure 3.18 *Stereoviews showing the relationship between the stacks shown in Figure 3.17 for 3-XXVII. (a) Triclinic structure, translational relationship. (b) Monoclinic structure, screw-axis relationship. (Adapted from Bernstein and Goldstein 1988, with permission.)*

crystal structures differ, of course, in that stacks are related by translation in the triclinic structure and by a screw axis in the monoclinic one, as shown in Figure 3.18, but the fact that these crystallize concomitantly would be consistent with the assumption that energetically, at least, these inter-stack interactions contribute in a less significant way to the total lattice energy than those within the stack.

3-XXVIII

It is of interest to note that another similar squarylium dye **3-XXVIII** has also been shown to concomitantly crystallize in a green monoclinic phase and a purple triclinic phase (Ashwell et al. 1996). The structure has been reported for the former, but crystals of the latter were not of sufficient quality to determine the crystal structure. **3-XXVIII** crystallizes in space group $P2_1/c$ with cell constants $a = 9.046$ Å, $b = 19.615$ Å, $c = 9.055$ Å, $\beta = 116.1°$, which, lacking a short axis, essentially precludes any molecular overlap of the type seen in Figure 3.17 for **3-XXVIII**. The crystal structures of **3-XXVII** and **3-XXVIII** are significantly different so that the similarity in the colors of the polymorphic behavior of the two compounds appears to be entirely coincidental, albeit unusual.

These studies suggest the existence of similar aggregates leading to the different polymorphs, as a function of thermodynamic and kinetic conditions, as investigated by Wojtyk et al. (1999).

3.7   Disappearing polymorphs

Although this chapter deals with the principles behind controlling the polymorph obtained—that is, in a sense controlling the interactions among molecules in the formation of crystals—in polymorphic systems, that control is not always easy to achieve, or to maintain. The loss of control is indeed disturbing, and might even call into question the reproducibility of chemical processes. However, in spite of the fact that crystallization as a technique and a process is taken for granted by most practicing chemists it is still very much the component of an art.

The appearance of a new polymorph in fact can change the environment in which the material is found, since there is now a competition between the new polymorph and the previously existing one(s). In fact, by Ostwald's rule (Section 2.3), it is very likely that the new polymorph will be thermodynamically more stable than the others, and hence will be energetically preferred, but its effect depends as well on the overlap of its occurrence domain with that of earlier existing polymorphs and kinetic factors of crystallization. The resulting situation that can develop is that there are crystal forms that are observed over a period of time but apparently are displaced by a more stable polymorph. I have termed this phenomenon "disappearing polymorphs," as described in two reviews of this topic (Dunitz and Bernstein 1995; Bučar et al. 2015). Many chemists remain skeptical about a subject that calls into question the criterion of reproducibility as a condition for acceptance of a phenomenon as being worthy of scientific inquiry (e.g., Thakur et al. 2015; Desiraju and Nangia 2016). Nevertheless, there are well-documented cases of such a phenomenon and a number of them are presented in the two earlier reviews.[2]

---

[2] Desiraju and colleagues echo an earlier comment of Feldman in a book review in *Physics Today* [Feldman, B. J. (2010) *Phys. Today*, **63**, 50] relating to the questionable validity of science that is not reproducible in every instance: "For example, the notion that a particular polymorph can suddenly

In the introduction to our 1995 paper we acknowledged the disconcertion this phenomenon may lead to about science and its practice (Dunitz and Bernstein 1995):

> What is disturbing about the phenomenon of disappearing or elusive polymorphs is the apparent loss of control over the process: we did the experiment last week and got this result and now we can't repeat it! This kind of statement can lead to raised eyebrows or even to outspoken expressions of disbelief.

As clearly stated in the first sentence below we addressed the principle of the scientific problem of recovering disappeared polymorphs in no uncertain terms in the last sentence of that paper. But we were not the first to do so by far, as the 1661 quotation from Robert Boyle (in a general context) at the start of this chapter testifies. Professor Dunitz recently commented in detail (Dunitz 2016):

> Once a particular polymorph has been obtained, it must *always* be possible to obtain it again; it is only a matter of finding the right experimental conditions. Why am I paying so much attention to this problem? Opinion seems to vary about the question of reproducibility in scientific experiments. In an article published a few years ago in *Physics Today* [the Feldman book review] the author takes a strong stand about the matter: "simply put, if an experimental group cannot reliably reproduce a result in its own laboratory, it should never submit the result for publication in a scientific journal, and no scientific journal should ever publish a paper based on such unverified experiments." In my opinion, this goes too far. If an experimenter finds that results cannot be reproduced, here or there, this indeed creates a problem. *Why are the results not reproducible?* The answers to such problems of non-reproducibility may often be trivial, but they may sometimes lead to the recognition of new factors that had gone unrecognized. Why keep the existence of such problems secret? What was disturbing about the phenomenon of disappearing polymorphs was just its non-reproducibility, the apparent lack of control over the outcome of the crystallization experiment. Silence, voluntary or imposed, about individual cases of puzzling, experimental non-reproducibility would not have benefited the progress of science.

Many other difficult questions arise when a situation described as a disappearing polymorph is encountered. Why did the new polymorph appear at all (often after years of no hint of its existence)? Why does a previously robust process no longer yield the crystal that had been obtained prior to the appearance of the new one? What crystallization parameters must be modified to obtain either the old or the new form exclusively and robustly?

As I have noted repeatedly in this chapter, there is a vast variety of conditions that can affect a crystallization, and for any polymorphic system it is indeed difficult to single out a particular factor that might dominate. However, it is certain

---

disappear and not appear again appears mysterious, but it has no real scientific basis and can only serve to muddle a complex issue."

that *seeding* can and does play an important role in determining the fate of a crystallization, especially one in which competitive processes can lead to individual polymorphs, or a mixture of them.

Intentional seeding is a common practice among chemists who wish to coax the crystallization of a compound from solution or from the melt: small crystals or crystallites of the desired material (seeds) are added to the system (e.g., Pavia et al. 1988; Shriner et al. 1997). In this way, the rate-limiting nucleation step, which may be extremely slow, can be accelerated. For this method to be applied it is of course necessary that a sample of the crystalline material is available; that is, the compound must have been already crystallized in a previous experiment. When polymorphic forms of a substance are known to occur, intentional seeding with one of the polymorphs is a useful and often the most successful way of preferentially producing it rather than the other form(s).

Unintentional seeding may also occur even if small amounts of the undesired polymorph are present as contaminants—in fact, in principle just one such seed is sufficient to act as a nucleation site.[3] Unintentional seeding is often invoked as an explanation of crystallization phenomena such as disappearing polymorphs which otherwise are difficult to interpret. I argue here in favor of such an explanation, although there is no consensus about the size and range of activity of such seeds, which have never been actually observed as such.[4]

Estimates of the size of a critical nucleus that can constitute a seed range from a few tens of molecules to a few million molecules (Mullin 2001). Even with a particle size of about 10^{-6} g (essentially the limit for visible detection) a crystal of a compound of molecular weight 100 contains approximately 10^{16} molecules, sufficient to make 10^{10} nuclei of the size at the large end of the estimated range.

Thus, on the basis of quantity and size required to play a role, once a crystal form exists in a certain locale the presence of seeds is almost always possible, indeed often unavoidable. One can think of local seeding, where the unintentional source may come from the experimentalists' clothing, a portion of the room, an entire room, a building, or even, with increasing degrees of improbability, increasingly larger environments.[5]

---

[3] It is well known among practicing chemists that it is often difficult to crystallize a newly synthesized compound. Subsequent crystallizations tend to be easier because of the presence of suitable seeds (Wiberg 1960).
[4] The lack of direct observation of a proposed object or phenomenon is not contrary to the scientific method. The directly observable world constitutes but a small portion of the range of sizes believed to exist (see, for instance, Morrison et al. 1982), but we consider many objects and phenomena outside of that range—modeled or understood on the basis of indirect evidence—as part of our well established scientific body of knowledge. The existence and understanding of atoms was in such a category for almost two centuries, although modern techniques are enabling us to "see" atoms, albeit with the aid of computer imaging. The atomic nucleus, being smaller by five orders of magnitude is certainly in the category of objects apparently quite well understood but never actually seen.
[5] On occasion, this has been expanded by some to make a claim for "universal seeding" which taken literally is obviously absurd. The universe is estimated to contain about a millimole ($\sim 10^{20}$) of stars, so one seed per star, amounting to *one seed per solar system*, would require about 100 kg of a compound of nominal molecular weight 100.

However it is important to bear in mind that in principle only one seed is required to initiate a crystallization so that indeed small amounts of material can play a significant role in a crystallization process. Moreover, other bodies such as specks of dust, smoke particles, and other small foreign bodies can act as nucleating agents in promoting crystallization. Invoking the presence of seeds to account for crystallization phenomena often generates skepticism, even though the presence of unseen dust and smoke particles as solid particles, water vapor as liquid, and unseen scents even as small as individual molecules are commonly recognized and acknowledged.[6] Three earlier short compilations of examples of disappearing polymorphs have been given (Woodard and McCrone 1975; Webb and Anderson 1978; Jacewicz and Nayler 1979), although with varying points of view about the nature of the phenomenon.

The fact that a particular crystal form may disappear does not doom it to chemical history. Indeed, its very existence means that it occupies a defined region of phase space, which in principle is accessible to reproduce this form, and I believe that once a particular polymorph has been obtained it is *always* possible to obtain it again; it is only a matter of finding the right experimental conditions. Redesigning a strategy to find those conditions is by no means a simple matter and may often require unconventional measures (e.g., Braga et al. 2000; Davey et al. 1993; Ludlam-Brown and York 1993). However, as proven by a number of recent case studies consideration of a number of experimental variables can assist in the design of such a strategy. Among these are careful observations using hot stage microscopy (Section 4.2), information on the relative thermodynamic stabilities (Section 4.3), the energy–temperature diagram (Section 2.2.3) and the enantiotropic or monotropic relationships among polymorphs (Sections 2.2.4 and 3.4), design of kinetic conditions to obtain the less stable form, consideration of the influence of seeding (both intentional and unintentional), judicious choice of solvent or tailor-made additives (Section 3.6) to inhibit a particular undesirable form, etc. A number of cases of the successful design and application of such strategies have been given (e.g., Bernstein and Henck 1998; Henck et al. 2001; Bombicz et al. 2003; Tidey et al. 2015) sometimes involving creative and sophisticated chemistry.

The important point is that the determination of the crystallization conditions for various polymorphic forms need not be a completely random process, as I have shown above. The combination of thoughtful observation with consideration

---

[6] A normal urban environment contains approximately one million airborne particles of 0.5 micron diameter or larger per cubic foot, the number being reduced by an order of magnitude in an uninhabited rural environment. A normal sitting individual generates roughly one million dust particles (≥0.3 μm diameter) per minute (for reference a visible particle is usually 10 μm or greater in diameter). Clean rooms for various purposes (e.g., surgical theaters, biological or pharmaceutical preparations, semiconductor fabrication) employ sophisticated technology to remove these particles and to prevent subsequent contamination (Thai HVAC 2001). Therefore the possible presence of seeds of even a newly formed crystal form in a laboratory, a manufacturing facility, or any location having been exposed to that form cannot be casually dismissed, indeed the presence would be hard to avoid.

of all the available chemical structures and thermodynamic information can provide extremely useful guidelines, if not for success, then at the very least for further experiments. Crystallization is almost never a surefire procedure, especially when one is trying selectively to produce a particular polymorph, or one that has proven consistently or suddenly elusive, even after unequivocally establishing its existence.

3.8 A final note

I have attempted to give here an overview of the realm of the solid form screen. For those beginning a solid form search for the first time, and even those with some experience, it is highly recommended to study the reviews cited in designing and executing the experimental protocol. In particular, approaches to overall strategy, along with sage advice on potential problems and limitations, can be found in those accounts, for instance that of Florence (2009) in Figure 3.10, with many, but not all, of the techniques for searching for solid forms.

Finally, a word of advice based on personal experience and long-time membership in the community of organic solid-state chemistry: much can be gained by careful thought and consideration of the results of each attempt at crystallization, even those that fail.

4

Analytical techniques for studying and characterizing polymorphs and polymorphic transitions

You can observe a lot just by watching.

Yogi Berra

4.1 Introduction

Since polymorphs represent different crystal structures, essentially every physical or chemical property may vary among the polymorphic structures of a material. A consequence of this is the fact that virtually any technique that measures the properties of a solid material may in principle be used to detect polymorphism and to characterize the similarities and differences among polymorphic structures. Some techniques are more sensitive to the differences in *crystal structure* or molecular environment, as opposed to *molecular structure*, and in many cases these are to be preferred in detecting and characterizing polymorphs.

The intent of this chapter is to survey the most commonly employed of these analytical techniques, with an eye to demonstrating how they are being used to discover and characterize polymorphs. There is considerable literature associated with each of these techniques and in the sense of being able to cover all the details and ramifications of each technique I do not intend to be comprehensive here. Rather, we will limit ourselves to describing briefly and qualitatively the principles of the technique, when necessary, to providing some leading references for further reading and entry to the literature, and to giving some examples of the application of the technique to the characterization and study of polymorphs. Since the publication of the first edition there have been many technological developments and examples of applications of these techniques individually or in combination. In particular, two multi-authored volumes (Hilfiker 2006; Brittain 2009b) incorporating contributions by recognized experts in the various analytical methods, with emphasis on the polymorphism of pharmaceuticals, contain a number of very

Polymorphism in Molecular Crystals. Second Edition. Joel Bernstein. © Joel Bernstein 2020. Published in 2020 by Oxford University Press. DOI: 10.1093/oso/ 9780199655441.001.0001

informative reviews. A useful summary of many examples employing a combination of techniques may be found in Chieng et al. (2011).

The fundamental nature of this chapter is to provide the basics of the analytical techniques used to investigate, identify and characterize polymorphic systems. Those basics have not changed since the publication of the first edition and most of the examples are still relevant; hence much of the earlier content has been retained with references to noteworthy technical advances or enlightening examples. Because every technique provides different information and some techniques will not distinguish among a particular set of polymorphs, the importance of utilizing a wide variety of techniques for the identification and characterization of a polymorphic system cannot be overemphasized (Threlfall 1995). A good example of the application of this principle can be found in Bannigan et al. (2016).

4.2 Optical/hot stage microscopy

Hot stage microscopy provides a rapid method for screening substances for the existence of polymorphism, and is highly recommended as one of the techniques to be employed in the initial investigation and characterization of a solid material; a great deal of information can be obtained in a minimum amount of time with a very small amount of material. In spite of the utility of this technique, it has not enjoyed the widespread use it deserves. Microscopy was much ignored during the middle decades of the twentieth century, due, in part at least, to the subjective nature of the observations and "measurement" and to the lack of practitioners to pass the knowledge and insight on to their scientific progeny. However, its utility and importance are being increasingly recognized (McLafferty 1990; Streng 1997; Bernstein and Henck 1998; Nichols 2006) facilitated to a great extent by widespread use of digital photography for recording optical events and the ability to publish the photographic record in color, even if limited to supplementary material in publications. Such a record can be invaluable to subsequent investigators and should be greatly encouraged in publishing circles. Three of the traditional centers of activity in chemical microscopy (Cornell University, University of Innsbruck, and McCrone Associates) have provided a number of excellent literature sources for becoming acquainted with the polarizing microscope equipped with a hot stage and its use in studying polymorphism (Kofler and Kofler 1954; McCrone, 1957; Kuhnert-Brandstätter 1971, 1982; Chamot and Mason 1973; McCrone et al. 1978; Cooke 1998). The use of the microscope declined, in part at least, in an age of increasing quantification, due to the lack of quantitative measures that can be attached to visual observations. Photographic records of microscopic observations were more in the realm of geologists and biologists than that of chemists. However, the ready availability and relatively low cost of video recording with the capability of digitally capturing individual images or sequences of images bode well for the future of chemical/thermal microscopy in the study of polymorphism.

The most general use of the microscope for the study of polymorphism is simply for observing the homogeneity or diversity of a crystalline sample. Variations in size, shape, or color may indicate the presence of polymorphism and the need for further examination (see, for instance, Figure 3.13). As noted in Section 2.4.1, differences in crystal habit are not necessarily indicative of polymorphism. Further physical characterization of the individual crystals may involve measurements such as optical constants (Hartshorne and Stuart 1964; Wahlstrom 1969; McCrone et al. 1978; Nichols 2006) or interfacial angles (Winchell 1943; Porter and Spiller 1951; Terpstra and Codd 1961).

By far the most useful accessory to polarized light microscopy is the hot stage originally invented by Lehmann (1877a, 1877b, 1888, 1891, 1910) (Figure 4.1) and now used in the versions developed by Kofler and Kofler (1954) and the Mettler Company in 1968 (Woodard 1970; Julian and McCrone 1971; see also Kuhnert-Brandstätter 1971, 1982). Melting point determination on small samples is arguably the most obvious application for hot stage microscopy. With practice true melting points (i.e., the temperature at which the solid and its melt are maintained at equilibrium) can be readily determined. Melting point is perhaps one of the longest used numerical physical constants for characterizing solids, although as Borka (1991) has pointed out different polymorphs can have similar melting points. An outstanding example is that of trimorphic D-mannitol (Burger et al. 2000) for which the melting points are: Form I 166.5 °C; Form II 166 °C, and Form III 150–8 °C incongruently. A variety of other properties may be examined and studied with this versatile tool, which should be the first option for any characterization of a solid material, especially in the search for polymorphs.

The systematic use of the hot stage microscope employing mainly the Kofler hot stage was originally described in the book by Kofler and Kofler (1954) (the founders of the Innsbruck group).[1] McCrone's book (1957) and its 1965 revision follow a very similar approach, although he pointed out that different experimental strategies guided the Innsbruck group and those in the US. In Innsbruck most of the observations were made during heating, while McCrone's were made during cooling.

A typical example of a solid–solid transition resulting in the observation of two observed melting points is given in Figure 4.2. The use of polarized light allows the ready detection of discontinuous changes in polarization colors during the heating process, the discontinuity in any property being symptomatic of a phase change. It is worth noting here that there may be optically observed phase changes that are not detectable by other analytical techniques such as differential scanning calorimetry (DSC).

---

[1] In addition to a thorough description of hot stage techniques this book contains a very useful chapter with a detailed description of experiments that may be done (including the specific substances that best illustrate the phenomenon) to characterize molecular crystals, including polymorphism. Unfortunately the book is out of print and hard to obtain. At the time of this writing a 1980 English translation was available through McCrone Associates, Chicago, IL.

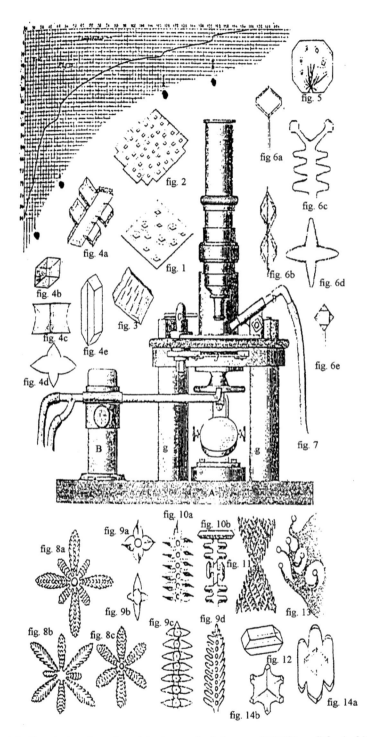

Figure 4.1 *Frontispiece from the original paper by Lehmann (1877b) on "physical isomerism" showing his hot stage microscope and various shapes of crystals observed. Another noteworthy feature of this figure is the time versus temperature cooling curve for ammonium nitrate in the upper left-hand corner, showing four inflection points that indicate polymorphic transitions. (Reprinted with permission from Zeitschrift für Kristallographie.)*

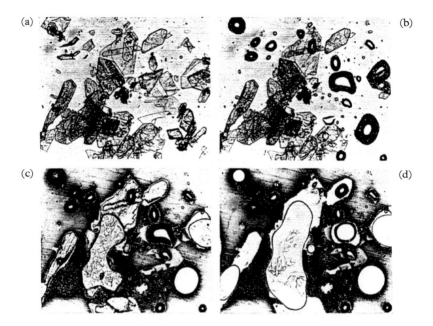

Figure 4.2 *Melting behavior of sulfathiazole. (a) At 175 °C, which is the melting point for commercially available Form II. (b) Some Form II crystals melting at 175 °C, while others transform to the stable modification. (c) At 200 °C, at which point some of the stable form begins to melt. (d) The equilibrium melting point for the higher melting form. (From Kuhnert-Brandstätter 1971, with permission.)*

The potential for carrying out studies below room temperature on the microscope cold stage has been facilitated by technological advances and should be increasingly employed. For instance, in one recent study, probenecid, a material that exhibited a theoretical propensity for polymorphism (Galek et al. 2009), failed to exhibit any polymorphic behavior over a seventy year history and a recent "traditional" attempt at a polymorph screen. However, a hot/cold stage study revealed two single-crystal-to-single-crystal transitions, one above room temperature and a second below room temperature (Nauha and Bernstein 2015), which could then be studied and characterized by additional analytical methods.

The determination of the index of refraction of the melt (a procedure readily performed on the microscope (McCrone 1957; Kuhnert-Brandstätter 1971) can serve as a confirmation of the existence of polymorphs, since all polymorphic forms must melt to the same liquid. Such studies provide preliminary, but often key and detailed information on the temperature range over which polymorphs exist, the degree of stability of metastable forms with respect to such factors as temperature, time, mechanical or thermal shock, the presence of impurities, etc.

The determination of the enantiotropic or monotropic nature of a polymorphic transition (see Sections 2.2.4 and 4.3) can be determined using the hot stage

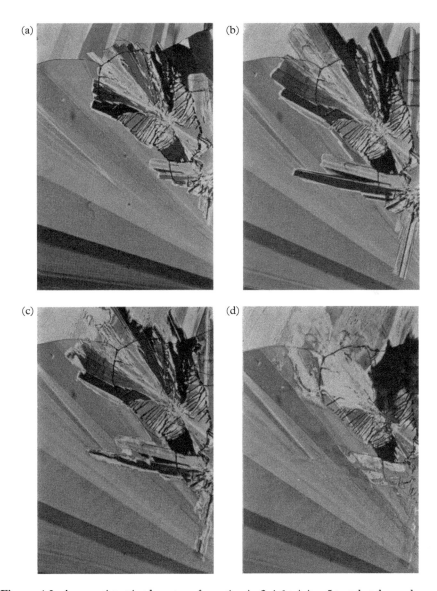

Figure 4.3 *An enantiotropic phase transformation in 2,4,6-trinitro-5-tert-butyl-m-xylene as observed on the hot stage microscope. Form II is stable at room temperature and the thermodynamic transition point is at 84 °C. (a) Room temperature stable Form II, the coarse crystals at upper right embedded in aggregate of Form II. (b) On heating Form II grows at the expense of Form I. (c) At 84 °C the transformation can be halted. (d) Above 84 °C, Form I is stable and has grown at the expense of Form II. (From Kuhnert-Brandstätter 1971, with permission.)*

microscope, as demonstrated in Figure 4.3. In a rarely used but potentially very powerful technique, transition temperatures and melting points can be determined by studying the behavior of the corresponding modifications in a two-component system and the determination of their melting points (Lautz 1913; Kofler and Kofler 1954; Kuhnert-Brandstätter 1971). A particular two-component method involves the preparations of the unstable modifications in mixtures with a second component with which the material under investigation may or may not be isomorphous. The fact that the stability often increases with the presence of impurities allows the preparation of the melting point diagram from mixtures and the extrapolation of the melting point. As examples, Kofler and Kofler (1954) cite the work of Brandstätter (1947) on Form II of 1,3-dinitrobenzene (with 1,2,4-chlorodinitrobenzene) and those of Francis and Piper (1939) and Phillips and Mumford (1934) on the unstable forms of methyl esters of carboxylic acids.

An additional indirect method employs the determination of the eutectic temperature with an appropriate test substance. It requires that the system does not form mixed crystals with the test substance, that it is monotropic, and that the equilibrium curves in the phase diagram be nearly parallel (Kofler and Kofler 1948). An elegant application of the eutectic method to determine the relative stabilities of a series of polymorphs of **4-I** was reported by Yu et al. (2000; see Section 5.8). Hot stage microscopy can also provide a wealth of information on the nature of polymorphic transformations. For instance, it is possible to distinguish whether nucleation of a new phase takes place randomly throughout the crystal, at specific defects or crystal edges, and whether the phase change occurs throughout the crystal in a diffuse manner (Figure 4.4) or at a front which moves through the crystal (Figure 4.5).

4-I

Solvates and hydrates may be readily detected and desolvation may be readily distinguished from a phase change using thermomicroscopy. The appearance of turbidity within the crystal upon heating is a sign of solvent being driven off, but a much more conclusive test involves covering the crystal with silicone gel or paraffin oil, which trap the bubbles of released solvent, as shown in Figure 4.6.

By thermomicroscopy the sublimation behavior may be readily studied and conditions (e.g., temperature, amount of sample) may be varied to achieve different habits and even different polymorphic modifications (Section 2.4.1; Figure 4.7).

Figure 4.4 *Phase transformation of β-naphthol, showing the diffuse nature of product phase formation. (From Kuhnert-Brandstätter 1982, with permission.)*

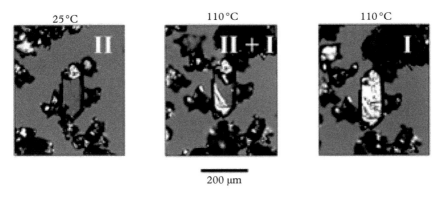

200 μm

Figure 4.5 *Phase transformation of Form II orthorhombic paracetamol to the monoclinic Form I. At 25 °C (left photo), the crystal in the center of the field is extinguished under crossed polarizers. The conversion takes place at 110 °C along a front which moves from lower left to upper right. The conversion is approximately 50% completed in the middle photo, and essentially complete in the right-hand one. (From Nichols 1999, with permission; see also Nichols 1998.)*

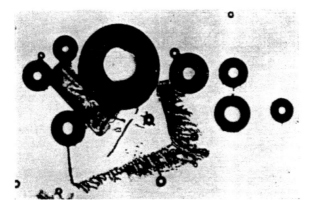

Figure 4.6 *Bubbles of solvent evolved from a crystal heated in a silicone gel preparation. (From Kuhnert-Branstätter 1971, with permission.)*

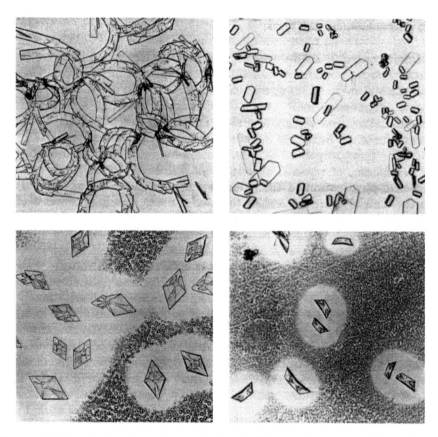

Figure 4.7 *Crystals of mephobarbital obtained under different sublimation conditions. All crystals are of the same polymorph but exhibit different habits. (From Kuhnert-Brandstätter 1982, with permission.)*

As noted in Section 3.5.3, in concert with his technique for surveying suitable solvents for crystallization McCrone (1957) has described techniques for carrying out crystallization on a microscope slide, a useful technique for preliminary studies for polymorphism, especially with limited sample quantities. This technique also allows subsequent characterization and/or reuse of the same sample for further crystallization experiments. One of the tests described for the relative stability of polymorphs (Section 4.12) involves the observation of competitive growth rates in a particular solvent (Figure 1.5). Such experiments may be easily followed and recorded on the microscope (at a variety of temperatures) as demonstrated in Figure 3.5.

A great deal of information on the polymorphic behavior of two-component systems may be obtained using the contact preparation method, originally developed by Lehmann (1888), and apparently rediscovered independently by A. Kofler

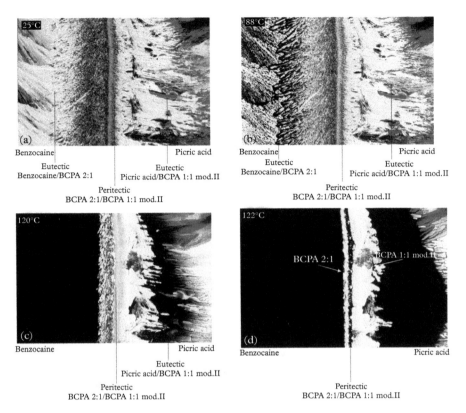

Figure 4.8 *Photographs of the microscope slide Kofler preparation showing the various phases of the benzocaine:picric acid (**BC:PA**) binary system. Top: (a) the photomicrograph of the recrystallized contact preparation of **BC** and **PA** at 25 °C. The interference colors are due to the use of crossed polarizers. The pure compounds are at the extremities of the preparation, while in the region where the original compounds have merged a number of different areas may be observed, due to the formation of different crystalline species combining the two components. Heating this preparation on the hot stage microscope to a temperature of about 88 °C shows (b) the eutectic melt of **BC** and the broad dark yellow crystals of a new compound. Due to the crossed polarizers, the isotropic melt appears black. At about 120 °C (c) on the left-hand side of the preparation **BC** is melted and the right-hand side the eutectic between **PA** and the remaining crystals of the **BC:PA** 1:1 complex melts; (d) the situation which is observed at 122 °C. PA is almost melted and in the middle of the preparation a eutectic melt appears. Thus, two chemically different kinds of complexes between **BC** and **PA** have been formed. The one on the right-hand side is the 1:1 complex while the small strip on the left side (the "**BC**" side) is a complex with the composition (BC)₂:PA. The former melts at 129 °C while the latter shows a melting point at 124 °C. (From Henck et al. 2001, with permission.) Overleaf The temperature–composition diagram derived in part from these hot stage observations. (From Henck et al. 2001, with permission.)*

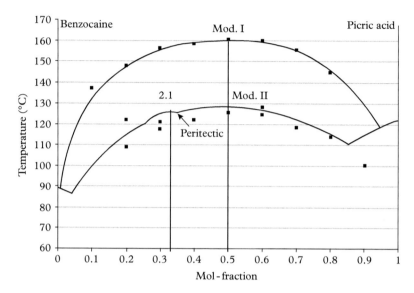

Figure 4.8 (Continued)

(1941, as noted by Kuhnert-Brandstätter 1971, 1982). The method involves melting the higher melting component between a microscope slide and cover slip so that about half of the intervening space is filled. After this melt is rapidly cooled the lower melting component is introduced at the free edge of the cover slip. Upon melting it flows under the slip until reaching the solid of the higher melting component. Preparations of this type allow one to determine the formation of one or more molecular compounds, of a eutectic, a mixing gap of the liquid phases or mixed crystals (Figure 4.8). The basic features of the melting diagram can be sketched out without the necessity for weighing, or the determination of time consuming and material consuming cooling-curve determinations; quantitative aspects of the diagram can be obtained from the thermal methods described in the next section. McCrone (1957) estimated that a simple binary eutectic diagram can be determined microscopically in 2–4 hours, while a ternary diagram might require two days, each additional polymorphic form doubling the time required. He also pointed out that Kofler's hot bench, or hot bar (Kuhnert-Brandstätter 1971,1982) can serve as a useful auxiliary in determining composition diagrams (Kofler and Winkler 1950a, 1950b).

4.3   Thermal methods

Whereas hot stage microscopy can be used to obtain qualitative information on polymorphic behavior, thermal analysis provides quantitative information about the relative stability of polymorphic modifications, the energies involved in phase changes between them and the monotropic or enantiotropic nature of those transitions. The two techniques are best used in conjunction, and indeed there is instrumentation for simultaneously carrying out the optical and thermal determinations.

Thermal methods are based on the principle that a change in the physical state of a material is accompanied by the liberation or absorption of heat. The various techniques of thermal analysis are designed for the determination of the enthalpy of the changes by measuring the difference in heat flow between the sample under study and an inert reference. These methods are all now commonly (and often mistakenly) referred to as DSC, since there are a number of ways of carrying out these experiments, each yielding slightly different information (McNaughton and Mortimer, 1975).

In the classical differential thermal analysis (DTA) system both sample and reference are heated by a single heat source. The two temperatures are measured by sensors embedded in the sample and reference. In the so-called Boersma system, the temperature sensors are attached to the sample pans. The data are recorded as the temperature difference between sample and reference as a function of time (or temperature). The object of these measurements is generally the determination of enthalpies of changes, and these in principle can be obtained from the area under a peak together with a knowledge of the heat capacity of the material, the total thermal resistance to heat flow of the sample, and a number of other experimental factors. Many of these parameters are often difficult to determine; hence DTA methods have some inherent limitations regarding the determination of precise calorimetric values.

On the other hand, in a genuine DSC instrument, sample and reference are each heated individually. A null balance principle is employed, whereby any change in the heat flow in the sample (e.g., due to a phase change) is compensated for in the reference. The result is that the temperature of the sample is maintained at that of the reference by changing the heat flow. The signal which is recorded (dH/dt, the heat flow as a function of time (temperature)), is actually proportional to the difference between the heat input into the two channels as a function of time (temperature), so that the integration under the area of the peak directly yields the enthalpy of the transition.

Thermogravimetric analysis (TGA) measures the change in mass of a sample as a function of temperature. It therefore provides information on the presence of volatile components, in the present context particularly solvents or water, which form the basis of solvates or hydrates respectively, as well as processes such as decomposition and sublimation.

Much of the literature on the thermal analysis of polymorphic systems up to 1995 has been cited by Giron (1995), along with many illustrative examples. Threlfall (1995), Kuhnert-Brandstätter (1996), and Perrenot and Widmann (1994) have also provided excellent discussions of many of the practical aspects of the application of thermal analysis to polymorphic systems. More recent surveys may be found in Craig (2006) and Bhattacharya et al. (2009). The following discussion draws much from these contributions.

Many typical features of a DSC of a polymorphic system are demonstrated for sulfapyridine in Figure 4.9. In the trace the starting sample is in an amorphous state, obtained by rapidly cooling (shock-cooling) the melt. This is one of the recommended procedures for detecting polymorphism in an uncharacterized sample. Initial heating results in a second order glass transition A to a supercooled liquid. A second order transition is characterized by a change in the heat capacity with no heat absorbed or evolved, and is recorded as a lowering of the baseline. The resulting unstable supercooled liquid can crystallize spontaneously upon heating, yielding in the sharp exotherm B, corresponding to an unstable phase in accord with Ostwald's rule (see Section 2.3). Further heating to ca. 145 °C leads to an (monotropic) exothermic solid–solid transition, denoted by C, resulting in a metastable phase. At D the latter melts at 175.30 °C and recrystallizes to the stable modification at E (at 177.30 °C), which in turn melts at 190.30 °C, at F.

Some of these changes can be followed optically, as evidenced by the microscope images in Figure 4.10. The quality of the thermal measurement, and consequently the amount of information that can be extracted from it depends on a number of experimental conditions. For instance, the heating rate is a crucial parameter, as demonstrated in Figure 4.11. On the same material, vastly different traces are obtained at different heating rates.

On the other hand, all of these contain information so that carrying out the measurement at a number of heating rates is always an advisable practice. This may not be a sufficient precaution, as evidenced in Figure 4.12. In spite of the fact that the heating rate is identical to that in Figure 4.9, this trace does not exhibit the melting of the metastable modification and subsequent crystallization into the stable form.

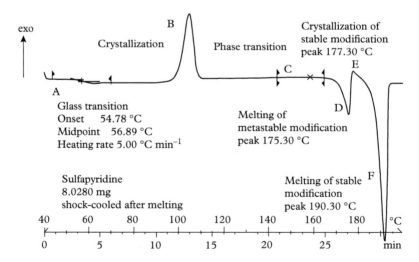

Figure 4.9 *Typical features of a DSC for the polymorphic system of sulfapyridine. (Reproduced with permission from Schwarz and de Buhr 1998.)*

Figure 4.10 *Photomicrographs of sulfapyridine at two temperatures. Left: ~110 °C, most of the field contains the metastable form that crystallized (peak B in Figure 4.9). Right: ~180 °C (between peaks E and F in Figure 4.9), most of the material has been converted to the stable form. (Reprinted from Schwarz and de Buhr 1998, with permission.)*

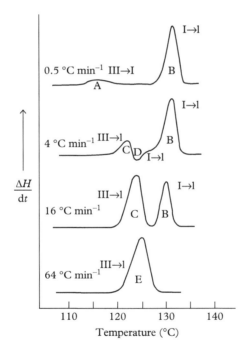

Figure 4.11 *A DSC measured at four different heating rates. At the slowest rate (0.5 °C min⁻¹) both a solid–solid phase transition (III → I) and the melting of the more stable phase I can be seen. At the fastest heating rate of 64 °C min⁻¹ modification III melts directly, but the heating rate is sufficiently fast to prevent the crystallization of form I. At 4 °C min⁻¹ form III melts (C) and recrystallizes to form I (D) which subsequently melts. At the intermediate rate of 16 °C min⁻¹ the system does not reach equilibrium, so the recrystallization of form I is masked by the direct melting of form III. (From Dr. D. Giron, with permission.)*

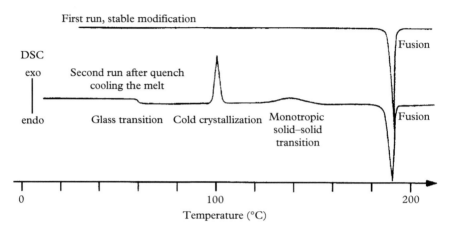

Figure 4.12 *DSC trace of sulfapyridine at 5 °C min⁻¹. (From Perrenot and Widmann 1994, with permission.)*

Other experimental factors that can influence the quality of the DSC measurement and the information that can be extracted from it are sample mass, particle size, the presence of impurities, the shape of the crystalline particles, and the presence of nuclei or seeds of various polymorphs. For the investigation of solvates the sample pan type also plays an important role (Giron 1995). Threlfall (1995) recommended routinely running both heating and cooling curves, while Perrenot and Widmann (1994) demonstrated the additional information that can be obtained by carrying out multiple heating runs on a particular sample.

In addition to providing information on the existence of polymorphs and the transformations among them, DSC measurements contain the quantitative information (complementary to optical microscopic observations) to aid in the preparation of the free energy–temperature diagram (Section 2.2.3). Since that diagram can be particularly useful in planning and executing the exploration of the solid form landscape the principles behind this process (Burger 1982b; Giron 1995; Grunenberg et al. 1996) are outlined here. The energy–temperature diagram and characteristic DSC traces for the monotropic case are given in Figure 4.13, while that for the enantiotropic case is given in Figure 4.14. The connections between the diagram and the traces are described in the figure captions. The principles are demonstrated for a system comprised of two polymorphic modifications, but of course, may be readily extended to more phases as demonstrated for five of the phases of aripiprazole in Figure 4.15 (Braun et al. 2009b). In practical terms the measured transition temperature for, say, a dimorphic enantiotropic system depends on the kinetic properties of the transition under investigation and is a function of the experimental conditions. The thermodynamic transition point (see Section 2.2.4) can be estimated if the melting points and the enthalpy of fusion of the two polymorphs are experimentally accessible (Yu 1995).

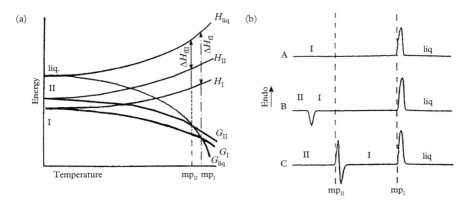

Figure 4.13 *Characteristic free energy versus temperature diagram (a) and DSC traces (b) for the monotropic relationship between polymorphs. The G_I and G_{II} curves do not cross below the melting points mp_I and mp_{II} indicated on the temperature axis. DSC trace A exhibits the melting of thermodynamically high melting Form I from which the value of ΔH_{fI} may be extracted. In trace B, the low melting modification II transforms monotropically to modification I which subsequently melts. The $\Delta H_{II \to I}$ gives a measure of the gap between the H_I and H_{II} curves at that temperature. In trace C, modification II melts followed by the crystallization of modification I. The melting and crystallization give ΔH_{fI} and ΔH_{fII} for I and II at the temperature of the process. (After Giron 1995, with permission.)*

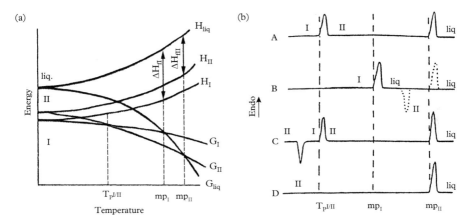

Figure 4.14 *Characteristic free energy–temperature diagram (a) and DSC traces (b) for the enantiotropic relationship between polymorphs. The G_I and G_{II} curves cross at the transition temperature $T_{I \to II}$ below their melting points T_{fI} and T_{fII} all indicated on the temperature axis. DSC trace A: at the transition temperature modification I undergoes an endothermic transition to modification II, and the heat absorbed is $H_{I \to II}$ for that transition. Modification II then melts at T_{fII}, with the accompanying ΔH_{fII}. DSC trace B: modification I melts at T_{fI} with ΔH_{fI} followed by crystallization of II with ΔH_{fII} at the intermediate temperature. Modification II then melts with details as above. DSC trace C: modification II, metastable at room temperature, transforms exothermically to modification I with $\Delta H_{II \to I}$ at that transition temperature. Continued heating leads to the events in trace A. DSC trace D: modification II exists at room temperature and no transition takes place prior to melting at T_{fII}, with the appropriate ΔH_{fII} (After Giron 1995, with permission.)*

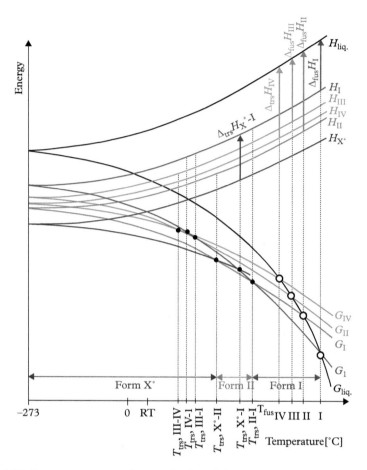

Figure 4.15 *Energy–temperature diagram for five of the forms of aripiprazole, including the original caption. (From Braun, D. E., Gelbrich, T., Kahlenberg, V., Tessadri, R., Wieser, J., and Griesser, U. J. (2009). Stability of solvates and packing systematics of nine crystal forms of the antipsychotic drug aripiprazole.* Cryst. Growth Des., 9, 1054–65. *With permission of American Chemical Society.)*

The principles of TGA are demonstrated in Figure 4.16 for glucose monohydrate, which is known to exist in an anhydrous form and as a monohydrate. The anhydrous form exhibits no weight loss on heating, while the monohydrate shows a weight loss of 7.1%, slightly below the expected value of 9.1% for a 1:1 molar ratio. TGA measurements are often accompanied by those of DSC, providing a great deal of information on the nature of the desolvation process, including heat of desolvation and subsequent transformations. For instance, the DSC measurement for glucose accompanying Figure 4.16 is given in Figure 4.17.

While the anhydrous form exhibits only melting at about 161 °C, the monohydrate shows a broad endothermic peak accompanying the dehydration. The resulting

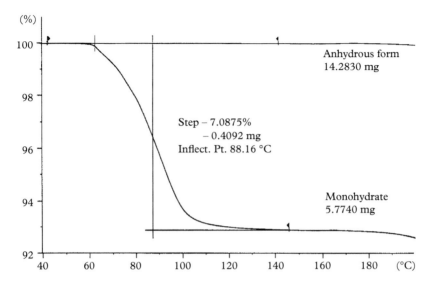

Figure 4.16 *TGA traces for the anhydrous and monohydrate forms of glucose. (Reprinted from Schwarz and de Buhr 1998, with permission.)*

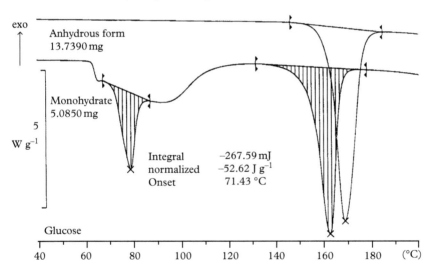

Figure 4.17 *DSC traces for the anhydrous and monohydrate forms of glucose. (Reprinted from Schwarz and de Buhr 1998, with permission.)*

anhydride melts at a temperature below that of the pure anhydrate, due, perhaps to incomplete crystallization or the presence of residual water. Of course, true polymorphism of hydrates and/or solvates may exist, and such cases will add complexity to the TGA and DSC traces. While other techniques are usually preferred for the quantitative analysis of mixtures of polymorphs there are some systems for which DSC has proven to be the preferred method (Carlton et al. 1996).

Mention should be made here of some of the recent developments in DSC technology. The first is oscillating, alternating, or modulated DSC (Barnes et al. 1993; Readings 1993; Dollimore and Phang 2000; Knopp et al. 2016). An oscillating time/temperature function is applied to the sample with simultaneous heating at a constant rate. The oscillation allows the application of Fourier transform techniques to the signal and its separation into two components: the reversible component contains the specific heat C_p, while the irreversible component contains kinetic information. The technique permits separation of thermal events that overlap. The irreversible ones, such as desolvation and crystallization will appear on the kinetic curve, while fusion will appear on the reversible one.

A second development is thermal scanning probe microscopy (Dai et al. 2012) which is an extension of atomic force microscopy (AFM) employing microthermal methods. It has been used, for instance, to detect two of the polymorphs of cimetidine (Sanders et al. 2000).

Solution microcalorimetry is another thermal method for the determination of the difference in lattice energy of polymorphic solids. The difference in heat of solution of two polymorphs is also the difference in lattice energy (more precisely lattice enthalpy), provided of course, that both dissolution experiments are carried out in the same solvent (Guillory and Erb 1985; Lindenbaum and McGraw 1985; Giron 1995; Royall and Gaisford 2016). The actual value for $\Delta H_{\text{I-II}}$ is independent of the solvent, as demonstrated in Table 4.1 for the two polymorphs of sodium sulfathiazole. Note also that the calculated heats of transition (ΔH_t) are virtually identical in spite of the fact that the heat of solution (ΔH_s) is endothermic in acetone and exothermic in dimethylformamide.

Since experimentally $\Delta H_{\text{I-II}}$ may be determined at a chosen temperature the technique allows the determination of the difference in lattice energy at a variety of temperatures. In addition the $\Delta H_{\text{I-II}}$ at one temperature (say T_1) may be used to calculate the same quantity at a different temperature, T_2, from the difference in the heat capacities of the two polymorphs:

$$\Delta H_{\text{I-II}}(T_2) = \Delta H_{\text{I-II}}(T_1) + (T_1 - T_2)\Delta C_{p(\text{I,II})}$$

Table 4.1 *Heats of solution ΔH_s (kcal mol^{-1}) measured at 25 °C and calculated heats of transition ΔH_t for the two polymorphic forms of sodium sulfathiazole in two different solvents (adapted from Lindenbaum and McGraw 1985)*

	Acetone	Dimethylformamide
$\Delta H_{s,\text{I}}$	2.853 ± 0.026	-1.113 ± 0.012
$\Delta H_{s,\text{II}}$	1.229 ± 0.016	-1.740 ± 0.023
ΔH_t (at 25 °C)	1.624 ± 0.042	1.627 ± 0.035

Some typical examples of these measurements are given by Lindenbaum et al. (1985). For quantitative analysis of mixtures, Botha et al. (1986) demonstrated how the percentage composition of two polymorphs may be determined essentially over the entire range of composition by measurements of the heats of solution of the pure polymorphs. Additional details on many aspects of the applications of solution microcalorimetry to the study of polymorphs and solvates may be found in the book by Hemminger and Höhne (1984), as well as the chapters by Grant and Higuchi (1990) and Brittain and Grant (1999).

4.4   X-ray crystallography

X-ray crystallographic methods, which reflect differences in crystal structure, in most cases can be definitive in the identification and characterization of polymorphs, and whenever possible should be included in the analytical methods utilized to define a polymorphic system.

In the application of X-ray diffraction methods to the study of molecular solids in general and polymorphic systems in particular a distinction is often made between powder methods and single crystal methods. Traditionally the former have been used for the qualitative identification of individual polymorphic phases or mixtures of phases, while the latter have been employed for the determination of the detailed molecular and crystal structure. Fortunately, the gap between these two sub-disciplines is being bridged, and much can be gained in the study of polymorphs by the cross fertilization of these two techniques. For instance, single crystal structure solution techniques are being applied to powder data to solve larger and previously intractable crystal structures (Andreev et al. 1997; Shankland et al. 1997; Tremayne et al. 1997; David et al. 1998; Chan et al. 1999; Putz et al. 1999) with subsequent refinement by Rietveld methods (Young 1993), and this field is growing very rapidly with increasingly powerful X-ray sources and computing resources. The more widespread application of synchrotron radiation to powder diffraction is also contributing significantly to closing this gap; in some cases revealing polymorphism not observed by laboratory scale X-ray sources (Sato 1999).

Two particularly elegant powder diffractions studies in the past decade on benzamide demonstrate the increasing sophistication of powder X-ray diffraction. Benzamide was the first molecular crystal to exhibit polymorphism, as described by Wohler and Liebig (1832). They noted that the compound crystallized from water initially appeared as fine needles, undergoing a transformation to rhombs over a period of minutes to hours. This phenomenon, readily observed on the microscope was cited as an example of concomitant polymorphism (Bernstein et al. 1999) since the needles and rhombs clearly coexist for a period of time during which they can be readily observed. While the crystal structure of the later appearing rhombs was reported by Penfold and White (1959), the crystal structure of the fine needles was solved only in 2005 using some elegant synchrotron

X-radiation techniques (David et al. 2005). Shortly thereafter a third polymorph was detected and its structure was determined by a no less sophisticated application of synchrotron powder diffraction combined with a number of analytical techniques described in this chapter (Thun et al. 2009). A recent contribution to the benzamide saga (Johansson and van de Streek 2016) is another demonstration of how the application of increasingly sophisticated techniques to iconic examples leads science forward.

Continuing developments in the field of electron diffraction of very small single crystals may also serve to bridge the gap between classical powder and single crystal diffraction techniques (Voigt-Martin et al. 1995; Dorset 1996; Dorset et al. 1998; Yu et al. 2000b).

The fundamentals of X-ray powder diffraction (XRPD) techniques are summarized in a number of texts and reviews (Azaroff and Burger 1958; Bish and Reynolds 1989; Jenkins and Snyder 1996). Recent developments in virtually all aspects of practice of XRPD are summarized in Chung and Smith (2000), Pecharsky and Zavalij (2009), Dinnebier and Billinge (2009), and Madsen et al. (2013).

The X-ray diffraction pattern from a solid results from the satisfaction of the Bragg condition ($n\lambda = 2d \sin\theta$), where λ is the wavelength of the X-ray radiation and d is the particular spacing between individual parallel planes. The condition can be satisfied when the angle θ between the incident radiation and that set of planes results in constructive interference. The XRPD pattern of a solid is thus a plot of the diffraction intensity as a function of 2θ values (or equivalently, d spacings) and may be considered to be a fingerprint of that solid. The values of the d spacings reflect the dimensions of the unit cell, while the intensities are due to contents of the unit cell and the way the atoms and molecules are arranged therein. As polymorphs comprise different solids with different unit cells and different arrangements of molecules within the unit cell they have different fingerprints—most often as different as the X-ray powder patterns of two different compounds. Thus XRPD is probably the most definitive method for identifying polymorphs and distinguishing among them. Typical experimental powder patterns for two polymorphs are given in Figure 4.18.[2]

XRPD patterns are often likened to fingerprints, since each solid—and hence each solid form—has a unique structure and a unique XRPD pattern. This is clearly manifested in the two patterns and recognized by simply viewing them. However, there are circumstances for which one might want to list characteristic features of

---

[2] There have been a number of reports of polymorphic systems in which the reader might be led to expect some similarity in the powder pattern of two polymorphs because they crystallize in the same space group. There is no physical basis for this expectation. Except for the systematic absences of certain reflections due to space group symmetry polymorphic structures in the same space group and different cell constants will have different powder diffraction patterns. On the other hand, polymorphs with similar cell constants but different space groups may exhibit some similarity in XRPD patterns, but these cases are very rare (*vide infra*).

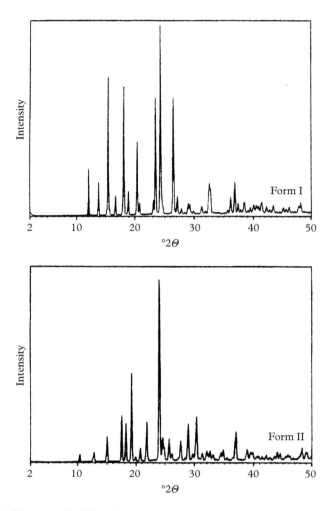

Figure 4.18 *X-ray powder diffraction patterns for the two polymorphs of paracetamol. (From Nichols and Frampton 1998, with permission.)*

the XRPD patterns of different forms in a tabular format. In that case a choice must be made on which, and how many peaks characterize and distinguish each powder pattern from its congeners. In our experience that choice is variable and depends very much on the crystal form landscape and one's familiarity with the system and the various forms. For instance, Figure 4.19 shows the XRPDs of the only two (so far) known forms of ranitidine hydrochloride. It is readily seen that the two most prominent peaks in Form 1, at ca. $2\theta \sim 17°$ and $22°$ appear at "windows" in the Form 2 XRPD, while the two prominent peaks in Form 2 at ca. $2\theta \sim 20°$ and $24°$ appear at "windows" in the Form 1 XRPD. Thus, these two pairs of peaks could readily be used to characterize and distinguish these two polymorphs.

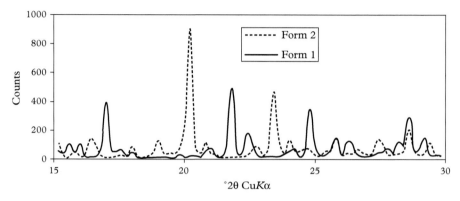

Figure 4.19 *X-ray diffractograms of the two polymorphic forms of ranitidine hydrochloride (from Agatonovic-Kustrin et al. 1999, with permission).*

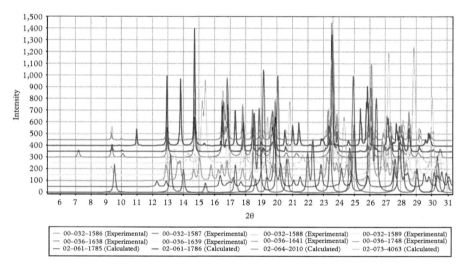

Figure 4.20 *Eleven XRPD patterns of cimetidine from the Powder Diffraction File. Some of these are calculated from the single crystal structure determination, while others were obtained experimentally (from Fawcett et al. 2009, with permission).*

When there are more forms the situation may on the surface appear to be more complex, but the reasoning is precisely the same—namely, the search for distinguishing characteristics in the various patterns. An example is given in Figure 4.20 for a number of forms of cimetidine. A few minutes of study clearly reveals that the density of peaks in the region $2\theta > 17°$ renders the determination of characteristic peaks rather difficult. In the lower angle regions there are clearly distinguishing characteristics, and while one or a few peaks in this region may not be absolutely characteristically distinguishing for a particular form those low angle peaks *in combination with* a relatively small number of higher angle peaks clearly can be distinguishing.

Single crystal X-ray crystallographic techniques are employed to determine details of the molecular and crystal structure of the solid—bond lengths, bond angles, intermolecular interactions, etc.—and are the source of some of the most precise metric data on these structural features (Dunitz 1979; Giacovazzo 1992; Glusker et al. 1994; Glusker and Trueblood 2010). The three-dimensional results are obtained from the collection of three-dimensional diffraction data. Those results may be used to simulate computationally the two-dimensional diffraction pattern to be expected from a powder of the same material. Such a calculated powder pattern may serve as a standard, unencumbered by impurities, the presence of other polymorphs, or the experimental difficulties enumerated below. The other side of this coin is the attempt to obtain the full three-dimensional crystal structure from the two-dimensional powder diffraction pattern, as noted above. Some of the important features of powder diffraction are discussed here. Those emanating from single crystal studies are alluded to in many other places throughout this work.

There are some reported cases of two genuinely polymorphic structures exhibiting very similar powder diffraction patterns. One of these cases, the two polymorphs of terephthalic acid (Davey et al. 1994), is shown in Figure 4.21.

These two patterns are remarkably similar, save for the 2θ region just above the large peak at ~30°. In such cases, there must be a structural explanation for the similarities in powder patterns. The two structures crystallize in the triclinic space group $P\bar{1}$ with different cell constants as given in Table 4.2 (Bailey and Brown 1967, 1984; Herbstein 2001).

As shown in Figure 2.14 and discussed by Davey et al. (1994) and Berkovitch-Yellin and Leiserowitz (1982), both structures are characterized by the formation of infinite hydrogen-bonded chains. The chains are organized into two-dimensional sheets that differ to some extent between the two polymorphs in the relationship between neighboring chains. The offset of neighboring sheets is more significant, and comprises the principal difference between the two structures. However, since the major contribution to the X-ray scattering is from the sheets, the similarity in structures is manifested in the similarity in powder patterns.[3]

---

[3] A cautionary note is in order here. In principle there is considerable freedom in choosing the axial system in triclinic unit cells, and in some higher symmetry ones as well (Hahn 1987). The convention calls for choosing the so-called reduced cell (Mighell 1976; Bauer and Tillmanns 1986; Hahn 1987), and there are a number of computer programs available for carrying out this cell reduction (e.g., LePage and Donnay 1976; Macicek and Yordanov 1992). For example, both the unreduced and reduced cells are presented in Table 4.2. Most crystallographic databases now routinely check that the recorded cell is in fact the reduced one, and all data collection programs on automated diffractometers do the reduction. However, many literature values have not been checked for the presence or absence of the reduced cell. Cell reduction can also lead to the recognition of a cell of different symmetry (e.g., Volume I of International Tables in 1965). As a result, a comparison of unit cell constants which leads to the conclusion that two structures are polymorphs may be in error. One way to check for identity or difference in polymorphs is to calculate the powder pattern, which is invariant to the choice of cell.

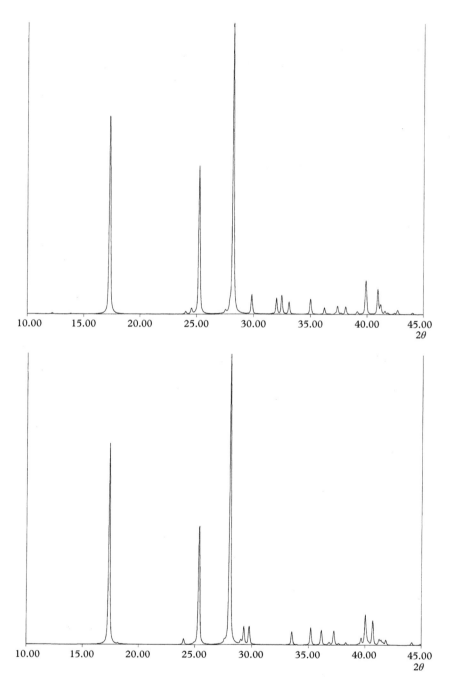

Figure 4.21 *Figure 4.21 X-ray powder diffraction patterns calculated from the crystal structures for the two forms of terephthalic acid. The structures are shown in Figure 2.14(a,b). Both polymorphs are triclinic, P$\bar{1}$, and are composed of very similar layers (which is essentially the plane of the paper) built up from linear chains of hydrogen-bonded molecules. The lateral offset of chains within a layer is also very similar in the two structures. The structures differ in the manner in which subsequent layers are offset from each other. (After Berkovitch-Yellin and Leiserowitz 1982, with permission.)*

Table 4.2 *Crystallographic cell constants for the two polymorphs of terephthalic acid*

	Reported cells[a]		Reduced cells[b]	
	Form I	Form II	Form I	Form II
a (Å)	7.73	9.54	3.76	5.027
b (Å)	6.443	5.34	6.439	5.36
c (Å)	3.749	5.02	7.412	6.991
α (°)	92.75	86.95	83.16	72.04
β (°)	109.15	134.6	80.87	76.03
γ (°)	95.95	104.9	88.53	87.09

[a] After Bailey and Brown (1967, 1984).
[b] Herbstein (2001). The cell constants for the reduced cell are based on those reported by Colapietro et al. (1984a) and Domenicano et al. (1990) for Form I and by Fischer et al. (1986) (by neutron diffraction) for Form II. Since they follow the convention for reporting reduced cells (Hahn 1987) they appear in a different order from the original; also by convention the cell angles are defined as acute.

Some other examples of very similar XRPD patterns have been noted. One of these is D,L-leucine (Mnyukh et al. 1975) for which the spectral data are clearly distinguishable. Another is caffeine (see Figure 4.28) (Suzuki et al. 1985; Griesser 2000), which is discussed in the next section. Karfunkel et al. (1999) also quoted the "surprisingly similar powder patterns" of some diketopyrrolopyrrole derivatives and have attempted to develop a model to describe the structural basis for similar powder patterns.

The preparation of samples for powder diffraction can lead to variations and inconsistencies among measurements on the same sample (e.g., Potts et al. 1994). Jenkins and Snyder (1996) summarized the possible causes for compositional variations between the original sample and that prepared for the diffraction experiment; grinding of the sample (generally required to reduce preferred orientation *vide infra*) can lead to amorphism, strain in individual particles, decomposition, solid-state reaction or contamination; the radiation used in the diffraction experiment can induce changes in the material, such as a solid-state reaction (e.g., polymerization), decomposition, or transformation to an amorphous state; the environment (humidity, temperature) can also effect the addition or loss of solvent, onset of reaction, decomposition, etc. All of these factors should be taken into account in determining and comparing powder patterns.

Perhaps the most pervasive problem influencing the intensities of powder diffraction lines is that of preferred orientation (Jenkins and Snyder 1996). Due to their non-spherical habit crystallites have a tendency to become oriented to efficiently occupy a minimum volume. For instance, plate-like crystals tend to lie flat on top of each other (Figure 4.22).

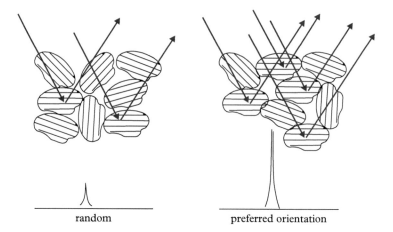

random   preferred orientation

Figure 4.22 *Schematic demonstration of preferred orientation for non-spherical particles. In the left-hand view the particles are randomly oriented so the diffraction condition is satisfied for a small number of crystals. In the right-hand view many of the crystals are in the same orientation, increasing the number of crystals for which the diffraction condition is satisfied and the intensity of the reflection from those planes. By the same token the intensities for many other reflections will be significantly reduced.*

Such orientational non-randomness will tend to diminish the intensities of those Bragg reflections that will not come into the diffracting geometry because of this preference for orientation of plates. A dramatic example of preferred orientation is shown for Form III of sulfathiazole in Figure 4.23, in which a comparison with the powder diffraction pattern calculated from the crystal structure indicates that the intensities of almost all of the expected diffraction peaks have been severely suppressed. The effects of preferred orientation can be reduced by grinding the sample prior to mounting, or spinning the mounted sample. Jenkins and Snyder note that one of the most effective ways of eliminating preferred orientation is by spray-dried preparation of the crystalline sample. Unfortunately, there may be considerable loss of control over the polymorphic form obtained in using this process.

It is important to make the distinction between the determination of *polymorphic identity* and *polymorphic purity*. The former is essentially a qualitative determination, asking the question, "Is a particular polymorph present in a given sample?" The latter is a question of quantitative analysis, and it is generally (though not always) assumed that the sample is *chemically* pure, so the analytical problem to be addressed is the determination of the relative amounts of different polymorphs in the sample. Recalling that different polymorphs are for all intents and purposes different solids, the determination of polymorphic purity is then no different in principle from quantitative determination of the composition of a mixture of solids. Such quantitative determinations comprise one of the traditional activities of

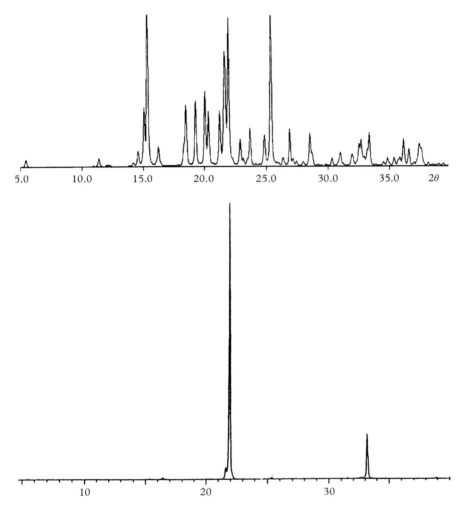

Figure 4.23 *The influence of preferred orientation on the experimental X-ray powder diffraction pattern of modification III of sulfathiazole. (a) Expected pattern calculated from the single crystal structure; (b) experimental powder pattern. (Adapted from Threlfall 1999, with permission.)*

analytical chemistry, especially when the materials are different chemical entities. In those cases a variety of different analytical methods may be employed, but generally one is interested in detecting and quantifying the undesired polymorphic forms in the presence of the desired ones, so that quantitative criteria such as *level of detection* and *level of quantitation* are invoked. More recently there have been an increasing number of instances in which it has been desired to measure the level of amorphous material in an essentially crystalline sample, or vice versa

(Madsen et al. 2011; Kern et al. 2012). In the case of polymorphic mixtures, or the determination of polymorphic purity, the choice of analytical method is considerably more restricted, and X-ray diffraction is one of the most definitive techniques.

The techniques of quantitative X-ray powder diffractometry were quite thoroughly summarized in a book of the same name by Zevin and Kimmel (1995). The basis of the technique is based on the assumption that the integrated intensity of a diffraction peak is proportional to the amount of the component (i.e., polymorph) present. Along with other factors that can influence the intensity noted above with regard to the determination of polymorphic identity, that intensity can also be severely affected by absorption of the incident radiation, for which appropriate corrections are available (Klug and Alexander 1974). Relative amounts of different polymorphs are determined by the relative intensities of a small number of (ideally) distinctive and preferably relatively strong peaks in the diffraction patterns, which means that the overlap of neighboring peaks is also a problem to be avoided or properly dealt with. Experimentally determined integrated intensities are also influenced by statistical errors, extinction effects, and other systematic aberrations, and all of these will affect the precision of the quantitative analysis. Of course, the usual precautions for quantitative analysis of mixtures must also be observed: the calibration and validation samples must be homogeneous and truly representative of the concentration level.

Many of these difficulties can be monitored and overcome with the use of standards, either internal or external (Zevin and Kimmel 1995). For the internal standard method, a known quantity of standard material is added to an unknown mixture, and the ratio of the intensity of the standard component is compared to a previously determined calibration curve to determine the mass fraction of the unknown (e.g., one or more of the polymorphic components). In the external standard method the entire composition of the unknown sample is determined simultaneously by comparing the measured intensities and respective calibration constants of reference intensity ratios (determined beforehand), which all must be with reference to the same reference standard.

As noted above, the traditional methods of quantitative analysis make use of one or a small number of non-overlapping peaks from the diffraction patterns of the different component (e.g., polymorphic) phases. With the advent of more powerful laboratory X-ray sources and synchrotron radiation, faster and more sensitive detectors, computer-controlled diffractometers, and the almost universal use of digitized data there is increasing use of the full diffraction pattern for quantitative X-ray diffraction analysis (Zevin and Kimmel 1995).

Qualitative and quantitative analytical applications of X-ray diffraction both require reference diffraction patterns to identify and quantify the different polymorphic modifications. Experimental powder patterns may be suspect for their use as standards as a result of experimentally induced errors or aberrations or the lack of polymorphic purity in the sample itself (which may even result from the sample preparation). The availability of full crystal structure determinations for any or all of the polymorphic modifications can considerably facilitate generation of standard powder patterns. A variety of public domain software is now available for calculating powder diffraction patterns from single crystal data (IUCr 2001; ICDD 2016).[4]

For example the calculated powder patterns for four of the polymorphs of sulfapyridine are given in Figure 4.24. The calculated pattern represents that of a pure sample. Using some of the more sophisticated programs to calculate the powder pattern, one can assume the absence of preferred orientation or alternatively some specified degree of preferred orientation. The line shape can be varied to match experimentally observed line shapes. If the crystal structures of all the polymorphic forms (and impurities) in the mixture are known, then the diffraction patterns of synthetically generated mixtures may be calculated for calibration or for use as benchmarks for experimental mixtures.

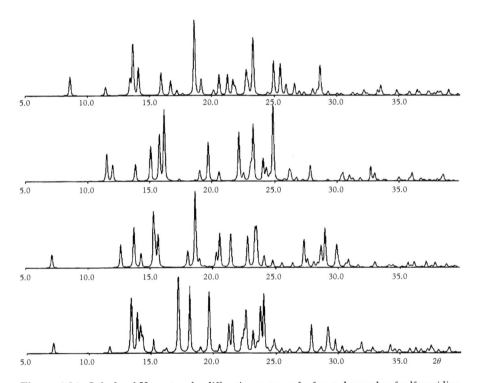

Figure 4.24 *Calculated X-ray powder diffraction patterns for four polymorphs of sulfapyridine.*

---

[4] Access to powder patterns generated from single crystal structures has been considerably facilitated by a cooperative arrangement between the Cambridge Crystallographic Data Centre (the repository of the experimental crystal structure data) and the International Centre for Diffraction Data (the repository for powder diffraction data). Either one of these organizations may be contacted for further details.

Many of the features of the use of X-ray diffraction in the analysis of a mixture of two unsolvated polymorphs (A and B) are demonstrated in a study by Newman et al. (1999). Under ambient conditions Form A has been shown to be thermo-dynamically more stable and the crystal structures of both forms have been deter-mined. They crystallize in different space groups with very different cell constants (Table 4.3); nevertheless, the laboratory generated powder patterns (Figure 4.25) are quite similar. According to Newman et al., the polymorphs are also not easily distinguishable by their DSC traces or from infrared (IR), Raman, and nuclear magnetic resonance (NMR) spectra. However, there are some clearly distinct features between their powder patterns measured using synchrotron radiation (Figure 4.26). Since manufactured lots appeared to contain both forms with

Table 4.3 *Crystallographic constants for two polymorphs*

	Form A	Form B
Crystal system	Orthorhombic	Monoclinic
Space group	$P2_12_12_1$	$P2_1/c$
a (Å)	10.260(3)	14.5878(3)
b (Å)	33.335(3)	14.111(3)
c (Å)	10.101(3)	18.101(3)
β (°)	–	111.85
Z	8	8

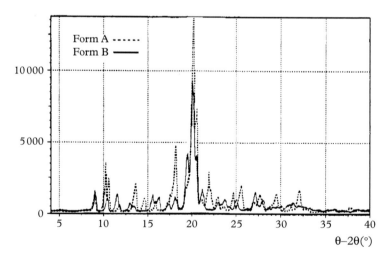

Figure 4.25 *Experimental X-ray powder diffraction patterns for Forms A and B of a polymorphic system (from Newman et al. 1999, with permission).*

(a)

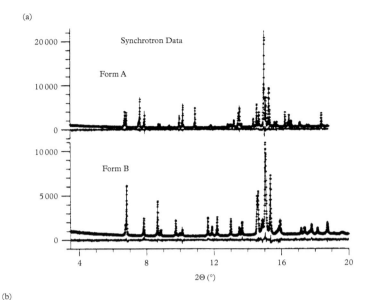

(b)

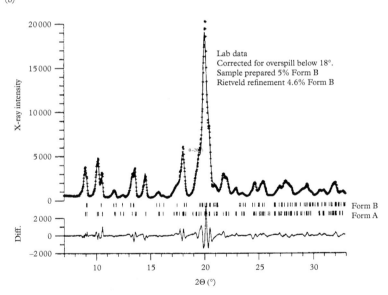

Figure 4.26 *(a) Synchrotron X-ray powder diffraction patterns for Forms A and B of a polymorphic system (from Newman et al. 1999, with permission). (b) A typical calibration run of a mixture of two polymorphs using Rietveld analysis. The calibration sample was prepared using 5% of Form B in a mixture of Form A and Form B. The upper trace shows the laboratory data for this sample. The next two rows indicate the positions expected for the diffraction peaks of Form B and Form A. The bottom trace shows the rms deviation resulting from the refinement of the combination of the full patterns for the two forms against the measured pattern. The best fit is obtained for a value of 4.6% Form B (from Newman et al. 1999, with permission).*

Form B generally as a minor component, it was necessary to develop a quantitative method to determine the amounts of the two forms in any batch. In developing the analytical protocol the authors followed the guidelines of the International Commission on Harmonization (1996), taking into account the specificity, working range, accuracy, precision minimum quantifiable limit, and robustness of the procedure.

Because of the general similarities in the diffraction patterns, and the lack of clearly resolvable distinguishing peaks, they employed the Rietveld method (Young 1993; Madsen et al. 2013). In the Rietveld method the entire experimental diffraction pattern for each solid phase is used as a basis for comparison. For structure determination using powder diffraction this comparison is made with a structural model used to generate a calculated pattern. In quantitative analysis of polymorphic phases, the known crystal structures are used to generate the standard diffraction patterns and these are then refined against the experimental powder pattern of the mixture to obtain the relative amounts of the polymorphs.

The development of the experimental procedure then involves the preparation of standard mixtures to prepare a calibration curve, with due care paid to corrections for particle size distribution, background, illuminated volume of sample, and preferred orientation. A typical calibration run is shown in Figure 4.26(b). Determinations on a series of similar "spiked" mixtures lead to the calibration curve in Figure 4.27. Analysis of the resulting data led to the determination of a minimum quantifiable limit of 5%, a working range of 5–50% Form B, and a relative standard deviation (RSD) of 16%. The method can be used to routinely monitor quantitatively the composition of mixtures of these two polymorphs for production lots, the control of processes, the stability of samples, and the monitoring and manipulation of process parameters to prepare each form or mixtures of them with predetermined proportions.

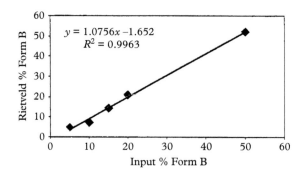

$y = 1.0756x - 1.652$
$R^2 = 0.9963$

Figure 4.27 *Calibration curve for mixtures of polymorphs A and B using laboratory X-ray data and Rietveld analysis. The value for the 50% mixture was obtained from synchrotron data (from Newman et al. 1999, with permission).*

Recently considerable progress has been reported in reducing the level of detection of small amounts of a solid chemical or polymorphic impurity, even in an intact formulated drug product (Yamada et al. 2011). The method involves optimizing the wavelength and path length of X-rays of the synchrotron radiation for the particular sample in order to gain maximum sensitivity. For a proof-of-concept experiment on fenoprofen calcium dihydrate a limit of quantification of 0.05% (RSD = 9.4%, n = 3) and a limit of detection of 0.02% (RSD = 17.3%, n = 3) were achieved, results which are approximately 102 times as sensitive as those obtained using conventional XRPD instruments. A European patent (Brescello et al. 2016) has been issued for a method that is claimed to be able to detect less than 0.005% of a minority polymorphic component (or alternatively, the absence of a non-prevailing polymorph), by employing the principle of optimization for each material.

It is important to point out here that each polymorphic system, indeed each polymorphic modification of a polymorphic system must be considered unique and must be individually characterized. For instance the 5% minimum quantifiable limit for the above system is well above the technical feasibility. Depending on the nature of the system, values at least as low as 1% are obtainable using quantitative X-ray analysis even by conventional methods (Tanninen and Yliruusi 1992).

Artificial neural network (ANN) theory (Zupan and Gasteiger 1991), which has been applied to other areas of quantitative chemical analysis (Bos and Weber 1991) has been shown to be useful in the quantitative analysis of mixtures of polymorphs using X-ray diffraction methods (Agatonovic-Kustrin et al. 1999, 2000). In studying the two polymorphs of ranitidine hydrochloride, and comparing the method to the more conventional method employing polynomial regression, the authors showed that the ANN methods yielded a smaller standard deviation and relative error, especially for the region of lower concentrations of Form 2, as low as 1–2%. The ANN methodology appears to be a potential alternative to the more traditional methods of quantitative analysis of mixtures, and the authors suggest extending the use to entire patterns (rather than a few selected peaks as in their study) as has been done with partial least squares methods in the quantitative analysis of mixtures using IR methods (see next section).

4.5   Infrared spectroscopy

IR spectroscopy is of course a standard technique for the characterization of compounds, in the current context of solid materials. Because it is based on the measurement of the vibrational modes generally of bonded atoms, with absorptions usually in the range 400–4000 cm^{-1} it is primarily a tool for investigating *molecular* properties rather than solid-state properties. Nevertheless, for over half a century (Ebert and Gottlieb 1952) it has been one of the most widely used methods for investigating the propensity of materials to form polymorphs (Kuhnert-Brandstätter and Junger 1976), including thermodynamic details such as transition points and number of components (Gu 1993). Fourier transform infrared (FTIR) spectroscopy is clearly the current method of choice, having replaced traditional grating and prism spectrophotometers (Krishnan and Ferraro 1982; Brittain 1997). Many of the developments since the first edition of this volume were very well summarized by Chalmers and Dent (2006), Brittain (2009a), and Heinz et al. (2009); all of these references contain numerous illustrative examples of the principles outlined here.

Since the characteristics of bonds or bonded atoms are monitored with this method, it is the perturbations of those vibrations due to variations in conformational or environmental factors among polymorphs that can lead to the differences in spectra. In general many molecular features are constant from polymorph to polymorph and the effect of environment on particular bonds and their vibrational manifestation may not be sufficiently large to be evident as differences in the IR spectra among polymorphs (Figure 4.28), for example, caffeine (Suzuki et al. 1985; Griesser 2000) and trovafloxacin mosylate (Norris et al. 1997), especially when presented in graphical form on a reduced scale. Hence IR spectra of different polymorphs quite often exhibit many similar features with differences showing up in a few specific regions, which may, however, provide distinctive markers for polymorphic characterization and distinction (Figure 4.29). The FTIR instrumentation and technique provides precise location of absorption bands, and that information, coupled with the IR assignments and comparisons for the various polymorphs, allows a means of characterization and comparison which is not available to the reader who is presented with a series of graphical spectra. Much information about the similarities and differences among polymorphs is often lost in such a graphical presentation.

An example of the preferred mode of presentation is in Figure 4.30 and Table 4.4 for two forms of lamivudine. While it is not always possible to make assignments of the bands to molecular bonding or functional groups, the tabulation of distinguishing peaks among polymorphic systems should be encouraged as standard practice. Evaluation of the statistical significance of similarities and differences in the spectra is considerably facilitated by inclusion of information on the experimental precision of peak location.

Since determination of the IR spectra of polymorphs is often carried out for the purpose of comparison, sample preparation becomes a particularly important factor in the experimental procedure. Threlfall (1995) reviewed many of the factors that must be taken into consideration in this regard, including labeling of polymorphic form (obvious, but in view of the confusion in polymorph nomenclature (see Section 1.2.3) still a potential problem), sample purity (both chemical and polymorphic purity), crystal size, crystal habit and orientation (Griesser and Burger 1993; Kobayashi et al. 1994), instability to pulverization and grinding (Farmer 1957; Hoard and Elakovich 1996; Threlfall 1999), solubility in the mulling medium or hydration (Kuhnert-Brandstätter and Riedmann 1989),

(a)

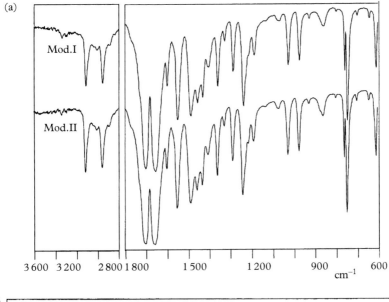

(b)

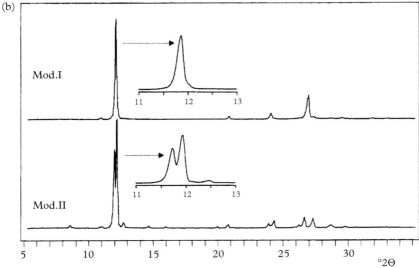

Figure 4.28 *(a) FTIR spectra of the two anhydrous modifications of caffeine, exhibiting virtually indistinguishable features. (b) X-ray powder diffraction patterns of the same modifications, showing considerable similarity, but distinguishable in the region 11°< 2θ< 13°. (From Griesser 2000, with permission.)*

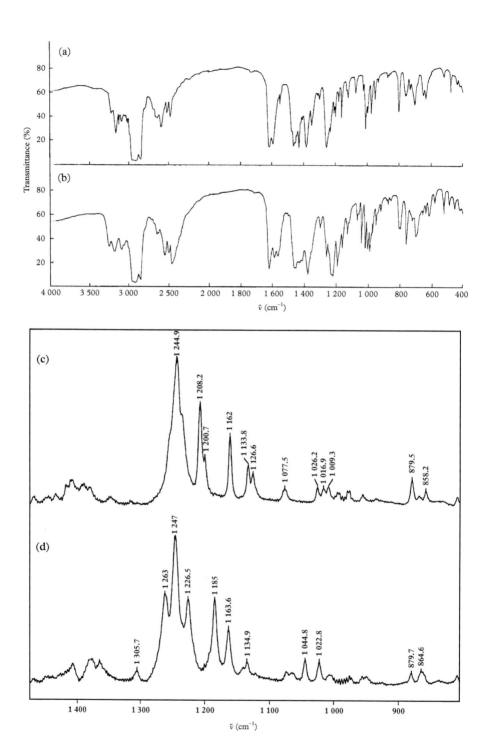

Figure 4.29 *IR spectra of (a) Form 1 and (b) Form 2 of ranitidine hydrochloride. The spectra are very similar (though not identical) in many respects, save one particularly distinguishing band at 1045 cm⁻¹ in the spectrum of Form 2 (from Cholerton et al. 1984, with permission). Detail of the (c) Form 1 and (d) Form 2 FTIR spectra of the two polymorphs of ranitidine hydrochloride, including the instrument-determined location of individual peaks (from Forster et al. 1998, with permission).*

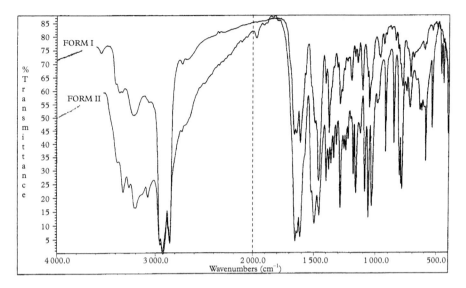

Figure 4.30 *FTIR spectra of the two forms of lamivudine. Note that the two spectra are presented as nearly overlaid as possible to facilitate comparison. In addition, the vibrational origin of significant peaks has been identified and the peak positions for the two polymorphs are compared in Table 4.4 (Harris et al. 1997) (figure from Lancaster 1999, with permission).*

Table 4.4 *IR assignments and frequencies for the two polymorphs of lamivudine (from Harris et al. 1997, with permission)*

Assignment	Form I	Form II
$\nu_{OH}{}^{a}$	3 545	—
ν_{NH}, ν_{OH}	3 404, 3 365, 3 341, 3 232	3 376, 3 328, 3 270, 3 201
ν_{CO}	1 662	1 652
ν_{CN}	1 643	1 636
ν_{CC}	1 613	1 613
ν_{CO}/δ_{OH}	1 053	1 060

[a] Of a hydration water molecule.

desolvation of a solvate (Burger and Ramburger 1979b), or instrumental variables (Free and Miller 1994).

A number of experimental alternatives to traditional IR transmission spectroscopy are suitable for overcoming some of these complicating experimental factors. In the technique of diffuse reflectance infrared Fourier transform spectroscopy (DRIFTS) (Neville et al. 1992; Hartauer et al. 1992) the sample is dispersed in a matrix of powdered alkali halide, a procedure which is less likely to lead to polymorphic transformations or loss of solvent than the more aggressive grinding necessary for mull preparation or pressure required to make a pellet (Roston et al. 1993). For these reasons, Threlfall (1995) suggests that DRIFTS should be the method of choice for the initial IR examination of polymorphs. He has also discussed the possible use of attenuated total reflection (ATR) methods in the examination of polymorphs and provided a comparison and discussion of the results obtained on sulfathiazole polymorphs from spectra run on KBr disks, Nujol mulls, and ATR.

In spite of many of the potential experimental pitfalls and difficulties (which should be viewed here as caveats rather than as deterrents), IR spectroscopy is still one of the simplest and most widely and routinely employed analytical tools in the study and characterization of polymorphs, especially when differences may be readily characterized and recognized. Some other modifications, developments and "hyphenated techniques" are worthy of note here, since they often considerably enhance the potential of the technique while reducing the drawbacks.

Photoacoustic spectroscopy is based on the absorption of modulated, rather than direct, radiation; hence it is independent of the linear absorption, thus eliminating attenuation due to excessive sample thickness or high extinction coefficients. Spectra can be measured on neat samples, without dilution (Vidrine 1982; Graham et al. 1985). This avoids the polymorphic changes that might be brought about by grinding, but particle size plays a role in such measurements (Rockley et al. 1984). An example of a study of a polymorphic system by photoacoustic spectroscopy on polymorphic forms sensitive to mechanical perturbations was given by Ashizawa (1989).

Just as polarized light enhances the utility of optical microscopy in the study of polymorphic systems, so polarization can be used in conjunction with IR spectroscopy. As I shall show later in greater detail (see Section 6.3.2) polarized spectroscopic methods provide detailed information on the *directional* properties which distinguish the spectral features of polymorphs. Thus, for instance, the directional properties of a polymorphic transformation of fatty acids (Kaneko et al. 1994a, 1994b, 1994c) and inferences about the differences in packing modes (Yano et al. 1993) have been investigated with polarized IR methods.

As with other analytical techniques previously widely used for polymorph characterization (i.e., polymorph identity), IR spectroscopy is being increasingly employed as a technique for quantitative analysis (e.g., polymorphic purity) (Patel et al. 2001). A typical example has recently been given by Bugay (1999) for cefepime dihydrochloride (Bugay et al. 1996). The question of polymorphic purity (more correctly in this case purity of crystal modification) involved in determining the amount of a dihydrate in a marketed monohydrate form. As seen in Figure 4.31a, the two modifications have quite distinct IR spectra in the 3000–4000 cm^{-1} range, and a mixture of 5% dihydrate in the monohydrate clearly shows how the peak of the former just below 3600 cm^{-1} may be used to quantify its presence. To extract quantitative information a calibration curve of known w/w mixtures was prepared in the working range of 1.0–8.0%, using diffuse IR reflectance. The statistical

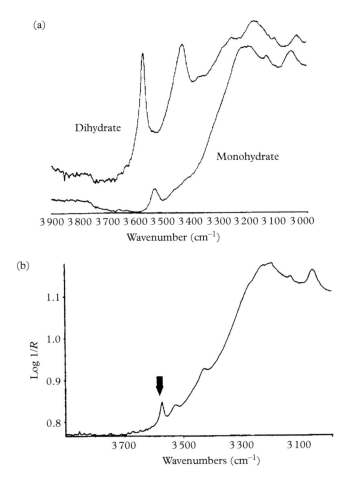

Figure 4.31 *(a) IR spectra for the monohydrate and dihydrate of cefepime dihydrochloride; (b) spectrum of a 5/95 w/w mixture of the dihydrate in the monohydrate. The arrow indicates the distinct dihydrate absorption used in the quantitative analysis (from Bugay 1999, with permission).*

analysis led to the graph shown in Figure 4.32, with a limit of detection of 0.3%, a minimum quantifiable level of 1.00%, and a good cross-validation with an independently developed assay based on XRPD (see Section 4.4).

For cases of the quantitative analysis of mixtures in which the distinction between two spectra is not as favorable or where unambiguous identification of many of the characteristic absorption peaks is desired (Zenith Laboratories, Inc. v. Bristol-Meyers Squibb Co. 1994), the method of partial least squares (Haaland and Thomas 1988) may be required. This technique originally developed for use in the near IR spectral region uses the entire (digitized) spectrum, rather than an individual peak or small number of peaks. (This approach is similar to the application of the Rietveld method in XRPD.) Some applications to the quantitative

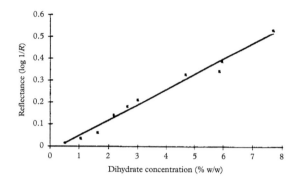

Figure 4.32 *Calibration curve for the presence of cefepime dihydrate in cefepime monohydrate determined by diffuse reflectance IR spectroscopy (from Bugay 1999, with permission).*

analysis of mixtures of polymorphs have been given by Geladi and Kowalski (1986), Hartauer et al. (1992), Tudor et al. (1993), and Jaslovzky et al. (1995) (the latter two with Raman spectroscopy), and it appears that this is a method that will see increasingly widespread use.[5]

4.6 Raman spectroscopy

IR and Raman spectroscopy are often grouped together, since both techniques provide information on the vibrational modes of a compound. However, since the two spectroscopic techniques are based on different physical principles the selection rules are different. IR spectroscopy is an absorption phenomenon, while Raman spectroscopy is based on a scattering phenomenon (Raman and Krishnan 1928). In general, IR energy is absorbed by polar groups, while radiation is more effectively scattered in the Raman effect by symmetric vibrations and nonpolar groups (Colthup et al. 1990; Ferraro and Nakamoto 1994). For a majority of molecules other than the simplest most symmetrical ones, the selection rules are not strictly obeyed, so both Raman and IR bands are likely be active for virtually all the bonds, but their relative intensities will differ, the more symmetrical ones giving higher Raman intensities, while the less symmetrical ones exhibit higher IR intensities (Lin-Vien et al. 1991). As with IR spectroscopy, Fourier transform methods have revolutionized the field and led to increasingly sensitive and sophisticated techniques. Cases where both IR and Raman spectra have been measured and compared have been given by Grunenberg et al.

---

[5] The desirability of using the entire spectrum (and hence every absorption peak) for both identification and quantification is particularly relevant in light of a recent court decision which required the plaintiff in a patent infringement litigation involving a particular polymorphic form to show that all the peaks in the patent were present in the defendant's accused infringing product (see Section 10.4.3).

(1996), Tudor et al. (1993), and Terol et al. (1994) (see Figure 4.33). In view of the similarity and overlap of the two techniques, the principles demonstrated and examples given in the previous section will not be repeated here. Rather, I will cite some relevant references, which can provide additional details as well as an entry into the specific literature.

A number of illustrative examples may be found in Wang et al. (2000), Ono et al. (2004), Bugay and Brittain (2006), Brittain (2009b, and references therein), and Dračínský et al. (2013). A typical example of the characterization of a polymorphic system by FT Raman spectroscopy has been given by Gu and Jiang (1995) while an application of the technique in the near IR spectral region to the polymorphic cimetidine system has been described by Tudor et al. (1991). The FT Raman technique has been compared to IR diffuse reflection spectroscopy in the study of the polymorphs of spironolactone (Neville et al. 1992), and the pseudopolymorphic transition of caffeine hydrate (i.e., loss of solvent) has been monitored using the technique (De Matas et al. 1998).

Some of the previous references contain descriptions of the use of Raman spectroscopy for quantitative analysis of mixtures of polymorphs. Additional examples may be found in Deeley et al. (1991), Petty et al. (1996), Langkilde et al. (1997), and Findlay and Bugay (1998).

One advantage of Raman spectroscopy over IR methods is that in general little or no sample preparation is required. This considerably facilitates the examination of samples in situ without undertaking experimental procedures for instance, grinding or pressing, which might induce polymorphic changes (e.g., Bell et al. 2000). One manifestation of this situation is the development of instrumentation that combines optical microscopy with Raman spectroscopy (e.g., Williams et al. 1994a; Webster et al. 1998). The initial applications of this technology were in quality control in industrial environments (Williams et al. 1994b; Hayward et al. 1994), but clearly the combination of microscopy and Raman spectroscopy can readily be applied to the analysis of polymorphic systems in a variety of environments, by facilitating the simultaneous optical and spectroscopic examination of multicomponent and/or heterogeneous samples. Typical systems would be the optical location and subsequent spectroscopic identification of individual polymorphs in a mixture of polymorphic forms, or the determination of the polymorphic form of the active ingredient embedded in the excipients of a formulated drug substance (Bugay 1999). In addition, because there is no need for sample isolation or preparation another important application in the employment of Raman spectroscopy is the identification and characterization of samples in high throughput crystallization experiments (Wall et al. 2007). In distinguishing between Raman spectra Mehuns et al. (2005) have suggested that a useful rule of thumb is that peak positions should differ by more than 1.6 cm^{-1}.

A comparison of the IR and Raman spectra for two polymorphs of an API is given in Figure 4.33.

A useful summary of IR and Raman methods applied to solid forms has been provided by Heinz et al. (2009) (Table 4.5).

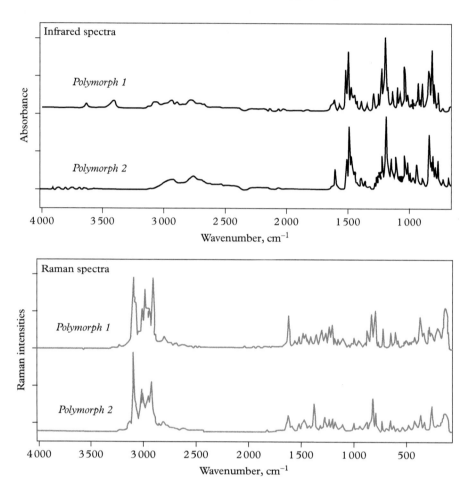

Figure 4.33 *Infrared and Raman spectra of A and E forms of an API are shown. This molecule contains three –OH groups, one NH, and one C=O. These three groups are characterized in the infrared spectra by the OH and NH stretching vibrations at about 3 425 and 3 300 cm⁻¹ and the C=O at ~1 720 cm⁻¹. The CH stretch vibrations at ~3 000 cm⁻¹ are more pronounced in the Raman spectra as are the ring modes in the 1 250–1 550 cm⁻¹ region. In the infrared and Raman spectra of this API shown there are major differences between the spectra of A and E forms. Two strong infrared bands between 1 050 and 1 150 cm⁻¹ characterize the A form, whereas these two bands are absent in the infrared spectra of the E form and are replaced by a very strong band at 1025 cm⁻¹. In the Raman spectrum, the E form has a very sharp band at 835 cm⁻¹, which does not appear in the spectrum of the A form. Moreover, the E form has a series of four strong–weak doublets in the Raman spectrum between 1 250 and 1 550 cm⁻¹. The A form also has a number of Raman bands in this same region, but the intensities are random. The A form has a doublet in the C–H stretching region just above 2 900 cm⁻¹, which is replaced by a single band in the spectrum of the E form. Thus, both the infrared and Raman spectra have a number of different patterns for identifying the two polymorphs of this API. (From Donahue et al. 2011, with permission. © 2018 CompareNetworks, Inc. All rights reserved.)*

Table 4.5 *Vibrational spectroscopic techniques used to analyze different solid-state transformations of active pharmaceutical ingredients. (From Heinz, A., Strachan, C. J., Gordon, K. C., and Rades, T. (2009). Analysis of solid-state transformations of pharmaceutical compounds using vibrational spectroscopy.* J. Pharm. Pharmacol., *61, 971–88. doi:10.1211/jpp.61.08.0001. With permission of John Wiley and Sons.)*

Vibrational spectroscopic technique	Information	Advantages	Disadvantages
FTIR spectroscopy	Intramolecular vibrations	Structural information on molecular level	Sample preparation for transmission or DRIFTS experiments can induce solid-state conversions
Modes:	*Crystalline polymorphs*: information about hydrogen bonding		
Transmission		No sample preparation required for ATR spectroscopy	
DRIFTS	*Solvates*: identification of solvent		Interference of excipients and environmental humidity
ATR	*Amorphous solids*: band broadening	Complementary structural information to Raman spectroscopy	
			Probes are not yet common
NIR	Overtones and combinations of intramolecular vibrations	No sample preparation required	Affected by environmental water and particle size
Modes:		Rapid measurements	
Transmission	*Crystalline polymorphs and solvates*: band splitting, changes in molecular symmetry	Use of probes	Low intensity
Reflectance		Possibility to measure through plastic containers	Subtle spectral differences between different solid-state forms

Continued

Table 4.5 Continued

Vibrational spectroscopic technique	Information	Advantages	Disadvantages
	Solvates: loss of solvent bands during dehydration, identification of different states of water in hydrates		
			Broad bands and overlapping spectral regions
	Amorphous solids: band broadening, lack of low-frequency bands		Poor fingerprint region and hence identification more complicated than with IR or Raman spectroscopy
Raman spectroscopy	Intramolecular vibrations	No sample preparation required	Local sample heating
Modes:	*Crystalline polymorphs and solvates*: band shifts indicate structural changes in the crystal lattice upon polymorphic conversions and hydrate formation/dehydration processes	Rapid measurements	Sample fluorescence
Backscatter		Use of probes	
Transmission		Not water sensitive, experiments in aqueous environment possible	
		Complementary structural information to IR spectroscopy	
	Amorphous solids: absence of spectral bands		

Vibrational spectroscopic technique	Information	Advantages	Disadvantages
		Possibility to measure through plastic containers	
Terahertz pulsed spectroscopy	Intramolecular vibrations, intermolecular vibrations and lattice vibrations	Information about crystal structure	Affected by water vapor
		Rapid measurements	Diffuse reflectance setup currently not available
Modes:	*Crystalline polymorphs and solvates*: hydrogen bonding, low-energy lattice vibrations	No sample preparation required for ATR measurements	
Transmission			Spectra currently difficult to interpret because vibrational modes are not yet fully understood
ATR			
Specular reflectance	*Amorphous solids*: bands disappear		

ATR, attenuated total reflection; DRIFTS, diffuse reflectance infrared Fourier transform spectroscopy; FTIR, Fourier transform infrared; IR, infrared; NIR, near infrared.

4.7 Solid-state nuclear magnetic resonance (SSNMR) spectroscopy

In terms of the structural features that are probed in various analytical methods, SSNMR may be looked upon as representing a middle ground between IR spectroscopy and XRPD methods. The former provides a measure of essentially molecular parameters, mainly the strengths of bonds as represented by characteristic frequencies, while the latter reflect the periodic nature of the structure of the solid. For polymorphs differences in molecular environment and/or molecular conformation may be reflected in changes in the IR spectrum. The differences in crystal structure that define a polymorphic system are clearly reflected in changes in the powder diffraction. Details on changes in molecular conformation or in molecular environment can only be determined from full crystal structure analyses, as discussed in Section 4.4.

SSNMR provides information on the environment of individual atoms. In essence the change in environment of any atom can arise from two factors, which usually are not separable in the interpretation of the SSNMR spectra, but are conceptually independent. Since different polymorphs are different crystal structures it is expected that the crystal environment of at least some atoms will differ from polymorph to polymorph (Section 2.4.2). In addition, since the molecular conformation may also vary among polymorphs (Section 5.5), the change in the environment of an atom due to conformational differences will also be reflected in the solid-state NMR (Levy et al. 1980; Strohmeier et al. 2001).

The theoretical basis and practical considerations for the application of SSNMR to the study of polymorphism may be found in a number of references, which themselves contain additional primary sources (for instance, Yannoni 1982, Fyfe 1983, Komorski 1986, Bugay 1993, Harris 1993; Brittain 1997, Byrn et al. 1999). Perhaps more so than many of the analytical techniques described in this chapter SSNMR is a particularly rapidly developing field with great potential in the investigation of polymorphic systems. It is not limited to a single nucleus (although most studies to date have concentrated on the ^{13}C nucleus) and it is being adapted for quantitative analysis of polymorphic mixtures and other multicomponent systems. Notable reviews since the publication of the first edition are those by Tishmack et al. (2003), Berendt et al. (2006), Harris (2006, 2007), Medek (2006), Offerdahl (2006), Geppi et al. (2008), Tishmack (2009), and Vogt et al. (2010, 2011).

The review by Geppi et al. 2008 in particular contains many well-described examples of application of SSNMR to problems of crystal forms, demonstrating not only the increasing sophistication of the various SSNMR techniques, but also their utility in investigating the details of structural differences in crystal forms and the nature of the transformations among them. In particular, the following topics of relevance here are described, with appropriate examples:

1. identification of different solid forms of an API in bulky samples;

2. identification and quantitation of API forms in either API mixtures or solid dispersions with excipients;

3. structural properties of API forms, involving conformational behavior, number of crystallographically inequivalent molecules in the unit cell, intra- and intermolecular interactions, internuclear distances;

4. stability of API forms;

5. structural behavior of the crystalline or amorphous forms of excipients in either bulky samples or solid dispersions;

6. dynamic behavior of API (usually related to inter-conformational motions) and/or excipient (for instance associated to the cross-linking degree in polymeric systems) forms;

7. physical or chemical interactions between APIs and excipients in solid dispersions;

8. average dimensions of API and excipient domains in solid dispersions on a 10–1 000 Å scale;

9. identification of impurities or degradation products;

10. solid-state chemical reactivity;

11. membrane protein structural properties and drug binding to membranes.

The original development of the basis for SSNMR by Schaefer and Stejskal (1976) through the combination of high power proton decoupling with magic angle spinning (MAS) and cross polarization SSNMR techniques for ^{13}C is demonstrated in Figure 4.34. A powdered crystalline sample measured by conventional

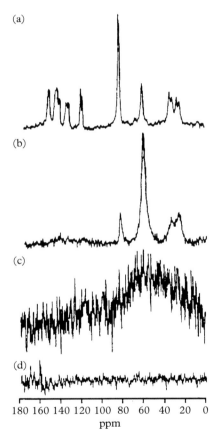

Figure 4.34 *Demonstration of the cumulative effects of techniques employed in ^{13}C CP/MAS solid-state NMR spectroscopy on a sample of dideoxyinosine. (From Bugay 1993, as modified by Byrn et al. 1999.)*

solution-phase pulse techniques gives an essentially featureless spectrum. The use of high power proton decoupling, MAS, and cross polarization leads to a spectrum with resolution similar to that obtained from solution, and the potential for extracting a great deal of chemical and structural information.

Perhaps the first application of this technique directly to polymorphic systems was by Ripmeester (1980), and Threlfall (1995) provided a useful early review supplemented more recently by many reviews cited above. For the study of polymorphic systems solid-state NMR has a number of advantages. The signal is not influenced by particle size, which may eliminate the complications of possible polymorphic transformations due to the grinding required in, say IR and XRPD techniques. The intensity of the signal is directly proportional to the number of nuclei producing it, so that the (qualitative) presence of mixtures of crystal modifications may be recognized (by noting the presences of overlapping spectra) or the quantitative composition of polymorphic mixtures (or, say, an active ingredient in a pharmaceutical preparation) may be determined. The lack of requirements for sample preparation means that investigations for the presence of polymorphism or polymorphic transformations may be easily carried out during any stage of the development or processing of a solid material. The required sample size is generally small, and in most cases is recoverable. There is at least one caveat, however: the high spinning rates required for the MAS technique generate considerable mechanical stress and local heating which may be sufficient to lead to polymorphic transformations or decomposition during the measurement.

In general, the similarity between the chemical shift in solution and solid-state ^{13}C cross polarization MAS (CP/MAS) spectra allow for the assignment of the peaks in the latter. Byrn et al. (1988) compared the range of chemical shifts for individual atoms in five crystalline forms of prednisolone with those obtained

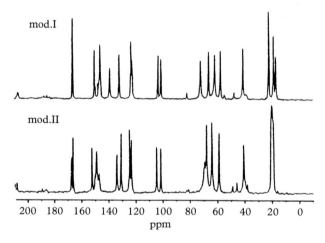

Figure 4.35 *Solid-state ^{13}C CP/MAS NMR spectra of the two modifications of nimodipine (from Grunenberg et al. 1995, with permission).*

from solution. The variation in the chemical shift from polymorph to polymorph
enables one to recognize significant differences in the atomic environment among
polymorphs. While the SSNMR spectra of two or more polymorphs may be considered fingerprint identifiers (e.g., Figure 4.35), identification of the individual
peaks and presentation of their chemical shifts in tabular form (e.g., Table 4.6)
provides considerably more information and is to be encouraged.

Table 4.6 *Comparison of the chemical shifts for the SSNMR spectra of the two polymorphs in
Figure 4.35*

	CDCl$_3$ (ppm)	Modification I (ppm)	Modification II (ppm)
C-1	148.15	150.43	148.86
C-2	123.28	123.39	124.54
C-3	150.09	150.43	152.31
C-4	134.72	139.06	133.82
C-5	128.61	132.13	130.67
C-6	121.28	122.63	123.10
C-7	40.10	47.70	40.76
C-8	103.00	101.26	104.51
C-9	145.38	148.26	147.76
C-10	144.60	146.20	147.01
C-11	103.81	103.31	101.35
C-12	166.62	166.44	166.03
C-13	62.99	62.09	63.95
C-14	19.39	17.46	20.28
C-15	19.54	19.12	20.28
C-16	167.13	166.44	167.29
C-17	67.37	66.45	67.86
C-18	22.12	22.49	20.28
C-19	21.80	22.49	20.28
C-20	70.54	72.36	69.30
C-21	58.83	57.89	58.78

In the characterization of polymorphs SSNMR can provide important crystallographic information, even in the absence of single crystal samples for full structure determination. Since the technique is a probe of the environment of a molecule in a crystal, differences among polymorphs in the number of molecules in the asymmetric unit are expected to be manifested in the solid-state spectra. Multiple molecules in the unit cell in principle lead to splittings for individual atomic peaks, since chemically equivalent atoms in crystallographically inequivalent molecules can have different surroundings and hence potentially different chemical shifts. (Similar environments or accidental redundancies can still lead to overlapping peaks.) A rather dramatic example of such a difference is presented in Figure 4.36 for lamuvidine. Form 2 crystallizes in a tetragonal structure with one molecule in the asymmetric unit ($Z' = 1$), yielding a rather straightforward spectrum. On the other hand, Form 1 crystallizes in an orthorhombic structure with five molecules in the asymmetric unit. Some peaks are merely broadened in going from Form 2 to Form 1 (due to near overlap), while others are split into multiple peaks.

Care must be exercised in interpreting such spectra in the absence of other confirming evidence for the polymorphic purity of a sample. For instance, the spectrum of a sample containing two polymorphs of a substance will appear as the sum of the two (Figure 4.37; see also Harper and Grant 2000), which might otherwise be interpreted as another form with multiple molecules in the unit cell.

In a crystal structure in which a molecule occupies a crystallographic special position, some chemically equivalent atoms can become crystallographically equivalent, meaning that they have identical environments in the crystal. Such a situation will reduce the number of lines observed in the solid-state NMR spectrum. For example in Figure 4.38, for Form I of 5,5′-diethylbarbituric acid ($Z' = 1$), peaks for all eight carbon atoms can be identified; in Form II the molecule lies on a two-fold axis that passes through carbon atoms 2 and 5, so that there are only five crystallographically unique carbon atoms and five unique peaks in the NMR spectrum.

As noted above, the characterization of polymorphs by SSNMR is by no means limited to ^{13}C spectra. For instance, Bauer et al. (1998) showed that ^{15}N CP/MAS could easily distinguish between the two polymorphs of irbesartan **4-II** and could be used to study the difference of the tautomeric behavior in the tetrazole ring of the two polymorphic forms. Variable temperature spectra indicate that the ring in Form B is involved in an exchange process that does not occur in Form A (Figure 4.39).

In addition to the identification of crystal modifications, SSNMR has been used to monitor reactivity and phase changes in different polymorphic forms. For instance, Harris and Thomas (1991) followed the photochemical conversion of formyl-*trans*-cinnamic acid with ^{13}C SSNMR (see also Section 6.4). Variable temperature techniques have been used to study the interconversion of

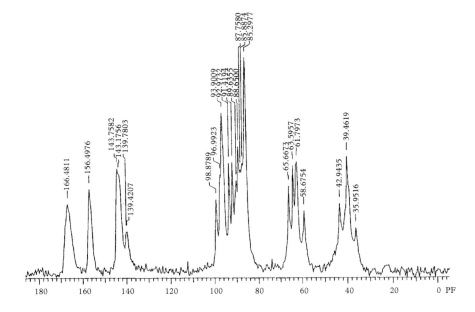

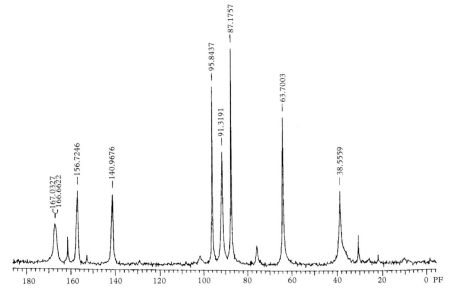

Figure 4.36 *Solid-state ^{13}C CP/MAS NMR spectra of the two modifications of lamuvidine. Lower, form 2, tetragonal, $P4_32_12$, $Z = 8$, $Z' = 1$; upper, form 1, orthorhombic, $P2_12_12_1$, $Z = 20$, $Z' = 5$ (from Lancaster 1999, with permission)*

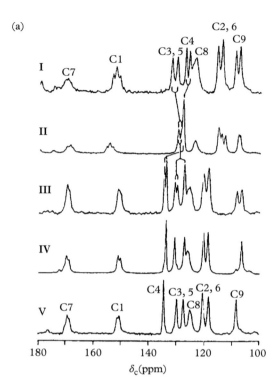

(a)

I C7 C1 C3, 5 C4 C8 C2, 6 C9

II

III

IV

V C7 C1 C4 C3, 5 C8 C2, 6 C9

180   160   140   120   100
δ_c(ppm)

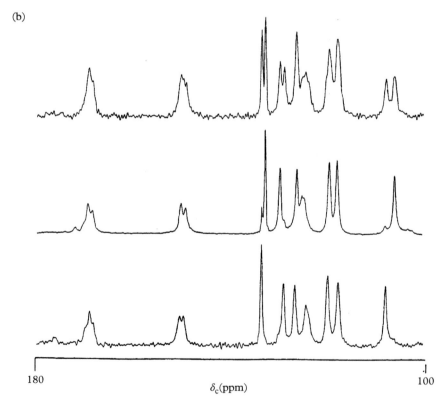

(b)

180   δ_c(ppm)   100

four polymorphic modifications of sulfanilamide (*p*-aminobenzenesulfonamide), including interpretation of at least some of the molecular motions during the course of the transformation (Frydman et al. 1990). A similar combination was augmented with colorimetric techniques to study the co-existence of two phases in the cause of a phase transition (Schmidt et al. 1999). Of course differences between unsolvated and solvated or hydrated crystal modifications may also be readily characterized by the SSNMR technique, as was done with the anhydrous and monohydrate forms of oxyphenbutazone (Stoltz et al. 1991). Due to the availability of the crystal structures for both modifications the SSNMR results could be interpreted directly in terms of the different atomic environments, especially for the differences in hydrogen bonding in the presence and absence of the water of hydration. SSNMR has been increasingly combined with XRPD, for instance to monitor changes in the crystal modification of the monohydrate of neotame (*N*-(3,3-dimethylbutyl)-L-phenylalanine methyl ester) (Padden et al. 1999), which is the most stable form of the compound under ambient conditions. Other modifications can be generated under vacuum, and the original monohydrate can be regenerated by exposure to moisture. XRPD monitoring of these processes indicated that no structural changes had occurred, but the changes in the ^{13}C SSNMR spectra suggested the presence of many forms during the reconversion process. Considerably more complex substances, such as starches, often distinguished as "A" and "B" forms, can yield useful structural information from SSNMR studies (Veregin et al. 1986).

The SSNMR technique has also been applied to characterize polymorphic systems of organometallic complexes. In one early application Lockhart and Manders (1986) studied three polycrystalline methyltin(IV) compounds. They were able to isolate one of the pure polymorphic modifications and to determine the presence of other modifications by SSNMR techniques.

A comparison of the environment of atoms in different crystal modifications necessarily requires the assignment of peaks to individual carbon atoms. As noted above, this assignment may be complicated in the solid state by the variation of chemical shifts from their solution values and by the presence of crystallographically inequivalent molecules in the asymmetric unit. A number of methods were used by Zell et al. (1999, and references 34–41 therein) to overcome these difficulties. These authors used two-dimensional SSNMR with high spinning speed and high ^{1}H decoupling power on uniformly labeled (Szeverenyi et al. 1982) samples of two of the three crystal modifications of aspartame. For the first time they were able to follow the connectivity between resonances of the crystallographically inequivalent molecules in a polymorphic modification of a compound. The further

---

Figure 4.37 *Three ^{13}C CP/MAS NMR spectra of sulfathiazole. (a) Spectra for genuine sample of each of the five polymorphs (from Apperley et al. 1999, with permission). (b) Effect of a mixture of polymorphs. The lower two spectra are of pure polymorphic forms, while the upper one is a mixture of the two lower ones (from Threlfall 1999, with permission).*

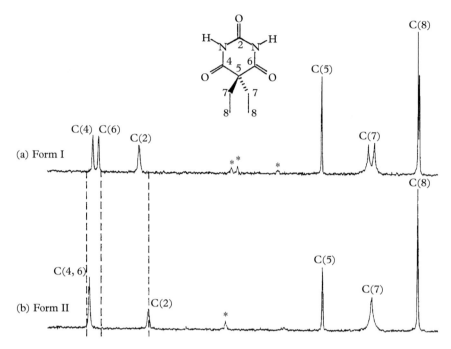

Figure 4.38 *^{13}C CP/MAS NMR of two of the polymorphs of 5,5′-diethylbarbituric acid. The atomic numbering is shown corresponding to the peak with the same number. (a) Form I; (b) Form II. (From Navon et al. 2001, with permission.)*

development of such techniques can considerably aid in providing tools for the detailed comparison of the molecular environment of molecules in different polymorphic forms in the absence of full crystal structure analyses. Since that environment includes intramolecular interactions, such information can also be used to build conformational models which are useful starting points for computational methods used for developing models for crystal structures and crystal structure solution (see Section 5.7 and 5.9). In this particular study Zell et al. assigned the peaks for three crystallographically independent molecules in the asymmetric unit of one of the aspartame crystal modifications.

There is also a growing use of SSNMR in the quantitative analysis of polymorphic mixtures (e.g., Harris 1985; Suryanarayanan and Wiedmann 1990; Bugay 1993; Stockton et al. 1998). An example was given by Bugay (1999) for Forms A and B of a polymorphic developmental drug substance. For both developmental and regulatory purposes it was required to quantitatively monitor the presence of Form B in Form A. The SSNMR spectra of the two forms are given in Figure 4.40. In the region of 10–20 ppm each polymorph exhibits a large, essentially single peak offset from the other polymorph. This region was therefore used for the calibration of known mixtures and assay of the unknown mixtures

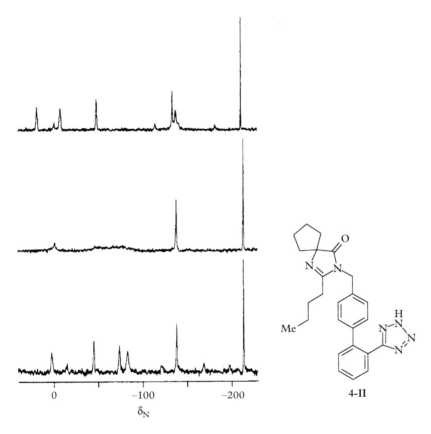

Figure 4.39 *¹⁵N CP/MAS spectra of irbesartan 4-II. Top, Form A at 295 K; middle, Form B at 295 K; bottom, Form B at 253 K (from Bauer et al. 1998, with permission).*

(Figure 4.40). The assay had a 5% w/w detection limit for Form B in Form A, and could be carried out on 100 mg and greater potency tablets, an important advantage of the technique being that bulk samples could be assayed without having to consider issues of sampling protocol.

The application of SSNMR to solid-state problems has fostered the discipline of NMR crystallography, as documented in a monograph with that title (Harris et al. 2009) and a number of recent examples. Crystal structures have been determined for flexible molecules such as lisinopril that did not yield crystals for single crystal structure determination by combining high resolution SSNMR with XRPD methods (Miclaus et al. 2014). Crystal structure prediction (CSP) methods (Section 5.9) and density functional theory calculations were combined with experimental powder NMR crystallography to successfully determine the crystal structure of a drug molecule of molecular weight 422, at the time the largest molecule to which these methods have been applied (Baias et al. 2013). In a third example the spectroscopic distinction of the α and β forms of Cu(II)(phthalocyanine),

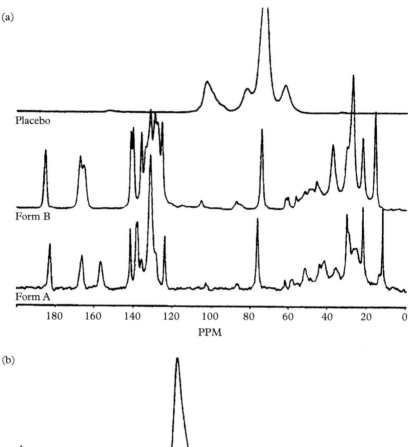

(a)

Placebo

Form B

Form A

| 180 | 160 | 140 | 120 | 100 | 80 | 60 | 40 | 20 | 0 |

PPM

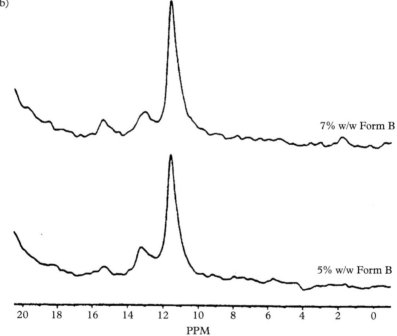

(b)

7% w/w Form B

5% w/w Form B

| 20 | 18 | 16 | 14 | 12 | 10 | 8 | 6 | 4 | 2 | 0 |

PPM

Figure 4.40 *(a)* ^{13}C *CP/MAS NMR spectra of two forms of a developmental pharmaceutical compound discussed in the text. (b) The spectra of two mixtures of Form A spiked with known quantities of Form B. The peak at ca. 15 ppm is indicative of Form B. (From Bugay 1999, with permission.)*

two of the six known polymorphs of this widely used semiconducting blue pigment, was accomplished using very-fast magic angle spinning (VFMAS) at 20 kHz, as confirmed by ab initio chemical shift calculations and marked differences of the ^{13}C and ^{1}H relaxation times (Shaibat et al. 2010).

Finally, electron density calculations were combined with SSNMR to confirm the presence of the rare enol form of barbituric acid in polymorph IV or barbital, whereas the keto form was shown to exist in polymorph II (Badri et al., 2014).

In situ monitoring of a crystallizing solution has been shown to lead to the discovery of new polymorphs, a technique that may be particularly useful in the survey of the solid form landscape discussed in the previous chapter. In one of the initial reports this led to the discovery of a new polymorph of 1,10-dihydroxydecane/urea and two new polymorphs of methydiphenylphosphine oxide (Hughes et al. 2012). In a no less impressive experiment by the same group, in situ solid-state ^{13}C NMR was employed to show that the α polymorph of glycine is obtained from methanol/water solution from the initial transient formation of the β polymorph (Hughes and Harris 2010).

There have also been significant advances in the quantitative analysis of solid forms. *Polymorphic purity* for a fluorine-containing compound has been quantitatively determined in batches of API using ^{19}F SSNMR (Barry et al. 2012). Similarly, as a proof-of-concept experiment levels of amorphous form as low as 1–5% in a formulated product of atorvastatin could be detected with a combination of ^{19}F SSNMR and chemometrics (Brus et al. 2011).

The presence of chlorine in a material, either as a substituent or as the hydrochloride salt (~50% of all pharmaceutical salts) provides another opportunity for utilizing SSNMR in the investigation of polymorphic materials (Hamaed et al. 2008). Hildebrand et al. (2014) have pioneered the use of ^{35}Cl SSNMR spectroscopy in characterizing solid forms. The ^{35}Cl electric field gradient and chemical shift tensor can be extracted from the spectra and used as fingerprint markers to uniquely identify individual phases, particularly those of hydrochloride salts. The method is particularly suitable for sampling the results of high throughput crystallization experiments. In another advance with this nucleus, ^{35}Cl SSNMR can be used in combination with dynamic nuclear polarization (DNP) to characterize APIs in dosage forms with chlorine concentration as low as 0.45% (Hirsch et al. 2016).

The use of sodium nuclei has also been developed for SSNMR studies on polymorphic systems. For instance, triple quantum ^{23}Na MAS spectroscopy has been employed to distinguish among the sodium cations in the three polymorphs of sodium acetate (Xu and Harris 2008).

It was demonstrated that DNP-enhanced SSNMR could be used to characterize the polymorphs and solvates of theophylline, although some of the samples underwent polymorphic transitions during sample preparation. Revised sample preparation protocols, including cryogrinding in an inert atmosphere and change of impregnating radical-containing solution overcame those problems; subsequent two-dimensional correlation experiments enabled structural characterization of the solid forms (Pinon et al. 2015).

The application of sophisticated new pulsing, cross-polarization, and DNP techniques has enabled the combined ^{14}N and ^{15}N SSNMR fingerprinting of polymorphs based on their unique chemical shift and quadrupolar NMR parameters, as well as the identity of impurity phases (Veinberg et al. 2016).

In short, the variety of these techniques and their increasing precision hold great promise for the application of SSNMR to many of the questions and problems associated with polymorphic systems.

4.8 Electron microscopy

Electron microscopy can provide extremely useful information at magnification orders of magnitude greater than optical microscopy. The two principle experimental techniques are *scanning electron microscopy* (SEM) and *transmission electron microscopy* (TEM).

SEM can achieve spatial resolution better than 1 nm and can provide considerable detail on topography, thus revealing, for instance, crystal habit, surface defects, etc. For instance, a comparison of the optical and SEM images of the two forms of paracetamol shown in Figure 4.41 demonstrates their similarity. The differences

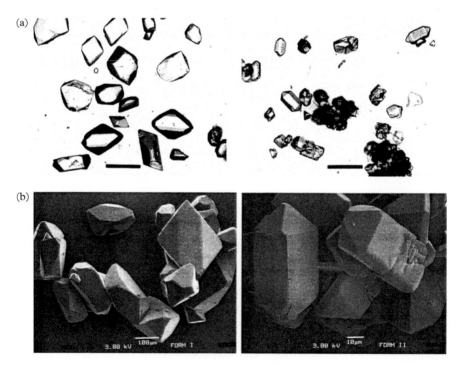

Figure 4.41 *Comparison of optical and SEM images of polymorphs of paracetamol grown from IMS. (a) Optical photomicrographs of Forms I (left, scale bar = 250 μm) and Form II (right, scale bar = 100 μm) (from Nichols 1998, with permission). (b) SEM images of Form I (left) and Form II (right) (from Nichols and Frampton 1998, with permission).*

in habit (monoclinic and orthorhombic; Section 2.4.1) which are clear from both techniques may not be obvious in the optical case for samples of small grain size.

The differences in morphology between the two polymorphs of ranitidine hydro-chloride (Figure 4.42) were used to account for the differences in filtering and dry-ing behavior. Form 1 crystallizes in poorly shaped plate-like particles that tend to agglomerate and adhere thus blocking the filter medium and reducing the effective-ness of suction for filtering and drying. On the other hand, Form 2 tends to form more needlelike crystals which form a much more granular porous layer on the filter medium, considerably increasing the filtering and drying efficiency. These differ-ences in filtering and drying were the principal advantages for the granting of the quite lucrative patent on Form 2 (Crookes 1985) (Section 10.4.3).

TEM has been developed to achieve even atomic resolution for the direct visu-alization of such atoms as light as carbon (Meyer et al. 2008). There are a number of imaging techniques available for enhancing the details of the image. One limita-tion for the molecular crystals of interest here, most of which have relatively high vapor pressures, is the experimental requirement of maintaining a vacuum in the sample chamber. However, the use of an electron beam for irradiating the sample means that TEM experiments can be carried out simultaneously with electron diffraction, a combination that has found considerable use with many non-molecular materials. Until recently, the limited use on molecular crystals has also been due, to a great extent, to the difficulties associated with the preparation of appropriately thin samples required for electron transmission, as well as sample damage caused by the electron beam.

However, in a promising technological advancement with this technique, some of these problems have been recently reduced, which has been demonstrated by the investigation and characterization of the molecular crystals of theophylline, paracetamol, and aspirin, as well as other pharmaceutical salts and co-crystals as demonstrated for theophylline in Figure 4.43 (Eddleston et al. 2010). Since elec-tron microscopy and electron *diffraction* both require the irradiation by an electron beam, such experiments can often be carried out sequentially on the same sample, as shown in Figure 4.43. The electron diffraction can often be used to determine crystal for samples that are much smaller than those required by X-rays, with sam-ple size approaching the picogram regime. One experimental solution to the sub-limation problem is the application of automated electron diffraction tomography (Kolb et al 2011). It has recently been demonstrated, however, that combining computational CSPs with electron diffraction allows structure solution with a limited amount of data providing greater possibilities for structure determination by electron diffraction on these small samples. The technique was developed and demonstrated by Eddleston et al. (2013) on the pharmaceutical compounds paracetamol, *scyllo*-inositol, and theophylline.

The increased magnification can provide details on the initial stages of crystal growth and distinctions in habit among polymorphs.

In addition to the characterization of habit and surface features, the structural symbiosis between two crystalline modifications can be readily studied by SEM.

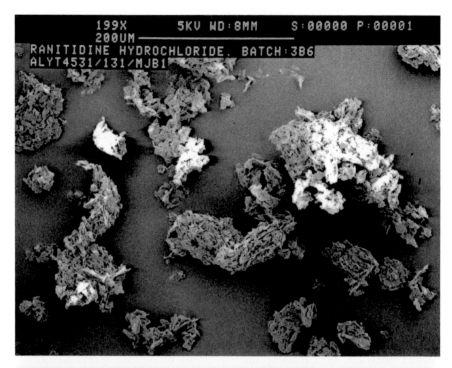

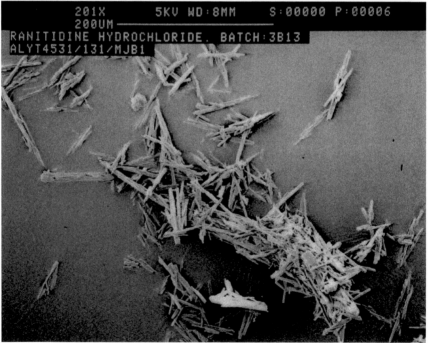

Figure 4.42 *SEM images of the two forms of ranitidine hydrochloride. (a) Form 1; (b) Form 2 (courtesy of GlaxoSmithKline).*

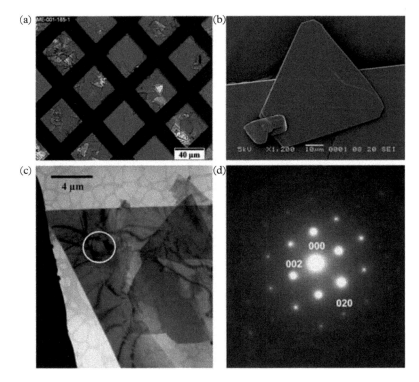

Figure 4.43 *Four views of form II of theophylline. (a) Polarized light microscopy image of theophylline crystals prepared on a copper sample support grid. (b) SEM image of triangular crystals of form II of theophylline crystals, (c) TEM image of overlapping triangular crystals of form II of theophylline. The dark lines running across the crystals are bend contours. (d) Electron diffraction pattern from the circled region in image (c). (From Eddleston, M. D. Bithell, E. G., and Jones, W. (2010). Transmission electron microscopy of pharmaceutical materials.* J. Pharm. Sci., **99**, 4072–83. *With permission of Elsevier.)*

Figure 4.44 shows the growth of the dihydrate modification of carbamazepine on the surface of the anhydrous form. Such information can considerably aid in the understanding of the process of transformations among crystal modifications and the development of robust procedures for the selective preparation of a desired crystal modification.

Aminobenzoic acid has been the subject of some of the revealing studies of the potential of SEM in the study of polymorphism. The *ortho* and *para* isomers are both at least dimorphic. For instance, Garg and Sarkar (2016) demonstrated with the aid of SEM the control of the polymorphism of p-aminobenzoic acid by isothermal anti-solvent crystallization. The two polymorphs of p-aminobenzoic acid are designated α and β, shown in Figure 4.45 together with the commercially available sample (at least from one source). [This perhaps should serve as a caveat for those carrying out any studies that might depend on or be influenced by polymorphism to exercise caution in using commercial samples "as received,

Figure 4.44 *SEM image of the growth of the fibrous needles of the dihydrate of carbamazepine on the larger faces of the monoclinic anhydrous modification (from Rodriguez-Hornedo and Murphy 1999, with permission).*

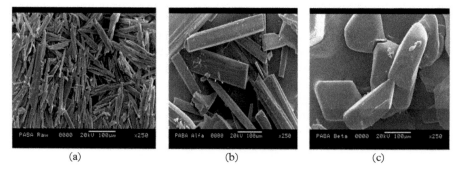

(a)   (b)   (c)

Figure 4.45 *SEM images of* p-aminobenzoic acid. *(a) Commerically obtained, (b) experimentally obtained* α *form, (c) experimentally obtained* β *form. (From Garg, R. K. and Sarkar, D. (2016). Polymorphism control of* p-aminobenzoic acid by isothermal anti-solvent crystalliza-tion. J. Cryst. Growth, **454**, 180–5. With permission of Elsevier.)*

without any further purification."] The α needles undergo a solvent-mediated transformation to the β form (Hao et al. 2012).

For the *ortho* isomer, Simone et al. (2016) have demonstrated the use of SEM in studying the effect of an additive on the induction time and crystal shape of Forms I and II. Such studies can provide detailed information on the structural consequences of changes in crystallization conditions. The influence of hydroxy-propyl methylcellulose (HPMC) on the nucleation of Form I is demonstrated in Figure 4.46, while the influence on growth is demonstrated in Figure 4.47.

On the other hand, the influence of HPMC on Form II grown from 10% iso-propyl alcohol (IPA) in water is demonstrated in Figures 4.48 and 4.49.

These studies clearly demonstrated that the additive inhibits nucleation of Form I but not of Form II since the transformation time from Form II to Form I is increased in the presence of HPMC. The addition of HPMC results in habit changes in both forms and for Form I the habit strongly depended not only on the

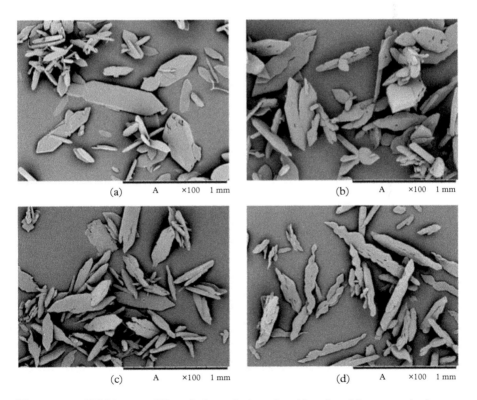

Figure 4.46 *SEM images of Form I of* o-*aminobenzoic acid nucleated from water in the presence of different concentrations of HPMC: (a) 0 ppm, (b) 5 ppm, (c) 10 ppm, and (d) 15 ppm. (From Simone. E., Cenzato, M. V., and Nagy, Z. K. (2016). A study on the effect of the polymeric additive hydroxypropyl methylcellulose (HPMC) on morphology and polymorphism of* ortho-*aminobenzoic acid crystals.* J. Cryst. Growth, **446**, *50–9. With permission of Elsevier.)*

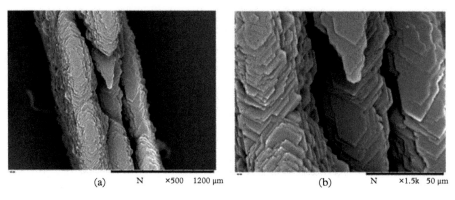

(a)   N   ×500   1200 μm   (b)   N   ×1.5k 50 μm

Figure 4.47 *SEM images of crystals of* o-*aminobenzoic acid Form I grown in water in the presence of 50 ppm of HPMC in solution. (From Simone. E., Cenzato, M. V., and Nagy, Z. K. (2016). A study on the effect of the polymeric additive hydroxypropyl methylcellulose (HPMC) on morphology and polymorphism of ortho-aminobenzoic acid crystals.* J. Cryst. Growth, **446**, *50–9. With permission of Elsevier.)*

amount of HPMC in solution but also on the solvent used. The irregularity of Form I increased with the concentration of HPMC additive in aqueous solution, while those grown from 10% IPA and water exhibited a flat prismatic shape, but increased in width with increasing concentration of HPMC.

Sullivan and Davey (2015) carried out a detailed description of the morphology of p-aminobenzoic acid using a variety of techniques, highlighted by SEM. This study beautifully demonstrates the mutual complementarity of optical and electron scanning microscopy as well as successes and limitations of the current theoretical methods for the prediction of crystal morphology, using the Bravais–Friedel–Donnay–Harker (BFDH) combination in the Mercury package (Macrae et al. 2008) and the attachment energy prediction method (Materials Studio 2013). For reference, the optical micrographs of samples of the α form from various solvents and the β form obtained from water are given in Figure 4.50. One optical micrograph and three SEM images of the α form are given in Figure 4.51. This is followed by a series of remarkably detailed SEM images of the α form demonstrating the differences in morphology obtained from different solvents (Figure 4.52). Since the experimentally determined morphologies did not match those computationally predicted the authors modified the calculations by taking into account some non-structural features (e.g., desolvation) to account for these differences. These results are shown in Figure 4.53. These kinds of studies yield crucial information on the nature of the crystal growth process and significantly increase our ability to control the polymorph obtained as well as the size and shape of the particles that are obtained from a crystallization experiment. As SEM becomes more available there is clearly great potential in the application of these methods to many vexing problems in the preparation and control of polymorphs.

(a)   A   ×100   1 mm

(b)   A   ×100   1 mm

(c)   A   ×100   1 mm

Figure 4.48 *SEM images of o-aminobenzoic acid nucleated in 10% isopropanol and water at three different concentrations of HPMC: (a) 0 ppm, (b) 3 ppm, (c) 15 ppm. (From Simone, E., Cenzato, M. V., and Nagy, Z. K. (2016). A study on the effect of the polymeric additive hydroxy-propyl methylcellulose (HPMC) on morphology and polymorphism of ortho-aminobenzoic acid crystals.* J. Cryst. Growth, **446**, *50–9. With permission of Elsevier.)*

Figure 4.49 *SEM images of Form I crystals of o-aminobenzoic acid obtained after complete transformation of Form II nucleated in 10% isopropanol and water solutions at different concentrations of HPMC: (a) 0 ppm, (b) 3 ppm, (c) 15 ppm. (From Simone, E., Cenzato, M. V., and Nagy, Z. K. (2016). A study on the effect of the polymeric additive hydroxypropyl methylcellulose (HPMC) on morphology and polymorphism of ortho-aminobenzoic acid crystals. J. Cryst. Growth,* **446**, *50–9. By permission of Elsevier.)*

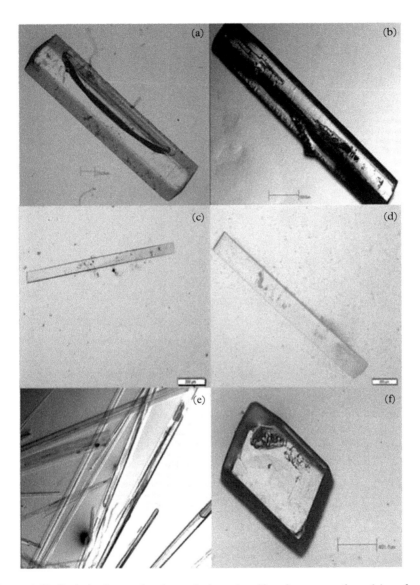

Figure 4.50 *Optical micrographs of p-aminobenzoic acid: α form grown from (a) methanol, (b) isopropanol, (c) acetonitrile, (d) ethyl acetate, (e) water; (f) β form grown from water. (From Sullivan, R. A. and Davey, R. J. (2015). Concerning the crystal morphologies of the α and β polymorphs of p-aminobenzoic acid.* CrystEngComm, *17, 1015–23. By permission of Royal Society of Chemistry.)*

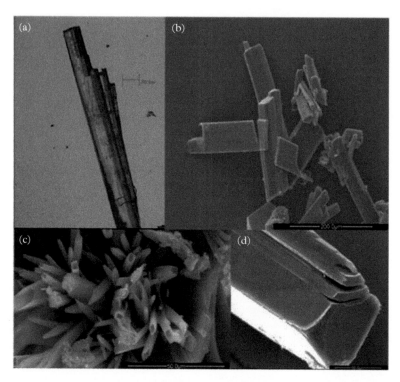

Figure 4.51 *Micrographs of α form grown by slow evaporation: (a) ethanol (optical micrograph), (b) nitromethane (SEM), (c) ethyl acetate (SEM), (d) acetonitrile. (From Sullivan, R. A. and Davey, R. J. (2015). Concerning the crystal morphologies of the α and β polymorphs of p-aminobenzoic acid.* CrystEngComm, **17**, 1015–23. *By permission of Royal Society of Chemistry.)*

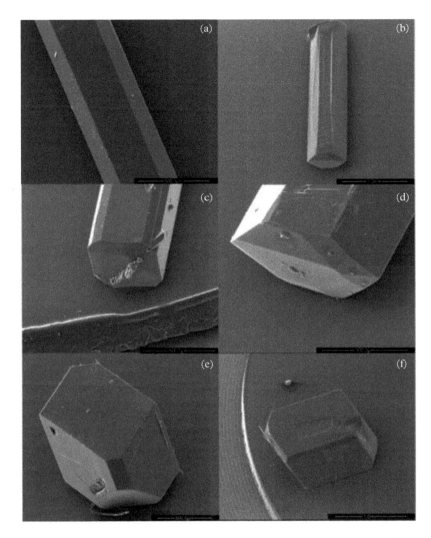

Figure 4.52 *SEM images of* p-*aminobenzoic acid grown by slow evaporation. α form: (a) methanol, (b) ethanol, (c) ethyl acetate, (d) ethanol; β form: (e, f) water. (From Sullivan, R. A. and Davey, R. J. (2015). Concerning the crystal morphologies of the α and β polymorphs of* p-*aminobenzoic acid.* CrystEngComm, *17, 1015–23. By permission of Royal Society of Chemistry.)*

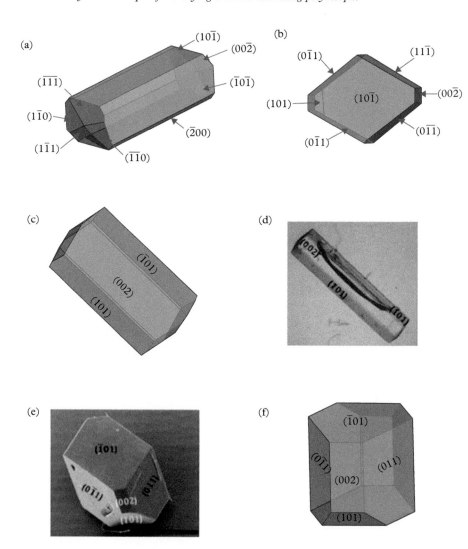

Figure 4.53 *Calculated and experimental morphologies for p-aminobenzoic acid. (a, b) Initial calculated growth morphologies. (c, d) Modified and experimental morphologies of the α form. (e, f) Experimental and modified calculated images of the β form. (From Sullivan, R. A. and Davey, R. J. (2015). Concerning the crystal morphologies of the α and β polymorphs of p-aminobenzoic acid. CrystEngComm, 17, 1015–23. By permission of Royal Society of Chemistry.)*

4.9   Atomic force microscopy (AFM) and scanning tunneling microscopy (STM)

As demonstrated in the previous section, surfaces play a crucial role in determining the polymorphic outcome of a crystallization. Hence, the study of the structure of surfaces and their interaction with a crystallizing material can provide detailed information on the nature of the process. That information may be used to design experiments and processes for the study of nucleation and the selective control of the growth of a particular crystal modification (Palmore et al. 1998). These AFM and STM techniques (and rapidly developing spinoff modifications of them) provide that information (Frommer, 1992; Ward, 1997). For instance, tapping mode atomic force microscopy (TMAFM) was employed to characterize the surface morphologies of the polymorphs of an organic radical, *p*-nitrophenyl nitronyl nitroxide (Fraxedas et al. 1999a). The use of templates to induce new solid forms described in Chapters 3 and 7 has also been applied to two-dimensional structures in a controlled fashion (Cicoria et al. 2008). The key tool in following the growth of these nanotemplates and the subsequent formation of the ordered organic structures is STM. For example, two different two-dimensional hydrogen-bonded structures of trimesic acid, described as "chicken-wire" and "flower," have been reported using this technique (Lackinger et al. 2005; Kampschulte et al. 2006). A recent review (Jones et al. 2016) describes in detail the production and characterization of polymorphic substrate-induced phases (SIPs) of 16 compounds, most of them semiconducting organic compounds. Currently, these new phases, which are generally a number of layers thick as opposed to surface induced monolayers, are studied primarily by incidence X-ray diffraction, but the techniques in the title of this section will likely also be employed.

4.10   Density measurements

The "density rule" proposed by Burger and Ramberger (1979a) (see Section 2.2.10) is useful for determining the relative stability of polymorphic forms. Hence the measurement of density or differences in density can provide the experimental information for making such determinations. A number of the techniques for carrying out these determinations are rather time-consuming and fraught with possibilities for error, but they should be included in the potential armory of techniques for studying polymorphs.

The density may be measured by flotation, pycnometry, or by volumenometry (Bauer and Lewin 1972; Andreev and Hartmanová 1989; Richards and Lindley

1999). Differences in density between polymorphs normally do not amount to more than 1–2% of the density of one of the forms, so that both high precision and high accuracy are required for the measurements to be meaningful. In addition, virtually all the errors that can be encountered in the experimental procedures of the density tend to lower the experimentally determined value. Helium pycnometry, based on the volume occupied in solid voids (Keng 1968), provides potentially the most precise measurement of density; commercial instruments utilizing this technique are available.

A reliable computational alternative is the calculation of the density from the experimental crystallographic unit cell constants, routinely obtained from single crystal structure determinations (e.g., Glusker et al. 1994), but also available from the indexing of powder patterns (e.g., Jenkins and Snyder 1996). An added advantage of these "indirect" methods is that the unit cell constants are usually accompanied by estimated standard deviations, so that the statistical significance of differences and similarities in the densities may be properly evaluated. As a caveat, Herbstein (2000) has shown that the standard deviations of the unit cell constants are often underestimated, suggesting a higher precision than the experiment warrants. With the increasing automation of both single crystal and powder diffractometry, and the accompanying sophistication of the appropriate software, comparative studies of the temperature dependence of the density of polymorphs can provide useful information on the phase relationships.

Dilatometry is one of the older classic methods for the determination of transition points between solids (Drucker 1925). The dilatometer usually consists of a large bulb connected to a capillary and filled with an inert liquid. Volume changes as a function of temperature or resulting from a solid–solid transition may be determined by changes in the volume of the inert liquid. Haleblian (1975) cites the use of dilatometry in the characterization of polymorphs and more recent advances in the miniaturization of chemical instrumentation (e.g., Jakeway et al. 2000; Krishnan et al. 2001), and the high precision associated with that miniaturization can lead to a renaissance of the use of some of these classic techniques and their applications to the study of polymorphism. A number of commercial instruments for convenient dilatometric measurement are currently available.

4.11 New technologies and "hyphenated techniques"

The potential application of chemical and physical techniques for the study of polymorphs is virtually limitless. The combination of techniques into a single apparatus in order to extract the maximum information from a single experiment and a single sample is particularly applicable to polymorphic systems. The joining of visual methods with spectroscopic and calorimetric measurements has already been mentioned. These can be expanded further, as witnessed, for instance, by a triple combination of microscopy, calorimetry, and microphotometry (Cammenga and Hemminger 1990; Lin 1992; Rustichelli et al. 2000) on SMATCH spectroscopy (simultaneous mass and temperature-change FTIR) (Timken et al. 1990). The combination of TGA with FTIR allows simultaneous quantitative analysis of weight changes during thermal processes with the IR identification of the decomposition products (e.g., solvent) resulting from those processes (Materazzi 1997), and the amalgamation of thermal methods with a variety of other techniques (Cheng et al. 2000) is very promising in research on polymorphic materials.

Optical microscopy, including both hot/cold stage techniques has been combined very effectively with both Raman (Smith et al. 2015) and FTIR instrumentation. This allows the examination and characterization of small regions of the solid sample during or following thermal events. As noted earlier a major advantage, especially in the case of Raman spectroscopy, is the lack of any need for sample preparation. Often an area chosen optically may be studied spectroscopically without moving the sample. Small areas may be measured in a serial fashion which allows for spectral mapping of the surface. Confocal Raman microscopy can probe areas of 250 nm in diameter and 1.7 μm in depth (Everall 2000; Smith et al. 2015).

One important technique for which progress has been somewhat slower is the combination of XRPD with other techniques, although some promising beginnings have been reported (Klein et al. 1998). Earlier reports on the potentially very useful combination of XRPD and DSC (Fawcett et al. 1989) have recently been proven although not yet commercialized (Berthold et al. 2011), in spite of some earlier successful results in characterizing polymorphic systems (Fawcett, et al. 1986; Fawcett 1987; Landes et al. 1990; Loisel et al. 1998). It is hoped that through the increasing sensitivity and speed of detectors, combined with more powerful and convenient laboratory X-ray sources, powder diffraction will be conveniently combined with optical, thermal, and spectroscopic methods in single unit instruments capable of more comprehensive studies of polymorphic systems, especially the metastable phases of those systems. The capability of measuring such (often fleeting) phases in situ is clearly a distinct advantage for the complete characterization of a polymorphic system. The convenience of being able to simultaneously measure and follow a number of physical properties must be tempered with an often unavoidable compromise in the final single instrument, namely the necessity to compromise on some or all of the ideal specifications and limitations of a stand-alone system for a single technique.

4.12 Are two samples polymorphs of the same compound?

Historically, most polymorphs have been discovered by serendipity, rather than as the result of a systematic search (Tutton 1922; Cholerton et al. 1984;

Chemburkar et al. 2000; Lommerse et al 2000). On the one hand this testifies to the general unpredictability of polymorphism, but on the other hand it also testifies to the intellectual curiosity and powers of observation and analysis of the discoverers. What generally characterizes polymorphs of a compound is that some or all of their properties will differ. As McCrone (1957) has noted, "[t]he evidence for a polymorphic transformation on heating is a discontinuous change in physical properties, usually most apparent as changes in polarization colors and extinction positions." This reflects Ehrenfest's thermodynamic classification (Jaeger 1998) that a phase change between two polymorphs is characterized by a discontinuity in the heat capacity versus temperature curve. Thus, in principle, an investigator armed with the knowledge that a particular compound is chemically pure, can employ almost any sufficiently sensitive physical measurement to determine if two crystalline samples of that compound may be polymorphs or not. As noted earlier, however, the full characterization of the polymorphic behavior should involve as many techniques as possible (e.g., Chiang et al. 1995). Previous sections outline many of these physical measurements and provide examples of the distinctions between polymorphs. In addition to those, however, McCrone's (1965) answer to the question posed here is worthy of repetition, since it demonstrates the principles and the relatively simple and straightforward techniques that are involved.

First, it is necessary to eliminate the possibility of tautomerism or dynamic isomerism by determining that the two materials give identical melts. X-ray diffraction can determine that crystal strain (which can be mistaken for polymorphism) is not a factor. Then, McCrone (1957, 1965) suggests the following additional tests for polymorphism, using the polarizing microscope equipped with a hot stage. The two samples are polymorphs if:

- they have different crystal properties (axial ratios, X-ray diffraction, indices of refraction, melting point, etc.) and they can be converted into each other through a solution or solid phase transformation;
- they differ in all crystal properties (as determined say, by the methods outlined throughout this chapter), but both melt to a liquid with the same refractive index and same temperature coefficient of the refractive index. In addition, if the two solids are mixed and melted to the melting point of the lower melting material, and the temperature is maintained at the melting point, in the case of polymorphism the two will both melt (if the melting points are very similar) or will crystallize as the higher melting form (whose melting point would be above that temperature). The persistence of a mixture of melt and solid at the melting temperature of the lower melting form indicates the presence of two different compounds rather than polymorphs;

- the crystal properties are different, yet the mixed fusion between them is identical;
- seeding of a fused sample of one of them with crystals of each of the two, and observation under cross polarizers on a polarizing microscope, indicates that one of the phases is growing through the other phase. If there are fronts between different areas of the growth of a solid that do not undergo change with time, then the two samples are likely the same material;
- mixing crystals of the two on a dry microscope slide (beneath a cover slip) and allowing a saturated solution of one of them to run under the cover slip leads to a phase transformation;
- heating crystals of both in proximity on a hot stage leads to a solid–solid transformation of one, followed by melting of both. Also, if on continued heating, one form melts and then can be induced to solidify by seeding with the other form, and then melted upon further heating, then the two are polymorphs;
- during heating of a mixture there is a vapor phase transition.

4.13   Concluding remarks

Since polymorphism is a structural manifestation of kinetic and thermodynamic factors in crystallization effects and leads to a variation of physical and chemical properties its characterization and investigation require a multidisciplinary approach. Different analytical techniques and methods provide a variety of structural and thermodynamic information, often complementing each other and adding to overall understanding of the relationships among the various phases. For the investigator the more information available the greater the control over the polymorph obtained and over the resulting properties; the message here is that the broadest range of the analytical techniques discussed in this chapter, many of which are summarized in Table 4.7 (Chieng et al. 2011), should be employed in the study of polymorphic systems.

Table 4.7 *Summary of methods used to characterize solid forms. (From Chieng, N., Rades, T., and Aaltonen, J. (2011). An overview of recent studies on the analysis of pharmaceutical polymorphs. J. Pharm. Biomed. Anal., **55**, 618–44.With permission of Elsevier.) The references numbers in the left-hand column are from the original publication and should be ignored here*

Modulated temperature differential scanning calorimetry (MTDSC) [143–145]	See DSC - Separation into reversing and non-reversing heat flow (i.e., more information available)	- Improved clarity of small (i.e., T_g) and overlapping thermal events	- More experimental variables (i.e., amplitude and period setting) - Relatively long data acquisition time - Interpretation of the 'separated' thermograms not always straightforward - Interpretation of the 'separated' thermograms not always straightforward
Thermogravimetric analysis (TGA)/ dynamic vapour sorption (DVS) [146.147]	- Transitions involving either a gain or a loss of mass - Decomposition temperature - Use in conjunction with Karl–Fischer titration	- Amount of solvate/hydrate in a sample - Experimental set up is straightforward - Small sample size (~3–10 mg)	- Interference with water-containing excipients - Sample is destroyed during analysis - Unsuitable for materials that degrade at low temperatures
Isothermal microcalorimetry (IMC) [148–152]	- Heat change in a reaction e.g., enthalpy relaxation of amorphous material (direct measurement), heat of crystallization	- High sensitivity - Qualitative and quantitative analyses - Stability study directly under the storage condition - Non-destructive method	- Low specificity (i.e., interpretation of data can be difficult) - Large amount (50–500 mg) of sample required

Solution calorimetry (SC) [148,149,153]	- Heat change in a reaction, e.g., heat of solution (main), heat of wetting, heat capacity of liquids, heat capacity of solids (mixture method)	- Qualitative and quantitative analyses	- Low specificity (i.e., interpretation of data can be difficult) - Large amount (15–200 mg) of sample required - Sample cannot be recovered - Long measurement time
Microscopy			
Polarized light microscopy (PLM) [13,115,154–156]	- Crystallinity (birefringence) - Morphology, colour and crystal habit	- Small sample size (microgram amount) - Easy to use - Very little sample preparation	- Quantitative information not available
With hot/cryo/freeze drying stage	- Complementary information on phase transition/physical changes in frozen state	- Temperature variability	- Careful sample preparation is required to avoid contamination with the thermal contact liquid
Scanning electron microscopy (SEM) [13,156]	- Topographical properties	- Higher resolution than light microscopy - Small sample size (microgram amount)	- Requires sample preparation and stage condition setup (vacuum setting)
Bulk level/other			
Karl–Fischer titration [157,158]	- Water content (adsorbed or hydrate) - Use in conjunction with TGA/DVS	- High sensitivity - Rapid analysis	- Sample needs to dissolve in the medium - Sample size of > 50 mg is preferred

continued

Table 4.7 Continued

Brunauer, Emmett and Teller (BET) method [159,160]	- Surface area of the samples (the BET equation is an extension of the Langmuir equation, for multilayer adsorption)	- Analysis is simple and straightforward - Non-destructive method	- Degassing step is required to remove adsorbed water or gas molecules - Small sample size (50–100 mg); sample amount may need to be adjusted depending on the surface area of sample
Density (gas pycnometer) [13]	- True density of the sample by dividing the known mass with the measured volume	- Analysis is simple and straightforward - Non-destructive method	- Degassing step (purging with helium) is required to remove adsorbed water or gas molecule - Sample size of >50 mg is preferred

5

Computational aspects of polymorphism

Can crystal structures be predicted? No!
Gavezzotti (1994a)

5.1 Introduction

To a great extent, the attraction and appeal of X-ray crystallography as an analytical tool for molecular structure determination is due to the precision with which that determination can be made. Bond lengths and bond angles with estimated standard deviations of a few thousandths of an angstrom or tenths of a degree, respectively, are difficult to obtain as routinely by other methods (Dunitz 1979; Glusker et al. 1994). Moreover, the accumulated experience and data of almost a century of such determinations have provided evidence for the limited variability of such parameters, so that characteristic values may be determined and used as benchmarks for additional studies (Allen et al. 1987; Orpen et al. 1989). But molecular structure is defined as well by torsion angles, and these tend to exhibit considerably greater ranges of values, for both electronic and stereochemical reasons. For instance, in a particular grouping of the four atoms that define a torsion angle, the variability can be manifested in different angles as that group appears in different molecular settings. In the context of the subject of polymorphism we are particularly concerned with the cases in which a molecular conformation as defined essentially by the torsion angles can and does vary from one polymorph to another. Such cases are of intrinsic interest for determining the characteristic value or ranges of values of a torsion angle, but more important, they provide particularly useful cases for investigating the interplay of intra- and intermolecular energetics.[1] Such studies involve the application of computational methods, which is the overriding theme of this chapter.

---

[1] Kitaigorodskii (1970) suggested four strategies for investigating the general problem of the influence of crystal forces on molecular conformation: (1) comparison of compounds in the gaseous and crystalline states; (2) comparison of the geometry of crystallographically independent molecules in the same crystal; (3) analysis of the structure of a molecule whose symmetry in a crystal is lower than that of the free molecule; (4) comparison of molecules in different polymorphic modifications. The

Polymorphism in Molecular Crystals. Second Edition. Joel Bernstein. © Joel Bernstein 2020. Published in 2020 by Oxford University Press. DOI: 10.1093/oso/ 9780199655441.001.0001

5.2   Molecular shape and energetics

As noted previously, molecular shape is defined by three different types of molecular parameters: bond lengths, bond angles, and torsion angles. Variations in the geometry of a molecule are then simply defined as changes in these parameters: bond stretching or compression, bond bending or deformation, and bond twisting or torsion, and these distortions are characterized by certain energy domains. Typical force constants for bond stretching (in 10^5 dyne cm^{-1}) range from 4.5 for the single bond in ethane to 15.7 for the triple bond in acetylene (Brand and Speakman 1960). Bond angle bending is less expensive. Mislow (1966) showed that for many carbon bond angles the following empirical relationship for the potential energy holds:

$$V_\theta \sim 0.01(\Delta\theta)^2 \, \text{kcal}^{-1} \, \text{deg}^{-2} \qquad (5.1)$$

where θ is the bond angle. Thus, an angular distortion of 10° involves approximately the same amount of energy as the distortion of a single bond by about 0.05 Å.

Torsional changes involve the rotation about the bond axis; the barrier to rotation about the single bond in ethane is about 2.8 kcal mol^{-1}, which is approximately the difference between the *trans* and the *gauche* conformations. I previously gave estimates of the energy "cost" of these distortions, based on the principles of molecular mechanics (Bernstein 1987). These were being compared with values derived from *ab initio* calculations at the MP2/6-31G* level by Hargittai and Levy (1999), as summarized in Table 5.1. Thus, as a rule of thumb, bond stretching is roughly one to two orders of magnitude more expensive energetically than rotations about single bonds, with bond angle deformations falling in the intermediate range.

Table 5.1 *Comparison of the estimates of energy "costs" resulting from distortions of molecular geometric parameters.[a] (From Hargittai and Levy 1999, with permission.)*

	Estimates from molecular mechanics	Calculated using *ab initio* methods
Bond length, C—C	0.1 Å, 14	0.02 Å, 0.6
Bond angle, C—C—C	10°, 10.3	2°, 0.4
Torsion angle, C—C—C—C	10°, 0.9	5°, 0.2

[a] For each entry the first number is the distortion from the nominal equilibrium value, and the second number is the (average) estimated energy involved in that distortion, in kJ mol^{-1}.

advantages and disadvantages of these various approaches have been considered elsewhere (Bernstein (1984, 1992).

5.3   Intermolecular interactions and energetics

On the intermolecular level, in molecular crystals one is dealing potentially with a variety of interactions, most of them considerably weaker than those involved in chemical bonding. A crystal structure corresponds to a free energy minimum that is not necessarily the global minimum. The existence of a number of energetically closely spaced local minima is the thermodynamic rationale for the possible appearance of polymorphism, assuming that these minima are kinetically accessible. While intuitively it is convenient to associate a local minimum with a situation in which attractive interactions are maximized, Dunitz and Gavezzotti (1999) have pointed out that:

> ...molecules in the bulk of a crystal are held together by mutual attraction, but are clamped in their places—in contrast to molecules in a liquid—mainly by resistance to compression, in other words, by repulsions that oppose the disentanglement of interlocking molecules and thus hinder any displacement from their equilibrium positions and orientations.

At any rate, a minimum does represent a situation in which attractions and repulsions are balanced (Brehmer et al. 2000). The nomenclature of these intermolecular interactions is quite variegated and the terms are not always clearly defined or distinguished from one another (Gavezzotti 2007). Some in common usage include van der Waals interactions, London forces, dipole–dipole interactions (and higher terms), dispersion forces, steric repulsion, hydrogen bonds, charge-transfer interactions (also called donor–acceptor interactions), electrostatic interactions, exchange repulsion forces, etc.

There has been some convergence of thought about the use of these terms, at least among those who deal with "crystal forces," especially from the computational point of view. Generally, the intermolecular interactions fall into three classes: (a) non-bonded, non-electrostatic (van der Waals, London, etc.); (b) electrostatic (coulombic); (c) hydrogen bonding. The lines of distinction between these general classes are not always particularly sharp so, for instance, hydrogen bonding has been treated by a combination of the first two general types of interactions (Hagler and Lifson 1974; Hagler et al. 1974; Jeffrey and Saenger, 1991). Another important distinction is that the non-bonded and electrostatic interactions are generally treated as isotropic, although it has long been recognized (Starr and Williams 1977a, 1977b; Pertsin and Kitaigorodskii 1987) that a more realistic physical representation requires the inclusion of anisotropicity into that treatment (Price 2000; Stone 2013; Reilly and Tkatchenko 2015). Hydrogen bonds, by their very nature, are directional and anisotropic.

The non-bonded (van der Waals or London) forces are generally weak interactions between uncharged atoms or molecules. The separation of a molecule from its crystal environment requires overcoming all the attractive forces acting on it. The energy involved in this process may be viewed as the sublimation enthalpy of the crystal. Thus the magnitude of the sum of forces acting on the molecule or the energies involved in the interactions of individual atoms of the molecule with atoms of surrounding molecules may be estimated from the sublimation energy of those molecular crystals in which other interactions are essentially absent. For most molecular crystals the experimental sublimation enthalpy is roughly in the range 10–25 kcal mol^{-1} (Acree and Chickos 2016, 2017). For a molecule with an intermolecular "coordination number" of 8–12 (Kitaigorodskii 1961; Braga et al. 1990; Gavezzotti 2002), the interaction energy thus is about 1–2 kcal mol^{-1} per molecular neighbor. A sub-class of these interactions, which are generally more anisotropic in nature, is the charge-transfer (i.e., π–π or σ–π) type, for which the energies involved rarely exceed 5 kcal mol^{-1} (Foster 1969; Steinmann et al. 2012). The electrostatic interactions can vary over a much wider range, depending on the distance and on the degree of polarization of the molecule or parts thereof, which computationally is manifested in the assignment of partial atomic charges to the various atoms. Intuitively, electrostatic interactions for, say, hydrocarbons might seem to be negligible; however, for many organic crystals they have been shown to comprise a significant part of the total lattice energy (Williams 1974; Cox and Williams 1981; Popelier et al. 1989). Hydrogen bond strengths are generally estimated to be in the range 1–10 kcal mol^{-1} (Jeffrey and Saenger 1991; Desiraju and Steiner 1999).

These intermolecular interactions all fall on the low end of the scale of energies required to bring about distortions of molecular geometry. This already suggests that if the crystal environment has any influence on the molecular geometric parameters, then those parameters most likely to be affected will be the torsion angles around the single bonds, rather than distortions in bond angles or bond lengths which require substantially larger energies to bring about significant changes. Hence, I will concentrate here on the changes in torsional parameters which may be due to the influence of the crystal environment. In spite of these arguments, the possibility of perturbations in bond lengths and bond angles brought about by intermolecular interactions should not be totally ruled out. Such situations are well documented and have been discussed in considerable detail (Colapietro et al. 1984a, 1984b, 1984c; Domenicano and Hargittai 1993; Wolff 1996; Wagner and Englert 1997; Cruz-Cabeza et al. 2012).

As Dunitz (1979) has pointed out, a crystal structure yields information about the "preferred conformations" (*sic* plural) of a molecule and that any arrangement of atoms or conformation "cannot be very far" energetically from the equilibrium structure of the molecule. Thus a number of conformations of a molecule may be energetically equivalent, or nearly so (e.g., Dobler 1984). An important consequence of this fact is that different conformations may appear in different crystal structures or, for that matter, in the same crystal structure when the number of molecules in the asymmetric unit exceeds one. Certain crystal packing motifs are more favorable than others (Kitaigorodskii 1961; Desiraju 1989; Wolff 1996; Steed and Steed 2015; Brock 2016), which may lead to a predominance of one conformation in a crystal structure, while in solution a number of different conformations may be present, including, most likely, the one(s) in the crystal structure(s). It is important to emphasize here that the energetic situation representing the combination of the crystal and molecular structures is a local minimum which is not necessarily unique, and at energies not far from the global minimum there may exist a number of possible molecular geometries of very nearly the same energy for both the molecular conformation and the crystal structure. It is this proximity of molecular conformational energies that makes possible the existence of different molecular conformations in a single crystal structure when there is more than one molecule in the asymmetric unit (i.e., $Z' > 1$), or of the same molecule in different crystal structures. The energetic justification for the existence of different conformations in different polymorphic structures comes from the fact that the differences in lattice energy among different polymorphic forms can be expected to be in the range 1–2 kcal mol^{-1} (Kitaigorodskii 1970; Kuhnert-Brandstätter 1971), although computed values for known structures exceeding that range have been reported (e.g., Buttar et al. 1998; Stockton et al. 1998; Starbuck et al. 1999; Cruz-Cabeza and Bernstein 2014). Gavezzotti (1991) has estimated that the differences in *total* energy (including molecular and lattice terms) will be of the order of 1 kcal mol^{-1}. Allen et al. (1996) have shown through a statistical analysis of the CSD that torsion angles associated with an intramolecular strain energy exceeding ca. 1 kcal mol^{-1} are relatively rare occurrences in crystal structures, including, of course, polymorphic structures, an observation recently confirmed by Cruz-Cabeza and Bernstein (2014). From the estimates of the magnitudes of intermolecular interactions, this is clearly in the range of energy required to bring about changes in molecular torsional parameters about single bonds, but it is generally not sufficient to significantly perturb bond lengths and bond angles. Therefore, for those molecules that do possess torsional degrees of freedom, various polymorphs may exhibit significantly different molecular conformations. This, then is the energetic rationale for the phenomenon of *conformational polymorphism*, a term apparently initially coined by Corradini (1973), and adopted more widely by Panagiotopoulos et al. (1974) and others (Bernstein and Hagler 1978; Cruz-Cabeza and Bernstein 2014). Corradini's definitions of a number of situations for the arrangement in a crystal of molecules of the same chemical composition but different conformations are illustrated in Figure 5.1.

Conformational polymorphism is the existence of different conformers of the same molecule in different polymorphic modifications. Additionally, Corradini originally specified that the conformers are nearly isoenergetic, and also included as an example, racemic and optically active crystals for the case of chiral molecules. Since both the general definition of polymorphism and chemists' general understanding of conformation are fraught with nuances and special complicating circumstances, I prefer the recently quantified definition of conformational polymorphism here (Cruz-Cabeza and Bernstein 2014), as discussed in detail in Section 5.10.5, without Corradini's additional qualifications.

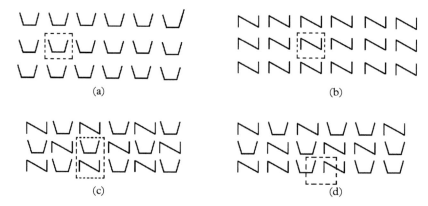

Figure 5.1 *Schematic illustration of three possibilities for arrangement of molecules of the same chemical composition but different conformations in the crystal. For each case two conformations represent symbolically cisoid and transoid dispositions around a single bond. The rectangle defined by a broken line represents one possible choice of the unit cell. (a, b) Conformational polymorphs; (c) conformational isomorphism; (d) conformational synmorphism (from Bernstein 1987, with permission, as adopted from Corradini 1973, with permission).*

Corradini also defined *conformational isomorphism* as the existence of different conformers of a molecule in the same crystal structure, a situation which can arise when there is more than one molecule in the asymmetric unit ($Z' > 1$). In such a case, each crystallographic site in the unit cell is always occupied by the same conformer and the ratio of conformers is defined by whole numbers as in Figure 5.1(c).

Conformational synmorphism describes the situation in which different conformers of a molecule are distributed randomly throughout the crystal lattice. Such a (rarely documented) situation usually exists when two or more conformers have similar overall molecular shapes. Thus, at any particular molecular site a number of conformations may be adopted, the relative population being determined by the relative intermolecular and intramolecular energies involved, as in Figure 5.1(d).

5.4   Presenting and comparing polymorphs

For the overall understanding of molecular and crystal structures and the relationship between them, the qualitative visualization of similarities and differences often plays an important role and can lead to considerable insight and understanding, for instance regarding the nature of particular interactions, or the geometric nature of the "reaction coordinate" for the transition between phases (Bürgi and Dunitz 1994). To facilitate the visual comparisons among all the polymorphs the crystal structures to be compared should be plotted on the same *molecular* reference plane in the *same molecular orientation.* Such a plane might be a phenyl ring or other suitably rigid structural unit and should be chosen so that significantly different torsion angles emanate directly from that unit or within one bond from it. This will serve to highlight the similarities and differences in molecular conformation and facilitate visual comparison. The crystal structure with which these molecular conformations are associated can then be readily compared by generating the molecules (usually 8–14) surrounding the reference molecule. It may be difficult to produce an understandable packing diagram that includes molecules directly above or below the reference molecule. In that case, an additional view rotated by 90° about the original in-plane horizontal or vertical axis may be required. Also, the initial view may be improved (for instance by elimination of some overlap) by *slight* (e.g., up to 5°) rotations from the initial reference orientation. Many of the currently available software packages for plotting molecular and crystal structures have the facilities for generating such figures. For instance, the Mercury plotting package of the CSD software is very conveniently designed for this purpose, and has the added advantage of being readily accessible from the CSD search software routines.

5.5 Some early examples of conformational polymorphism

The number of examples of conformational polymorphism has increased significantly since the first formal recognition of the phenomenon (Corradini 1973; Panagiotopoulos et al. 1974; Bernstein and Hagler 1978) and a more recent review (Bernstein 1987; Cruz-Cabeza and Bernstein 2014). Primarily for historical reasons in this section, I will give a brief overview of the situation, in light of the diversity of the systems studied, to highlight some of the aspects of structures exhibiting conformational polymorphism that generated considerable subsequent interest and activity. Many of the earlier examples were chosen on the basis of qualitative recognition of different conformations rather than quantitative criteria (see Section 5.10) but are still classic ones. The increasing ease and proliferation of crystal structure analysis, along with the growing awareness and importance of polymorphism has prompted the study of many polymorphic systems, providing a fascinating variety of examples of conformational polymorphism. As I will try to demonstrate here with a necessarily limited number of examples, any class of compounds with conformational degrees of freedom may exhibit conformational polymorphism, and those seeking examples for particular compounds, functional groups, structural motifs, etc. should be able to readily identify them by making use of the extensive data and software available on the CSD, and criteria given in Section 5.10.

Often, other physical methods (e.g., Chapter 4) may be used to predict the existence of conformational polymorphs, in the absence of full crystal structure analyses, for example, *n*-propyl acetate (Ogawa and Tasumi 1979), 1-bromopentane

(Ogawa and Tasumi 1978), 2-(4-morpholinothio)benzothiazole (Guzman and Largo-Cabrerizo 1978), and dibenzo-24-crown-8 (Stott et al. 1979; Weber et al. 1998). The early prediction of conformational polymorphism based on the IR spectra (Tomita et al. 1965) was realized in the trimorphic iminodiacetic acid **5-I** system, the details of which were determined from the full crystal structures (Boman et al. 1974; Bernstein 1979). Of the four independent molecules in the three structures, two are nearly identical; however, there are torsional differences of up to 30° about C–N bonds. In two cases the hydroxyl hydrogen of one carboxyl group is *trans* to the nitrogen, while in two cases it is eclipsed.

$$HO_2C \overset{\overset{\displaystyle H_2}{\underset{}{C}}}{} \overset{\overset{\displaystyle }{\underset{\displaystyle H}{N}}}{} \overset{\overset{\displaystyle H_2}{\underset{}{C}}}{} CO_2H$$

5-I

Because of its sensitivity to differences in molecular conformation and molecular environment, solid-state NMR can be a particularly useful technique to recognize conformational polymorphs. Often, detail can be extracted to characterize structurally the polymorphs which may be useful in the absence of crystal structure analyses (Smith et al. 1998; McGeorge et al. 1996).

Information on space group and crystallographic site symmetry may also be sufficient to provide evidence for conformational polymorphism. For instance, in one of the forms of 2,2′,4,4′,6,6′-hexanitroazobenzene **5-II** (with space group $P2_1/a$ and $Z = 2$) the molecule must lie on a crystallographic inversion center, requiring the two phenyl rings to be in parallel planes, while in the second form the molecule lies on a crystallographic general position ($P2_1/a$ and $Z = 4$), making no restrictions on the molecular conformation in the crystal. These implications of molecular and crystallographic symmetry are borne out in the conformations observed in the crystal structures (Graeber and Morosin 1974), as seen in Figure 5.2. This is an example in which the identity of space group for the two forms clearly is not manifested in any similarity in molecular or crystal structure, which should be considered the general case, as noted earlier.

5-II

In contrast to the cumulative effect of a number of small differences along a chain, a large conformational difference can be generated by a single rotation, for instance in a dithiahexyl anthracene derivative (Reed et al. 2000) and in the triclinic and monoclinic modifications of vitamin A acid **5-III**, a member of

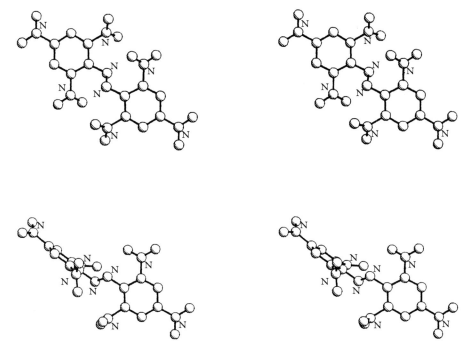

Figure 5.2 *Stereoviews of the molecular conformation observed in the two forms of 5-II. Upper, P2₁/a and Z = 2 structure; lower, P2₁/a and Z = 4 structure. In both cases the molecule is plotted on the plane of the phenyl ring on the right-hand side of the molecule (from Bernstein 1987, with permission).*

the family of visual pigments. The cyclohexenyl ring has essentially the same conformation in both forms (Stam and MacGillavry 1963; Stam 1972). However, as seen in Figure 5.3 the triclinic form exhibits an s-*cis* conformation about the exocyclic bond and clearly a non-planar molecular conformation. The monoclinic form is s-*trans* about the same bond and the stereoview reveals a much more planar structure.

5-III

The very rich structural chemistry of organometallic compounds also provides many examples of conformational polymorphism. For instance, cyclic ligands often have low barriers to rotation so that conformational variations between

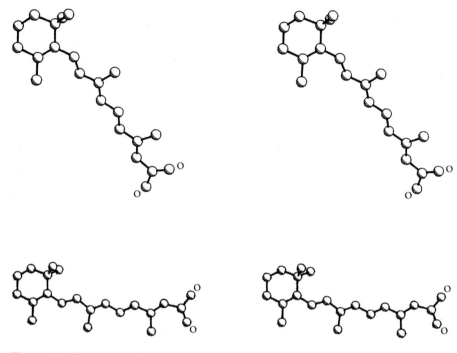

Figure 5.3 *Stereoviews of the two forms of vitamin A acid* **5-III**. *Upper, triclinic form; lower, monoclinic form. In both cases the view is on the plane of the three lower atoms of the cyclohexenyl ring (from Bernstein 1987, with permission).*

polymorphs might not be unexpected. Riley and Davis (1976) reported the structures of the dimorphic sandwich compound **5-IV**, a system in which the molecular geometry is determined by the crystallographic site symmetry for both polymorphs (Figure 5.4). In the triclinic form, the molecular site symmetry is C_i, requiring the "*trans*" orientation of the two pyridyl rings, while in the orthorhombic form, the molecule lies on a twofold axis (site symmetry C_2) leading to the "*gauche*" orientation.

5-IV

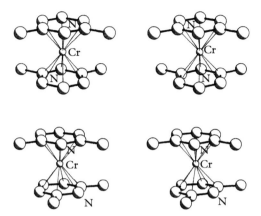

Figure 5.4 *Stereoviews of the two conformers of bis(2,6-dimethylpyridine) chromium 5-IV. Upper, triclinic; lower, orthorhombic. Carbon labels have been deleted for clarity. Note that in this case an identical reference plane would not have been a good choice for comparison. Instead, I have chosen to view structures nearly along the N⋯p-carbon axis of the upper pyridyl ring, with that axis tilted slightly downward, and the two methyl groups pointing toward the viewer (from Bernstein 1987, with permission).*

Variation of the coordination geometry about the metal atom is also exhibited in polymorphic systems. **5-V** crystallizes in four modifications (von Stackelber 1947; Lingafelter et al. 1961; Clark et al. 1977), two of which are green and two of which are brown. In the brown γ form the Cu atom is 0.5 Å from the plane of the naphthalene part of the ligand and there are dimers leading to a 5-coordinate (4 + 1 tetragonal pyramid) arrangement of ligands about the metal with an oxygen atom from the centrosymmetric dimer pair providing the fifth coordination site. In a second brown modification (β) (Clark et al. 1975) the corresponding distance from the plane is 0.63 Å and the metal is 6-coordinate (4 + 2, pseudo-octahedral). One of the two green forms (δ) (Martin and Waters 1973) exhibits a

5-V

5-VI

4-coordinate square-planar geometry with the Cu being only 0.09 Å from the ligand plane.

In another example, Kelly et al. (1997, 1999) showed that **5-VI** exists in two forms. Crystallization from hot acetonitrile yields blue crystals with square-planar coordination about the metal. Slow evaporation from methylene chloride/petroleum ether yields green needles with pseudo-tetragonal coordination about the metal.

5.6   What are conformational polymorphs good for?

Advances in chemical understanding are often based on making generalizations about systems, identifying the exceptions to those generalizations and studying the reasons for those exceptions. In arriving at the initial generalizations, one would like to minimize the number of variables considered, so that relationships can be clearly defined with cause and effect unambiguously assigned. Conformational polymorphs provide almost ideal systems for such an approach to the study of structure/property relationships, since the number of *chemical* variables is reduced to zero, the role of substituent being eliminated by the very choice of polymorphic systems. Thus changes in properties may be directly related to changes in structure. The subject of utilizing polymorphs in general and conformational polymorphs in particular to study structure/property relationships is treated in detail in Chapter 6. In conformational polymorphs in particular, differences in molecular structure *must* be due to differences in crystal environment; hence, these are ideal systems for the study of the relationship between the two. Furthermore, the identity of the molecular component among polymorphs, coupled with the differences in conformation and the relatively small, but important energy differences among them noted in Sections 5.2 and 5.3 make these systems at the same time excellent and often demanding test cases for the development and use of computational methods and parameters used in the calculation of lattice energies. These aspects of conformational polymorphism are treated in subsequent sections of this chapter.

5.7   Computational studies of the energetics of polymorphic systems

As discussed in Chapter 2, at a given temperature and pressure, the energies of polymorphs correspond to different Gibbs free energies. The relative stabilities of those polymorphs are then the differences in those energies (ΔG). These are expressed as:

$$\Delta G_{T,P} = \Delta U_{T,P} + P\Delta V - T\Delta S \qquad (5.2)$$

For the computational evaluation of these Gibbs free energy differences, in principle it is necessary to calculate all three terms on the right-hand side of equation (5.2). The third term is often neglected, since differences in lattice-vibrational energies are usually very small (Gavezzotti and Filippini 1995). The $P\Delta V$ term can be calculated but may be neglected for normal pressures due to the small value of ΔV. Hence $\Delta U_{T,P}$ is usually taken as an estimate for $\Delta G_{T,P}$. The main error in this approximation is the assumption of small entropy differences, which may not be correct in all cases of polymorphic phase transformations. As Dunitz has pointed out, as the temperature is increased through a transition point, the polymorph stable at 0 K may become metastable and transform to another polymorph. For the new, more stable polymorph, the entropy increase with temperature will always be faster than for the less stable polymorph (Dunitz 1996).

Therefore, most computational methods are aimed at calculating ΔU. The semi-empirical methods were originally developed by Kitaigorodskii (1970, 1973) and much of the basic theory and developments through 1985 were summarized by Pertsin and Kitaigorodskii (1987). The basic model is built on the atom–atom potential method and assumes that the lattice energy can be obtained by a sum of pairwise van der Waals interactions ($V(r_{ij})$) calculated using a Lennard-Jones (equation (5.3)) or Buckingham (equation (5.4)) potential:

$$V(r_{ij}) = \frac{B}{r_{ij}^n} - \frac{A}{r_{ij}^6} \quad n = 9,12 \tag{5.3}$$

$$V(r_{ij}) = B' \exp(-C'r_{ij}) - \frac{A'}{r_{ij}^6} \tag{5.4}$$

A coulombic term, $q_i q_j / r_{ij}$, is now normally included, especially since Williams (1974) showed that even for hydrocarbon crystals such a contribution may approach 30% of the total energy.

The individual partial charges on atoms, q_i, q_j, may be estimated from bond dipole moments (Hagler et al. 1974), quadrupole moments (Hirshfeld and Mirsky 1979) or charge densities obtained from molecular orbital calculations, and are assumed to be isotropic. The assignment of a partial charge to a particular atom from calculations remains an approximation and the derivation of that single charge is still a subject of investigation (Cornell et al. 1995; Masamura 2000). However, both experimental observations (Sarma and Desiraju 1985) and theoretical evidence (Williams and Hsu 1985; Price and Stone 1992) indicate that anisotropic considerations may be required for such calculations (Price 2000), and as computing power increases these will become more widely used.

The limitations of these methods should be understood. Their application requires the determination of the empirical parameters in equations (5.3) and (5.4), as well as the partial atomic charges in the coulombic term. The former are usually parameterized from experimental solid-state data such as vibrational frequencies or sublimation enthalpies, which in themselves contain some experimental uncertainties and variability from system to system. The partial atomic charges can and do vary with the choice of basis set for the calculations from which they are derived (*vide infra*). The function chosen and the complete set of parameters are often collectively termed a "force field." Ideally, one would like to develop a "universal" force field, but given the diversity of chemical systems there are doubts as to whether this is attainable.[2] Furthermore, the computations generally do not take temperature into account, so the temperature for a particular calculation is often undefined. If the parameterization is based on *ab initio* calculations, the computed energies correspond to 0 K and neglect zero-point vibrational effects.

In spite of these caveats, there is intense activity in the application of these methods to polymorphic systems and considerable progress has been made. Two general approaches to the use of these methods in the study of polymorphism may be distinguished. In the first, the methods are utilized to compute the energies of the known crystal structures of polymorphs to evaluate lattice energies and determine the relative stabilities of different modifications. By comparison with experimental thermodynamic data, this approach can be used to evaluate the methods and force fields employed. The other principal application has been in the generation of possible crystal structures for a substance whose crystal structure is not known, or which for experimental reasons has resisted determination. Such a process implies a certain ability to "predict" the crystal structure of a system. However, the intrinsically approximate energies of different polymorphs, the nature of force fields, and the inherent imprecision and inaccuracy of the computational method still considerably limit the efficacy of such an approach (Lommerse et al. 2000). Nevertheless, in combination with other physical data, in particular the experimental X-ray powder diffraction (XRPD) pattern, these computational methods provide a potentially powerful approach to structure determination. The first approach is the one applicable to the study of conformational polymorphs. The second is discussed in more detail at the end of this chapter.

For the lattice energies, a good test of the computational methods and parameters is a comparison with the sublimation energy, which may be measured fairly readily when sufficient quantities of material with suitably high vapor pressures are available (Daniels et al. 1970; Acree and Chickos 2016, 2017). Alternatively (although with considerably less precision), they may be estimated from the sublimation energies of analogous model compounds and group contributions (Bondi 1963; Cardozo 1991; Arnautova et al. 1996). These latter empirical methods generally cannot be used to estimate the differences in energy for

---

[2] There is a question whether such an approach is even practical or realistic. If a force field is viewed as a tool for calculating the energetics of real or possible crystal structures, then one might imagine using different tools for different systems, a strategy very successfully adopted by Neumann et al. Just as a carpenter's toolbox contains a variety of tools for a variety of tasks, the computational chemist should probably have (and use) different force fields for different chemical systems. For instance, one would not expect a force field developed and parameterized on peptides to be effective or reliable on aromatic hydrocarbons, or vice versa, at the precision demanded for polymorphic systems.

polymorphic modifications, although some promising attempts have been made in that direction (Gavezzotti and Filippini 1995). Unfortunately, and perhaps not surprisingly, the literature contains very few experimental determinations of the sublimation energies of more than one member of a polymorphic system and filling that gap remains a challenge.

5.8   Some exemplary studies of conformational polymorphism

The study of **5-VII**, termed ROY by the investigators (Yu et al. 2000a; Yu 2010) for the red, orange, and yellow colors of the various polymorphs, demonstrates quite a few of these aspects. Although ten polymorphs are now known for ROY (Figure 5.5) (Chen et al. 2005), this study covers the original six that coexist at room temperature. They are readily distinguished by their color and morphology, as shown in Figure 5.5. All six can be obtained from methanolic solutions, occasionally crystallizing as concomitant polymorphs (see Section 3.6) from a supercooled melt. In spite of considerable effort to determine conditions for the selective crystallization of individual polymorphs by solution methods, the apparent similarity in thermodynamic stability often led to concomitant crystallization or rapid conversion. Due to thermal instability of two of the forms [YN (yellow needles) and ORP (orange red plates)] it was possible to determine the melting points of only four of the forms, which fall in the range 106.2–114.8 °C. Nevertheless, the authors succeeded in determining the crystal structures of all six forms.

5-VII

The energetic relationships among the polymorphs were established by combining a number of techniques, providing a prototypical example of how this can be done. The pure melting endotherms of four forms (R, Y, OP, and ON) could be determined by DSC (at 10 °C min^{-1}; Figure 5.6). Each one exhibits homogeneous melting, without any solid–solid transformation. A DSC trace of a mixture of the four (at a slower heating rate, inset Figure 5.6) indicates subsequent recrystallization from the supercooled melt. An additional DSC experiment on YN (at 20 °C min^{-1}; Figure 5.7) shows the exothermic conversion to Y and (trace amounts of) R prior to melting. In other reported experiments, YN also converted to ON and ORP converted to ON or Y.

The data obtained from the DSC measurements could be used to determine free energy and entropy differences from expressions developed earlier by Yu (1995):

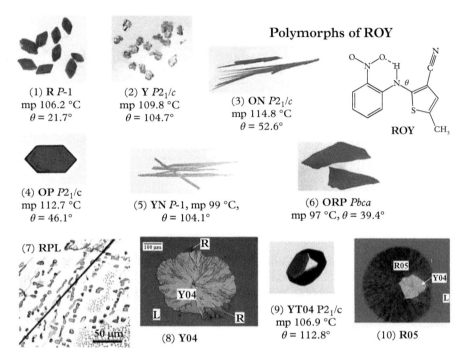

Polymorphs of ROY

(1) R *P*-1
mp 106.2 °C
$\theta = 21.7°$

(2) Y *P*2$_1$/*c*
mp 109.8 °C
$\theta = 104.7°$

(3) ON *P*2$_1$/*c*
mp 114.8 °C
$\theta = 52.6°$

ROY

(4) OP *P*2$_1$/*c*
mp 112.7 °C
$\theta = 46.1°$

(5) YN *P*-1, mp 99 °C,
$\theta = 104.1°$

(6) ORP *Pbca*
mp 97 °C, $\theta = 39.4°$

(7) RPL

(8) Y04

(9) YT04 *P*2$_1$/*c*
mp 106.9 °C
$\theta = 112.8°$

(10) R05

Figure 5.5 *Photomicrographs of the ten polymorphs of ROY 5-VII, showing the different color (R = red, Y = yellow, O = orange) and morphology (P = plates, N = needles, T = tabular). The first six are the subject of the detailed discussion in the text. (From Yu, L. (2010). Polymorphism in molecular solids: an extraordinary system of red, orange and yellow crystals. Acc. Chem. Res., 43, 1257–66. With permission of American Chemical Society.)*

$$(G_j - G_i)_{Tmi} = \Delta H_{mj}(T_{mi} - T_{mj})/T_{mj} + \Delta C_{pmj}[T_{mj} - T_{mi}\ln(T_{mj}/T_{mi})], \qquad (5.5)$$

$$(S_j - S_i)_{Tmi} = \Delta H_{mi}/T_{mi} - \Delta H_{mj}/T_{mj} + \Delta C_{pmj}\ln(T_{mj}/T_{mi}), \qquad (5.6)$$

where T_{mi} and T_{mj} are the melting points of i and j, ΔH_{mi} and ΔH_{mj} are their enthalpies of melting, and ΔC_{pmj} is the heat capacity change upon melting j. The subscript indicates that $(G_j - G_i)$ and $(S_j - S_i)$ are evaluated at T_{mi}.

In many cases, data from DSC measurements are sufficient to determine the phase relationships among polymorphic modifications (Grunenberg et al., 1996). However, due to the complexity of this system, the authors employed an older, but unfortunately rarely used technique of melting and eutectic melting data (McCrone 1957) to determine the stability relationship between the Y and ON forms. The procedure involves the preparation of mixtures of known composition of each polymorphic form under investigation with a series of common

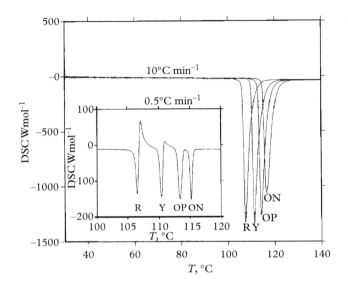

Figure 5.6 *DSC measurements on individual samples of polymorphs R,Y, OP, and ON of **5-VII** (see caption for Figure 5.5 for definition of terms), recorded at 10 °C min⁻¹, each exhibiting homogeneous melting. Inset: DSC trace of a mixture of the four modifications, recorded at 0.5 °C min⁻¹, showing better separated melting endotherms and exotherms resulting from crystallization from supercooled melts (from Yu et al. 2000, with permission).*

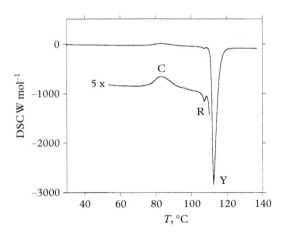

Figure 5.7 *DSC trace (at 20 °C min⁻¹) of polymorph YN of **5-VII** (see caption of Figure 5.5 for definition of terms) showing an exothermic conversion (noted as C) and subsequent melting (noted as R, trace amount) and Y. The area under C gives an estimate of the enthalpy difference between YN and Y (from Yu et al. 2000, with permission). The authors attribute the differences between these traces and that in Figure 5.6 to the difference in heating rate.*

reference materials (McCrone 1957) to form eutectics that reduce the melting point by different degrees. For **5-VII** such mixtures were prepared with acetanilide, benzil, azobenzene, and thymol, and the resulting melting data are shown in Figure 5.8.

The data from such traces may be used to calculate $(G_j - G_i)$ at the eutectic melting temperatures by the relationship (Yu et al. 2000a):

$$
\begin{aligned}
x_{ej}(G_j - G_i)_{Tei} &= \Delta H_{mej}(T_{ei} - T_{ej})/T_{ej} + \Delta C_{pej}[T_{ei} - T_{ej} - T_{ei}\ln(T_{ei}/T_{ej})] \\
&+ RT_{ei}\{x_{ej}\ln(x_{ej}/x_{ei}) + (1 - x_{ej})\ln[(1 - x_{ej})/(1 - x_{ei})]\}
\end{aligned}
\tag{5.7}
$$

where T_{ei} and T_{ej} are the melting points of i and j with a common reference compound, x_{ei} and x_{ej} are the eutectic compositions, ΔH_{mei} and ΔH_{mej} the enthalpies of eutectic melting, ΔC_{pej} the heat capacity change upon melting of the jth reference compound eutectic, and R is the ideal gas constant.

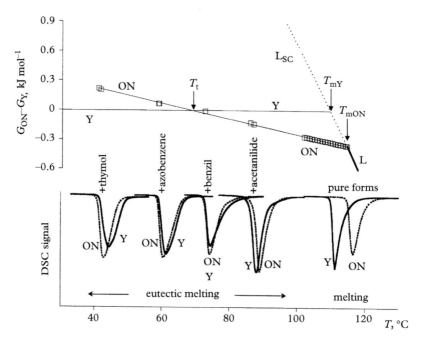

Figure 5.8 *Melting and eutectic data for determining the stability relationship between the Y and ON modifications of 5-VII. Lower: melting endotherms of the two modifications as pure crystals, and in the eutectics with four different reference compounds. Upper: ΔG versus T relationship for the two forms based on data derived from equations (5.5)–(5.7). (T_{mY} and T_{mON} indicate melting points of pure forms, T_t the transition temperature, L indicates the liquid phase, and L_{SC} indicates the supercooled liquid phase. (From Yu et al. 2000, with permission.)*

The eutectic melting data for Y and ON with the reference compounds are given in Figure 5.9, together with the thermodynamic data derived from equations (5.5)–(5.7). From the relationship between the melting points of the eutectics in the lower part of the diagram, it is seen that at higher temperatures ON is more stable than Y, with a stability reversal at ~70 °C. The upper part of the diagram summarizes the free energy differences between ON and Y. The lines for Y and ON terminate at the line for the liquid.

Similar data obtained for other pairs of polymorphic modifications are presented in Figure 5.9. The thermodynamic data and enantiotropic or monotropic relationship between modifications are given in Table 5.2, summarizing the relative stability of the phases from room temperature to the melting points (save the difficult to obtain ORP form).

The authors have made some interesting comments regarding these data in particular and the relative importance of energy and entropy contributions to the relative stability of polymorphs in general. As noted in Section 2.2.2, the relative stability of polymorphs depends on the free energy difference ($\Delta G = \Delta H - T\Delta S$) between them. The relative importance of the two terms on the right can be measured by the ratio between them (say $T\Delta S/\Delta H$). As seen in Figure 2.5, at absolute

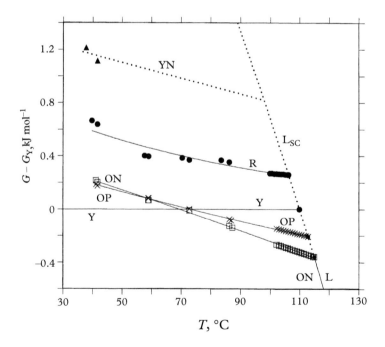

Figure 5.9 *Stability relationships between polymorphs of* **5-VII**, *constructed from melting and eutectic data as in Figure 5.8. The symbols have the same meaning as in Figure 5.8 (from Yu et al. 2000, with permission).*

Table 5.2 *Summary of thermodynamic relationships among modifications of* **5-VII**.[a]
(From Yu et al. 2000a, with permission.)

	Forms	ΔH, kJ mol^{-1}	ΔS, J K mol^{-1}	T_t, °C
Y = LT	ON = HT	2.6	7.7	70
Y = LT	OP = HT	1.9	5.3	72
Y = S	R = MS	1.4	3.0	c
Y = S	YN = MS	3.0[b]		c

[a] Enantiotropic systems: LT = low temperature form, HT = high temperature form. Monotropic systems: S = stable form, MS = metastable form.
[b] From YN–Y enthalpy of conversion.
[c] $T_t > T_m$ (virtual transition temperature of monotropic polymorphs).

zero ($T = 0$), $\Delta G = \Delta H$ and $T\Delta S/\Delta H = 0$. At a transition temperature between two polymorphic phases, $\Delta G = 0$ so the ratio $T\Delta S/\Delta H = 1$. Above a transition temperature this ratio will be > 1. Applied to some of the polymorphs of **5-VII**, for example, for the pair Y–R at the melting point of R the ratio is 0.85, which means that while Y is the more stable form at that temperature, the entropy is an important contributor to the free energy. Other similar comparisons based on the data in Table 5.2 strengthen the notion of the importance of entropy in the consideration of thermodynamic relationships among polymorphs.

Having established the thermodynamic relationships among (most of) the polymorphs, the authors then carried out the X-ray crystal structure determination of all six of them. The most significant conformational variations are due to rotations about the N–C (thiophene) bond, ranging from 21.7° in the R polymorph to 104.7° in the Y polymorph. The nitro group is essentially planar with the phenyl ring in Y (−1.8°) and shows a maximum out-of-plane rotation of 18.7° in the OP modification, while the rotation of the phenyl group out of the C-N-C plane varies from −150.0° to −175.2°. Hence there are experimentally significant conformational differences among all of the molecules in the six polymorphs.

The authors postulated that these differences in conformation should be manifested in the frequency of the CN stretch, which would also provide a measure of the molecular conformation in solution. The appropriate spectra, given in Figure 5.10, clearly reflect the differences in molecular conformation, and indicate that both the melt and solution contain a mixture of the conformations found in the six polymorphic structures.

Essentially following the procedural stages outlined in Section 5.7, the authors then computationally investigated the energetics of the conformations observed in the six polymorphic forms. *Ab initio* calculations at the RHF/6-31G* level were used to explore the conformational energy profile about the N–C (thiophene) bond (see also Section 5.5). The authors also considered the role of the (computed) molecular dipole moment (which changes with molecular conformation)

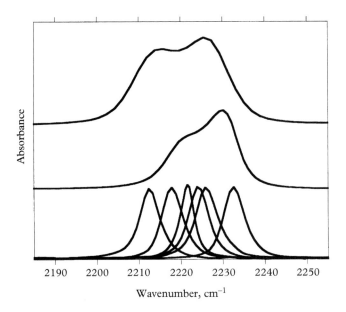

Figure 5.10 *Infrared absorption spectra in the region of the CN stretch for **5-VII**. Bottom: solid-state spectra of forms (left to right) R, ORP, YN, ON, OP, and Y. Middle: solution spectra in CCl$_4$ (0.90 mM). Top: supercooled melt at 22 °C (from Yu et al. 2000, with permission).*

in the stabilization of crystal energies (albeit while noting the evidence against the role of dipole moments in determining stability and space-group preference (Whitesell et al. 1991)). The combined results, presented in Figure 5.11, indicate that a number of the structures are not in the conformational energy minimum (at ±90°), and that those away from the minimum tend to have higher dipole moments. Both the *syn* planar and *trans* planar conformations are maxima, the former having a higher dipole moment.

Using the YN form as a reference, the authors then made a comparison of the computed conformational energies and the experimentally determined crystal energies, including noting the type of hydrogen bonding present in each of the structures (Figure 5.12).

The lowest energy modification (Y) is also unique in exhibiting an inter-molecular hydrogen bond. The authors attribute the ordering for the other four polymorphs (again not including ORP) as due to two factors: favorable packing geometry of more planar conformers, and greater electrostatic interactions between larger dipoles. While these explanations are reasonable and consistent with other data, some additional evidence, provided say by some lattice energy calculations might strengthen them even more.

The detailed study of the conformational polymorphism of ROY **5-VII** by Yu et al. is still one of most comprehensive reported to date (see also Section 5.10.4). Similarly wide ranging studies, albeit with different emphases on the particular

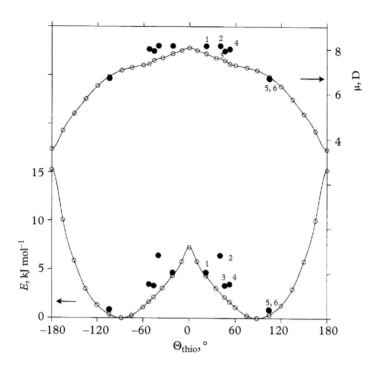

Figure 5.11 *Conformational energy (bottom, left scale) and dipole moment (top, right scale) calculated at the RHF/6-31G* level for 5-VII. Solid circles correspond to conformers observed in the six crystal modifications: 1-R, 2-ORP, 3-OP, 4-ON, 5-Y, 6-YN (from Yu et al. 2000, with permission).*

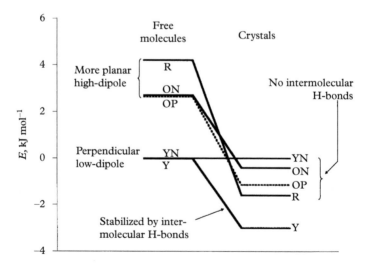

Figure 5.12 *Comparison of the computed conformational energies and experimentally determined crystal energies for 5-VII (see text for discussion of the generation of these data points). Modification YN is the reference point (E = 0) for both comparisons (from Yu et al. 2000, with permission).*

phenomena studied and the techniques employed have been appearing with increasing frequency and breadth, indicating concerted attempts to characterize and understand polymorphic systems, and these studies bode well for the increasing amount of information and understanding to be gained from such investigations (Campeta et al. 2010; Lopez-Mejias et al. 2012; Zeidan et al. 2013; Patel et al. 2015).

With regard to these lattice energy calculations, it is important to reiterate the necessity of optimizing the positions of hydrogen atoms as part of the computational protocol. While the hydrogen atom positions are not precisely determined by X-ray methods (and are often included in structure determinations in positions assumed from chemical and geometric principles), their location on the molecular periphery means that they make significant contributions both to non-bonded intramolecular interactions and to intermolecular interactions. It is therefore crucial to properly locate and optimize the hydrogen atom positions at the start of any lattice energy computational procedure.

5.9   The computational prediction of polymorphs

5.9.1   Early developments

As described throughout this chapter, once the general features of molecular crystals became apparent through X-ray crystallographic methods (e.g., Pauling 1960, first edition 1939), there were attempts to model the structures, based on the notion of relative constancy of intermolecular contacts for specific atoms and the tendency to minimize the amount of voided space in a crystal.[3] This prompted the pioneering work by Kitaigorodskii in the late 1940s, summarized in his 1955 book (in Russian) (English translation: Kitaigorodskii 1961), which included the design and construction of a mechanical "structure seeker" for physically modeling possible structures using hard sphere molecular models. Many of Kitaigorodskii's basic ideas have been incorporated into today's increasingly more sophisticated computational approaches, but the original book still contains many very useful insights into the nature of packing regularities in molecular crystals.

The ultimate goal in this endeavor is to be able to solve (see Timofeeva et al. 1980 for a review of early approaches to this problem), or predict a crystal structure on the basis of the molecular structure alone. Clearly, for polymorphic systems, or potentially polymorphic systems, that goal includes some additional challenges: how many polymorphs can be expected to exist, what is the structure of each, and how might one proceed experimentally to obtain each of the

---

[3] As noted in the Preface, the interest and activity in polymorphism, especially the prediction of polymorphic structures, have seen remarkable growth since the publication of the first edition. In order to provide background and historical perspective, especially with regard to developments since the publication of the first edition, this section has been retained as in the original, albeit with some minor editorial adjustments.

polymorphs? As in most areas of chemistry, the modeling approaches range from empirical to *ab initio*, each with its advantages, limitations, successes, and failures. In this section, I will attempt to review those approaches as they relate to polymorphic structures in this dynamic and rapidly developing field (Gdanitz 1997, 1998; Verwer and Leusen 1998; Lommerse et al. 2000; Mooij 2000; Pillardy et al. 2000). While the emphasis here is on molecular crystals, similar efforts are being made on crystals of polymers (Ferro et al. 1992; Boyd 1994; Aerts 1996).

All the programs that attempt to predict the number and structures of polymorphs of course begin with a molecular structure. For some a very rough estimate of the molecular shape and size is sufficient, while for others even precise details may not suffice, as described in the previous section. Moreover, it is usually not obvious a priori how much detail is required in the molecular model. At any rate, if the molecule does have conformational flexibility then choices must be made for the conformations of the starting model. Most allow for the simultaneous variation of molecular conformation and crystal structure. For instance, an early contribution to the effort to combine intramolecular and intermolecular searches is the successful inclusion of the rotation of hydroxyl groups together with the search for possible crystal structures (van Eijck and Kroon 1999).

At the empirical end of the spectrum of methods employed for the theoretical/ computational generation of possible polymorphic crystal structures is the approach taken by Gavezzotti (1991, 1994a, 1996, 1997) and co-workers (Gavezzotti and Filippini 1995; Gavezzotti et al. 1997; Filippini et al. 1999), as well as by Holden et al. (1993). Trial structures with fixed molecular geometry are generated (Gavezzotti 1994a) using (singly, and in combination) the most commonly appearing symmetry elements in molecular crystals: inversion centers, twofold screw axes, glide planes, and translation (sometimes appearing as centering) (Kitaigorodskii 1961). Of the 230 mathematically possible space groups, statistics indicate (Donohue 1985; Bauer and Kassner 1992) that organic crystals tend to crystallize in one of the following with a high probability: $P\bar{1}$, $P2_1$, $P2_1/c$, $C2/c$, $P2_12_12_1$, and *Pbca*. The generated trial structures with the highest packing coefficients and substantially cohesive packing energy are accepted for further study, which involves optimization of the lattice energy by a suitable program (e.g., Williams 1983). The resulting structures must then be screened by the user, either visually, or with the aid of some quantitative criteria to choose reasonable structures. Of course, one quantitative measure is the lattice energy which should be at (or near) a minimum for the expected polymorphic structures. In many cases, however, the number of low-energy structures within the (rather narrow) energy range of ~10 kJ mol^{-1} is much too large to permit a rational choice of a reasonable number of expected polymorphic structures even for relatively simple molecules (van Eijck et al. 1995; Mooij et al. 1998).

Another approach (common to many strategies of structure generation and "polymorph prediction") is to attempt to match the powder diffraction pattern calculated from the generated trial structure with an experimental one, although in principle this can no longer be justified as a truly predictive procedure.

In a strategy with a similar philosophy, Perlstein (1994a, 1994b, 1996) outlined an "*aufbau* process" for building up trial structures, following the approach to understanding crystal packing originally advocated by Kitaigorodskii (1961). The process proceeds from one- and two-dimensional motifs using some common patterns revealed in the CSD (Perlstein 1992), which are then built up into a number of possible crystal structures, all of which are potential polymorphic structures (Perlstein 1999).

The CSD also provides information on the coordination spheres of molecules and the symmetries associated with them (e.g., Braga and Grepioni 1992; Allen and Kennard 1993, Dunitz 1996; Motherwell 1997). This information was used by Holden et al. (1993) with a program called MOLPAK to generate trial structures on the basis of the densest packing of such coordination geometries. Using these model structures, other workers then minimized the resulting lattice energies. For instance, the group of Price applied the multipole electrostatic model for electrostatic contributions together with the Buckingham empirical atom-atom potentials in a program called DMAREL (Willcock et al. 1995) to successfully generate a number of crystal structures, including polymorphic ones containing rather complex hydrogen bonding patterns and to address a number of questions about competing forces (Coombes et al. 1997; Price and Wiley 1997; Aakeröy et al. 1998; Potter et al. 1999).

A more purist approach, which in principle ignores the accumulated structural data and associated structural trends, involves placing each independent molecule with a defined conformation at a certain location (X, Y, Z) with a certain orientation (ϕ, θ, φ) in a unit cell defined by its six parameters $(a, b, c, \alpha, \beta, \gamma)$. Each additional independent molecule in the asymmetric unit is defined with six additional location and orientation parameters. Each variation of conformation for one or all of the individual molecules constitutes a new starting model.

The starting models are then optimized by a variety of techniques in the search for the global energy minimum (for the computationally most stable structure) and energetically neighboring structures (for possible less stable polymorphs). One of the fundamental problems in this process is the definition of the space group. In the procedure described in the previous paragraph, the space group would be $P1$, rarely encountered for molecular materials. One solution is to add additional molecules, each with six parameters, which clearly requires additional computer resources. A computationally more efficient approach is to check for the origin of symmetry as a result of the packing optimization process, a procedure incorporated into a number of standard programs for analyzing the results of crystal structure determinations (e.g., Platon program, Spek 1990). Of course, another solution is to adopt part of the earlier related strategy and incorporate accumulated information on space group frequencies, at the risk of not generating a true structure in a less commonly encountered space group. For instance, one of the two polymorphs of acetamide **5-VIII** crystallizes in space group $R3c$ (Deene and Small 1971), one of the polymorphs of 5,5′-diethylbarbituric acid **5-IX** crystallizes in $R\bar{3}$ (Craven et al. 1969), and one of the modifications of aspartame **5-X** crystallizes in $P\bar{4}$. Additional considerations that must be included in developing a general strategy for generating trial structures include the possible presence of more than one molecule in the asymmetric unit (van Eijck and Kroon 1999), the location of a molecule on a crystallographic symmetry element, and the presence of solvent molecules. Furthermore, Dunitz et al. (2000) have shown that entropy differences among the computed polymorphs can be of the same order as differences in packing energy at 300 K (usually the temperature of greatest practical interest, if not of computational rigor). This suggests that a means for including lattice-vibrational entropy should also be included in these computational algorithms. The importance of the inclusion of dynamic effects into atom–atom computational procedures was demonstrated by Rovira and Novoa (2001).

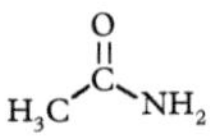

5-VIII

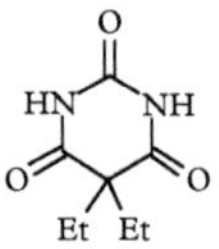

5-IX

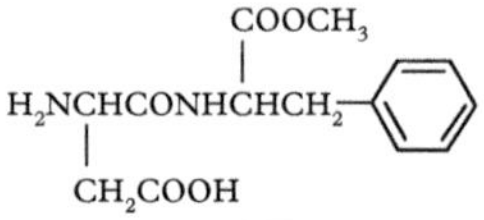

COOCH$_3$

H$_2$NCHCONHCHCH$_2$–⟨phenyl⟩

CH$_2$COOH

5-X

In spite of these limitations there has been impressive activity and progress in generating possible crystal structures, and polymorphs, as documented by the six (to date) CCDC blind tests (see Section 5.9.2).

5.9.2   Subsequent activity

Much of the current and increasingly intense computational activity has been summarized by Beran (2016) and Price (2008b, 2014).

Beginning in 1999, in a project that may be unique to the community of chemical and computational crystallography, the CCDC began a series of so-called "blind tests" to periodically bring members of this community together to benchmark and monitor the progress in the computer generation of crystal structures. The rules are quite simple. Private and public invitations announce the beginning of a particular "round," with the specification of the molecules for study. The crystal structures for the 5–6 candidate molecules have been determined but not published, so the participants are not aware of the solutions to the problems. In the early blind tests the compounds were chosen with various levels of molecular complexity and flexibility, space group limitations, limited number of molecules in the crystallographic asymmetric unit, and even lack of polymorphism (if it had not been determined that the material was polymorphic). With a time frame of nominally a year participants could employ any computational techniques to predict the structures of those compounds. The results would be sent to the CCDC preceding a meeting in which all participants would present their top three results for each of the compounds on which they had worked. A correct structure for any one of the three was considered success. The results were discussed and jointly published to allow the larger community to follow progress in the field (Lommerse et al. 2000; Motherwell et al. 2002; Day et al. 2005, 2009; Bardwell et al. 2011). A total of 21 compounds were studied in the first five blind tests.

The sixth blind test was concluded in Cambridge in October 2015 (Reilly et al. 2016). The rules were somewhat altered from previous blind tests, reflecting considerable progress in the field and an expansion of the narrower goal of structure prediction to a more thorough investigation of the methodology of the calculations and the means for ranking and interpreting the results. For instance, each participant could submit up to 100 ranked predicted structures in order of likelihood. In addition, they could submit a second list of 100 ranked predicted structures that could be re-ranked by the submitting group or another group, the latter to encourage those groups studying ranking algorithms.

The molecules for study were chosen with the following criteria (verbatim from the literature report):

(1) Rigid molecules, with functional groups restricted to C, H, N, O, halogens, S, P, B; one molecule in the asymmetric unit; up to 30 atoms.

(2) Partially flexible molecules with two to four internal degrees of freedom; one molecule in the asymmetric unit; up to about 40 atoms.

(3) Partially flexible molecule with one or two internal degrees of freedom as a salt; two charged components in the asymmetric unit, in any space group; up to about 40 atoms.

(4) Multiple partially flexible molecule (one or two degrees of freedom) independent molecules as a co-crystal or solvate in any space group; up to about 40 atoms.

(5) Molecules with four to eight internal degrees of freedom; no more than two molecules in the asymmetric unit, in any space group; 50–60 atoms.

Molecule (2) (number XXIII in the continuing blind test series) **5-XI** is particularly relevant for our discussion here. It had five known crystal structures (*A–E*). The participants were informed that for forms *A*, *B*, and *D* $Z' = 1$ and for forms *C* and *E* $Z' = 2$, and that slurrying experiments had determined that the most stable polymorphs at 257 K and 293 K both have $Z' = 1$; they were also apprised of the crystallizations conditions: "including slow evaporation of acetone solution and ethyl acetate–water mixture." The test for success in calculating a model that matched the experimentally determined crystal structure was carried out by a CSD program, the Crystal Packing Similarity Tool (Chisolm and Motherwell 2005) (see also Section 5.10 on comparing structures).

5-XI

On the molecular level the molecule formally has five rotatable bonds. However, the conformation as drawn facilitates the formation of an N–H⋯O hydrogen bond, significantly reducing the freedom of rotation around two formal single bonds. While in principle this could reduce molecular conformational flexibility, a priori there is nothing to prevent that hydrogen bond from forming; hence a full crystal structure prediction (CSP) search would require taking those two degrees of freedom into account as well. One of the two molecules in the asymmetric unit exhibited signs of disorder, but not to a sufficient extent to exclude that structure as a valid target.

Twenty-five groups participated in the sixth blind test, but many were selective on which of the five test molecules they decided to work. Some of that choice is due to the very high demands for computer resources, specifically CPU time. Hardware employed ranged from desktop PCs to massively parallel machines at national computing centers. For all the submissions more than 40 million CPU hours were used on a variety of machines, ranging from 274 hours to over 30 000 000. For the groups that submitted solutions to all five compounds the number of CPU hours ranged from 1720 to 1 220 000.

As noted in the previous section the first step in the overall process is determination of the energetic landscape of the molecular conformation. As in all stages, the participants chose a variety of methods that generally fall into the following:

(1) *ab initio* calculations of the molecule in the gas phase (for flexible molecules of the type of XXIII the energetics must be mapped for all five

torsion angles to determine preferred conformations, and the ease with which the molecule can transfer from one to another);

(2) conformational preferences for specific torsion angles obtained from appropriate searches in the CSD (Bruno et al. 2004);

(3) exploration of molecular flexibility using molecular mechanics and appropriate force fields (Kolossváry and Guida 1996);

(4) perturbations on systematic grid searches (Goto and Osawa 1989, 1993).

These preliminary mappings of conformational space were often refined using more sophisticated *ab initio* calculations.

The next step in the overall process is the generation of trial crystal structures, and the strategies chosen are even more diverse. One of the first decisions to be made is the choice of space groups from among the 230 possibilities. But, as noted earlier, most molecules crystallize in a relatively small group of ~20 space groups, so most CSP practitioners limit themselves to roughly that number. (This of course always leaves open the possibility that the true structure in a rarely occurring space group will not be computationally predicted.) A second consideration at this juncture is the choice of Z', the number of symmetry independent molecules in the crystallographic asymmetric unit. $Z' = 1$ is the most common, but molecules lying on special positions lead to $Z' < 1$ and a significant number of structures have $Z' > 1$, some even with fractional values for those molecules occupying crystallographic special positions. The blind test does reveal the known Z' value to the competitors.

The wide variety of methods employed for generating trial structures is symptomatic of the vitality of the research in this area. Every trial structure requires the input of up to six cell constants (fewer for higher symmetry space groups) and the coordinates of the atoms for the initial location(s) of the molecules according to space group symmetry.

Many of the latest blind test participants employed random or quasi-random methods for trial structures. An improvement on the totally random generation is the Sobol' sequences (Sobol' 1967), which are quasi-random low discrepancy sequences that cover the structure landscape space generated in a more uniform fashion, thus lowering the possibility of missing an important starting model.

Another method is Monte Carlo simulated annealing (Brünger 1991) that is based on starting with a randomly generated critical state from which a minimum is obtained. From each starting state a different minimum may be found and the ultimate solution is expected to be in the ensemble of those minima. An additional variation on the Monte Carlo method employed for generating starting structures is known as parallel tempering (Earl and Deem 2005).

A different approach is based on systematic grid searches in multidimensional space (Hammond et al. 2003). Still another approach is the genetic algorithm (Łużny and Czarnecki 2014). Finally, in an approach used by one participant, a

CSD search was carried out to match the shape of the candidate molecule with the shapes of molecules in the CSD. Those structures in the database were then used as starting structures for possible solutions.

The result of any of the searches based on these starting strategies is an ensemble of possible crystal structures. The cardinal challenge at this point is the need to rank these proposed structures in order to find the most likely structure, and in the case of polymorphs, the most likely structures. The strategies in that evaluation process are even more variegated than the generation of trial structures, but essentially all are based on the fundamental assumption that the thermodynamically most probable possible computed polymorph is that with the lowest energy, but other confirming factors (e.g., highest density or packing efficiency) may be taken into consideration as well. But one of the limitations is that for most algorithms the energy—actually in most cases the enthalpy rather than the free energy—is calculated for a state system at 0 K, which of course is far from ambient conditions. The free energy must include contributions from the zero point energy and vibrational contributions to both the enthalpy and the entropy, with disorder—not uncommon in real systems— providing another contribution to the free energy. A further—and perhaps the major—challenge is that, as noted previously in this chapter, differences among lattice energies for polymorphs generally fall within the range 1–10 kJ mol^{-1} and often at the low end of that range. This places a heavy burden on the ability to accurately compute energies, and perhaps more important, differences in energies. However, an increasing number of groups did attempt to use free energies in determining their final ranking. The extensive variety of approaches to this ranking attests to the challenges, the activity, and the current lack of reliable and robust methods for achieving that ranking. This arguably represents the major challenge to CSP in general and polymorph prediction in particular.

The variety of methods for calculating the energies of the computed structures are summarized in Reilly et al. (2016), and considerable detail is given by each of the 25 participating groups in the extensive supplementary deposited information. The reader is encouraged to consult those for sources. While the details of all the strategies and methods employed for energy evaluation and ranking are beyond the scope of this volume, some general aspects of the process are worthy of note.

One strategy employed by a number of groups was a two-tiered approach. The initial energies were computed using methods that required less computational resources, and then those in the low-energy range were recalculated with much more sophisticated methods for the final ranking. In a number of instances those sophisticated methods included distributed multipole electrostatics (Stone 2005; Price et al. 2010) and a number of different dispersion-repulsion potentials for the intermolecular energies, while *ab initio* methods were employed for intramolecular energies (Kazantsev et al. 2011; Hapgood et al. 2015). In this and in recent past blind tests, the group of Marcus Neumann (Neumann et al. 2008) has had considerable success with empirical potentials tailor made for the compound in question, and some other groups have adopted various aspects of that approach (e.g., van Eijck et al. 2001).

Another notable development from this sixth blind test is the increasingly widespread use of density functional theory (DFT) in energy calculations of solid-state properties. These were employed by nearly half of the participants albeit with considerable variety in the precise treatment. Notable among these, because of the indication of their importance for molecular materials is the inclusion of many-body van der Waals effects (Reilly and Tkatchenko 2015); there are indications that these may be particularly important for treating and understanding polymorphic systems (Marom et al. 2013).

The results of the sixth blind test were summarized as follows. Of the twenty-five groups that submitted responses, seven attempted all the structures, while fourteen attempted only some of them. Four groups re-ranked the solutions presented by others based on their own algorithms. It is difficult to give a numerical or statistical description of the results that provides a single measure of success or progress, say, on a percentage basis. However, the following can be noted. Except for Form E of the polymorphic system (discussed in more detail below) which presented additional difficulties due to disorder, all the experimental structures were generated by at least one group. In many cases a number of groups succeeded and except for the polymorph above one group did succeed in generating all the experimental structures. Five of the structures were ranked as the lowest energy by a number of methods, but not consistently by a single method. Calling the possibility of polymorphism into play the authors noted that the higher ranked members of the predicted structures could represent yet undiscovered forms, since the compounds in question have not been the subject of the type of solid form screen described in the previous chapter.

How did the blind test participants fare on the polymorphic system? Before proceeding to a summary of the results it is well to recall the relevance of this exercise in the exploration of the crystal form landscape. In that case nothing is known about the landscape when the exploration begins. In the case of the blind test, the participants knew there were five forms; they knew that three had $Z' = 2$ (not simply $Z' \neq 1$), while the other three had $Z' = 1$; they knew that none of the five contained water or solvent; they did not have to determine the experimental conditions to *prepare* any forms. In other words this was not an exercise in crystal form landscape exploration, or even, for that matter *polymorphism prediction*. The object of the treasure hunt was known; it was only a matter of whether it could be found.

The pentamorphic target compound XXIII had been a Pfizer/Warner-Lambert drug candidate (Augelli-Szafran et al. 2005; Simons et al. 2009). Slurrying experiments (see Sections 3.6.1 and 7.10) had shown that Form A is the stable form at 257 K, while Form D is the stable form at 293 K. The details of similarities and differences in molecular conformation among the five polymorphs are described in Reilly et al. (2016).

Fourteen of the groups attempted solutions of the three $Z' = 1$ polymorphs (Forms A, B, and D), and three groups re-ranked structures. The lingering general reluctance of many groups to even attempt the $Z' > 1$ cases testifies to the considerable leap in difficulty involved. From the eighteen attempts Form A was found among the top 100 four times, Form B was found among the top 100 ten times and Form D was found among the top 100 three times. Two groups (Neumann, Leusen, and Kendrick and Day et al.) were both successful for all three structures. It will be recalled that the criteria for success did not require that the correct structure be among the top three for an attempt to be considered successful. Rather it required that the correct structure be among the top 100.

In practical terms knowledge of the relative stability is crucial for drug development, especially in terms of the specific form chosen for production. Only a few of the results ranked the experimental structure as one of the ten most stable structures, with Form A (most stable at 257 K) having the best rank of twenty-three (results of Day et al.). Contrary to the experimental evidence, a number of groups predicted Form B to be the most stable (of A, B, and D), appearing as the highest ranked for the submission of Price et al. A significant number of low-energy results predicted crystal structures in which the conformations were incorrectly predicted as compact rather than extended. All the experimental polymorphs exhibited the common $R_2^2(8)$ carboxylic acid dimer, which also appeared in the majority of the low-energy computed structures, so that the main differences in the latter were in molecular conformations and packing.

Of all twenty-five submissions only four attempted the $Z' = 2$ for two of the polymorphs of **5-XI**. One group that listed all $Z' = 1$ and $Z' = 2$ structures ranked together had the calculated C polymorph ranked sixth; no group successfully predicted polymorph E.

The sixth blind test represents considerable progress over the fifth that preceded it by five years. More complex systems, from both the molecular and crystallographic points of view, are possible to attempt by at least some of the practitioners. The group of Marcus Neumann has had particular success in the past few blind tests, albeit requiring significant computational resources. A number of other groups are not very far behind.

However, there are still many major limitations in the current methodology. The ideal in CSP is to be able to compute the expected structure—or energetically ranked structures in the case of (unknown) polymorphic systems—simply on the basis of a two-dimensional molecular structure diagram. This would include any possible space group, any value of Z', and in the ultimate solution, the presence of solvent molecules in any stoichiometric or non-stoichiometric ratio (i.e., variable solvates). The difference in temperature between that of the calculation (usually at 0 K) and (usually) ambient temperature is still a challenging aspect, as is the role of solvent and kinetics in producing or inhibiting the growth of a particular crystal form. And perhaps the major problem of computing reliable lattice energies and reliable differences in lattice energies required for ranking of solution is still quite far from solution; witness the variety of approaches under development and used to determine those rankings. Thus, for the foreseeable future, while the CSP calculations will provide an increasingly valuable tool in understanding polymorphism as well as possible solutions for structure determination from XRPD data, the ultimate solution to the problem is still not in sight.

5.10   The computational comparison of polymorphs

5.10.1   Introduction

In Section 2.4.2 I described the recommended methods for the *visual* comparison of the crystal structures of polymorphs and the use of the graph set approach for comparison of the hydrogen bonding networks in polymorphs. Both of those approaches are indeed valuable tools for the understanding of the differences and similarities among polymorphs. Since the publication of the first edition there have been many significant developments of the *computational tools* for the comparison of polymorphs in addition to the methods outlined in the previous section. Just as the application of a variety of analytical techniques is recommended for greater understanding of a polymorphic system, so the utilization of a number of these computational techniques can considerably further that understanding.

5.10.2   Comparison based on geometry criteria

Intuitively, the most natural method of comparing structures is based on geometric criteria. Indeed, a number of groups have applied such methods to comparing crystal structures in the search for and verification of polymorphs.

Chisolm and Motherwell (2005) base their algorithm COMPACK on interatomic distances, thus avoiding the need to use space group and unit cell parameters. For a reference molecule the program constructs a coordination shell based on nearest neighbors. The cluster is of variable size; the authors defaulted to fourteen molecules for molecular structures, but for the study of conformational polymorphism (see Section 5.10.4) the cluster size was generally increased to twenty. Neighboring molecules are those found within the van der Waals radius plus 2.0 Å. Having defined this search structure the task of determining whether it is the same as, or different from another structure (i.e., whether two structures are polymorphic or not) then the computational task involves searching the three-dimensional coordinates of the second structure for a match within a specified tolerance. The authors indicate a tolerance of ±15%, but that value is clearly adjustable to determine the confidence level of polymorphic identity. In the conformational polymorphism study the match was determined by calculating the *rmsd* over all the atoms in the two structures.

The program is currently incorporated into the CSD software library.

5.10.3   Comparison of crystal structures using Hirshfeld surfaces

Developed by the group of Mark Spackman (McKinnon et al. 2004, 2007a, 2007b; Spackman and Jayatilaka 2009) based on ideas originally put forward by Hirshfeld (1977), the fundamental concept behind this development is that a single diagram (which is actually possible to represent in a number of different ways) can be used to describe the details of the surroundings of a molecule in a crystal. Since by definition different polymorphs are different crystal structures, the surroundings of a molecule must be different in some way from that same molecule in a different crystal structure. The generation of packing diagrams with the guidelines given in Chapter 2 might aid in the understanding and interpretation of the similarities and differences between polymorphic structures. As I shall show, in spite of the crucial information that these packing diagrams can provide they are often difficult to decipher and understand. On the other hand, the various Hirshfeld diagrams, in particular the so-called "fingerprint," provide a single representation usually with distinct features, and the finely honed ability of humans to recognize and distinguish among patterns facilitates the recognition of difference and the ability to distinguish among polymorphs. I will provide some of the basic principles on which the generation of these diagrams are based and demonstrate their utility using the earlier cited example (Sections 2.4.2 and 4.4) of terephthalic acid.

The Spackman et al. application of the Hirshfeld method is based on partitioning molecular crystals into "molecules" for electron density integration (Spackman and Byrom 1997). The method takes advantage of the difference between the spherical electron density of individual atoms, used, for instance in most crystal structure refinements, and the non-spherical real electron densities of atoms as bonded in molecules. The Hirshfeld atoms, that is, the aspherical atomic density functions (ADFs), are obtained by a Hirshfeld's "stockholder" partitioning method. Hirshfeld's method partitions an electron density into atomic fragments using a weight function. The weight function for an atom at a point in space is calculated as the ratio of two parts: the spherical atomic density of that atom, and the promolecular atomic density. The molecular Hirshfeld surfaces developed by Spackman then partition a crystal into regions in which the electron distributions from a sum of *spherical atoms* for the molecule (called the promolecule) exceeds the contribution from the corresponding sum over the whole crystal (the perturbations from non-sphericity). Hirshfeld defined a weighting function $w(\mathbf{r})$ for a particular molecule in the crystal structure as

$$w(\mathbf{r}) = \sum_{a \in molecule} \rho_a(\mathbf{r}) / \sum_{a \in crystal} \rho_a(\mathbf{r}) = \rho_{promolecule}(\mathbf{r}) / \rho_{procrystal}(\mathbf{r}) \qquad (5.8)$$

where $\rho_a(\mathbf{r})$ is a spherically-Hartree-Fock atomic electron density function centered on nucleus **a**. From this definition it follows that the volume within which the promolecule dominates the procrystal electron density is the region for which $w(\mathbf{r}) \geq 0.5$. The Hirshfeld surface is thus defined by $w(\mathbf{r}) = 0.5$.

The cardinal point for using this approach to compare the structures of polymorphs is that the surface is unique for each structure because the electron density of the molecule, especially in the region of the surface of the molecule in each crystal structure is unique. Thus, the Hirshfeld surface *is defined only* for the molecule in the bulk crystal, specifically *intermolecular interactions*; and it represents the interplay between different atomic sizes in bonded atoms and intermolecular contacts. In addition to the atomic electron distributions, the size and shape of the surface depend on the molecular geometry and the crystal geometry. The Hirshfeld surface fills the crystal space in non-overlapping molecular regions in which no single molecule dominates the electron density. Crystallographers are familiar with packing diagrams using space-filling models. These are based on Pauling's (1960) definition of van der Waals radii together with Kitaigorodskii's (1970) notion of packing coefficients and generally fill 65–80% of the crystal volume. The *non-overlapping* Hirshfeld surfaces are much more closely packed than in Kitaigorodskii's model and fill 95% or more of the crystal volume. It turns out that the computational basis for generating the Hirshfeld surface is rather simple, since the function—essentially derived from the electron density—is everywhere smooth and differentiable. Finally, as I demonstrate in the following discussion the properties of the surfaces and the information that may be obtained from them are easily color coded and mapped to be readily read, compared, and interpreted.

Before moving on to some examples of the use of Hirshfeld surfaces in the study of polymorphic systems, a few more details on their generation are required. Two geometric properties related to the *shape* of the surface may be defined: (1) the *curvedness*, which is a measure of *how much shape* the surface displays, and (2) the *shape index*, a measure of *which* shape is exhibited by the surface. Finally, the *distance properties* are defined with the aid of the diagram in Figure 5.13.

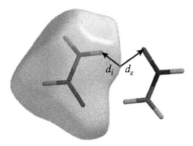

Figure 5.13 *Demonstration of the distance properties of the Hirshfeld surface. For each point on the surface two distance properties are defined:* d_e, *the distance to the nearest nucleus outside (external to) the surface, and* d_i, *the distance to the nearest nucleus internal to the surface (from Spackman 2004, with permission).*

Many of the characteristics and utility of the Hirshfeld surfaces can be demon-strated with the surfaces for naphthalene and one of the polymorphic forms of terephthalic acid, in Figure 5.14. It is readily seen that although the two molecules are similar in size their surface qualities differ significantly. The surface for naphthalene is more rounded, reflecting the relatively weak and non-directional interactions present in the crystal, while the upper and lower surface for terephthalic acid is much more abruptly terminated, reflecting the existence of strong hydrogen bonds at the two ends. The almost flat upper surface (nearly parallel to the plane of the page) is indicative of $\pi \cdots \pi$ stacking between adjacent ribbons of hydrogen bonded molecules.

The "curvedness" surface provides information on the close contacts between two molecular Hirshfeld surfaces. The "shape index" surface provides informa-tion on which shapes appear on the surface; for instance, in terephthalic acid the region shown indicates a concave (blue) region complementary to a convex (red) region. On the other hand, the Hirshfeld surface for naphthalene has a rather broad concave region due to C–H$\cdots \pi$ contacts. The d_e surface represents the dis-tance from that surface to the nearest nucleus in another molecule; the most dis-tant regions are represented in blue, with increasingly shorter distances appearing in green and then red.

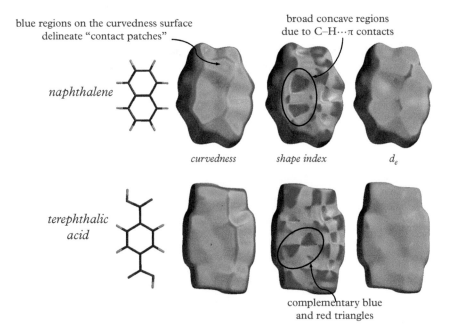

Figure 5.14 *Demonstration of the three surface representations of Hirshfeld surfaces for naph-thalene (NAPHTA10) and terephthalic acid (TEPHTH03). The scales are identical for the two molecules (from Spackman 2004, with permission).*

Another graphical representation of the molecular environment is the "2D finger-print plot" (Figure 5.15), which is probably the clearest and most demonstrative way of comparing polymorphic structures. First of all, it is abundantly clear that even a casual inspection reveals significant differences in these two figures, and demonstrates that each crystal structure indeed has a unique fingerprint. It is important to point out that all of the fingerprint plots are plotted in the same way, and on the same scale: d_i is on the x axis while d_e is on the y axis, both running from 0 to 2.4 Å. Therefore, the surroundings of any molecule *in any structure* may be readily compared.

What are the important features in these plots? The wings in the naphthalene are due to C–H$\cdots\pi$ interactions that clearly dominate the structure. The "probos-cis" in the region where $d_i \approx d_e \approx 1.2$ Å is due to H$\cdots$H contacts. The terephthalic acid fingerprint plot is dominated by two long donor and acceptor spikes which are characteristic of a hydrogen bond. The red area in the region $d_i \approx d_e \approx 1.8$ Å is due to the strong contribution made by $\pi\cdots\pi$ stacking.

The utility of Hirshfeld surfaces in structural polymorphism can be demonstrated by examining the terephthalic acid in greater detail using this graphical-analytical tool. In Section 4.4, I demonstrated how some similar structure features of polymorphic structures can lead to similar (but not identical!) XRPD patterns. In that demonstration I pointed out the importance of using reduced cells to com-pare the cell constants, and, in order to become visually oriented to compare the structures, the importance of plotting them on the same molecular reference plane in the same orientation. The CSD now contains five room-temperature structure determinations entries for terephthalic acid; the cell constants are given in Table 5.3 together with the CSD Refcodes and the form labels as given by the authors. TEPHTH and TEPHTH07 appear to be the same structure, while the space group of TEPHTH13 clearly differentiates it from all the others. TEPHTH01 has been labeled as "form II" by the authors and also appears to be different from TEPHTH12.

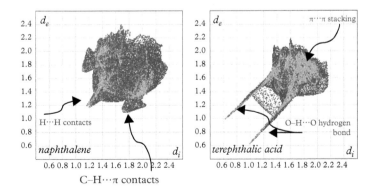

Figure 5.15 *Two-dimensional fingerprint plots from the Hirshfeld surface (from Spackman 2004, with permission).*

Table 5.3 *Comparison of cell constants for the five room-temperature structures of terephthalic acid in the CSD (from Spackman 2004, with permission)*

	TEPHTH "form I"	TEPHTH01 "form II"	TEPHTH07 "form I"	TEPHTH12	TEPHTH13
Space group	$P\bar{1}$	$P\bar{1}$	$P\bar{1}$	$P\bar{1}$	$C2/m$
Z'	0.5	0.5	0.5	0.5	0.25
$a(\text{Å})$	7.730	9.54	7.761	5.027	8.940
$b(\text{Å})$	6.443	5.34	6.439	5.360	10.442
$c(\text{Å})$	3.749	5.02	3.760	7.382	3.790
$\alpha(°)$	92.75	86.95	91.47	115.72	90.0
$\beta(°)$	109.15	134.65	109.45	101.06	91.21
$\gamma(°)$	95.95	104.90	95.81	92.91	90.0

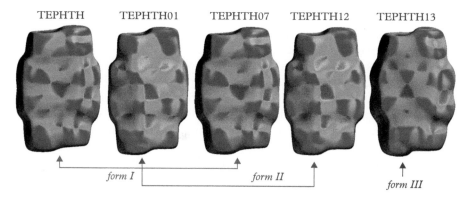

Figure 5.16 *Hirshfeld surfaces for five structures of terephthalic acid in the CSD (from Spackman 2004, with permission).*

While the cell constants are different, it must be recalled that for a triclinic structure the choice of the unit cell is arbitrary except that for $P\bar{1}$ the origin is on an inversion center, so they could, in fact, be the same structures. The Hirshfeld surfaces for the five structures are shown in Figure 5.16. The identity of TEPHTH and TEPHTH07 is confirmed, and TEPHTH01 and TEPHTH12 are clearly also identical.

How are these differences reflected in the fingerprint plots? Since two pairs of structures have already been identified, it is necessary to examine the fingerprints of the three remaining unique structures. TEPHTH13 clearly differs from the other two, but on close inspection the subtle differences between TEPHTH and TEPHTH01 are recognizable. Of course there is some error in the experimental determination of these structures, but as Figure 5.17 clearly demonstrates, the subtle features of all of these plots are reliable identifiers and distinguishers among structures.

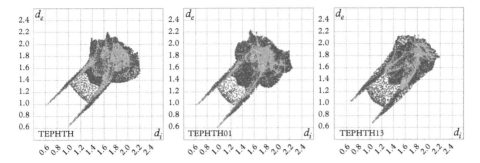

Figure 5.17 *Fingerprint plots for the three unique structures of terephthalic acid (from Spackman 2004, with permission).*

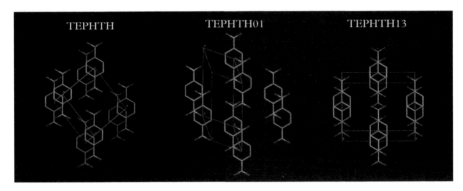

Figure 5.18 *Packing diagrams of three polymorphs of terephthalic acid. Note that the three structures are plotted as suggested in Chapter 4, namely on the same molecular plane (a phenyl ring) and oriented in the same direction, with the para axis vertical (from Spackman 2004, with permission).*

Some features of the diagrams are worthy of note. All three show spikes characteristic of the cyclic ($R_2^2(8)$ in this case) hydrogen bond motif. The spikes for TEPHTH13 are shorter (terminating at greater distance on the grid) representing longer hydrogen bonds. In TEPHTH and to a much lesser degree in TEPHTH01 there are red points at $d_i \approx d_e \approx 1.8$ Å that indicate the earlier noted $\pi\cdots\pi$ stacking; there is only a slight hint of it in TEPHTH01. For TEPHTH13 there is a red line at $d_i + d_e \approx 3.8$ Å; this corresponds to the short crystallographic c axis of 3.79 Å (see Table 5.3).

All of these features can also be identified in the packing diagrams, which are compared in Figure 5.18. It is readily seen that all three have ribbons of infinite bonded chains (running vertical in the diagrams), and that TEPHTH and TEPHTH13 have significant $\pi\cdots\pi$ overlap, in contrast to TEPHTH01. It is worthy of note here that in spite of the ready availability of programs for generating crystal packing diagrams the preparation of a set like this, especially for more complex structures can be a difficult and time-consuming task, while the

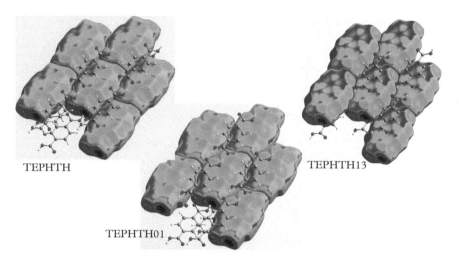

TEPHTH

TEPHTH13

TEPHTH01

Figure 5.19 *Demonstration of the "cluster" option of* CrystalExplorer *for three polymorphs of terephthalic acid (from Spackman 2004, with permission).*

fingerprint plot is conveniently and readily generated from the *CrystalExplorer* program (Wolff et al. 2012).

An additional mode of representation of the crystal structure—and hence an additional means for comparing structures—is available through the same *CrystalExplorer* package. The "cluster" option produces essentially a packing diagram of the Hirshfeld surfaces, on which is superimposed a ball and stick representation of the structure, as shown in Figure 5.19. This provides often enlightening information on the matching protrusions and voids on neighboring molecules as well as the role of specific interatomic interactions.

In addition to determining the presence or absence of polymorphism among a set of crystal structures, the Hirshfeld surfaces may be used to determine or to verify the number of molecules in the asymmetric unit (Z') of a crystal structure (e.g., Bernstein 2011). For example, the compound 4,4-diphenyl-2,5-cyclohexa-dienone (HEYHUO) **5-XII** was reported to crystallize in four different forms A, B, C, and D (Kumar et al. 2002). It was noteworthy that it was reported that two of the forms, B and C, crystallized in the same space group $P\bar{1}$ with $Z' = 4$ and $Z' = 12$, respectively. This system was also cited as an example of *conformational polymorphism* (Nangia 2008) (see Section 5.10.4). The fingerprint plots for the 19 presumed crystallographically unique molecules are given in Figure 5.20.

5-XII

A comparison of the fingerprint plots for Form C, presumably with $Z' = 12$, readily reveals the following equivalences: $(1 = 2 = 3; 4 = 5 = 6; 8 = 9 = 12; 7 = 10 = 11)$. Thus, there are four triplets of molecules with identical surroundings, which means that there are only four crystallographically independent molecules in the asymmetric unit, and $Z' = 4$ for this structure. Moreover, further comparison of the four unique fingerprints of Form C are identical to Form B $(C1 = B1; C4 = B2; C7 = B3; C8 = B4)$. Thus, this system is trimorphic rather than tetramorphic, and exhibits not 19, but seven crystallographically different molecules.

The tetramorphic benzidine **5-XIII** system provided another test example for the application of Hirshfeld surfaces to determine the identity or uniqueness of polymorphs. For a simple molecule the structures were particularly unusual in their Z' values (1.5, 3, and two with 4.5), the last being particularly rare in the CSD. For the four polymorphs the result is 15 crystallographically distinct molecules (Rafilovich and Bernstein 2006). Proof of the uniqueness of the two $Z' = 4.5$ structures (which was also strongly indicated by a significant difference in the length of the unique monoclinic b axis for the two structures) and indeed for all 15 molecules is readily observed in the fingerprint plots (Figure 5.21). In fact, these two examples, involving the comparison of quite a few fingerprint plots, rather dramatically demonstrate the role of a human being's finely honed sense of pattern recognition in the scientific process (Bernstein 2017). That human ability has been augmented to a quantitative level by the application of automated pattern-matching algorithms and evaluation criteria, including Hirshfeld fingerprints, by Parkin et al. (2007) and Collins et al. (2010). Although it does not deal specifically with polymorphism, this paper contains an enlightening set of fingerprints for the (non-polymorphic) structure of anthracene at six temperatures in the range 94–295 K and the corresponding XRPD patterns at those temperatures. The former series indicates a subtle shift of the entire fingerprint towards higher d_i values, while those for d_e do not exhibit such a shift. The shifts in the XRPD patterns are somewhat less subtle. These are factors that should be considered when comparing these two representative characteristics at different temperatures.

$$H_2N - \langle\!\!\!\!\!\rangle\!-\!\langle\!\!\!\!\!\rangle - NH_2$$

5-XIII

In this section I have attempted to give an introduction to the Hirshfeld surfaces and their powerful utility in investigating and understanding polymorphic structures. Further details on the method and applications may be found in the following publications, in addition to those cited in the opening paragraphs (Spackman and McKinnon 2002; McKinnon et al. 2007a, 2007b; Spackman et al. 2008; Turner et al. 2011).

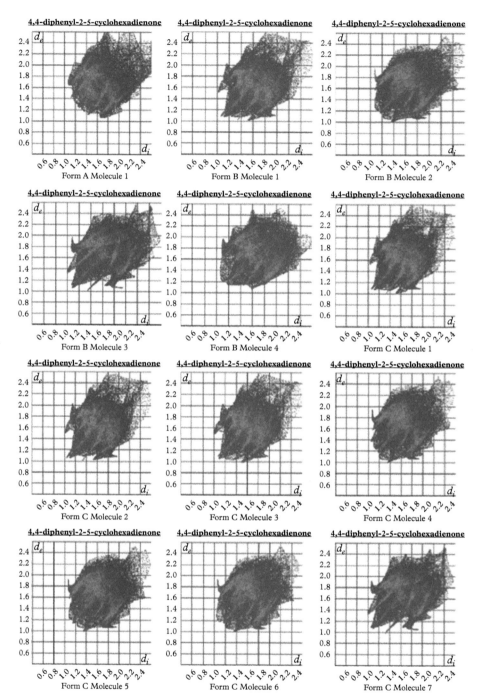

Figure 5.20 *Hirshfeld fingerprint plots for the 19 apparently crystallographically independent molecules in the four reported polymorphs of HEYHUO **5-XII**. See text for the analysis of identities and analysis that led to the reduction from 19 to 7 crystallographically independent molecules. (From Bernstein, J. (2011). Polymorphism – a perspective. Cryst. Growth Des., **11**, 632–50. With permission of American Chemical Society.)*

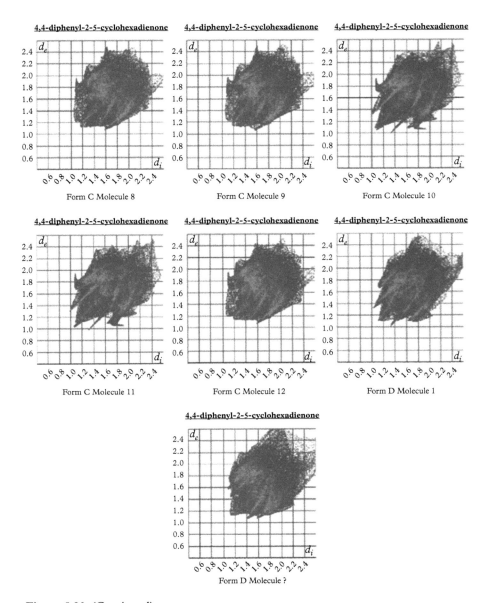

Form C Molecule 8

Form C Molecule 9

Form C Molecule 10

Form C Molecule 11

Form C Molecule 12

Form D Molecule 1

Form D Molecule ?

Figure 5.20 (Continued)

5.10.4 Conformational polymorphs

The original work on conformational polymorphs in the 1970s (Corradini 1973; Panagiotopoulos et al. 1974; Bernstein and Hagler 1978) was based on a qualitative monitoring and review of the literature, citing and using for examples structures that qualitatively and subjectively appeared to exhibit different conformations. Subsequently, the underlying and continuing fundamental questions in designating

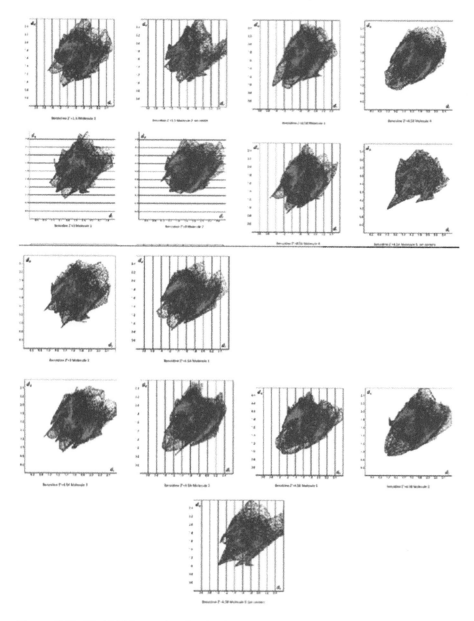

Figure 5.21 *Hirshfeld fingerprint plots for the 15 crystallographically independent molecules in the four reported polymorphs of benzidine* **5-XIII**. *The molecular labeling corresponds to that in the original structure determination (Rafilovich and Bernstein 2006). (From Bernstein, J. (2011). Polymorphism – a perspective.* Cryst. Growth Des., *11, 632–50.With permission of American Chemical Society.)*

a system as an example of *conformational polymorphism* was what physically constituted differences in conformation. That remained essentially undefined until quite recently (Cruz-Cabeza and Bernstein 2014). I briefly review the developed methodology and the major findings of that methodology as applied to validated polymorphic systems in the CSD.

Following the selection of true polymorphs using the COMPACK program (Section 5.10.2) the flexibility of the molecule is evaluated by specifying the number of rotatable bonds (R-bonds). That degree of flexibility (DOFlex) is the sum of (1) the number of acyclic R-bonds, (2) the number of groups attached to triple bonds, and (3) the number of aliphatic cycles that could potentially also change their geometry.

Molecular geometries were compared by mapping the atoms of one molecular conformation onto that of a second molecular conformation (from another polymorphic structure) using the Tormat program (Weng et al. 2008). In addition to providing a minimum *rms* deviation between atomic positions, the procedure also involved calculating the maximum difference for any torsion angle of a given pair of molecules ($\Delta\theta$).

Following a rather detailed evaluation procedure a total of 2770 crystal structures were selected for 1297 molecules for the study of conformational polymorphism. The question of replacing the qualitative judgment with working definition of conformational polymorphs still remained. This was resolved by distinguishing between the concepts of "conformational change" and "conformational adjustment" as illustrated in Figure 5.22.

Consider two polymorphs A and B. The energies of the experimentally determined molecular structures in the two polymorphs are represented on the gas phase potential energy curves by the points A_{crys} and B_{crys}. The two molecular conformations are located in the vicinity of different minima in the potential energy curve. Suitable molecular energy minimizations (in the case of this study, DFT including dispersion corrections at the B97-D/ccpVTZ level using the Gaussian 09 package (Gaussian, Inc., Wallingford, CT) would bring the experimental structures to the two minima. That process is termed "conformational adjustment" and all experimental molecular geometries lying within a single energy well are similarly related.

To proceed from a conformation in one energy well to another requires crossing an energy barrier; molecules so related in different energy wells undergo a "conformational change" and thus are termed conformational polymorphs. Hence the working definition requires the calculation of the potential energy landscape, although in many cases the calculation of only a few points will be required to locate the barrier. Using these principles it can be asked if a general parameter or physical constant can be defined for the determination of the distinction between conformational adjustment and conformational change. To make that determination the relative frequencies of conformational change and conformational

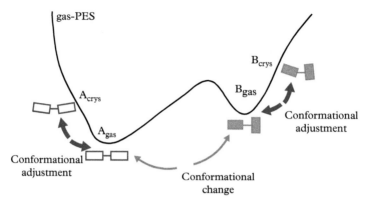

Figure 5.22 *Demonstration of the concepts of "conformational change" and "conformational adjustment." (From Cruz-Cabeza, A. and Bernstein, J. (2014). Conformational polymorphism.* Chem. Rev., *114, 2170–91.With permission of American Chemical Society.)*

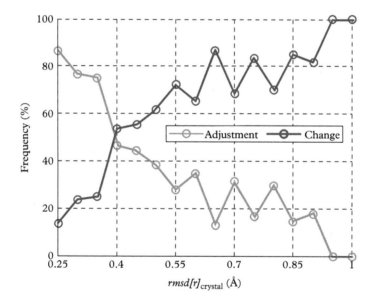

Figure 5.23 *Relative frequencies of conformational change and adjustment as functions of* rmsd[r]$_{crystal}$ *(452 pairs of conformations). (From Cruz-Cabeza, A. and Bernstein, J. (2014). Conformational polymorphism.* Chem. Rev., *114, 2170–91. With permission of American Chemical Society.)*

adjustment were plotted as a function of *rmsd[r]*$_{crystal}$ for 452 pairs of conformations. The results are shown in Figure 5.23.

This sampling of 452 pairs of conformations can probably be considered representative, and from the graphs in Figure 5.23 it is possible to estimate an approximate *rmsd[r]*$_{crystal}$ cutoff of 0.375 Å that may be used as a criterion for distinguishing between conformational change and adjustment. I note that this is considered an approximation rather than an absolute criterion, since there will no doubt be some outliers. It is also worthy to recall that the 0.375 Å criterion was obtained from data contained in the CSD. Most of the molecular structures have between ten and fifty non-hydrogen atoms; thus it is possible that this cutoff should be modified for larger molecules. It was found that this cutoff will place ~90% of the pairs with the correct conformational relationship across the entire range of $rmsd[r]_{crystal}$ values. Hence the geometric criterion may be used with some confidence in place of the computer-intensive energy calculations.

A similar approach was taken towards evaluating the torsional variations of conformational adjustment and conformational change, as seen in Figure 5.24. The data presented in Figure 5.24 for the comparison of two different conformations (with flexible R-bonds only, no aliphatic rings or triple bonds) can be summarized as follows. (1) For $\max(\Delta\theta) < 25°$, they are related by conformational adjustment only. (2) For $25° \leq \max(\Delta\theta) < 45°$ they are related by conformational adjustment with a probability of 92%. (3) For $45° \leq \max(\Delta\theta) < 95°$, it is not possible to make a prediction and the R-bonds need to be analyzed individually. (4) For $\max(\Delta\theta) \geq 95°$, they are related by conformational change with a probability of 99%. A very good linear relationship was found for and $rmsd[r]_{crystal}$ and $\max(\Delta\theta)$ up to the value

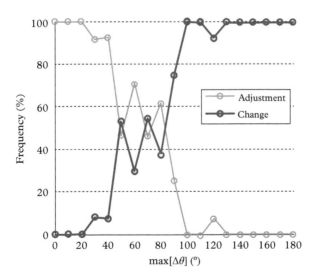

Figure 5.24 *Relative frequencies of conformational change and adjustment as functions of max($\Delta\theta$) for a subset of 267 DFT-d-optimized pairs of conformations containing flexibility due to acyclic bonds only. (From Cruz-Cabeza, A. and Bernstein, J. (2014). Conformational polymorphism.* Chem. Rev., **114***, 2170–91.With permission of American Chemical Society.)*

of 0.375 Å for the latter, with considerably large standard deviations for the correlation at higher values.

When energy calculations are not feasible, it may nevertheless be possible to determine if a situation represents one of conformational polymorphism, as demonstrated by the decision tree in Figure 5.25. If the conformations in A and B in the absence of energy calculations have one or more torsion angles with $\Delta\theta \geq 95°$ they are conformational polymorphs. Otherwise, the conformational parameters must be examined. If there is at least one conformational feature in clearly distinct areas of the distributions for conformations in A and B, then they are conformational polymorphs. If not, then a program such as the CSD's *Mercury* can be used to calculate an $rmsd[r]_{crystal}$ value for the two molecules. If that value exceeds 0.375 Å then the two are conformational polymorphs (within the confidence level given above).

What are the typical energetics for conformational adjustment and conformational change? The results of DFT-d calculations on 452 pairs of conformations are given in Figure 5.26. The histogram for conformational adjustment quantifies the conformational energy penalty involved in calculating a gas-phase energy for a molecular conformation observed in a crystal structure. Slightly less than 50% of

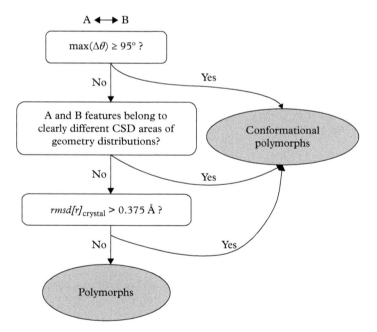

Figure 5.25 *Decision tree to identify the conformation relation between the molecules in two polymorphs A and B with structural data only (e.g., in the absence of energy calculations). (From Cruz-Cabeza, A. and Bernstein, J. (2014). Conformational polymorphism.* Chem. Rev., *114, 2170–91. With permission of American Chemical Society.)*

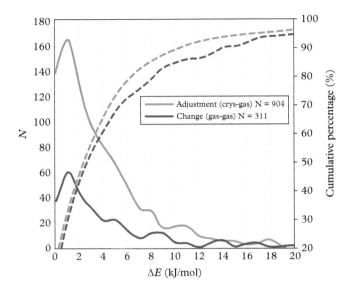

Figure 5.26 *Histograms of conformational adjustment (* N = 904*) and conformational change*
(N = 311*) energies (1 kJ mol⁻¹ per bin). Dashed lines illustrate the cumulative percentages of the*
distributions. (From Cruz-Cabeza, A. and Bernstein, J. (2014). Conformational polymorphism.
Chem. Rev., *114, 2170–91.With permission of American Chemical Society.)*

the crystal conformations differ in energy from the computed gas-phase conformers by less than 2.5 kJ mol⁻¹, with the percentages rising to 70 and 90 for 4.5
and 10.5 kJ mol⁻¹, respectively. These numbers give a measure of the differences
in molecular energy that can be expected from the fact that they are observed in
the solid rather than the gas phase.

The histogram for conformational change in Figure 5.26 demonstrates how
different conformers of a given relative energy are seen in different polymorphs.
Conformational change is also more common as $\Delta E_{\text{gas-gas}}$ between the two gas-
phase conformers becomes smaller. Over 50% of the pairs related by conformational
change differ by less than 3.5 kJ mol⁻¹, with the percentages rising to ~70 and ~90
for energies of 5.5 kJ mol⁻¹ and 14.5 kJ mol⁻¹, respectively. I note that most of the
conformational pairs differing by more than 14.5 kJ mol⁻¹ consist of a very stable
conformer with a strong intermolecular interaction and a second conformer for
which such an intermolecular interaction is broken. Some of the outstanding
examples are given in Figure 5.27.

The iconic polymorphic ROY system can be examined to see how these principles apply in practice. As seen in Figure 5.27, the conformation can be defined
by essentially one rotatable bond with the torsion angle θ, which varies from 21.7°
to 112.8°. Which conformations are related by conformational change and which
are conformational polymorphs? The potential energy surface for this torsional

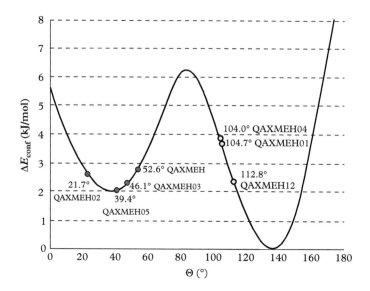

Figure 5.27 *Gas-phase potential energy surface of ROY as a function of the torsion angle Θ calculated using the B97-D/cc-pVTZ basis set. The experimental conformations found in the seven determined ROY polymorphs are plotted as red and yellow circles. (From Cruz-Cabeza, A. and Bernstein, J. (2014). Conformational polymorphism. Chem. Rev., 114, 2170–91. With permission of American Chemical Society.)*

parameter is also shown in Figure 5.27 with the locations of the various polymorphs on that curve. It is clear that the four molecules with the low torsion angles are members of a group related by conformation change, and similarly for the three molecules with the high torsion angles. On the other hand these two groups comprise conformational polymorphs.

Some additional aspects of conformational polymorphism have also been noted. Many of these are based on statistics of the sample reported in 2015, but it seems unlikely that the overall trends will change significantly with the increasing numbers of reported structures. Of the polymorphic structures chosen for study in the CSD 89.2% had two polymorphs, 8.8% had three, and only twenty-six molecules have four polymorphs or more (Figure 5.28).

It is important to recall that these data are based on reported crystal structure determinations that have been carefully vetted to this study. A survey of reported polymorphism in, say, pharmaceutically important compounds, would reveal a very different picture. For example, seventy-eight crystal forms (including solvates and hydrates) have been reported for atorvastatin (Jin 2012) and at least twenty-five forms have been reported for olanzapine. The current record holder in the CSD is flufenamic acid (Lopez-Mejias et al. 2012) with nine structures in the CSD.

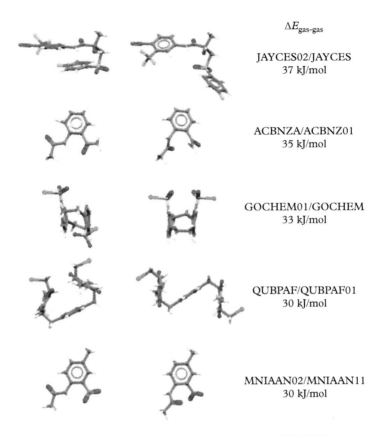

$\Delta E_{\text{gas-gas}}$

JAYCES02/JAYCES
37 kJ/mol

ACBNZA/ACBNZ01
35 kJ/mol

GOCHEM01/GOCHEM
33 kJ/mol

QUBPAF/QUBPAF01
30 kJ/mol

MNIAAN02/MNIAAN11
30 kJ/mol

Figure 5.28 *Examples of gas-phase conformers (denoted by REFCODEs) related by high $\Delta E_{gas\text{-}gas}$ energy differences. The more stable conformers are shown on the left. (From Cruz-Cabeza, A. and Bernstein, J. (2014). Conformational polymorphism. Chem. Rev., 114, 2170–91. With permission of American Chemical Society.)*

Some of the more dramatic examples of conformational polymorphism are given in Figure 5.29.

In addition to the points made above, I note the following features of conformational polymorphism. Rotatable bonds of different nature exhibit different propensities for conformational change and adjustment and polymorphic molecules with the higher propensity tend to exhibit conformational polymorphism. As a consequence of this observation it is also true that polymorphic molecules, prone to both change and adjust, are likely to display a rich polymorphic landscape. Finally, conformational polymorphs might be harder to crystallize than non-conformational polymorphs because the conformer might be less accessible, and

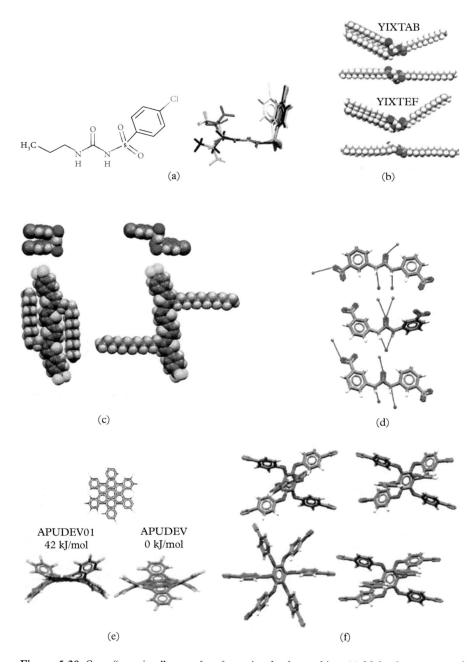

Figure 5.29 *Some "notorious" cases of conformational polymorphism. (a) Molecular structure of chlorpropamide, and the five different conformers (the record) found in polymorphs BEDNIG12/02/03/06/05. (b) Bent and linear conformations of two saturated triglycerides YIXTAB and YIXTEF. (c) Combinations of (top) bis(4-bromobenzenesulfonyl)amine as in polymorphs*

the resulting structures might differ considerably in properties, especially those that depend on molecular structure (see Chapter 6).

5.10.5   Comparison of energetic environment

Since the energetic environment of a single reference molecule must vary from one polymorphic structure to another, the similarities or differences in that energetic environment can be used to determine the identity of polymorphic structures. Dunitz and Gavezzotti have developed a method for examining and comparing the energetic environment (Dunitz and Gavezzotti 2005a; Gavezzotti 2007a).

The method is based on selecting a reference molecule in the crystal structure and calculating the *molecule–molecule* interaction energies for the surrounding molecules. The intermolecular distances are based on those calculated between the centers of mass of the molecular pairs. In the Gavezzotti approach the resulting interaction energy is calculated using the PIXEL semiclassical density sums (SCDS) method (Gavezzotti 2002, 2003). The PIXEL method, based on integration over the molecular charge density, partitions the energy into coulombic, polarization-dispersion, and repulsion contributions, which makes it particularly suitable for molecular crystals.

To compare, say, two polymorphic structures the analysis requires using the crystallographic information (cell constants and space group and symmetry operators) to generate the twelve nearest neighbors of the reference molecule based on the centers of gravity. The molecule–molecule interaction energy is then calculated preferably, according to Gavezzotti, using the PIXEL algorithm, but molecular mechanics may also be adequate. In addition to a numerical value for the energy (associated with a molecule–molecule distance) the symmetry relationship of the molecular pair is noted. The values are then ranked in decreasing order; those above 3–5% of the total are usually sufficient to provide characterizing data for the polymorph. The numbers may be compared for two or more polymorphs, but a graphical representation usually reveals the identity of two polymorphs or lack thereof. The graphical comparison of the two polymorphs of benzidine with $Z' = 4.5$ is given in Figure 5.30a. It is readily seen that there is no overlap among the points, confirming that they are different structures. The graphical comparison for the two polymorphs (B (HEYHUO01 $Z' = 4$) and C (HEYHUO02 $Z' = 12$))

YABKUI and YABKUI01 an oligothiophene as in polymorphs KISQEJ and KISQEJ01. (d) Conformations of 1,3-bis(m-nitrophenyl)urea found in alpha (SILTOW), beta (SILTOW01), and delta (SILTOW11) polymorphs that are also concomitant polymorphs. The red lines indicate termini of hydrogen bonds. (e) Left, high-energy and right, low-energy conformations adopted by a fluorinated benzocoronene molecule with REFCODE APUDEV. (f) Four different conformations found in the four known conformational polymorphs of the hexa-host molecule hexakis (4-cyanophenyloxy)benzene. (All from Cruz-Cabeza, A. and Bernstein, J. (2014). Conformational polymorphism. Chem. Rev., **114**, *2170–91. With permission of American Chemical Society.)*

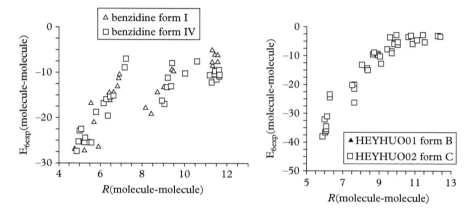

Figure 5.30 *Distance/energy plots for (a) the two polymorphs of benzidine with Z' = 4.5; (b) two polymorphs of 4,4-diphenyl-2,5-cyclohexanedione. (From Bernstein J., Dunitz, J. D. and Gavezzotti, A. (2008). Polymorphic perversity: crystal structures with many symmetry-independent molecules in the unit cell.* Cryst. Growth Des., **8**, *2011–18. With permission of American Chemical Society.)*

is given in Figure 5.30b. In this case there is essentially a complete overlap, indicating that they are one and the same crystal structure. (Recall that the Hirshfeld fingerprint plots clearly indicated that $Z' = 4$ for Form C, rather than the reported value of 12.)

5.10.6 Partitioning of lattice energy

The CSP methods described above are all based to some extent on the calculation and subsequent ranking of lattice energies. As described in Section 5.7, this is a single number that represents contributions from a variety of interactions, generally calculated on the basis of the sum of interactions from each atom in the molecule. While these "global" energies are useful, or even essential in determining the most likely computed structure(s) they provide no information on the various contributions to the total energy, either in terms of the nature of the interactions or specific atoms in the molecule for which contributions to the total may play a major role in determining the structure. Moreover, in terms of the origin and understanding of a polymorphic system such details can be particularly useful for, say, developing strategies to search for additional solid forms.

The information required to meet these demands can be obtained by partitioning of the total lattice energy in a number of ways. First, the contribution of all the interactions from each atom to the total lattice energy can be extracted from the global total energy. This provides information, for instance, on whether the energy contribution from one or a small number of atoms dominates total energy. In much of the current literature such observations are based simply on the presence of short intermolecular distances, assuming, often incorrectly, that short distances are necessarily equivalent to increased stabilization energy (Jeffrey and Saenger, 1991; Dunitz and Gavezzotti 2005a). This partitioning of the energy can provide detailed information on the distribution of the contributions of each atom to the total lattice energy.

For instance, one very early attempt at such a partitioning of lattice energies calculated on the basis of semi-empirical atom–atom potentials (Bernstein and Hagler 1978) on a dimorphic pair of a dichlorobenzylideneaniline. The initial working assumption was that the Cl···Cl interactions dominated the structures and that the difference in their energetic contribution in the two structures could help to explain the origin of the polymorphism. The partitioning led to the conclusion that in neither case was the lattice energy dominated by the contribution from a single atom, but rather was comprised of small contributions from all of the atoms. This is not inconsistent with the current views of Dunitz and Gavezzotti (2005a).

A further refinement of this approach is to partition the energy contribution on each atom according to the nature of the interaction that leads to that energy. A very useful computational and graphic tool, based on a concept described as *energy frameworks,* has recently been developed to facilitate and present that partitioning (Turner et al. 2015; Dhananjay et al. 2016). This tool is an extension of the Hirshfeld surfaces described earlier.

The energy partitioning is based on the development of the efficient calculation of the four individual components of the total energy:

$$E_{tot} = E_{ele} + E_{pol} + E_{dis} + E_{rep} \qquad (5.9)$$

where the subscripts on the right-hand side represent the *electronic, polarization, dispersion,* and *exchange-repulsion* energy respectively (Turner et al. 2014). The individual energy components between molecular pairs can then be represented as cylinders joining the centers of mass of the molecules, with the radius of the cylinder proportional to the magnitude of the interaction energy. The resulting networks comprise an "energy fingerprint" of the structure, displaying the relative strengths of the features of molecular packing in different directions. The four contributors may be color coded to aid in comparison of the role of each in a particular structure (e.g., Figure 5.31) or in comparing polymorphic structures (Figure 5.32).

Variations on this energy framework approach have been presented by Shishkin et al. (2012a, 2012b) using an energy vector approach and Bond (2014) who applied similar techniques to the output of Gavezzotti's *PIXEL* program. These energy frameworks can be expected to provide a great deal of insight into the subtleties of the similarities and differences in interaction in polymorphic structures in particular, and more generally in organic solid-state chemistry (Dunitz and Gavezzotti 2012).

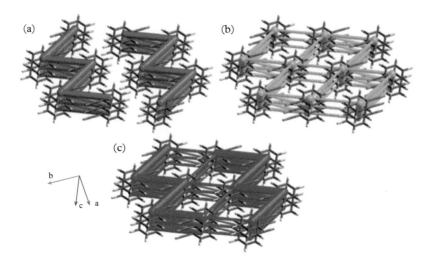

Figure 5.31 *Example of use of energy frameworks for 4-fluorobenzonitrile: (a) electrostatic component in red; (b) dispersion component in green; (c) total interaction energy in blue. (From Turner, M. J., Thomas, S. P., Shi, M. W., Jayatilaka, D., and Spackman, M. A. (2015). Energy frameworks: insights into interaction anisotropy and the mechanical properties of molecular crystals. Chem. Commun., 51, 3725–38. With permission of Royal Society of Chemistry.)*

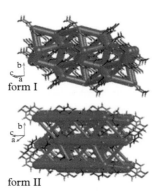

Figure 5.32 *Comparison of the total energy frameworks for the two polymorphs of paracetamol. (From Turner, M. J., Thomas, S. P., Shi, M. W., Jayatilaka, D., and Spackman, M. A. (2015). Energy frameworks: insights into interaction anisotropy and the mechanical properties of molecular crystals. Chem. Commun., 51, 3725–38. With permission of Royal Society of Chemistry.)*

5.10.7   Comparison of X-ray powder diffraction patterns

As noted in Chapter 4, the XRPD of a solid serves as a fingerprint of that solid. Hence, comparison of the XRPD patterns of different solid samples of a compound serves as an excellent test of the presence or absence of polymorphism among the samples examined. Such a comparison very often involves a visual inspection of the patterns in question, which requires the application of "fuzzy logic" in making a judgment of identity or lack thereof (Bernstein 2017; Rouvray 1995, 1997). No two XRPD patterns—indeed, no two measurements of any sort, especially based on point for point digital data—are ever exactly identical, so one must rely on knowledge and accumulated experience with the technique and often with the material in question to make this distinction. That kind of judgment is part and parcel of the training and practice of chemistry.

The comparison of two or more XRPD patterns is, of course, amenable to computer methods, and considerable efforts have been made to facilitate that mode of comparison (Karfunkel et al. 1993). This computational comparison of XRPD patterns is, in principle, similar to the comparison of spectra (De Gelder et al. 2001).

Specific applications to XRPD have been developed by Hofmann and Kuleshova (2005) who proposed a similarity index for automated comparison of powder diagrams. The basis for comparison is the *integrated* powder patterns rather than the digitally recorded experimental patterns. The measure of similarity is an index defined as the mean difference between a normalized simulated pattern and an observed integrated powder pattern. Since the basis of comparison is a simulated pattern generated from a result of a CSP study or a structure solution attempt from a powder pattern, the method is not immediately applicable to the comparison of two or more experimental powder patterns.

For a variety of experimental reasons (e.g., preferred orientation, presence of amorphous material, problems with signal-to-noise ratio, unavoidable strong background, variable and/or broad peak shapes) peak positions in XRPD patterns are often difficult to locate consistently. Gilmore et al. (2004) and Barr et al. (2004) used full pattern profiles to overcome these problems in attempting to compare diffraction patterns. In a qualitative proof-of-concept experiment on twenty-one pharmaceutical examples five polymorphs were expected. The data yielded six well differentiated clusters, which suggested that one of them may be a mixture, a question that was successfully settled by resorting to quantitative analysis.

Another application of the full peak profile pattern matching of experimental XRPD patterns (and other spectral data) has been described by Ivanisevic et al. (2005). This method was applied in a wide-ranging screen of pharmaceutically acceptable co-crystals of carbamazepine that yielded twenty-seven unique solid forms of the compound and eighteen coformers (Childs et al. 2008). The computational technique for comparing experimental powder patterns takes into consideration effects of preferred orientation and/or systematic peak shifting. Matching algorithms are based on peak position and/or envelope matching, and can qualitatively identify mixtures. A test case on ~200 capillary crystallizations of buspirone hydrochloride yielded five unique crystal forms, an amorphous form, and three groups of mixtures (Park et al. 2007). As of this writing the software, *PatternMatch,* is a proprietary SSCI in-house program.

It is worthy of note that the ICDD also has proprietary software (Integral Index) based on full pattern comparison for the automated comparison of experimental and calculated XRPD patterns (Faber and Blanton, 2008).

Finally, the CCDC developed software for comparison of (mainly) calculated XRPD patterns from the deposited structural data in the CSD in order to sort out true polymorphs from the multiple reported structures (van de Streek and Motherwell 2005).

The above discussion hints at the difficulty of automating the comparison of experimental XRPD patterns, and indeed there are still few published accounts of such studies, due mainly to the exigencies of the measurements. One of the groups of authors have commented on the situation noted in the opening of this section: "The best algorithms are still inferior to manual human matching, primarily due to the common problems with lab-generated XRPD data that are difficult to address in fully automated programs…performance of the algorithm was evaluated on six different data sets and shown to be comparable to that of a trained human expert (though still somewhat worse on average)" (Ivanisevic et al. 2005).

6

Polymorphism and structure–property relations

Polymorphic modifications of a given compound show significant differences in chemical behavior.

Cohen and Schmidt (1964)

There are two strategies for studying structure/property relationships: maximize the number of observations or minimize the number of variables.

Willis (1986)

6.1 Introduction

The design and preparation of materials with desired properties are among the principal goals of chemists, physicists, materials scientists, and structural biologists. Achieving that goal depends critically on understanding the relationship between the structure of a material and the properties in question. To be most effective, studies of structure–property relationships generally require systematically eliminating as many as possible of the structural and composition variables that play the most important role in determining the particular property under investigation. For molecular materials, a typical strategy might involve, for instance, a methodical variation in the mode or type of substitution on one part of the molecule in order to test a particular hypothesis about the causal relationship between structures and properties. Variations in substituents, while they do often result in changes in structure, and the corresponding changes in properties, also lead to perturbations in the geometric and electronic structure of the molecules in question. In such cases, changes in properties cannot always be correlated directly with changes in structure. The existence of polymorphic forms provides a unique opportunity for the investigation of structure–property relationships, since by definition the only variable among polymorphs is that of structure, and one of the most effective strategies for studying structure–property relations has been to follow the behavior of a physical property through a polymorphic phase change. The study of the thermodynamics, kinetics, and mechanism of phase transitions by most or all of these analytical or physical techniques is a discipline in itself (e.g., Verma and

Polymorphism in Molecular Crystals. Second Edition. Joel Bernstein. © Joel Bernstein 2020. Published in 2020 by Oxford University Press. DOI: 10.1093/oso/ 9780199655441.001.0001

Krishna 1966; Rao 1984, 1987; Bayard et al. 1990; Rao and Gopalakrishnan 1997), and is not addressed specifically here; rather, the polymorphic systems are considered as points in the multidimensional phase space representing structures with different packing and possibly different molecular conformations for the utilization of the study of structure–property relations. For a polymorphic system, differences in properties among polymorphs must be due only to differences in structure. As a corollary to this principle, a constancy in properties for a polymorphic system indicates a lack of structural dependence on that property, at least within the limitation of the structural variation through that series of polymorphic structures.

For molecular materials, studies of structure–property relations fall into two broad categories. In the first category, the properties under investigation are due to strong interactions between neighboring molecules, and one wishes to study the changes in bulk properties resulting from differences in the spatial relationships among molecules in the crystal. In the second category, we seek information on variations in properties related to differences in molecular structure, generally, for reasons noted in the previous chapter, in molecular conformation. Cases of conformational polymorphism, discussed in detail in the previous chapter, also provide opportunities for the study of the influence of crystal forces on molecular conformation. This chapter is devoted to describing representative examples of structure–property studies from both of these categories. In view of the increasingly extensive literature it is not possible to be inclusive in this regard, but it is hoped that the description of a number of case studies will be sufficient to demonstrate the utility and advantages of choosing polymorphic systems, when possible, for the investigation of structure–property relationships. In addition to providing information on the structure–property relationships, many of these studies also reveal the experimental strategies and procedures required to obtain the polymorphic modification(s) with the specifically desired properties (Bernstein 1993).

There obviously have been many advances in all aspects of this subject since the publication of the first edition. For example, see the recent review by Chung and Diao (2016) on organic electronic materials. In another recent aspect, the use of substrate-induced polymorphic structures described in Section 7.4.6 has led to the study of a variety of structures and properties, as reviewed by Jones et al. (2016). Some of the most notable examples not noted in those two reviews have been included in this chapter. Much of the material from the previous edition remains, since it still demonstrates the principle of extracting knowledge and understanding from the study and comparison of properties of polymorphic materials.

6.2 Bulk properties

6.2.1 Electrical conductivity

Molecular materials are traditionally considered to be electrical insulators, but the discovery in the early 1970s of metallic conductivity in crystals of the π molecular complex of tetrathiafulvalene (TTF) **6-I** and tetracyanoquinodimethane (TCNQ) **6-II** (Coleman et al. 1973; Ferraris et al. 1973) led to a revolution in thinking about these materials in particular, and the potential for organic materials in general, as the basis for the next generation of electronic components (Wudl 1984; Williams et al. 1985), as well as the Nobel Prize in Chemistry for the year 2000. In contrast to the vast majority of previously known π molecular complexes that crystallize with plane-to-plane stacks of alternating donors and acceptors (Herbstein 1971, 2005) (Figure 6.1), the complex of **6-I** and **6-II** crystallized with segregated stacks of molecules along the same crystallographic axis (but not mutually parallel molecular planes), each stack containing only one type of molecule. This structural feature was shown to be a necessary condition for electrical conductivity in these materials, although the mixed mode of stacking is generally considered to be the thermodynamically preferred one (Shaik 1982).

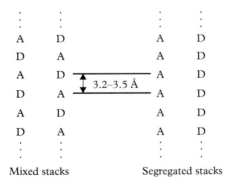

6-I   6-II   6-III

Proof of the relative stability of the mixed and segregated stack motifs, and a recipe for obtaining crystals of the latter, came with the discovery of a pair of polymorphic 1:1 complexes of **6-II** with **6-III** (Bechgaard et al. 1977; Kistenmacher et al. 1982). The red, transparent, mixed-stack form of the complex is a semiconductor, while the black, opaque structure with segregated stacks is a conductor (Figure 6.2).

This finding demonstrated that the presence of segregated stacks is a necessary condition for electrical conductivity. Reflecting the relative stabilities for the two stacking modes noted above, crystals of the red semiconductor form were obtained from a thermodynamic or equilibrium crystallization: equimolar solutions of the

```
     ⋮    ⋮              ⋮    ⋮
     A    D              A    D
     D    A              A    D
     A    D  ⎯⎯⎯⎯⎯        A    D
              ↕ 3.2–3.5 Å
     D    A  ⎯⎯⎯⎯⎯        A    D
     A    D              A    D
     D    A              A    D
     ⋮    ⋮              ⋮    ⋮

   Mixed stacks       Segregated stacks
```

Figure 6.1 *Schematic diagram of the mixed-stack and segregated-stack motifs for packing of π molecular charge-transfer complexes (from Bernstein 1991b, with permission).*

(a)   (b)

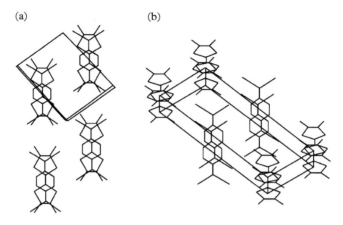

Figure 6.2 *Views of the two polymorphic structures of **6-II**:**6-III**. In both cases the view is on the plane of the tetracyanoquinodimethane molecule **6-II**. (a) The red, transparent, mixed-stack complex, a semiconductor; (b) the black opaque, segregated-stack complex, a conductor (from Bernstein 1991b, with permission).*

donor and acceptor in acetonitrile are mixed and allowed to evaporate slowly. On the other hand, crystals of the black form were obtained from a kinetic or non-equilibrium crystallization: hot equimolar solutions of the donor and acceptor in (the same) acetonitrile solvent are mixed and cooled rapidly. Some microcrystals of the resulting black powder are then used as seeds to obtain larger crystals of the segregated-stack black form.

Non-equilibrium crystallization methods, in particular electrochemical techniques, have become standard procedure for obtaining crystals of organic conductors, in part because of the ability to control and reproduce the crystallization conditions (Williams et al. 1991). However, there are examples of cases of polymorphic charge-transfer complexes in which a mixed-stack structure and a segregated-stack structure were obtained concomitantly under nearly equilibrium conditions of slow diffusion, although, not surprisingly, under these conditions, the segregated-stack form was obtained "with much difficulty" (Nakasuji et al. 1987). The segregated-stack modification exhibits metallic conduction, while the mixed-stack form is a semiconductor (Imaeda et al. 1989). Similar principles may be applied and similar observations have been made when the structural units contain organometallic complexes (Legros and Valade 1988; Cornelissen et al. 1992, 1993; Almeida and Henriques 1997).

The utilization of polymorphism in understanding the connection between structure and electrical conducting properties is perhaps most poignantly represented by the so-called "ET" salts of **6-IV** (Bechgaard et al. 1980). Generically designated $(ET)_2X$, they generated intense activity because some exhibited the highest then known T_c values for organic superconductors. Because the characterization of the solid-state properties is the essence of the investigation of these materials, they were intensively studied, and many were found to be polymorphic (Kikuchi et al. 1988; Williams et al. 1991; Schlueter et al. 1994; Saito 1997). For $X = I_3^-$, there are at least fourteen known phases, and in such cases learning to define and control the growth conditions leading to a specific desired polymorph or the avoidance of an unwanted polymorph is one of the challenges facing workers in this field. A common method of crystal growth for these compounds is electrocrystallization. Experiments can extend to periods up to months, with a number of polymorphs appearing on the same electrode, in many cases with indistinguishable colors or crystal habits. Only the physical characteristics (conductivity, spectral response, X-ray diffraction pattern, etc.) can be used to distinguish them, and often the identification of the various phases requires the examination and characterization of each individual crystal in a batch (Kato et al. 1987; Wang et al. 1989). For example, Kobayashi et al. (1987) identified the presence of the α, β, γ, δ, τ, and κ polymorphs of $(ET)_2^+ I_3^-$ in the same experiment. Yoshimoto et al. (1999) showed by solution-mediated transformations that the β form is more stable than the α form in the temperature range 0–50 °C, leading to the selective and controlled growth of large crystals of each form.

6-IV

As an example of the type of variation observed, the β and κ phases can be compared (Figure 6.3) along with the schematic packing motifs of other polymorphic forms. The former, apparently favored by thermodynamic crystallization conditions (e.g., low current density) is a centrosymmetric triclinic structure with one formula unit of the salt in the unit cell. The symmetry arguments require that the anion lies on a crystallographic inversion center, and that the donor ET molecules all be parallel, as shown in Figure 6.3. The structure is thus characterized by stacks along the diagonal of the unit cell, with strong intermolecular S···S interactions between stacks. T_c for this phase is 1.4 K. The κ phase may be obtained, concomitantly with the α and θ phases, in a tetrahydrofuran solution under N_2 with a mixed supporting electrolyte of $(n\text{-}C_4H_9)_4NI_3$ and $(n\text{-}C_4H_9)_4NAuI_2$ at 20 °C and constant current. It is characterized by the formation of dimers, crystallizing in a centrosymmetric crystal structure with one layer per unit cell in the $P2_1/c$ space group, and exhibiting a T_c at 3.6 K.

Understanding the structure–property relationship in these materials is crucial to the rational development of organic conductors and superconductors with increasingly higher T_c values. The plethora of polymorphic structures can easily lead the unwary investigator astray, but it provides an opportunity not available in many other systems for isolating the structural characteristics required for a very specific physical property. Williams et al. (1991) have also pointed out that the *isostructural* series of salts are important for the information they can yield. In this

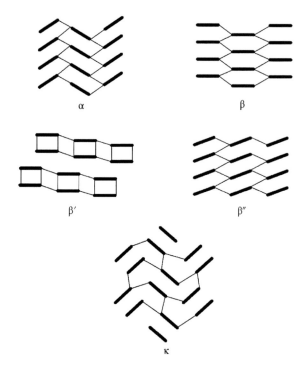

Figure 6.3 *Schematic representation of the network of ET* **6-IV** *molecules in five phases of the salt* $(ET)_2^+ I_3^-$. *Thick lines indicate ET molecules, and thin lines indicate short intermolecular S⋯S contacts. (A. Kini, personal communication).*

case, the *structural* parameter is kept nearly fixed (or only slightly perturbed) and the effect of *chemical* perturbations can then be evaluated. For instance, the ET salts tend to crystallize with the same structure, so the effect on the properties of varying the anion can be studied in a series of isostructural salts.

While polymorphic structures can reveal subtleties about the structure–property relationship, they do not necessarily provide the keys to this understanding, in particular when the structural changes are more subtle than their manifestations in particular physical properties. The study of the system of 1,6-diaminopyrene **6-V** and chloranil (CA) **6-VI** is edifying in this regard since it indicates both the utility and some of the limitations of studies of polymorphic systems. **6-V** is a relatively strong donor that was the subject of an early study as a component of potential electrically conducting complexes (Kronick and Labes 1961; Kronick et al. 1964). Its complex with **6-VI** is polymorphic, with a dark brown α form and green β form. They may be obtained concomitantly (see Section 3.6) from benzene solution by both slow cooling and "prolonged" slow evaporation (Inabe et al. 1996). Structurally, both forms are mixed-stack complexes, and on the basis of the previous discussion might be expected to be electrical insulators. They do differ in crystal structure, however, as can be seen in Figure 6.4. In the

(a)

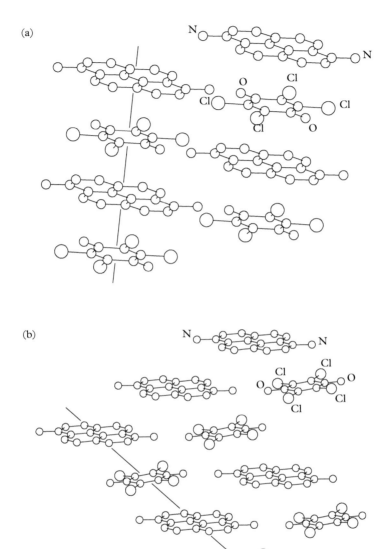

(b)

Figure 6.4 *Packing diagrams of the two forms of the complex formed by **6-V** and **6-VI**. (a) α form. The vertical stacks are along the c crystallographic axis. (b) β form. The mixed donor/acceptor stacks are along the diagonal. The structure also exhibits donor/donor and acceptor/acceptor interactions (after Bernstein 1999, with permission).*

α form the stacks along the *c* axis are vertical, leading to a significant donor–acceptor overlap reminiscent of the earlier mentioned mixed-stack structure. In the β form the *c*-axis stacks are offset, leading to a reduction in the donor–acceptor overlap and an increase in both donor–donor and acceptor–acceptor overlap. Reiterating the meaning of the concomitant crystallization in practical terms, according to the authors the two polymorphs are nearly isoenergetic at room temperature.

6-V   6-VI

   Upon increasing the temperature of the α phase, the electrical conductivity rises by approximately seven orders of magnitude, although the onset of that increase varies from crystal to crystal. The same phenomenon is observed upon formation of a pellet prepared under pressure for conductivity measurements. In spite of the change in conductivity, the IR spectra are identical for the solids before and after the change. The full X-ray crystal structure determination of the highly conducting material indicated that no statistically significant structural change takes place. Additional physical measurements (ESR, solid-state CP/MAS NMR) on both the "pristine" α form and the "low resistance" α′ form do not provide any clue as to the source of the vastly increased electrical conductivity (Goto et al. 1996). This system provides a poignant reminder that some physical phenomena are based on subtle or small perturbations which are not directly observable in structure determinations or are beyond the detection limits of other physical techniques. The closing statement of the authors that "no ordinary conduction mechanism can rationalize the low resistivity value [of the low resistance α′ form]" is a challenge to our ingenuity in measuring and interpreting physical phenomena.

   An increasing variety of recent conductivity studies on polymorphic single component systems have shed considerable light on this structural connection (Purdum et al. 2016). The substances studied were dimorphic chlorinated naphthalene diimides **6-VII** with the aim of understanding the connection between semiconductivity and structure. The authors studied the kinetics of the transformation between the triclinic and monoclinic forms, but also found that the thin films of the triclinic form consistently outperformed the thin films of the monoclinic form, which is in contrast to the conductivity performance of the single crystals of the two forms. They suggested that the structure with the smaller in-plane anisotropy is more beneficial for efficient lateral charge transport in polycrystalline devices.

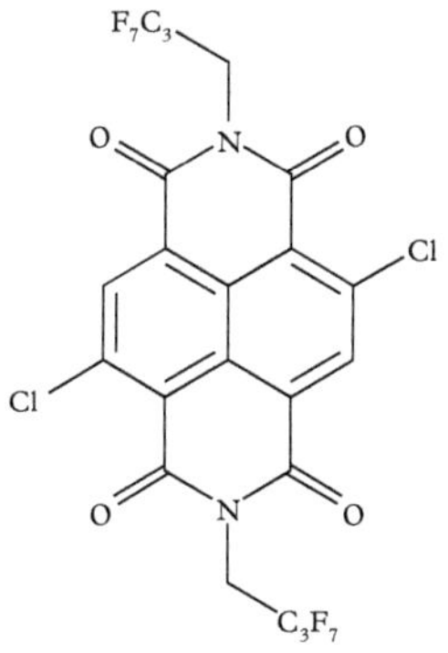

6-VII

Similarly, Hiszpanski et al. (2014) and Hiszpanski and Loo (2014) found two previously unreported polymorphs of hexa-cata-hexabenzocoronene (HBC) **6-VIII**, and characterized all three, including "tuning" of the crystal structure, and the resulting out-of-plane orientation of thin films, to study the manifestations of the performance as thin-film transistors. They found that molecular geometry and polymorphism are equally important—the field-effect mobility changes by an order of magnitude by changing one of them. The same group found that substituting hydrogen with fluorine in HBC affects the polymorphic behavior: two of the hydrocarbon polymorphs are obtained, but the preference is altered with increasing fluorine substitution (Hiszpanski et al. 2017).

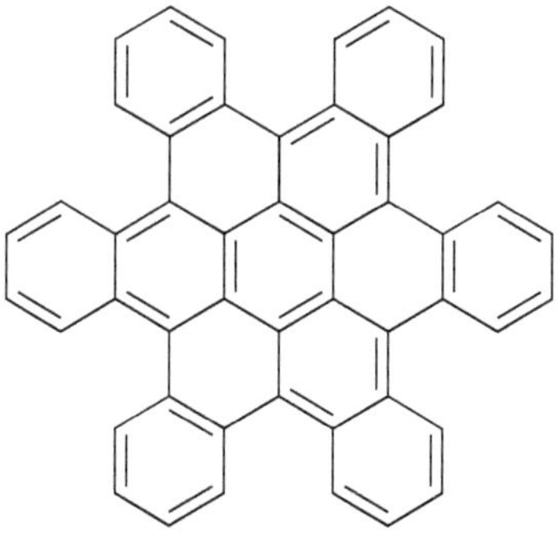

6-VIII

The nanoconfinement method described in Section 4.11 for directing polymorphism crystallization has been applied to the study of structure–property relations (Diao et al. 2014). A flow-enhanced crystallization technique was used to identify existing polymorphs, prepare a new form, and to selectively prepare individual forms of **6-IX** at ambient conditions. The molecular packing was

correlated with charge transport properties using charge carrier mobility measurements and quantum mechanical calculations.

6-IX

The eventual application of organic field-effect transistors depends on the ability to prepare layers with controlled structures. A recent report on the preparation
of blends of dibenzo-TTF **6-X** and polystyrene **6-XI** demonstrated that such
control could be achieved and also resulted in the first preparation of a thin film
of the α form (Galindo et al. 2017).

6-X   **6-XI**

6.2.1.1   *The neutral/ionic transition in charge-transfer complexes*

One particular phenomenon associated with mixed-stack charge-transfer complexes that has received considerable attention is the neutral-to-ionic (N–I) transition. As noted, the vast majority of π charge-transfer complexes crystallize in the
mixed-stack motif (Herbstein 1971). Except for cases in which the ionization
potential of the donor is very low and the electron affinity of the acceptor is very
high, the ground state of these complexes may be formally described as stacks of
alternating neutral donors and acceptors. At the other extreme, in the formal definition of the ionic phase, each donor will have lost an electron and each acceptor

will have gained one, and the structure will now consist of alternating anions and cations. In the range between these two extremes, there is a group of compounds forming mixed-stack complexes which can exist in both the ionic (I) and neutral (N) states, with transitions between them (Torrance et al. 1981; Jacobsen and Torrance 1983). In principle, this may also be considered a case of polymorphism, albeit with the definitional complication of whether the components are indeed the same for both phases. However, both dissolve to give the same solution, although the equilibrium distribution between ionic and neutral species may depend on the polarity of the solvent.

$$\cdots \mathbf{D^\circ A^\circ D^\circ A^\circ D^\circ A^\circ D^\circ A^\circ D^\circ A^\circ D^\circ A^\circ D^\circ A^\circ} \cdots$$ Neutral phase (N)

$$\cdots (\mathbf{D^+ A^-})(\mathbf{D^+ A^-})(\mathbf{D^+ A^-})(\mathbf{D^+ A^-})(\mathbf{D^+ A^-}) \cdots$$ Ionic phase (I)

The prototypical N–I system is formed by TTF **6-I** and CA **6-VI**, although others have also been reported (Katan and Koenig 1999). The nature of the structural change is represented in Figure 6.5. At room temperature and atmospheric pressure, the neutral (monoclinic, $P2_1/n$) phase is stable. The asymmetric unit is composed of one half of each of the centrosymmetric donor and acceptor molecules.

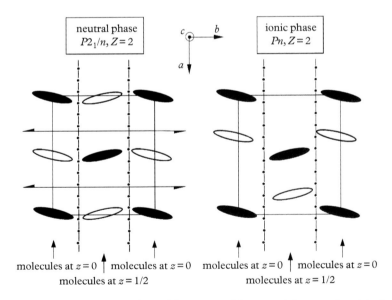

Figure 6.5 *Schematic representations of the structural features of the neutral and ionic phases of the complex of tetrathiafulvalene (TTF)* **6-I** *and chloranil (CA)* **6-VI***. In both cases the view is on the ab plane. In the neutral phase the molecules lie on crystallographic inversion centers at (0, 0, 0) and (0, 1/2, 0), while in the ionic phases those points are no longer crystallographic inversion centers (based on Cailleau et al. 1997, with permission).*

The crystal structure generated by the space group symmetry (with each molecule lying on a crystallographic inversion center) leads to the mixed-stack structure with parallel stacks of regularly spaced alternating donor and acceptor molecules along the *a* crystallographic axis. Although formally represented as the neutral complex, the degree of charge transfer was determined to be approximately 0.3 (Girlando et al. 1983; Jacobsen and Torrance 1983; Tokura et al. 1985). This means that over the space-averaged structure approximately 30% of the TTF donors have lost electrons to the CA acceptors.

At 81 K, the first-order transition to the ionic state takes place leading to a degree of charge transfer of approximately 0.7, indicating that 30% of the donors and acceptors are still neutral. The structural change involves the retention of the glide plane and virtual loss of the screw axis (with concomitant loss of the inversion center), the space group being thus reduced to *Pn* (Le Cointe et al. 1995b), and the asymmetric unit being one full molecule each of TTF and CA. The lowered space group symmetry lifts the restrictions that led to the regular spacing and the structure is now characterized by pairs of molecules, rather than regular stacks, as indicated by the parentheses in the schematic notation for the ionic structure. The change in the degree of charge transfer is also reflected in the molecular geometry of the molecular moieties (Le Cointe et al. 1995b) as well as in the ^{35}Cl nuclear quadrupole resonance (NQR) response (Gallier et al. 1993; Le Cointe et al. 1995a).

The (*P,T*) phase diagram for the system has been prepared (Figure 6.6) using a combination of data from neutron scattering, NQR, vibrational spectroscopy,

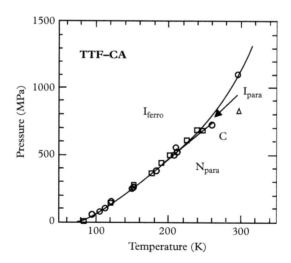

Figure 6.6 *(P, T) phase diagram of the TTF:CA system. Experimental points were obtained by a variety of methods, including neutron scattering O, NQR □, and conductivity Δ. C indicates the estimated critical point. N_{para}, I_{para}, and I_{ferro} indicate the neutral paraelectric, ionic paraelectric, and ionic ferroelectric phases, respectively (from Cailleau et al. 1997, with permission).*

and condu100tivity measurements (Cailleau et al. 1997). The neutral phase is para-electric, while the ionic phase has paraelectric and ferroelectric regions. The nature and mechanism of the phase transition has been reviewed and treated, as well as a number of other physical properties (Le Cointe et al. 1996; Cailleau et al. 1997; Horiuchi and Tokura 2008).

In closing this section I again refer the reader to the excellent review on subsequent developments of many aspects of electrical conductivity and polymorphism in molecular crystals by Chung and Diao (2016).

6.2.2   Organic magnetic materials

Preparing magnetic materials from organic molecules is no less a challenge than preparing electrically conducting organic materials, although activity in this field started somewhat later. The subject was reviewed by (Kinoshita 1994; Miller and Epstein 1994, 1996; Miller 1998; Kahn et al. 1999), and activity has increased significantly in recent years. The study of magnetism has traditionally been closely but indirectly related to that of polymorphism, in particular with the changes in magnetic behavior resulting from phase changes. The area has been dominated very much by physics and physical measurements and the emphasis has been phenomenological, focusing more on the temperature of phase transitions and the macroscopic nature of the changes in magnetic behavior associated with those phase transitions, rather than on the structural basis for those changes or differences in behavior. For examples of some early studies on the magnetic properties of polymorphic transition metal complexes see, for instance, Boyd et al. (1981), Scheidt et al. (1983), and Decurtins et al. (1983). The development of molecular magnetic materials by chemists expanded the field considerably to include an interest in the fundamental individual molecular basis for potential magnetic behavior, as well as the geometric and spatial requirements of the crystalline solids comprised of those molecular bases for different magnetic behavior (Kahn et al. 1999).

A design requirement of magnetism in molecular materials is the presence of unpaired electrons on the molecular species. The magnetic behavior is then determined by the environment of the electrons and the nature of the interactions among them. Thus, polymorphic structures can provide particularly useful information about the changes in magnetic behavior resulting from changes in molecular environment, and a number of case studies of magnetic behavior of polymorphic materials demonstrate the utility as well as some of the problems inherent in such studies.

One of the early purely organic materials to have been shown to undergo a well characterized ferromagnetic transition is the azaadamantane derivative **6-XII** (Dromzee et al. 1996). The material is dimorphic, and both forms appear concomitantly by evaporation from a diethyl ether solution at room temperature, suggesting that they are thermodynamically nearly equivalent. In fact, the authors noted that they did not succeed in determining the conditions for selectively crystallizing either form. The monoclinic α form undergoes a ferromagnetic transition at 1.48 K (Chiarelli and Rassat 1991). Views of the α and β forms are shown in Figure 6.7. It can be seen that there is a subtle difference in the environment of the oxygen atoms on which (at least formally) the unpaired electrons are located (as indicated in **6-XII**), and the crystal structure analysis suggests that the oxygen is disordered in the β phase. As yet, no detailed analysis of the structural basis for the magnetic behavior has been given, although Miller (1998) has noted that in the monoclinic phase the intermolecular NO· distances are shorter than in the orthorhombic phase. This is a system that can potentially provide additional useful information.

Me

Me

Me

O·

Me

N

O·

Me

N

6-XII

The sensitivity of magnetic properties to changes in polymorphic structure is demonstrated in the trimorphic system of decamethylferrocene **6-XIII** with TCNQ **6-II**. Two forms were originally prepared by Miller et al. (1987). The "thermodynamically favored" form is prepared from an acetonitrile solution refrigerated at −35 °C for more than two weeks, or electrochemically generated in

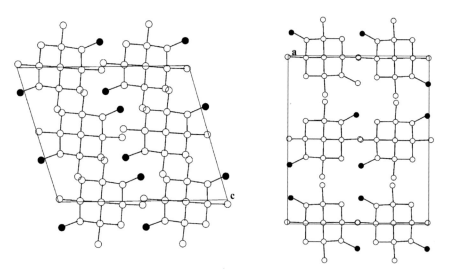

Figure 6.7 *Representation of the packing of the two forms of 6-XII. Oxygen atoms are denoted by solid circles. Left, the monoclinic α form; right the orthorhombic β form (from Dromzee et al. 1996, with permission).*

24 h. It appears as purple plates which behave as a paramagnetic material. The crystal structure contains "dimeric" units (Figure 6.8) which are stoichiometrically represented as $[Fc^*]_2^{+}[TCNQ]_2^{-}$, $(Fc^* = Fe(C_5Me_5)_2)$ and form the building blocks of a classic herring-bone structure of a molecular crystal (Desiraju 1989). The "kinetically favored" phase is prepared from warm solutions of donor and acceptor in dichloroethane or acetonitrile, from which the salt rapidly precipitates, yielding air-sensitive chunky green crystals. In this case, the basic building block is a one-dimensional mixed-stack chain (hence referred to as the "1-D" phase by the authors; Figure 6.8), and the crystals are metamagnetic. Form III was obtained by recrystallization of the complex from acetonitrile at $-20\ °C$. It is also purple in color but "upon close inspection" (Broderick et al. 1995) is distinctly different from Form I, appearing as parallelepipeds rather than as plates. It is an air-sensitive ferromagnetic material, which crystallizes in one-dimensional chains as in Form II but exhibits a slightly different color. There are a number of structural differences, though. The two Cp* groups within a donor are staggered in Form II, and eclipsed in Form III. Within a stack in Form II, all the acceptors are oriented in the same direction; in Form III they form a zigzag arrangement along the *b* axis. The inter-stack arrangements are also different. This system was revisited in 2013 to study the pressure dependence of the magnetic behavior of the two polymorphs (DaSilva and Miller, 2013).

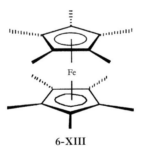

6-XIII

A number of additional polymorphic molecular materials exhibiting variations in magnetic behavior warrant mention here. The stable free radicals **6-XIV** and **6-XV** are known to be at least dimorphic (Banister et al. 1995, 1996; Palacio et al. 1997a, 1997b) and tetramorphic (Turek et al. 1991; Nakazawa et al. 1992), respectively. Both forms of **6-XIV** may be obtained by sublimation, the thermodynamically stable β form being produced slowly, while the triclinic α form is generated via a more rapid sublimation. They differ in the dihedral angle between the rings ($\alpha = 32.2°$; $\beta = 57.8°$) (Banister et al. 1995, 1996). Both structures are characterized by the formation of head-to-tail chains, all parallel in the β structure, but antiparallel in the α structure. The former undergoes weak ferromagnetic ordering below 35.5 K, whereas the latter orders antiferromagnetically below 8 K; attempts have been made to attribute these differences in behavior to the structural features distinguishing the two modifications.

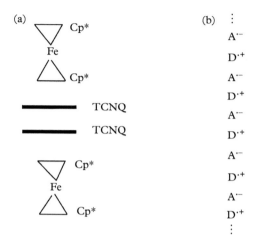

Figure 6.8 *(a) Schematic representation of the "dimeric" units which are the structural building blocks for the herringbone motif in the crystal structure of Form I of the complex* [Fc*]$_2^+$[TCNQ]$_2^-$. *(b) Schematic representation of the mixed stacks of anion and cation radicals in Forms II and III of the complex (from Bernstein 1999, with permission).*

6-XIV

6-XV

The four forms of **6-XV** have very different crystal structures (Miller 1998), and correspondingly different magnetic properties. At room temperature, the thermodynamically stable orthorhombic β form is obtained by slow evaporation from benzene at 20 °C. Other solvents and crystallization temperatures are used to generate the three additional forms (α: 1,1,2-trichloroethane at ~30 °C; γ: aceto-nitrile or chloroform at 65 ± 5 °C; δ: chlorobenzene at 132 °C). The α and δ phases are paramagnetic (Allemand et al. 1991; Tamura et al. 1993), while the γ phase orders antiferromagnetically at T_N = 0.65 K (Kinoshita 1994) and the β-phase undergoes a ferromagnetic transition T_c = 0.6 K (Tamura et al. 1991). Three polymorphs (α, β, and δ) of the compound have been prepared as a thin film (Fraxedas et al. 1999a) and the conditions and mechanism of the transition of the α to the β form have recently been studied in some detail (Fraxedas et al. 1999b).

Further studies on polymorphic derivatives of **6-XV** have been reported by the group of Veciana (Catala et al. 2006; Feher et al. 2008). In the former case the α forms of the pure radical **6-XVI** exhibits only antiferromagnetic interactions, while the β form shows competing ferro- and antiferromagnetic interactions. These interactions were then related to the hydrogen bonding networks which resulted in different relative orientations of the spin carriers. The latter case deals with a trimorphic uracil derived nitronyl nitroxide. Again, differences in hydrogen bonding characterize the three structures: one form contains a hydrogen-bonding tape with two molecules stacked in the tape; a second form exhibits layer-forming tetramers. Both are paramagnetic.

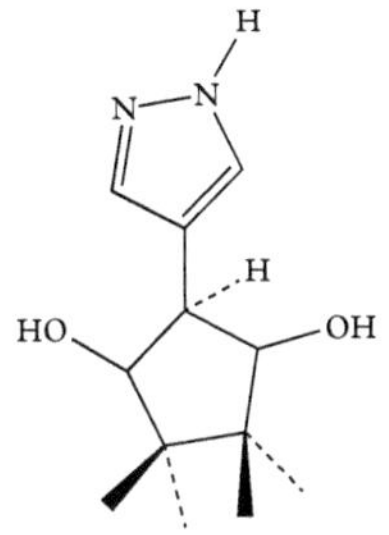

6-XVI

The closely related molecule **6-XVII** has been shown to be dimorphic (Sugawara et al. 1994). Both forms can be crystallized from diethyl ether, but at different temperatures: the bluish-purple blocks of the α phase above 4 °C, and the blue needles of the β phase below 0 °C. Selective crystallization can be accomplished by seeding. The α modification, structurally characterized by both intramolecular and intermolecular hydrogen bonds, is a three-dimensional ferromagnet. On the other hand, the β form exhibits intermolecular hydrogen bonds and π–π stacking, and its magnetic behavior was interpreted by the ferromagnetic ST model, which was accompanied at lower temperatures by a weak antiferromagnetic interaction (Matsushita et al. 1997). The authors concluded that the hydrogen bond is not only a structural feature, but also plays a role in the transmission of spin polarization between molecules. The structures and properties of different polymorphic forms, as well as the role and potential use of hydrogen bonds in generating structures of these (and other) materials as organic magnets was reviewed by Veciana et al. (1996).

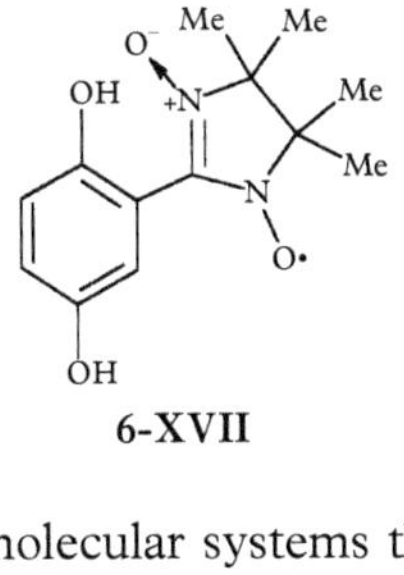

6-XVII

Two additional dimorphic molecular systems that have been studied for their magnetic behavior include the copper complex [Cu(cyclam)(TCNQ)$_2$] **6-XVIII** (Ballester et al. 1997) and the gold complex **6-XIX** (Leznoff et al. 1999). The common structural feature in the former is the presence of parallel chains of copper macrocyclic units linked by dimeric $[(TCNQ)_2^{2-}]$ units which are coordinated to the metal. The chains are connected by neutral TCNQ units. In one polymorph the nature of the connection leads to a two-dimensional network while in the second modification a three-dimensional network is generated. The magnetic behavior of the two forms is similar, but not identical. In both there is strong antiferromagnetic coupling between the organic radicals, and the dimeric $[(TCNQ)_2^{2-}]$ anion behaves diamagnetically below room temperature. On the basis of analysis of the distance between neutral and anionic TCNQ species (~0.2 Å shorter in Form 2 than in Form 1) the authors suggest that the formally neutral TCNQ is less so (i.e., more changed) in the former than in the latter, indicating more delocalization and hence accounting for a lower antiferromagnetic coupling between radicals in accord with the magnetic susceptibility measurements.

6-XVIII

6-XIX

In the gold complex **6-XIX**, there are differences in the packing modes that are particularly manifested in the intermolecular contacts involving the radical-bearing aminoxyl N-oxide fragments. This leads to differences in the (N–O)···(N–O) interactions between the two structures (α and β), which results in differences in the magnetic behavior. Both exhibit intermolecular antiferromagnetic interactions. The β modification shows no maximum in that behavior, the magnetic susceptibility rising with decreasing temperature. On the other hand, the α form goes through a maximum in the magnetic susceptibility at 3.5 K, a behavioral effect which could be successfully modeled based on the observed crystal structure.

A number of other systems involving magnetic properties of organic polymorphs are worthy of brief note here, to demonstrate the variety of issues that have been raised, if not always resolved, and the kinds of opportunities that polymorphic systems provide for investigating these systems.

Compound **6-XX** (tanane or TEMPO) has a high-temperature tetragonal phase that transforms at around 14 °C to an orthorhombic phase, which is both ferroelastic and ferroelectric, and a monoclinic phase has also been reported and prepared (Capiomont et al. 1972; Jang et al. 1980). Structural data (Capiomont et al. 1981) were used to model the transition (Legrand et al. 1982).

6-XX

Banister et al. (1996) prepared an organic free radical **6-XXI** not based on NO. It is concomitantly dimorphic by sublimation, and varying the conditions can lead to either the centrosymmetric $P\bar{1}$ α form (Banister et al. 1995) or the non-centrosymmetric *Fdd2* β form (Banister et al. 1996). The result is that in the α structure neighboring chains are antiparallel, while in β they are all parallel. Due to these symmetry considerations the α modification cannot exhibit a so-called "spin-canting" mechanism (Carlin 1989) which can account for the transition from a low-dimensional antiferromagnet to a weakly ferromagnetic state. However, these restrictions are absent in the β modification, and this mechanism, based to a large extent on the structural distinction between the two polymorphs, has been used to account for the fact that this β form is the first example of an open-shell molecule to exhibit spontaneous magnetization above liquid helium temperature (36 K).

6-XXI

Finally, the stable free radical diphenylpicrylhydrazyl **6-XXII** is the classic reference standard for ESR measurements, with solid samples showing stability over as long as thirty years (Yordanov and Christova 1997). Despite the common use of this material and a number of crystallographic studies, the polymorphic and solvate behavior apparently still have not been fully characterized. A variety of crystal modifications, both solvates and solvent-free, have been reported (Weil and Anderson 1965). Among the solvent-free modifications were included an amorphous form with a melting point (*sic*) of 137 °C, an orthorhombic form with a melting point of 106 °C (Williams 1965), and a triclinic form with melting point 128–129 °C (Williams 1965). Williams (1967) reported the structure determination of a benzene solvate, which was repeated by Boucherle et al. (1987). The structure reported by Kiers et al. (1976) was originally thought to be the triclinic form of Williams (1965) but turned out to be monoclinic (*Pc*) and to be an acetone solvate. A monoclinic form has been reported by Ellison and Holmberg (1960) and another modification, with similar but apparently not identical cell constants and unidentified space group has been studied by Prokop'eva and Davidov (1975). The solid-state magnetic properties, presumably of the orthorhombic form, were studied by two groups (Burg et al. 1982; Boon and Vangerven 1992) but the apparent structural variety and the variation of magnetic properties with crystal structure still have not been fully elucidated. Also worthy of note here is the fact that both a yellow and an orange modification of a 2,2-di(*p*-nitrophenyl)hydrazine derivative of **6-XXII** crystallize concomitantly from dichloromethane, with very similar molecular conformations (Wang et al. 1991). The color difference (*vide infra*) has been attributed to a difference in molecular packing.

6-XXII

Which of these structural factors, if any, determine the differences in magnetic properties? What kinds of experiments can be designed to make these distinctions? Are there additional polymorphs that reveal the subtleties of structural differences and physical properties? There are many questions, challenges and opportunities for discovery and understanding in systems of this type that exhibit distinguishable structures and properties, and it is clear that the growing awareness of the valuable information gained by studying polymorphic materials will lead to increasing study and understanding of these systems (Miller 2011).

Perhaps the other side of the coin in the connection between polymorphism and properties is whether polymorphs can be generated, or specific forms preferentially generated by the application of an electric or magnetic field on the growth medium. For instance, the use of electric fields in poling polymers is well known (e.g., Day et al. 1974). However, evidence for the influence of a magnetic field on polymorphic structure has been scarce, with a few exceptions. Honjo et al. (2008) demonstrated the influence of a magnetic field on the production of different polymorphs of 2,2′:6′,2″-terpyridine while magnetic levitation was used to separate mixtures of polymorphs (Atkinson et al. 2013). More recently Potticary et al. (2016) serendipitously obtained a new polymorph of coronene **6-XXIII** by the application of a magnetic field; the new β form is stable in the absence of the magnetic field and exhibits significantly different optical and mechanical properties from the previously known α form.

6-XXIII

6.2.3 Photovoltaicity and photoconductivity

The oil crisis of the middle 1970s generated a great deal of activity in the search for alternate energy sources, including organic photovoltaic and photoconducting materials. The activity slackened off during the 1980s, but has increased recently with the threat of global warming. The development of new technologies to generate and structurally characterize materials, surfaces and structures of bulk and monomolecular thick films, often at the atomic level, has led to a resurgence of activity in this field (Law 1993), that has continued unabated, for instance in the evolution of advanced photocopier and laser printer technology. While much direct information is becoming available, as with any correlation of structure and properties, the information provided by polymorphic systems can still prove very useful. Many of the materials studied for photovoltaic or photoconductive properties are also used as pigments and dyes, and the more general discussion of the polymorphic behavior of that class of compounds is found in Chapter 8. I will concentrate here on those two related photophysical phenomena.

Much of the early and continuing activity on photovoltaic materials was centered about squarylium **6-XXIV** (e.g., Zheng et al. 2015) and cyanine dyes, such as for example those with formula **6-XXV** (Merritt 1978; Morel et al. 1984; Law 1993), but there has been continuing and intensive work on other promising classes of candidate materials such as azo pigments and perylene pigments (Borsenberger and Weiss 1993; Würthner et al. 2016) as well as a variety of other more specific molecular systems (Youming et al. 1986; Law et al. 1994). In spite of the increasing interest in these two photoeffects and the effort that has been made in developing and studying these materials, there have been few direct studies of photovoltaic or photoconducting behavior as a function of polymorphism. This is in part due to the lack of proclivity of many of these materials to crystallize at all, to say nothing of crystallizing as polymorphs, so that relatively few crystal structures have been determined. Also, since the potential applications would involve nano-dimensional devices, much of the effort in the study of the materials has been concentrated on films, composites, etc., although there is also evidence for multiplicity of structural motifs in these media as well (Ashwell et al. 1997; Wojtyk et al. 1999; Stawasz et al. 2000).

6-XXIV   6-XXV

Nevertheless, among the well-studied photoconductive molecular materials, polymorphism is most prevalent in the intensively studied phthalocyanines (e.g., Vasseur et al. 2012). There are also some examples from the squaraines, apparently a few instances among the perylenes, but as yet apparently no reported examples among the photoconducting azo dyes. This situation is bound to change as intensive work continues in this field.

In Section 3.6.3 I discussed the structural aspects of the dimorphic green and violet forms of the concomitant crystallizing squarylium dye **6-XXIV** (R = C_2H_5, R′ = OH) (Bernstein and Goldstein 1988). For another derivative of **6-XXIV** (R = C_4H_9, R′ = H), Ashwell et al. (1996) have reported the crystal structures of two polymorphs, which coincidentally are also green and violet. There are no studies of the photoconductivity of either of these structurally characterized polymorphic systems, although Tristani-Kendra and Eckhardt (1984) did carry out a thorough spectroscopic investigation of the nature and manifestation of the intermolecular coupling on the solid-state optical spectra of the two forms of the former. These investigations also provide information on the important question of the direct structural relationship between dye aggregates in solution and macroscopic crystals, a question that has been addressed by Marchetti in a number of spectroscopic studies (Marchetti et al. 1976; Young et al. 1989).

Law (1993) has reviewed the widespread polymorphism in phthalocyanines and its relationship to their photoconductivity and use in xerographic applications. These are best demonstrated by the prototypical copper phthalocyanine CuPc (**6-XXVI**), whose structural chemistry is discussed in detail in Section 8.3.3.1. The polarized absorption spectra of five forms are given in Figure 6.9.

6-XXVI

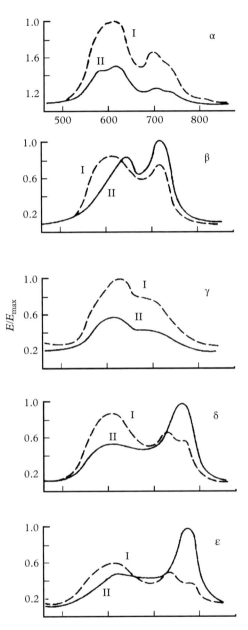

Figure 6.9 *Polarized absorption spectra of the five polymorphs of CuPc. The directions of the polarized light are perpendicular (I, - - -) and parallel (II, ——) to the long axis of the pigment aggregates (adapted from Sappok 1978, with permission).*

One outstanding feature of these spectra is the rather intense red-shifted band at ~770 nm in the δ and ε modifications compared to the solution absorption at ~678 nm (Law 1993). This band has been interpreted as arising from an intermolecular charge-transfer interaction, the latter being required for high photoconductivity. In fact, it has been reported that the xerographic performance of the ε form is superior to the other crystalline modifications (Yagishita et al. 1984; Enokida and Hirohashi 1992).

Law (1993) pointed out the difficulties in comparing the photoconductivity data of different compounds and even of different polymorphs. Many of them arise from the variations in processing procedures and conditions in various laboratories, and the possibility of polymorphic changes during processing. Moreover, the direct relationship between the crystal structure and the photoconductivity has not been firmly established, because of difficulties caused by impurities, crystal size, etc. Nevertheless, the importance of that relationship is clearly recognized, and the results on photoconductive polymorphic phthalocyanines that have been intensively studied for their xerographic response, that is, **6-XXVI** (M = 2H), **6-XXVI** (M = TiO), and **6-XXVI** (M = VO) (and derivatives thereof), led Law (1993) to state that "the precise stacking arrangement is important." There are plenty of examples to support this view indirectly, among them the data cited above for CuPc—namely, the common feature of all photosensitive phthalocyanines is the presence of the red shifted absorption, found in the aggregate and absent in solution. The structural conclusions are that the photoconductivity is indeed a solid-state phenomenon requiring strong intermolecular interactions, short (~3.5 Å) interplanar distances, and large charge mobility upon excitation (i.e., the presence of charge-transfer states). There is clearly much still be learned here, which can be revealed by combined structural and spectroscopic studies on the various polymorphs.

6.2.4 Nonlinear optical activity and second harmonic generation

The transformation of data transmission systems from analogue to digital means, more specifically to optically-based digital systems, led to intense interest in the development of optical switching devices. Molecular materials comprise a promising group of potential candidate systems for those switching devices (Kolinsky 1992) and much work was done in preparing and characterizing potentially useful compounds for these applications (Munn and Ironside 1993; Zyss 1994).

The principal structural requirement for second order nonlinear effects in assemblies of molecules is the lack of a center of symmetry, and considerable efforts have been expended in trying to induce potentially useful molecular entities to crystallize in non-centrosymmetric or polar crystals (Curtin and Paul 1981; Etter et al. 1991; Radhakrishnan 2008). As demonstrated in the following, this is a necessary, but not sufficient condition for obtaining nonlinear effects.

The existence of a polar crystal requires that it be composed of polar molecules. While polar molecules might tend to form chains in which the dipoles are aligned (or approximately aligned) in a head-to-tail fashion, or even form sheets with the dipoles aligned in one direction, achieving the third dimension of aligned dipoles is a major challenge: in many cases the packing in the third dimension leads to cancellation of the overall dipole of the first one or two dimensions and the resulting crystal will not exhibit a nonlinear effect. Hulliger et al (2012) have demonstrated with Monte Carlo calculations that the formation of "mono domain" polar crystals developing from a single seed suffer from a fundamental growth instability.

Nevertheless, considerable experimental efforts have been made to overcome and control this tendency, with a variety of approaches taken. For instance, Holman et al. (2001) and Ward (2005) suggested the design of polar solid-state host frameworks with adjustable size, shape, and chemical character that would force polar guest molecules into a constant orientation within the lattice. Wang et al. (2009) obtained a series of crystalline compounds with non-centrosymmetric polar packing directly from a one-pot, three component synthesis. In an analogous structural—but totally different chemical—situation, a series of seven ball-shaped tetrameric assemblies of tetramethylammonium halides with uranylsalophen complexes all crystallized in non-centrosymmetric space groups. The tetragonal molecular symmetry is either retained in a crystal structure of $I\bar{4}$ space group symmetry or lost in a structure with space group symmetry $R3c$ or $I\bar{4}3d$ (Cametti, et al. 2010). Saha et al. (2006) demonstrated that halogen···halogen interactions may be used to increase the probability of obtaining non-centrosymmetric packing arrangements over the nominal ~15% frequency for achiral molecules. Sasaki et al. (2016) showed that type II halogen···halogen interactions could be used to bond layers formed by a combination of charge-assisted hydrogen bonds and van der Waals interactions of alkyl chains.

In the more traditional investigations, the variety of crystallization experiments has led to a number of polymorphic structures, and with them information about the relationship between the properties of these materials and their structures, as well as useful guidelines for attempting to obtain the desired non-centrosymmetric crystal structures.

The seminal work in this regard is still that by Hall et al. (1986). This study summarizes work on three widely studied molecules whose crystals have been shown to have nonlinear optically active molecules, **6-XXVII–6-XXIX**, acronymically designated as PAN, FNBH, and MNP, respectively, and demonstrates the importance of space group symmetry as well as the details of packing in determining the nonlinear optical activity of molecular crystals. The first two are known to be dimorphic while four polymorphs have been reported for the third.[1]

---

[1] The crystallographic data for these polymorphic structures do not yet appear in the CSD, except for the inactive form of FNBH (CSD Refcode SEFBEK) (Aldoshin et al. 1988). I am grateful to Prof. M. Hursthouse (Southampton) for providing those data.

PAN
6-XXVII

FNBH
6-XXVIII

MNP
6-XXIX

The two known forms of PAN crystallize in the non-centrosymmetric space group $P2_1$. Both are grown from solutions of water and absolute alcohol. The polymorph that does not exhibit activity may be obtained with plate-like morphology by "equilibrium" methods such as programmed decrease of solution temperature or evaporation. The form exhibiting activity may be obtained as dendritic needle-like crystals by "non-equilibrium" methods, for instance rapid crystallization from a highly supersaturated solution. In general, dendritic growth often results from kinetic crystallization, since many nuclei are generated, and the high degree of supersaturation means that growth continues from them essentially unabated. Under less kinetic conditions, the larger crystals grow at the expense of the smaller ones. In fact, it was noted in this system that upon standing, solutions containing the nonlinear optically active dendritic crystals tended to convert to the inactive plate-like crystals, consistent with the relative stability implied by the growth conditions. Such a solvent-assisted conversion process also indicates that the inactive form is more stable. The crystal structures of the two forms are compared in Figure 6.10. It can be seen that in both forms the molecular dipole moment has a similar projection on the polar b axis. In the active form, the molecular dipole vectors of the two molecules in the unit cell are oriented so that they add to give a net component along the b axis, while the pseudocentrosymmetric arrangement of the molecular dipoles in the inactive form (albeit in the non-centrosymmetric space group) leads to an effective cancellation of the components along that axis. Note that the inactive form contains two pseudocentrosymmetrically related molecules in the asymmetric unit ($Z' = 2$), leading to an overall Z-value of 4. This is the crystallographic situation that permits the cancellation of the dipoles. With one molecule per asymmetric unit in the active form, the screw axis must generate a net dipole moment. Using the structural information and knowledge of the crystal faces that dominate the non-polar modification, Popovitz-Biro et al. (1991) designed a polymeric crystallization inhibitor for that form, resulting in the preferential crystallization of the polar polymorph (see Section 3.7).

In the case of **6-XXVIII** (FNBH) different crystallization conditions from absolute alcohol led to the two forms, equilibrium conditions leading to crystals of high optical quality of the nonlinear optically inactive form, while a kinetic crystallization from a highly supersaturated solution led to poor quality crystals of the

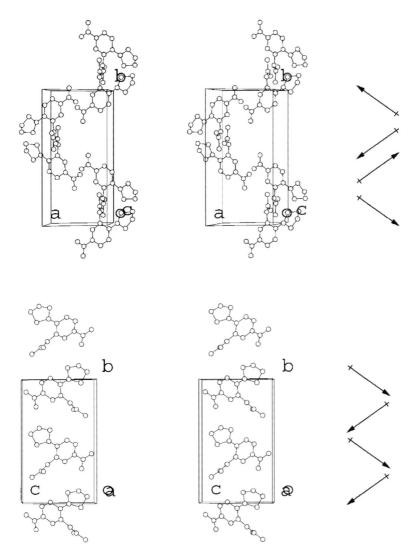

Figure 6.10 *Stereoviews of the two forms of 6-**XXVII** (PAN). Both views are oriented so that the* b *crystallographic axis is vertical. Approximate orientations for the molecular dipole moments are also shown (see text). Upper, crystal form exhibiting nonlinear optical activity; lower, inactive form.*

active form. Here, the space group symmetries clearly distinguish between the two forms. The inactive form crystallizes in the centrosymmetric space group $P2_1/c$, with eight molecules in the unit cell, the asymmetric unit being composed of an approximately centrosymmetrically related pair of crystallographically independent molecules, as shown in Figure 6.11. On the other hand the active form is in (the rarely encountered, at least for molecular crystals) space group $I4_1cd$, in which the c axis is the polar direction (Figure 6.11).

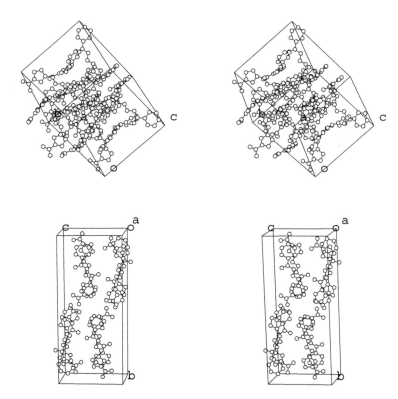

Figure 6.11 *Stereoviews of the two forms of **6-XXVIII** (FNBH). Upper, tetragonal form exhibiting nonlinear optical activity; lower, inactive monoclinic form, showing the presence of centrosymmetrically related dimers of crystallographically independent molecules.*

In **6-XXIX** (MNP) the four different forms may be obtained from different solvents under different rates of cooling and evaporation. The space groups for the two forms for which crystal structures have been reported (Forms 2 and 3) are again both non-centrosymmetric monoclinic $P2_1$. The needles of Form 2 obtained by rapid cooling are unstable in the presence of Form 3. The structures of both forms are shown in Figure 6.12. In Form 2, the asymmetric unit is again a pair of crystallographically inequivalent but pseudocentrosymmetrically related molecules. In Form 3, there are two such pairs. In both cases all the molecular dipoles are approximately perpendicular to the polar b axis, so that they essentially cancel upon the action of the twofold screw parallel to this axis. The necessary condition of a non-centrosymmetric space group is met, but the sufficiency condition of favorable orientation of molecular dipoles is not satisfied and both cases lead to a near cancellation of optical coefficients and a very small nonlinear effect.

These three examples clearly demonstrate the physical principle of the necessity for a polar structure to obtain nonlinear optical effects, the difficulty in obtaining

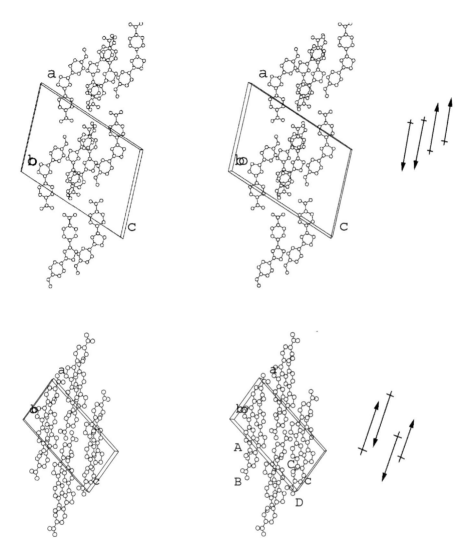

Figure 6.12 *Stereoviews of the two forms of* **6-XXIX** *(MNP). Approximate orientations for the molecular dipole moments are also shown (see text). Note that in both cases the molecular dipole moments are nearly perpendicular to the polar* b *axis. Top, Form 2, with two molecules in the asymmetric unit; bottom, Form 3, with four molecules in the asymmetric unit.*

that polar structure, and possible ways to overcome those difficulties and the utility of studying polymorphic structures to establish these principles. The crystallographically centrosymmetric arrangement is clearly preferred under equilibrium conditions, but even under kinetic conditions molecules apparently tend to arrange themselves with spatially opposing dipoles. The not-necessarily foolproof recipe for obtaining nonlinear optical activity would therefore strongly suggest the use of kinetic conditions of crystallization. There have also been recent attempts to associate particular molecular conformations with the tendency to pack in non-centrosymmetric structures (Huang et al. 1996) and a report of concomitant crystallization of potential nonlinear optical materials (Suponitskii et al. 2002).

The recognition of these structural features associated with nonlinear optical activity in fact led to the discovery of a new polymorph of *m*-nitrophenol **6-XXX**. The description of the material appears in Groth's (1919) compendium, indicating two melting points, but the interfacial angles determined by Barker (1908) and Steinmetz (1915) were mutually consistent, not hinting at polymorphism. However, in a 1934 determination of the index of refraction, Davies and Hartshorne (1934) noted "occasional individuals (of crystals) with more or less rhombic outlines" amid monoclinic prismatic crystals. In 1972, Shigorin and Shipulo (1972) reported that the compound exhibited a strong second harmonic generation (SHG), but Groth's morphological description indicated a centrosymmetric space group. In an attempt to resolve this discrepancy, Pandarese et al. (1975) carried out the crystal structure determination of the material described by Groth and others from the melt and from benzene. The crystals were indeed centrosymmetric in space group $P2_1/n$ leaving the nonlinear optical activity unaccounted for. With the dilemma not resolved, Pandarese et al. carefully reexamined many individual crystals from the benzene-grown batch and discovered that approximately 20% of them belonged to the non-centrosymmetric orthorhombic space group $P2_12_12_1$, thus explaining the source of the nonlinear effects. A method for purifying *m*-nitrophenol and growing crystals was developed by Wojcik and Marqueton (1989) and the structure of the non-centrosymmetric orthorhombic form was published by Hamzaoui et al. (1996).

6-XXX

Subsequent to the report by Hall et al. (1986), a number of other groups utilized the information on the kinetic vs thermodynamic crystallizations in polymorphic potential nonlinear optical materials to generate active materials. Black et al. (1993) utilized selective anisotropic solvent adsorption on specific crystal faces to favor the growth of morphologically polar crystals. Some additional reports of the study of crystal modification and nonlinear optical activity include those on anhydrous and hydrated sodium *p*-nitrophenolate (Brahadeeswaran et al. 1999), derivatives of 2-benzylideneindan-1,3-dione (Matsushima et al. 1992), straight-chain carbamyl compounds (Francis and Tiers 1992), benzophenone derivatives (Terao et al. 1990), a 1,3-dithiole derivative (Nakatsu et al. 1990), α-[(4′-methoxyphenyl)methylene]-4-nitrobenzeneacetonitrile (Oliver et al. 1990), and so-called "lambda-shaped molecules" (Yamamoto et al. 1992). Hall et al. (1988) followed the thermal conversion of the centrosymmetric ($P2_1/c$) form of 2,3-dichloroquinazirin to the non-centrosymmetric *Pc* form by monitoring the development of an SHG signal.

Consistent with the earlier observation, the centrosymmetric form was obtained under equilibrium conditions, while the non-centrosymmetric one could be obtained under more kinetic conditions.

A change of solvent of crystallization may also generate polymorphs with different SHG responses, as observed in 2-adamantylamino-5-nitropyridine (Antipin et al. 2001) in the trimorphic 4-methoxy-4′-nitrotolane system (Tabei et al. 1987) and in (4-pyrrolidinopyridyl)bis(acetylacetonato)zinc(II) (Anthony et al. 2003). In one case at least, it has been demonstrated that the non-centrosymmetric polymorph is preferred in crystallizations from non-polar solvents, while the centrosymmetric polymorph results from crystallizations from polar solvents (Sharma and Radhakrishnan 2000). In another case, a centrosymmetric and a non-centrosymmetric form were found to crystallize concomitantly (Timofeeva et al. 2000). The two forms have almost identical layer structures differing in the superposition of successive layers; hence they may be considered molecular polytypes (Verma and Krishna 1966). Serbutoviez et al. (1994) also demonstrated for two molecules that the addition of one equivalent of pyrrolidine to 2:1 ethanol/water solvent mixtures could also be used to generate SHG-active crystal modifications in zwitterionic substances. In the case of **6-XXXI** (R = SCH$_3$) two polymorphic monohydrates were obtained with and without the addition of pyrrolidine; upon heating, both converted to the same inactive anhydrate. For **6-XXXI** (R = H), the monohydrate obtained in the absence of pyrrolidine was inactive, the dihydrate obtained from the addition of pyrrolidine was active, and again, both converted to an inactive anhydrate upon heating.

6-XXXI

6.2.5 Chromoisomerism, photochromism, thermochromism, mechanochromism, etc.

Changes in the color of a substance resulting from perturbations which result in changes in structure or environment—so-called "chromogenic effects" (Nassau 1983)—have fascinated chemists for centuries (Kahr and McBride 1992). The observation that solutions of a pure substance could lead to concomitantly crystallizing crystals of different color led chemists in the first two decades of the twentieth century to coin the term *chromoisomerism* (Hantzsch 1907a, 1907b, 1908; Toma et al. 1994). The crystals may undergo changes in color due to exposure to light (photochromism), to changes in temperature (thermochromism) or to mechanical shock (mechanochromism). In some cases, the structural changes associated with the color change are at the molecular level, while in other cases they may be due to subtle or significant changes in the crystal structure; a combination of molecular and bulk structural changes is also possible (e.g., Reetz et al. 1994; Reed et al. 2000). If the molecular integrity is maintained, then such changes involve polymorphic systems. When the chemical nature of the molecular entity changes, one is again faced with the definitional difficulties of polymorphism (see Section 1.2). With those caveats in mind I describe a number of systems that demonstrate some of the principles and challenges in these systems involving changes in color. Many other systems described in this book exhibit color differences among polymorphs—in fact, that is often the reason for the recognition of the polymorphism—and those color differences are often noted, although the reasons for the differences in color are not further investigated or specifically described. In view of the fact that for some of the chromogenic phenomena thousands of chemical systems have been recognized and studied, the coverage here can serve only as a brief introduction to the important connection between these effects and polymorphism. I will concentrate here on systems chosen to study the specific physical phenomenon and, where possible, the attempts to relate the consequences of the physical perturbation to changes in structure and color. All of these involve monomolecular species or salts but they serve equally well as prototypes for polymorphic polymeric systems (e.g., Chunwachirasiri et al. 2000).

6.2.5.1 *Chromoisomerism*

The rather thorough study by Toma et al. (1994) of the red and green polymorphs of 9-phenylacridinium hydrogensulfate **6-XXXII** demonstrates many aspects of the difficulties and challenges in studying these closely related phenomena (along with an historical summary of the debate surrounding the origin of chromoisomerism). The two forms of **6-XXXII** crystallize concomitantly, although selective crystallization may be accomplished by appropriate seeding; they do not undergo thermal interconversion. The molecular structures found in the two modifications are essentially identical, with the torsion angle of the exocyclic phenyl ring differing by one degree, also in agreement with the geometry expected from computed molecular energetics. The hydrogen-bonding patterns are also identical $D_3^3(9) R_2^2(8)$ (see Sections 7.12 and 7.13), involving centrosymmetric dimeric hydrogensulfate units flanked by (also centrosymmetrically related) phenylacridinium units, although the packing of these units is quite different (Figure 6.13). In spite of the fact that this was a rather thorough study, it was not possible to determine the physical basis for the significant visible difference in color between the two polymorphs.

6-XXXII   **6-XXXIII**

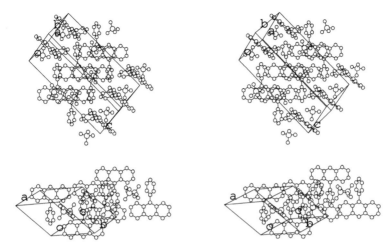

Figure 6.13 *Stereoviews of the red (upper) and green (lower) polymorphs of 6-**XXXII**, showing similarity in hydrogen bonding patterns, but differences in packing. In order to facilitate comparison, in both cases the view is on the best plane of a centrally placed acridinium moiety.*

Toma et al. refer also to the studies of **6-XXXIII** (Byrn et al. 1972; Yang et al. 1989; Richardson et al. 1990) as an example of a case in which the source of differences in color were explained. While these systems were both very thoroughly studied and structural changes were *correlated* with differences in color (for another excellent example, see, for instance, Desiraju et al. 1977), the *origin* of the differences in color, which is based on the details of the electronic structure, has for all intents and purposes remained elusive in all of the systems studied. One notable exception is indeed the case of **6-XXXIII** in which the color change could be associated with a proton transfer (Richardson et al. 1990).

A number of subsequent studies of chromoisomerism based on polymorphic systems attempt to determine the source of differences in color. The examples are by no mean comprehensive, but rather serve to portray the current state of the art on this fascinating subject. The substituted anthraquinone **6-XXXIV** is dimorphic, with a yellow, low melting triclinic form and an orange higher melting monoclinic form. The differences in color have been qualitatively attributed to the presence of discrete pairs of anthraquinone rings in one form compared to infinite lamellar stacks in the second form (Reed et al. 2000). Similarly, the trimorphic copper complex **6-XXXV** exhibits easily distinguishable concomitant light green, emerald green, and orange forms. The color differences are attributed to different π-π stacking: the emerald green and orange forms exhibit face-to-face (dimers) and slipped stacking (infinite chains) respectively, while the light green is devoid of these interactions (Peresypkina et al. 2005). The stable yellow form of dimorphic tetramesityldisilene **6-XXXVI** can be converted to the metastable orange form by laser illumination in the region 514–457 nm, with accompanying changes in the Raman spectra. The two forms exhibit different conformations but little information was obtained on the fundamental nature of these color differences (Leites et al. 2003). The differences in packing for the colorless and yellow forms of *N*-hydroxyphthalimide **6-XXXVII** have been described in considerable detail. The colorless form is nearly planar while the yellow form is less so and the intermolecular hydrogen bonding is different. No source for the lack of color in one form and the presence in a second form is provided (Reichelt et al. 2007). A similar type of study was carried out on the metastable green and stable yellow polymorphs of **6-XXXVIII** without arriving at a detailed explanation of the source of the difference in color (Miao et al. 2012). Finally, the tetramorphic system of the nicotinic acid derivative **6-XXXIX** displays different colors attributed by the authors to differences in molecular conformation and conjugation between two aromatic rings for three of the forms $Z' > 1$ (Long et al. 2008). This study, typical of those noted earlier in this paragraph are phenomenological, providing little information on the electronic basis for these color differences, and in the end the color in every case is due to an electronic transition—the absorption of light by electrons. The four polymorphs of a diketopyrrolopyrrole **6-XL** were prepared both in thin films and single crystals (Salammal et al. 2016). The optical spectra are clearly different. The authors attribute a change in the band gap from 1.75 to 1.5 eV during a phase change from the α form to the ω form to enhanced π–π interaction between neighboring molecules. They also attribute a change of 0.006 Å in the π–π stacking of the γ form to a reduction in hole mobility.

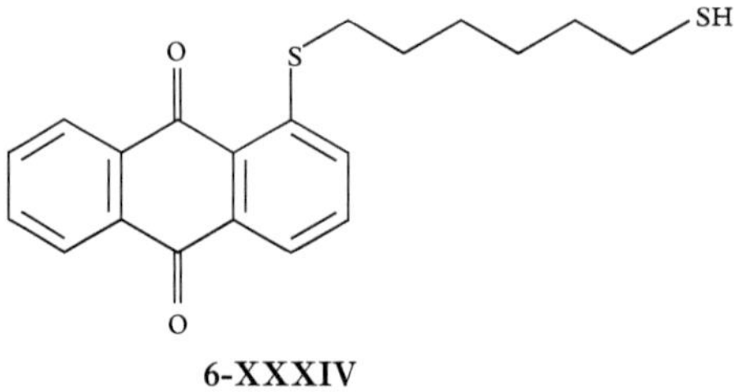

6-XXXIV

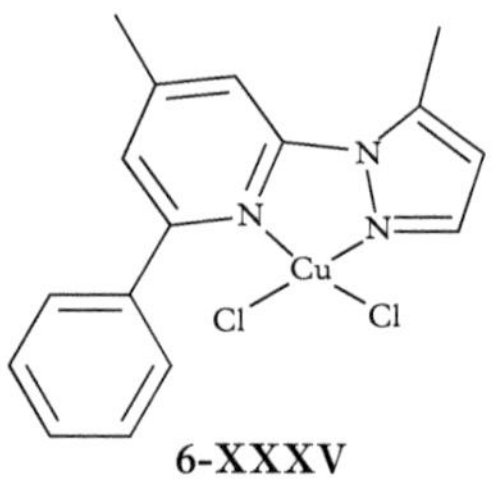

6-XXXV

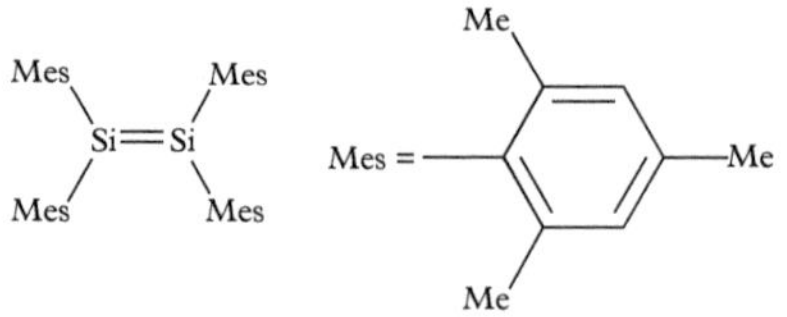

6-XXXVI

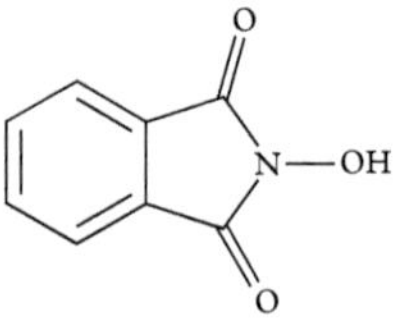

6-XXXVII

6-XXXVIII   6-XXXIX

6-XL

There are a number of reasons for the difficulties in determining the origin of color differences. First, the color of a crystal actually observed by a human being results from a variety of optical phenomena (scattering, absorption, reflection, etc.) (Nassau 1983), which are not generally simultaneously measured by any particular analytical optical or spectroscopic technique. Also, as noted for magnetic phenomena in Section 6.2.2, the origins of the changes in color may be beyond the detection limits of the techniques currently in use. The human eye is a rather sensitive detector of color and variations in color. While we can quite confidently determine differences in structure to less than one part per 100 (generally better than 0.01 Å for a nominal bond length of 1.5 Å) by a variety of analytical methods, the small differences in extinction coefficient or subtle changes in the overlap of absorption bands can lead to optically recognizable changes in shade or color which are often difficult to quantify experimentally.

6.2.5.2 *Photochromism*

The photochromic process involves the formation, destruction or change of color due to the action of light. The fact that such phenomena can be useful for optical devices information storage has made photochromism a subject of considerable interest (Maeda 1991; Weber et al. 1998; Bouas-Laurent and Durr 2001), and much of the literature has been reviewed by Kitagawa (2014).

For example, compound **6-XLI** can be obtained from xylene, toluene, or benzene as "superb, long, very dark blue-black needles with metallic luster." A "brilliant scarlet form" can be obtained from polar solvents such as acetic acid, DMF, or DMSO. The black form can be converted to the red form both photophysically and thermally (abruptly at 139–142 °C) (Begley et al. 1981). The two structures are shown in Figure 6.14. In the black form, the molecules form infinite stacks ("in the manner of roof tiles") in the *ac* plane with intermolecular contacts of 3.31 Å which the authors attribute to charge-transfer interactions. The stacking is absent in the scarlet form, and the molecular conformation is considerably more non-planar, making this pair of structures also conformational polymorphs. No mechanism for the conversion or color difference has been proposed.

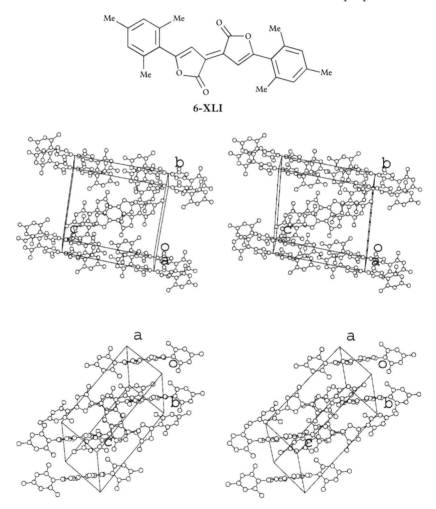

6-XLI

Figure 6.14 *Stereoviews of the two forms of 6-XLI. Upper, black form showing stacking of neighboring molecules; lower, scarlet form. In both structures the view is on the best plane of the essentially planar portion of the reference molecule, shown with its surroundings.*

In another example, the photochromic transformation was carried out in solution and the starting material and product were investigated in the solid in order to establish the structural nature of the process (Burns et al. 1988). The compound **6-XLII** is a formazan derivative that can adopt a number of conformations. The orange, light-stable form crystallizes from 1:1 ethanol–water, while the red form is obtained by dissolving the orange form in the dark and evaporating the solvent under vacuum. The crystal structures (Figure 6.15) clearly show that in the red form the molecule adopts the *syn*, s-*trans* **6-XLIIa** conformation, while in the light-stable orange form the *anti*, s-*trans* conformation **6-XLIIb** is adopted. The solution composition of isomers apparently depends on the light level. The source of the difference in the *solid-state* color has not been investigated.

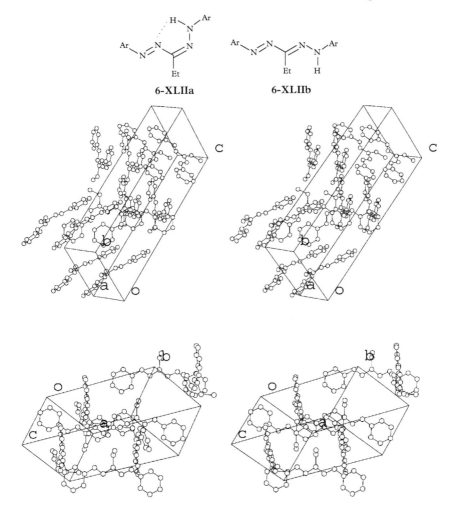

Figure 6.15 *Stereo packing diagrams of the red form (upper) and the light-stable orange form (lower) of 6-XLII. Both structures are plotted on the same reference plane which is the N–C(Et)=N atomic triplet in a molecule located near the center of the cell.*

A number of studies of photochromism and polymorphism have been based on the chemistry of anils, specifically salicylideneanilines, as described, for example, by Ohashi (2014). Taneda et al. (2004) described a dimorphic salicylideneaniline in which the two forms displayed similar photochromic behavior and attributed the difference in the thermal stability of the photochromic product to differences in the shape of the molecule and the cavity it occupies in the crystal. Another dimorphic salicylidene system reported by Safin et al. (2016) exhibited red and yellow crystals, the former transforming monotropically into the latter between 165 and 190 °C. Reversible photochromism is displayed exclusively by the red form upon radiation at 365 nm, with a slow thermal relaxation. This system is also the first in which a temperature induced *cis/trans*-keto equilibrium has been studied for an *N*-salicylideneaniline system.

A system based on diarylethene exhibits triclinic needles and monoclinic plates, grown from acetone at different temperatures (Iwaihara et al. 2015). A solvent mediated transformation at room temperature leads to the monoclinic form, which can be observed as a photochromic reaction at 78 °C that proceeds as a blue front through the colorless starting material due to UV radiation. All four polymorphs of a tetramorphic cyclopentene system undergo photocyclization reactions with UV ($\lambda = 370$ nm) or visible light ($\lambda > 500$ nm) with quantum yields around unity, but the reverse photoreactions varied by a factor of four (Morimoto et al. 2003).

6.2.5.3 *Thermochromism*

Thermochromism has been widely studied, and Nassau noted even in 1983 the existence of several thousand examples. Thermal transformations of molecular crystals have also been studied in some detail (Paul and Curtin 1973), and many of these are accompanied by changes in color. The salicylideneanilines (Cohen et al. 1964a, 1964b, 1964c) and the substituted salicylideneamines (Carles et al. 1987) have provided a source of a variety of compounds, exhibiting both photochromic and thermochromic crystal modifications. In spite of continuing interest and the determination of the structure of the β form and an α_1 polymorph, the detailed understanding of the basis and mechanism of the thermochromism and photochromism of this system remain enigmatic (Arod et al. 2007). I review some of the earlier examples in order to demonstrate the kinds of problems that may be encountered before citing some of the more recent reports that often employ a combination of sophisticated techniques in the search for an understanding of the system.

Reetz et al. (1994) reported the preparation of concomitant polymorphs of the 4,4′-bipyridinium salt of squaric acid **6-XLIII**. The major product is a monoclinic ochre modification, while "a few individual crystals" of a slightly darker triclinic form were also isolated and characterized. Upon heating to 180 °C the monoclinic form undergoes a single-crystal to single-crystal thermochromic change to a red crystal.

6-XLIII

The change takes place via a reaction front that moves rapidly (i.e., a few seconds) along the needle axis, which corresponds to a short (~3.8 Å) crystallographic axis in the starting material. The transformation is reversible, but is accompanied by hysteresis, occurring in the opposite direction only at 150 °C. Despite the fact that the process appears reversible there are some subtle changes that occur upon cycling. Upon returning to the low temperature ochre modification the relative intensities of the UV maximum (at 240 nm) and the charge-transfer band (at 390 nm) are different from the starting material. The locations of the lines in the X-ray powder diffraction pattern (reflecting the unit cell dimensions) are not perturbed, but the intensities (reflecting the unit cell contents and orientation) are altered. The same effects were observed when the transformation was carried out under hydrostatic pressure. The color change has been attributed to a proton transfer mechanism along the chains of the interleaved bipyridinium and squaric acid moieties.

The thermochromic transition of an organometallic compound **6-XLIV**, reported by Etter and Siedle (1983), exhibits a number of additional noteworthy features and riddles. Yellow triclinic needles obtained from hexane expand suddenly along the needle axis when heated to 90 ± 10 °C. If they are heated only on one face, they exhibit the thermosalient (i.e., "hopping crystal") effect (Steiner et al. 1993; Zamir et al. 1994; Lieberman et al. 2000; Naumov et al. 2015) (Section 6.2.6). X-ray photographs indicate that the original lattice is maintained at this point, but disorder is increased. Following the expansion, a red phase develops at one end of the needle and a front progresses along the needle as the temperature is raised. The transition is not reversible and at 157 °C the red crystals melt. The red material, identical by X-ray diffraction with the transformed phase, may be obtained independently from hot xylene. The crystal structures of the two forms exhibited very different packing (Figure 6.16). Etter and Siedle attempted to characterize the yellow to red transition structurally and thermodynamically, but noted then that significant questions remain unanswered, in particular why the color changes along a front *after* the thermal expansion.

6-XLIV

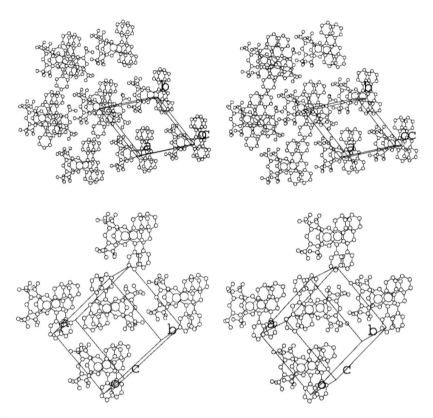

Figure 6.16 *Stereo packing diagrams of the low temperature yellow (upper) and high tempera-ture red (lower) forms of* **6-XLIV**. *In both cases the view is on the best molecular plane, which is approximately along the* c *axis, measuring 8.44 Å in the yellow form and 6.99 Å in the red form.*

Other systems raise similar questions. Crystals of the overcrowded molecule **6-XLV** undergo a reversible orange to red conversion repeatedly, apparently without any sign of visible degradation. Yet a phase change is not detectable by DSC measurements (Stezowski et al. 1993). Finally, in a remarkable system studied by Katrusiak and Szafranski (1996), guanidinium nitrate undergoes a

6-XLV

first-order phase transition at 296 K, followed by another continuous phase transition at 384 K. The lower temperature transition is accompanied by large crystal strain and a physical lengthening of the crystal by over 44%, which is not reported specifically to lead to physical movement of the crystal, but can hardly be ignored. The mechanism, based on structural and calorimetric data and lattice energy calculations, has been interpreted in terms of the rearrangement of the hydrogen-bonding patterns in the course of the transition.

A number of studies of thermochromic polymorphic systems have led to considerable understanding of this phenomenon, although, as in all of these instances each case is different and must be treated differently. As we have seen from a number of examples in this chapter, Schiff bases exhibit thermochromism as well as photochromism. Two of three polymorphs of a thermochromic benzylideneaniline derivative **6-XLVI** were synthesized in solvent-free conditions by mechanochemical syntheses (Zbačnik et al. 2015). Carletta et al. (2016) have proposed new criteria for the prediction of those properties based on the study of polymorphic and isomorphic co-crystals of *N*-salicylidene-3-aminopyridine **6-XLVII** with fumaric acid and succinic acid. Biedermann et al. (2006) apparently solved the century old riddle of thermochromism (and thus refuting eleven previously proposed models) in overcrowded bistricyclic aromatic enes by studying the deep-purple and yellow polymorphs of **6-XLVIII**. The deep purple form exhibits a twisted molecular conformation while the ambient temperature form has a folded conformation.

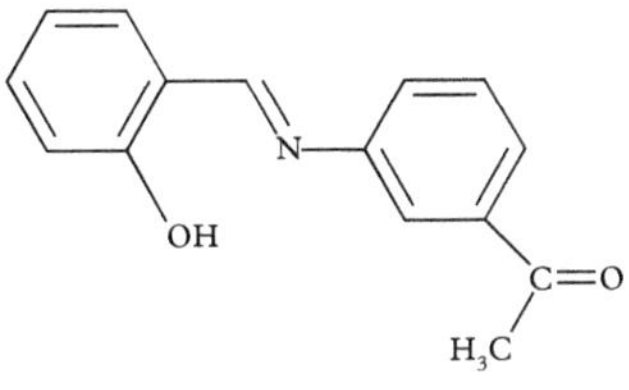

6-XLVI

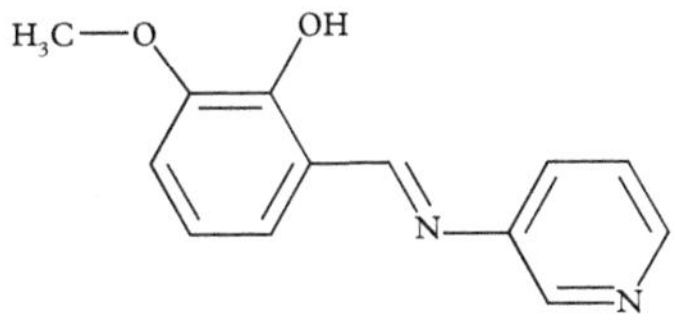

6-XLVII

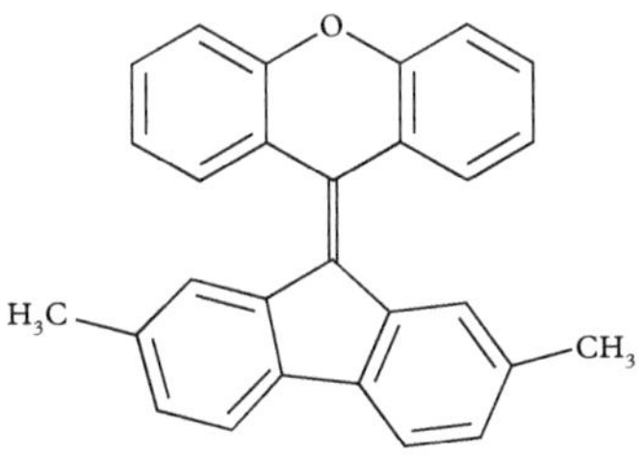

6-XLVIII

Two very thorough and enlightening studies by Naumov et al. (2009) and Lee et al. (2011c) demonstrate the extent of the information that may be obtained by studying polymorphic systems. The first seven polymorphs in the system based on **6-XLIX** (*n* = 2–5) exhibit what was designated "dual" thermochromism, that is, a gradual color change from yellow to orange at lower temperatures followed by sharper color change from orange to red at higher temperatures. Structurally the polymorphism is described by a "locking" and "unlocking" mechanism of the alkyl chains enabled by intramolecular hydrogen bonding.

6-XLIX (BDB*n*, *n* = 2–5)

It was shown that the lower temperature slow thermochromic change can be related to increased distance and weakened π–π interactions between stacked benzene rings leading to the red-shift of the absorption edge of the intramolecular charge transfer band responsible for the color. The higher temperature sharp color change accompanying the phase change results from an intramolecular proton transfer of one amino hydrogen to a nitro group. This mechanism in which these two factors have been elucidated distinguishes this system from many earlier thermochromic compounds based on either intermolecular π–π *or* thermal proton transfer.

The second system is based on a study of ten derivatives of 1,3-diamino-4,6-dinitrobenzene dyes **6-L** that exhibit a reversible color change. One of the three polymorphs of the trimorphic bisphenyl derivative goes from red to yellow—that is *negative thermochromism*, wherein the high temperature phase absorbs at higher energies than the low temperature phase.

6-L

6.2.5.4   *Mechanochromism, luminescence, and piezoluminescence*

These are closely related phenomena that may not warrant individual definitions. Mechanochromism describes a change in color resulting from a physical perturbation involving pressure such as crushing or grinding, while triboluminescence is defined as the generation of light due to friction. In a demonstration of the mechanochromic effect for polymorphic structures, **6-LI** was found to crystallize in four modifications, in some cases concomitantly, depending on the solvent and conditions; three structures have been reported (Mataka et al. 1996). Colors vary from colorless (the form whose structure has not been reported) through pale yellow to dark orange. They differ slightly but experimentally significantly (by a few degrees) in the rotations of the phenyl groups and the carbonyls and so may be considered conformational polymorphs, but the authors attribute differences in color more to intermolecular interactions than to conformational variations. Stereoviews of a pair of molecules in the three structures are shown in Figure 6.17.

6-LI

The mechanochromism resulting from grinding or crushing was monitored using visible fluorescence resulting from 365 nm incident radiation. The blue fluorescence is exhibited by the pale yellow form, which upon crushing becomes green with green fluorescence. The dark orange form which fluoresces yellow-green is also transformed to a green solid with green fluorescence. The chromatic changes are less dramatic for the yellow form, which shows a change in the shade of yellow and fluoresces yellow-green. Upon grinding, the colorless form with blue fluorescence turns yellow-green with yellow-green fluorescence. The X-ray powder patterns of the yellow and dark orange forms change as a result of the grinding or crushing; the pale yellow and colorless ones do not change. The original colors can be restored by washing the products or exposing them to solvent vapor. Except for this phenomenological description, no structural or physical basis for the mechanochromic effects has been proposed.

During the 1970s, there was rather intense activity surrounding the triboluminescent effect, which can readily be demonstrated by chewing Wintergreen Life Savers® in front of a mirror in a darkened room. Much of the work was summarized in a review by Zink (1978). The prevailing model at the time was that the triboluminescent effect required a non-centrosymmetric crystal (such as the sugars in the Life Savers®) which developed fluorescent charged states as a result of the friction generated by chewing. The study of a polymorphic system with a non-triboluminescent centrosymmetric form and a triboluminescent non-centrosymmetric form (Hardy et al. 1978) tended to confirm such a model. In subsequent work there were a number of reports of centrosymmetric crystals that nonetheless exhibited triboluminescence, which in some cases was attributed to disorder (e.g., Chen et al. 1998), meaning that local regions of the crystal may not have been strictly centrosymmetric. Sweeting et al. (1997) carried out a study of twelve 9-anthracene-carboxylic acids and their esters to correlate their

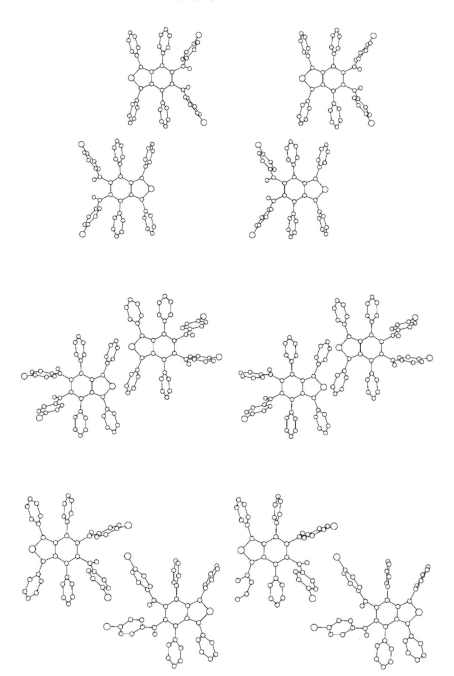

Figure 6.17 *Stereoviews of the molecular conformation and pairwise relationship of three of the four modifications of **6-LI**. All three structures are viewed on the same reference plane, which is the benzothiophene ring. Upper, pale yellow form; middle, light yellow form; lower, dark orange form.*

triboluminescent activity with crystal structure, purity, and photoluminescence. Their conclusion was that a non-centrosymmetric crystal structure is a necessary, but not sufficient condition for triboluminescence, with charge separation and recombination actually being responsible for the triboluminescent effect. Clearly, the identification and study of some non-disordered polymorphic structures could provide additional edification in this matter, and a considerable amount of work in this area has been reported in recent years.

The closely related phenomenon of chromic luminescence has been investigated on a number of polymorphic systems. For example, Zhang et al. (2010) studied the polymorph-dependent emission and mechanochromic luminescence of the sunscreen active material difluoroboron avobenzone **6-LII**. Subsequently, Li et al. (2014) reported the piezochromic luminescence of the imidazole derivative **6-LIII**, and demonstrated that the stacking mode and fluorescence can be "reversibly tuned" to lead to a piezochromic material by thermal and mechanical treatment. They have attributed this process to the perturbation of π–π stacking between anthracene planes in the crystal structure resulting from the physical treatment. Wang et al. (2016) prepared two monoclinic polymorphs of a tetraphenylethylene derivative, both displaying aggregation-induced emission (AIE): the centrosymmetric ($P2_1/c$) form exhibits mechanoluminescence, which is absent in the non-centrosymmetric *C2* form. The authors claim that they are able to interpret the difference based on the molecular packing in the two structures. The influence of intermolecular interactions has also been invoked in the interpretation of the peizochromism of a number of polymorphic substituted tetraphenylethylene derivatives **6-LIV** (Zhuang et al. 2016), while an earlier study on similar compounds attributed the differences in polymorphism-dependent emissions following grinding to intramolecular conformations (Qi et al. 2013). Another group termed this phenomenon "tribochromism" (Asiri et al. 2014): a fulgid derivative exhibited yellow or dark red crystals, and crushing of the former yielded the latter. The yellow crystals are comprised of molecules with a folded conformation while the red form contains molecules with a twisted molecular conformation. The authors showed that calculations of molecular and lattice energies could provide a basis for the difference in color.

6-LII

6-LIII

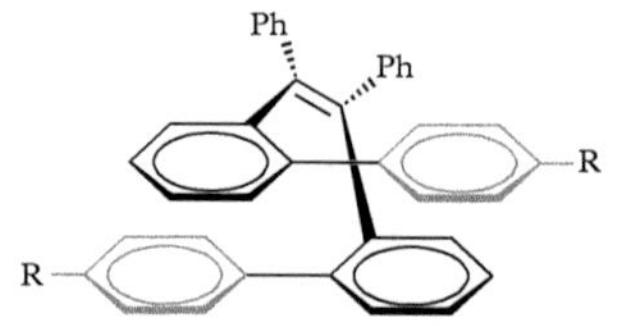

6-LIV

Yoon and Park (2011) generated phase changes between the two polymorphs of dicyanodistyrylbenzene by both thermal and mechanical means. The resulting differences in fluorescence were correlated in considerable detail with the different modes of dipole coupling associated with the different structures characterized by C–H⋯π and C–H⋯N interactions. In a similar study, switchable solid-state fluorescence in a salicylaldehyde azine derivative **6-LV** resulting from mechanical and/or thermal stimuli was reported by Chen et al. (2011). An impressive array of analytical techniques was used to investigate the system. The conclusion was that the differences between the fluorescence response in the two polymorphs originated from structures exhibiting different π–π interactions, manifested specifically in a local-dipole couple mechanism resulting from the situation with donor-planar-acceptor registry as the essential factor for the solid luminescence-switching properties. The study of a switchable and reversible thermochromic fluorescence of a naphthyridine derivative **6-LVI** has been reported (Fernández-Mato et al. 2013) with few details. The orthorhombic form exhibits a blue emission; following a thermally induced transformation to a monoclinic form and subsequent conversion to amorphous the emission turns cyan. The nanoprecipitation of the molecule **6-LVII** resulted in nano particles differing from the micro particles in morphology, structure, and luminescence (Anthony and Draper 2010). A new polymorph appeared (see Section 7.4.5) as one-dimensional nanowires. The different morphologies of the nano/microstructures allowed tuning of the luminescence over the range 604–519 nm.

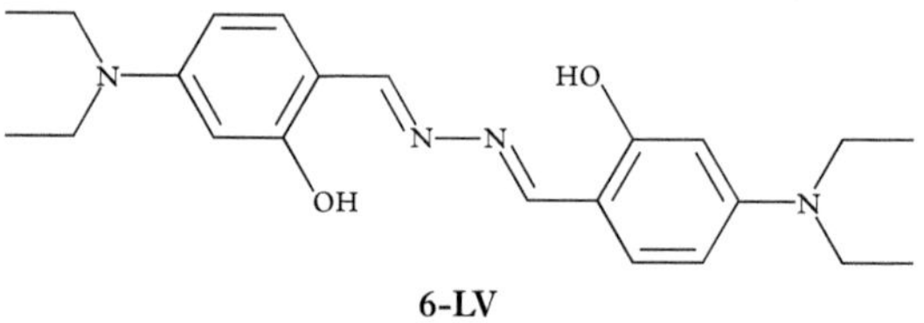

6-LV

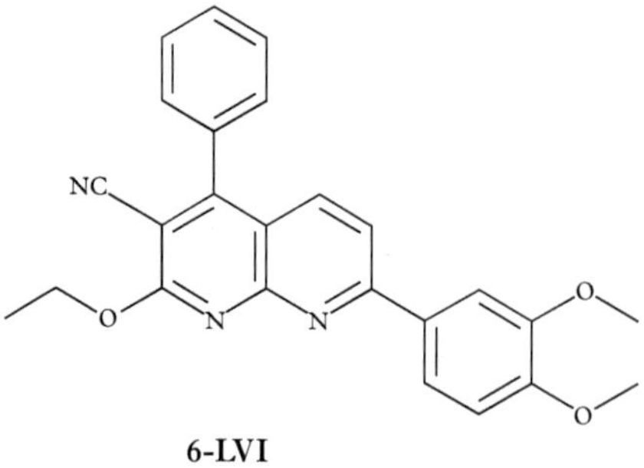

6-LVI

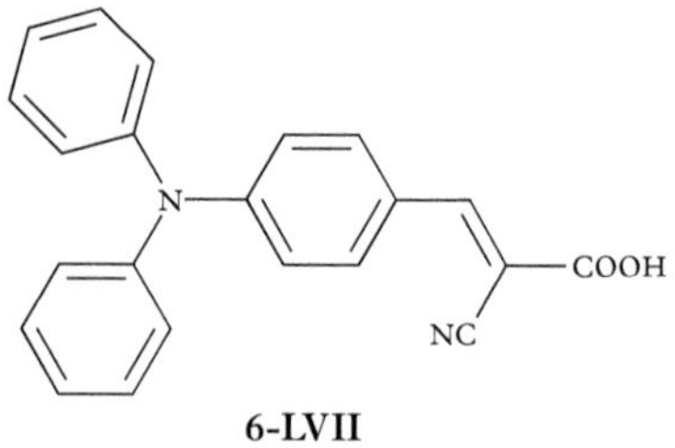

6-LVII

As a final example, a trimorphic pyrrolopyrrole derivative **6-LVIII** exhibits blue, cyan-blue, and yellow-green emission which can be modified by mechanical and thermal perturbations, as well as by fuming with chloroform (Ji et al. 2017). An attempt has been made to attribute the multicolor response to the molecular conformation and packing as revealed in the X-ray determined crystal structures.

6-LVIII

6.2.6 The thermo-photo-mechanosalient effect—"hopping" or "jumping" crystals

The field of thermally, photo, and mechanically induced phase transitions is rather vast, and while involving transformations among polymorphic forms, is beyond the scope of this work (see, e.g., Rao and Rao 1978; Bruce and Cowley 1981; Salje 1990). However, as polymorphs are simply different phases, there has been increasing interest in the study of the structural mechanism (in chemists' terms the "reaction coordinate") of these transitions in molecular crystals. As Dunitz (1991) has pointed out, there is no clear boundary between solid-state transitions and chemical reactions. Only a few of these investigations, all emphasizing different aspects of such studies, are noted here. They all have one feature in common; that is, the structures of the starting and product phases are known, and attempts are made to deduce the nature of the intra- and intermolecular motions during the course of the thermal transformation. These motions may lead to thermally induced reactions (Paul and Curtin 1973) for which the kinetics can be quite complex (Shalaev and Zografi 1996), and the current status of this aspect of solid-state reactions and the benefits that can be derived by a multidisciplinary investigative approach have been given in a review by Even and Bertault (2000). The development of time-resolved crystallography, especially with synchrotron radiation, promises to shed much light on the details of these reactions (Coppens and Fournier 2015; Ki et al. 2017).

The earlier occasional reports of crystals that jump in the course of these thermal phase transitions—the thermosalient effect—in organic crystals (Gigg et al. 1987; Ding et al. 1991; Steiner et al. 1993; Zamir et al. 1994; Lieberman et al.

2000; Skoko et al. 2010), and inorganic crystals (Crottaz et al. 1997) have been supplemented by a great deal of activity. The hopping or jumping of the crystal, visible on a macro scale, is usually associated with a thermally induced polymorphic phase change. In virtually all of these reports, the starting and final structures are known, some of them are reversible (e.g., Zamir et al. 1994), and considerable progress has been made in determining the precise relationship between the microstructural changes (detectable in the structure determinations) and the macrostructural manifestations of those changes (Naumov et al. 2015; Mittapalli, et al. 2017). The thermosalient event may be accompanied by desolvation of a solvated starting material (Shibuya et al. 2017). It has also been demonstrated that the properties of these potential transducers may be tuned by appropriate levels of chemical substitution (Nauha et al. 2016) and evidence is accumulating based on acoustic emission measurements that the thermosalient transitions in these molecular crystals are analogous to martensitic transitions in some metals (Panda et al. 2017). There are undoubtedly many more systems that exhibit thermosalient phase changes which can be readily observed on the hot stage microscope (see Section 4.2). Mechanosalient effects in terephthalic acid single crystals have been monitored by a combination of high speed optical analysis and serial scanning electron microscopy (Karothu et al. 2016). Determining the details of the reaction coordinate for any of these transitions can lead to a great deal of understanding of the nature of the polymorphic transition, especially when the cooperative effects of the transition can be defined (Van den Ende et al. 2016). Photosalient effects are no less dramatic and a number of examples have been reported. Seki et al. (2015) described the photoinduced shortening of an Au···Au separation that initiated a single-crystal-to-single-crystal phase change. All of these manifestations of polymorphic transitions due to various external stimuli can be harnessed into mechanical devices (Commins et al. 2016).

6.2.7 Other physical properties

As the awareness of the value of investigating polymorphic systems to gain understanding of the structural basis for differences in physical properties has spread, the variety of systems studied has increased. A sampling of those follows, again without any claims for completeness.

The difference in compressibility of Forms I and II of ranitidine hydrochloride (see also Section 10.4.3) were interpreted in terms of the crystal packing, the presence or absence of C–H···O interactions, and specific slip planes. These differences in structural features were also invoked to explain the difference in compressibility of the two forms (Upadhyay et al. 2013). Comparison of Young's modulus and yield stress in polymorphic pairs of carbamazepine, sulfathiazole and sulfanilamide have been reported by Roberts and Rowe (1996). More recently Mohapatra and Eckhardt (2008) have used Brillouin scattering to determine the complete elastic constant tensor for carbamazepine; that tensor may be used to calculate the linear compressibility and bulk modulus.

Saraswatula and Saha (2015) have reported on the thermal expansion properties of a polymorphic solvate in which one form exhibits biaxial negative thermal expansion while the second form exhibits the normal positive thermal expansion. Klapper et al. (2000) studied the variation of thermal expansion of the polymorphs of 4-methylbenzophenone between its two polymorphs, as well as determining the pyroelectric, dielectric, and piezoelectric coefficients for both forms.

6.3   Molecular properties

The manifestation of changes in molecular structure among polymorphic modifications as detected by a number of physical and analytical techniques was discussed in Chapter 4. The primary aim there was to describe how those techniques can be used to detect and characterize polymorphic forms. In this subsection, I wish to demonstrate how those techniques can be used to investigate structure–property relations on a molecular level utilizing the polymorphic nature of the compound. The distinction between these two approaches is not always sharp, as suggested also by the discussion of bulk properties. Again, the coverage here is not meant to be inclusive, but rather to demonstrate how the utilization of polymorphic structures can often provide the key to unraveling molecular structure–property relations, regardless of the analytical technique employed—often by combining crystallographic (e.g., symmetry arguments) and spectroscopic information.

6.3.1   Infrared and Raman spectroscopy

Since infrared (IR) spectroscopy is a standard and widely used tool in the search for and characterization of polymorphs, there are likely to be thousands of references to the use of the technique in connection with polymorphs. The vast majority of these deal with the determination of the IR fingerprint of a polymorphic modification. In this section, I wish to note a few cases in which the IR and Raman techniques were employed to obtain *chemical information* somewhat beyond the mere "fingerprint" identification of a particular crystal modification. For instance, in an early study Mathieu (1973) showed for a number of chiral compounds that it is possible to distinguish between a *dl* racemate and a conglomerate of *d* and *l* crystals by use of IR and/or Raman spectroscopy, even when it may not be possible to make such a distinction by physical or visual means.

As in many studies of polymorphism the combination of techniques is particularly effective for obtaining structure–property relationships (Yu et al. 1998). Combining IR with SSNMR and X-ray crystallography to study the two polymorphs of **6-LIX**, Fletton et al. (1986) were able to correlate the IR frequencies with the number of molecules in the unit cell as well as the intramolecular and intermolecular hydrogen bonding.

6-LIX   **6-LX**

Perhaps a classic case of the application of Raman spectroscopy on polymorphs involved the resolution of a controversy surrounding the correlation of geometry and stretching frequency near 500 cm^{-1} of the biologically important disulfide bridge, —S—S— (Lord and Yu 1970a, 1970b). Dithioglycolic acid **6-LX** may be considered a model compound for this important chemical linkage. In a series of papers, Sugeta and co-workers had suggested that bands showing large shifts to higher frequencies indicate one (for ~525 cm^{-1}) or two (for ~540 cm^{-1}) —C—C—S—S—torsion angles approximating the *trans* (~180°) conformation (Sugeta et al. 1972, 1973; Sugeta 1975). Van Wart and Scheraga (1976a, 1976b, 1977) had argued contrarily that the same frequency shifts were characteristic of considerably smaller torsion angles, in the range 20–50°.

The resolution of the controversy was provided by Nash et al. (1985) who prepared two polymorphs of **6-LX**. Form I was obtained only when an aqueous solution evaporated slowly (through an orifice) over a period of weeks. On the other hand, Form II was obtained by a more rapid evaporation of an aqueous solution or from a variety of organic solvents. The latter thus appears to be the kinetic form, the former the thermodynamic form. The two forms crystallize concomitantly when aqueous solutions are allowed to evaporate more rapidly. In Form I, the molecule lies on a crystallographic twofold axis, requiring equality of the two —CCSS— angles (−167.4°) and a value of −86.3° for the —CSSC— torsion angle. Molecules in Form II lie on crystallographic general positions, the two —CCSS— angles being −76.2° and −64.6°, with a —CSSC— angle of −92.3°. There are some other geometric parameters that differ by experimentally significant amounts. The variations in torsion angles render this system suitable for addressing the spectroscopic points of the controversy.

As expected, the Raman spectra of the two polymorphs also differ. In the region under dispute, Form I has a very strong line at 536 cm^{-1}, while the corresponding line for Form II appears at 510 cm^{-1}. Van Wart and Scheraga had reported a very strong line at 508 cm^{-1}, which means that they must have obtained Form II, not Form I as they contended. On the other hand, the 536 cm^{-1} value for Form I does correspond to that of Sugeta et al., confirming their conclusions. Nash et al. also performed normal coordinate calculations that were consistent with these findings, which were subsequently confirmed by Van Wart and Scheraga (1986).

6.3.2 UV/vis absorption spectroscopy

Most interactions of electromagnetic radiation with matter contain a geometric, as well as an energetic component. For visible and ultraviolet (UV) absorption this is because the fundamental relationship governing the absorption of light is the transition moment integral

$$f = \int \Psi \, \mathbf{e} \cdot \mathbf{r} \, \Psi^* d\tau \tag{6.1}$$

in which $\mathbf{r}$ is the transition moment *vector* defining the direction along which the electric vector $\mathbf{e}$ of the light must operate for the transition to occur from the ground state, defined by the wave function Ψ, to the excited state, Ψ^*. While many, indeed most, attempts at reconciling absorption spectral observations with theory concentrate on the energetics, the directional properties of the spectral features are certainly no less important, and in some instances may even be more critical than the energies in determining the extent to which experimental results correspond to theoretical models. In solution the directional properties are randomized, but crystals provide an essentially fixed matrix for orienting the molecules in a way potentially known through the crystal structure determination. It is not sufficient, however, to fix the molecules in a crystal structure. The direction of the electric vector of the incident light must be fixed by using polarized light. Then, by carrying out spectroscopic studies using polarized light on single crystals for which the crystal structure has been determined, and with a known orientation of the crystal with respect to the polarization of the electric vector of the incident light, the directional properties of the observed spectral transitions can be determined. These can then be compared directly with the symmetry properties of the wave functions Ψ, Ψ^* which form the basis of the electronic model or theory.

A complicating factor arises when the oscillator strength of the transition is large (e.g., has a large extinction coefficient) as it is in many materials of spectroscopic interest. In such cases it may not be possible to prepare single crystals or films of known orientation sufficiently thin to allow the passage of light for an absorption measurement. However, because of a complementary relationship between the absorption and reflection of light, it is possible to measure the polarized reflection spectra of single crystals, and these spectra may be converted to the equivalent absorption spectra through a Kramers–Kronig transformation (Anex and Simpson 1960; Anex and Fratini 1964; Anex 1966). The application of such techniques to single crystals of polymorphic materials can give direct and rich information on the origin of optical effects resulting from molecular as well as bulk processes.

The utility of such an approach is demonstrated on the molecular level with the case of benzylideneaniline **6-LXI** (R = H), which occupied spectroscopists for nearly three decades (Haselbach and Heilbronner 1968, and references therein). This material is isoelectronic with azobenzene **6-LXII** and stilbene **6-LXIII**, but its solution absorption spectrum differs significantly from them (Figure 6.18).

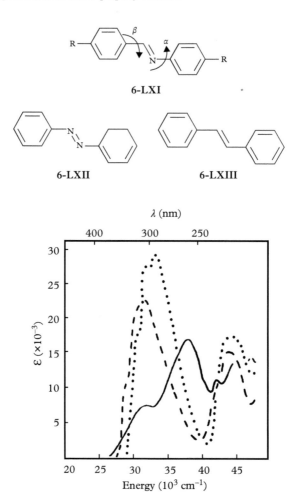

Figure 6.18 *Solution absorption spectra of benzylideneaniline* **6-LXI** *(R = H) (——), azobenzene* **6-LXII** *(- - - - - -), and stilbene* **6-LXIII** *(· · · · ·) in ethanolic solutions (from Bernstein et al. 1979, with permission).*

The difference in the absorption spectra between **6-LXII** and **6-LXIII** on the one hand and **6-LXI** on the other hand has been attributed to a difference in molecular conformation in solution: the former two were believed to be essentially planar on the average, while the latter is not. The non-planarity of **6-LXI** is due to the repulsion between the hydrogen on the bridge and one of the *ortho* hydrogens on the aniline ring. The bridge hydrogen is absent in **6-LXII** and the increased length of the —C=C— bond in **6-LXIII** alleviates that steric effect. For **6-LXI** the tendency towards non-planarity due to the hydrogen repulsion is balanced in part by the π-electron conjugation, leading to the minimum energy conformation of $\alpha \approx 50°$ (Bernstein et al. 1981).

The existence of two polymorphic structures of the dichloro derivative of **6-LXI** (R = Cl) (Bernstein and Izak 1976) provided an opportunity for the direct examination of the relationship between the molecular structure and the electronic spectrum. The metastable very pale yellow triclinic needle form exhibits a planar molecular conformation ($\alpha = \beta = 0°$) (Bernstein and Schmidt 1972) and the stable yellow orthorhombic form (with chunky rhombic crystals) exhibits a non-planar conformation ($\alpha = 25°$; $\beta = -25°$) (Bernstein and Izak 1976). Assuming that the two crystal structures merely serve to hold the molecule in the two different conformations (i.e., the "oriented gas" model), the absorption spectra should reflect the difference in conformation: that measured on the triclinic structure, with a planar conformation, should closely resemble the spectra of **6-LXII** and **6-LXIII**, while that for the orthorhombic structure, with the non-planar molecular conformation, should retain the characteristics of **6-LXI** (R = H) in solution.

Polarized normal incidence specular reflection spectroscopy on single crystals was used to investigate this system (Bernstein et al. 1979). Faces of the two polymorphs were chosen for study on the basis of a favorable projection of the long axis of the molecule (Figure 6.19), which was believed to be the polarization direction of the lowest energy transition (Bally et al. 1976). The reflection spectra

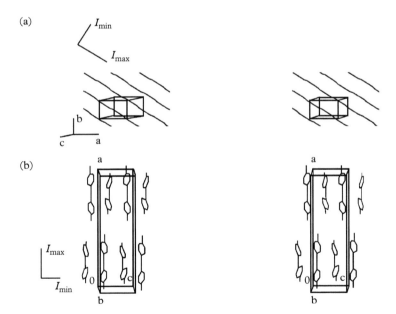

Figure 6.19 *Projections of the crystal structures of **6-LXI** (R = Cl) onto the faces studied spectroscopically: (a) the (001) face of the triclinic form; (b) the (110) face of the orthorhombic form. I_{min} and I_{max} are the extinction directions that were oriented parallel to the electric vector of the incident light for the measurement of the reflection spectra (from Bernstein et al. 1979, with permission).*

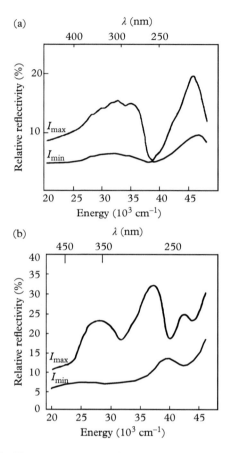

Figure 6.20 *Normal incidence polarized reflection spectra for the two forms of **6-LXI** (R = Cl): (a) the (001) face of the triclinic structure (spectra for light polarized parallel to the* I$_{max}$ *and* I$_{min}$ *directions, respectively); (b) the (110) face of the orthorhombic form (from Bernstein et al. 1979, with permission).*

with the light polarized nearly along the long axis of the molecule (Figure 6.20) clearly indicate a significant difference, with that of the triclinic form even exhibiting vibronic structure reminiscent of the stilbene and azobenzene solution spectra. The corresponding absorption spectra obtained through the Kramers–Kronig transform (Anex and Fratini 1964) are given in Figure 6.21. These can be considered equivalent to the absorption spectra that would be obtained if the measurement could be made through that particular crystal face.

The absorption spectra differ in the same manner as the reflection spectra. The similarity to the expected solution spectra is indicative of weak interactions between the molecules, hence confirming the assumption that these crystals may simply be considered to be oriented gases. The spectrum of the planar

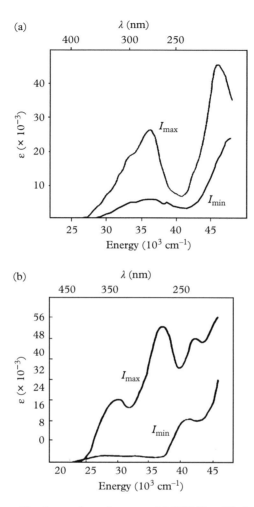

Figure 6.21 *Kramers–Kronig transformed spectra of **6-LXI** (R = Cl) obtained from the reflection spectra of Figure 6.20. The captions for (a) and (b) are as in Figure 6.20 (from Bernstein et al. 1979, with permission).*

molecule exhibits a single low-energy band, which as noted above, is punctuated by reproducible vibronic structure due to the superposition of the —C=N— stretching mode (~1350 cm^{-1}) on the long-axis transition. In the non-planar conformation found in the orthorhombic structure the conjugation of the π system is broken, so the long-axis transition associated with that system no longer exists and consequently no vibronic structure is expected or observed. These observations provide direct confirmation of the interpretations of spectroscopists on the origins of the spectral differences among benzylideneanilines, stilbenes, and azobenzenes. This system is a good example of how polymorphic structures, in particular those which are conformational polymorphs, can provide excellent systems for the study of the relationship between molecular structure and spectral properties. The assumption that these crystals behave as oriented gases is borne out by the similarity between the Kramers–Kronig derived absorption spectra and the solution spectra.

At the opposite extreme from the oriented gas model for molecular crystals, the neighboring molecules do interact with each other resulting in spectral properties of the bulk that differ considerably from those of the individual molecule. Interacting molecules of this type often tend to form aggregates even in solution, a phenomenon that has been exploited by the photographic industry for the tuning of the spectral response of silver halide emulsions (Herz 1974; Smith 1974; Nassau 1983). Aggregate formation can lead to the development of new, and often quite intense absorption bands (e.g., Jelley 1936), and the nature of both the aggregates and their spectral response has been a matter of long-standing interest and study (Würthner et al. 2011).

Polymorphism may be utilized to study this phenomenon of dye aggregation, employing similar experimental spectroscopic techniques as above. As noted and discussed earlier (Sections 3.6.3 and 6.2.3) the dye molecule **6-XXIV** (R = Et, R′ = OH) with the solution spectrum shown in Figure 6.22, has been shown to concomitantly crystallize as violet and green crystals (Tristani-Kendra et al. 1983). Of interest here are the spectral manifestations of those structural differences.

The molecule is essentially flat in both structures, so differences in the spectroscopic properties must be an expression of the intermolecular relationships in the two structures. The polarized normal incidence reflection spectra of the two crystals are given in Figure 6.23, and it can be readily seen that the spectra, representing extremes for quasi-metallically reflecting crystals (Anex and Simpson 1960;

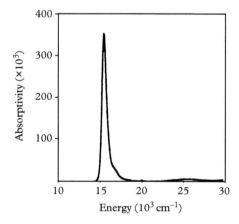

Figure 6.22 *Solution absorption spectrum of molecule **6-XXIV** (R = Et, R′ = OH) in methylene chloride (from Tristani-Kendra et al. 1983, with permission).*

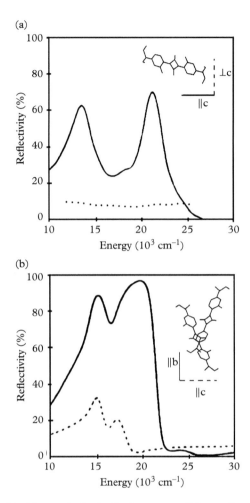

Figure 6.23 *Normal incidence polarized reflection spectra of the two forms of molecule 6-XXIV (R = Et, R' = OH). For each crystal modification there are two spectra measured with the light polarized along each of the two directions (the so-called principal directions), as indicated in the upper right-hand corner, which also shows the projection of the molecule(s) onto the crystal face studied. (a) Triclinic polymorph, (100) face; (b) monoclinic polymorph, (100) face (from Tristani-Kendra and Eckhardt 1984, with permission).*

Fanconi et al. 1969) are significantly different between the two forms for light polarized essentially along the long axis of the molecules. This must be a consequence of the difference in the interaction of a single molecule with its surroundings. The reflection spectrum of the band in the monoclinic form has been interpreted as being composed of two transition-dipole oscillators, while that of the triclinic form contains three, or possibly four, oscillators. The interpretation of these spectra was one of the first attempts to investigate the nature of dye aggregation by studying polymorphic systems (Tristani-Kendra and Eckhardt 1984; Yagi and Nakazumi 2008; Zheng et al. 2015).

Similar studies employing polarized specular reflection spectroscopy on single crystals have been carried out on a polymorphic cyanine (Sano and Tanaka 1986) dye and a polymorphic acridine (Mizuguchi et al. 1994) dye. As many as seven polymorphic oriented layers of a polymethine dye have been prepared and characterized, the different polymorphs serving as different models for states of aggregation (Dähne and Biller 1998a, 1998b).

6.3.3   Excimer emission

Förster and Kasper (1954) reported that in some cases the emission of light following excitation of a molecule can be attributed to fluorescence from a dimer rather than from the individual molecule that initially absorbed the radiation; the emitting moiety was termed an "excimer," and the determination of the relationship between its structure and emitting properties was a matter of considerable interest (Chandra et al. 1958; Murrell and Tanaka 1964; Smith et al. 1966).

The discovery of two excimer emitting polymorphs of the stilbene derivative **6-LXIV** (Cohen et al. 1975) provided an opportunity to determine directly the relationship between excimer structure and its emission properties. The structure of Form A was determined crystallographically and indicated a stacking arrangement, while the structure of Form B, inferred from the cell constants and the photochemistry consistent with the topochemical principles (Section 6.4), was a pairwise arrangement (Figure 6.24). The emission and excitation spectra for the two polymorphs are compared to that for a glassy ethanol solution in Figure 6.24. There are clear differences that the authors have semi-quantitatively attributed to differences in the potential energy curves of the ground state for the stack and pair structures (ca. 1000 cm^{-1}).

6-LXIV

In a later study, Theocharis et al. (1984) were able to make a direct determination of the relationship between intermolecular registry and excimer behavior through the study of two conformational polymorphs of bis(*p*-methoxy)-*trans*-stilbene **6-LXV**. Irradiation of the orthorhombic form with a planar conformation is likely to form excimers, which requires the approach of two non-parallel neighboring molecules. Apparently, relaxation to the original molecular positions following excimer emission may not be possible, leading to a topotactic transformation to a monoclinic form, in which the molecules are not planar, but are parallel, since they are related by translation.

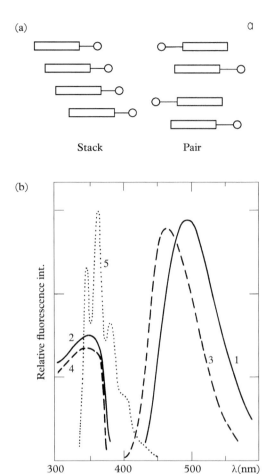

Figure 6.24 *(a) Schematic representation of the stack and pairwise structures for the two poly-morphs of **6-LXIV**. (b) Emission and excitation spectra of **6-LXIV** at 77 K. (1, 2) Emission and excitation spectra of modification A; (3, 4) emission and excitation spectra of modification B; (5) emission spectra of glassy ethanol solution (from Cohen et al. 1975, with permission).*

6-LXV

6.3.4   Excited state diffraction studies

As with analytical methods employed to characterize polymorphs, the combin-ation of techniques in studying structure–property relations is also seeing increased use. With the increasing power and speed of flash photo and synchrotron sources it is now possible to study the structural changes that take place in transient states, in particular in excited-state species (Coppens et al. 1998; Ozawa et al. 1998; Coppens 2011). In the early studies, using the combination of flash laser techniques and synchrotron radiation, Zhang et al. (1999) studied the excited state of 4,4′-dihydroxybenzophenone in two polymorphs of its complex with 4,13-diaza-18-crown-6. In its own crystal, the benzophenone molecules are in close proximity, leading to triplet–triplet annihilation. The formation of a complex leads to greater separation between individual benzophenone molecules, and the existence of two polymorphs means that the triplet lifetimes can be studied as a function of separation of carbonyl groups. The carbonyls are more widely separated in the triclinic modification than in the monoclinic one, leading to corresponding triplet lifetimes of 49.2 ± 0.5 and 44.2 ± 1.2 µs, respectively. The difference makes these promising candidates for diffraction studies of excited-state species. The technological advances since the previous edition have created extensive opportunities for X-ray diffraction studies of excited state species (Coppens 2011).

6.4 Photochemical reactions

In 1964, Cohen and Schmidt formulated the topochemical principles of reactions in the solid state, in particular photochemical reactions, very much on the basis of the different photochemical behavior of polymorphic substances (Hertel 1931). Their investigations arose from the rather long-standing controversy over the nature of the solid-state dimerization of *trans*-cinnamic acid **6-LXVI** (Störmer and Förster 1919; Störmer and Laage 1921; Stobbe and Steinberger 1922; Stobbe and Lehfeldt 1925; De Jong 1922a, 1922b). The material was long known to be polymorphic (Lehmann 1885; Erlenmeyer et al. 1907), and considerable light was shed on the problem by Bernstein and Quimby (1943), but it remained for the combined application of organic photochemistry and X-ray crystallography to these polymorphic systems to define the topochemical principles (Cohen et al. 1964d; Schmidt 1964).

6-LXVI

The principles derived from these experiments are summarized in Figure 6.25. In the β-type crystal, the molecules are arranged in a translationally equivalent manner along a short (≤ 4.2 Å) crystallographic axis, so that the necessarily parallel reactant double bonds are within an appropriate distance and in the proper orientation to undergo a [2 + 2] addition; the resulting photochemical product,

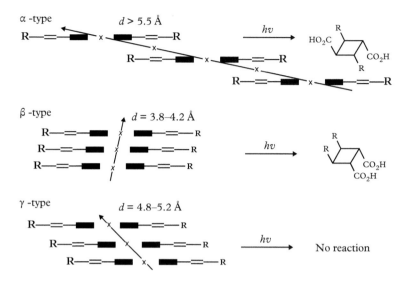

Figure 6.25 *The topochemical principles, as demonstrated by the photochemical behavior of polymorphic cinnamic acids.*

with molecular mirror symmetry, reflects the pre-registry of the reactant molecules in the β structure. In the prototypical α modification, neighboring molecules are related by an inversion center, which again results in a parallel registration of the reactive double bonds, but the crystallographic axis is approximately doubled to 7.5–8.0 Å. The photochemical [2 + 2] product also has inversion symmetry. In the absence of a short contact and/or suitable orientation of the neighboring reactive centers, the reaction does not take place; this is the case for γ-type crystals.

A great deal of solid-state chemistry has been carried out and understood using these topochemical principles (Schmidt 1971; Thomas et al. 1977; Lahav et al. 1979; Thomas 1979; Hasegawa 1986; Ramamurthy and Venkatesan 1987; Venkatesan and Ramamurthy 1991; Singh et al. 1994). Some polymorphism-based variations on this basic theme are worthy of note. Moorthy and Venkatesan (1994) studied the concomitantly crystallizing dimorphs of 4-styrylcoumarin **6-LXVII**, which has two reactive double bonds. The triclinic needles and the monoclinic prisms, which are easily distinguished and separated, yield two different photoproducts at different rates and with different yields (Figure 6.26) depending on which of the two double bonds participates in the [2 + 2] photoreaction. The potential limited variety of photoproducts from these systems depends on the polymorphic form, and the desire to control the nature of the product has led to attempts to generate polymorphs of styrylcoumarins with intermolecular relations that would lead to the preferred products (Vishnimurthy et al. 1996; Row 1999).

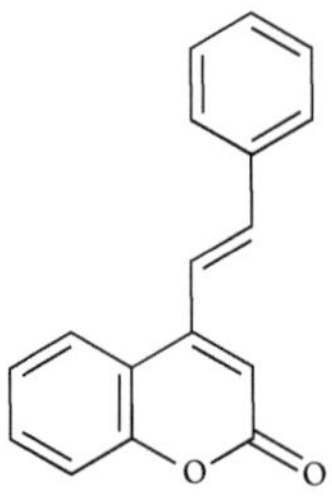

6-LXVII

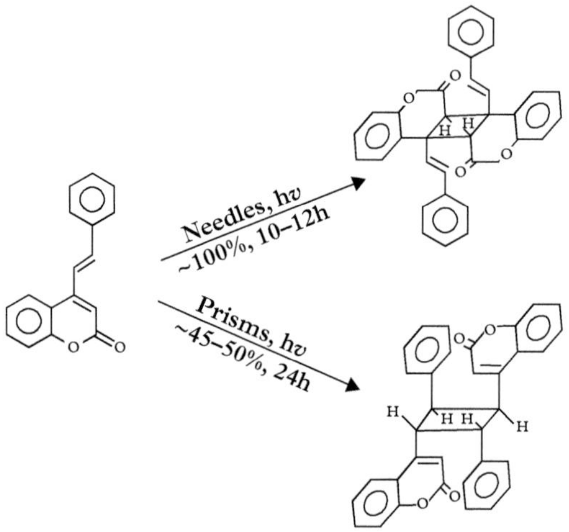

Figure 6.26 *Schematic diagram of the different photoreaction pathways for the triclinic needles and monoclinic prisms of 4-styrylcoumarin* **6-LXVII** *(from Moorthy and Venkatesan 1994, with permission).*

The topochemical principles were also applied to the interpretation of *intramolecular* photochemical reactions of polymorphic materials depending on polymorphic form. Recently these topochemical principles have been extended for elegant intramolecular synthetic reactions (e.g., Elacqua et al. 2012).

1,14-Cyclohexacosanedione **6-LXVIII** can be prepared as conformational dimorphs that undergo Norrish type II photochemistry which can be correlated with the molecular conformations in the two modifications, one leading to a *cis*-butanol, while the other leads to a *trans* product (Gudmunsdottir et al. 1996). Similarly, these principles can be invoked to understand the relative photochemical stability of two polymorphs of the active ingredient in a compound developed for the treatment of psoriasis (Lewis 2000).

6-LXVIII

In the study of the four conformational polymorphs of a cobaloxime complex, Sawada et al. (1996) were able to establish a quantitative relationship between the size of the reaction cavity for an intramolecular photoisomerization reaction and the rate constant for the process in three of the modifications. Such studies provide a great deal of direct detailed information about the relationship between the environmental influences on the mechanism of a reaction at essentially the atomic level, information which is much more difficult to obtain from studies on a single crystalline form.

6.5   Thermal reactions and gas–solid reactions

Thermally induced phase transformations and reactions in the solid state also provide a potentially rich source of information on structure–property relations and often have practical economic implications (e.g., Loisel et al. 1998). As noted earlier, in the past the investigation of such phenomena relied to a large extent on the determination of the structures (and properties) of the starting materials and products (e.g., Jameson et al. 1984; Theocharis and Jones 1984). Two reviews based on that principle surveyed the literature through the 1970s (Paul and Curtin 1973; Curtin et al. 1979), with a more recent one by Even and Bertault (2000).

   The current ability to control and program the crystal temperature, combined with increasingly rapid means for determining crystal structures, and the development of techniques for the study of surfaces that also reveal details of internal structure all serve to provide information that can be used to study stages along the reaction coordinate, rather than just at its extremities (Boese et al. 1987; Sanchis et al. 1997). Individual crystal structures determined as the reaction proceeds provide snapshots of the changes in molecular structure, and in combination produce a representation of the dynamics of the transition or reaction (e.g., Katrusiak 2000). As noted elsewhere in this book, such studies often indicate the lack of a clear distinction between polymorphic transitions and thermal reactions in the solid state (Dunitz 1991).

On the other hand, reactions between solids and gases involve two chemical species and such a distinction is much clearer. The reviews by Paul and Curtin (1975, 1987) remain two of the key early sources of information on the subject, and both contain a number of examples in which the study of polymorphic systems shed considerable light on the mechanisms by which such reactions do or do not proceed (Perrin and Lamartine 1990). More recent examples may be found in Braga and Grepioni (2003, 2005) and Braga et al. (2006). One of the important features of gas–solid reactions is the anisotropy of the reaction, which depends on the fact that on the various faces of a crystal different parts of a molecule may be exposed to the reagent gas, and the reaction will proceed at various rates (or not at all) depending on the functionality of the exposed part of the molecule. Similarly, different structures lead to different crystal habits, but also to different intermolecular relationships that may allow or prevent the initial surface reaction to proceed through the bulk of the crystal.

6.6 Pressure studies

Although pressure is one of the fundamental physical constants, the experimental challenges of pressure-dependent investigations compared with, say, temperature-dependent ones, is still rather daunting. This is in spite of the fact that some quite simple devices were developed for highpressure crystallographic studies on molecular crystals over a quarter of a century ago (Piermarini et al. 1969; Block et al. 1970; Bassett and Takahashi 1974). On the other hand the combination of pressure, temperature and crystallographic studies can lead to detailed understanding of the structural basis of the phase diagram and the mechanisms of the phase transitions (Szafranski and Katrusiak 2000). As with the "hyphenated" analytical techniques described in Chapter 4, the advantages to be gained by carrying out those measurements at a variety of pressures carries great potential for the study of polymorphic systems and their utilization for structure–property relationships (Szafranski et al. 1992; Busing et al. 1995; Boldyreva 1999).

It is natural to expect pressure to induce polymorphic phase changes, and considerable progress has been made recently in developing the techniques to make pressure-dependent crystallographic measurements more readily accessible (Katrusiak 1991). This has led to structural studies of pressure-induced phase transitions (Katrusiak 1990, 1995, 2008; Fabbiani et al. 2004). A number of recent structural investigations as a function of pressure have also been carried out by Boldyreva et al. (2000, 2008, and references therein) and Neumann et al. (2015). One of these was aimed at comparing the anisotropy of the pressure-induced lattice compression of two polymorphs of the complex salt $[Co(NH_3)_5NO_2]I_2$.

The two forms (I and II) are concomitant (Ephraim 1923), but show no evidence of a solid to solid phase transformation. In a rather detailed study over a number of pressures, Boldyreva et al. found a non-reversible transition from I to a new third phase (III), which appears when the hydrostatic pressure is applied in the presence of a methanol–ethanol–water mixture. Phase I is more compressible than II or III with increasing pressure. The structural distortion of the unit cells for all three polymorphs is anisotropic, with the major distortion occurring along the shortest axis, and on the molecular level these distortions could be related to hydrogen-bonding networks and iodine–iodine interactions.

Finally, I describe a system which is difficult to "pigeonhole" into one of the subsections of this chapter, in many ways reflecting the increasing interdisciplinary nature, variety and potential of polymorphism research. Užarević et al. (2009) described the concomitant crystallization of three forms of a methanolate of a molybdenum complex (methanol)*cis*-dioxo(*N*-salicylidene-2-amino-3-hydroxypyridine)molybdenum(VI). Solvent-free mechanochemical and thermal reactions led to the formation of two new complexes, which on subsequent exposure to methanol vapor led to the same form of the original complex.

6.7 A variety of emission phenomena

Since the first edition there has been a great deal of activity in the study of light emission in polymorphic systems (e.g., Varughese 2014). These studies involve many different aspects of the structure and emission phenomenon. Clearly, any attempt at covering all of them is beyond the scope of this work. However, even a brief catalog of the variety of topics covered should serve as another testament to the potential information content in the study of polymorphic systems. As noted earlier in this chapter the origin of the color observed is a function of many factors, but ultimately involves the absorption and/or emission of light by a molecule or a collection of molecules. In order to fully understand the source of the color, and moreover, the source of differences in color from different crystal structures, the fundamental nature of that absorption or emission process must be understood. As will become clear from what follows, determining that fundamental nature is a very challenging task. What is more often done in these studies is a phenomenological qualitative or even quantitative description of the similarities and differences in spectral responses without actually getting to the source of those differences. Nevertheless, the kinds of studies reported here, albeit briefly, are crucial steps to the ultimate goal.

One typical study of fluorescent polymorphs demonstrated the differences in the emission spectra for two polymorphs of (*Z*)-2-phenyl-3-(4-(pyridine-2-yl))arylonitrile **6-LXIX** (Percino et al. 2014), while another involved a series of 2,5-diamino-3,6-dicyanopyrazine dyes **6-LXX** in which the authors determined that intramolecular conformational changes rather than intermolecular exciton interactions were responsible for the color differences (Matsumoto et al. 2006).

6-LXIX   **6-LXX**

Cheng et al. (2016) described the emission properties of the red and orange polymorphs of **6-LXXI**, including the detection of multicolor amplified spontaneous emissions (ASEs). The orange form can be thermally transformed to the red form via a single-crystal-to-single-crystal process, during which the change in the emission can be monitored. The authors suggest that these combined phenomena may provide a development strategy for organic laser science and technology.

6-LXXI

Similar studies have been carried out on polymorphic organometallic compounds **6-LXXII** and **6-LXXIII** (Gussenhoven et al. 2008; Bai and Wang 2006)

6-LXXII   **6-LXXIII**

The relationship between orientational disorder, molecular conformation and the solid-state fluorescent properties were studied on a trimorphic (anthracene/naphthalene) system **6-LXXIV**, with the conclusion that the compound may be considered a tunable solid-state fluorescent material (Li et al. 2012).

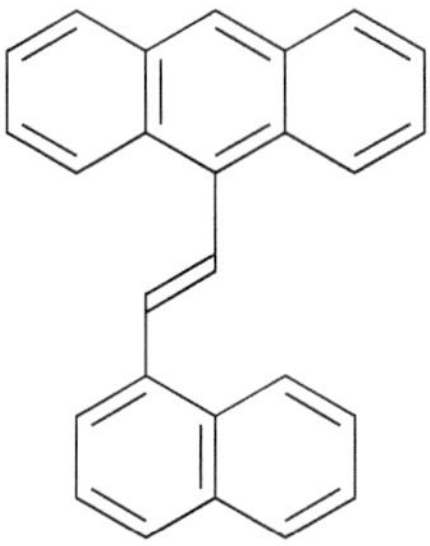

6-LXXIV

Polymorphism-dependent emission behavior was found to be reversible by grinding, heating, or exposure to solvent vapors for a series of bis(diarylmethylene) dihydroanthracenes **6-LXXV** (He et al. 2015). On the other hand the fluorescence (and the underlying electronic properties) of pentamorphic cyclic octadehydrotribenzo[14]annulene **6-LXXVI** were shown to be strongly affected by defects depending on the magnitude of the π-overlap of the molecular planes and defects in the structures. In particular, the strong fluorescence of one polymorph was attributed to those defects (Hisaki et al. 2011).

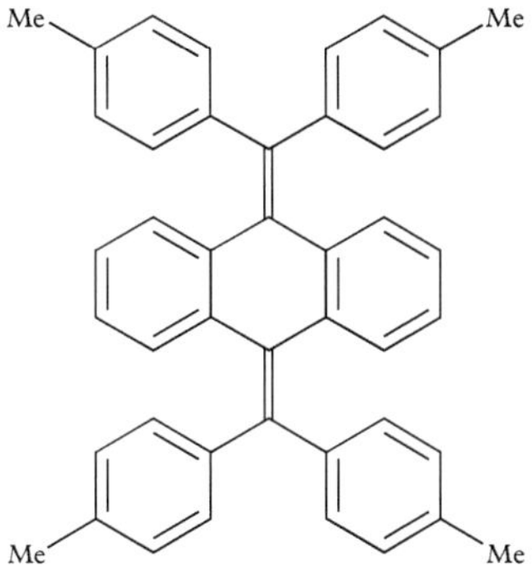

6-LXXV

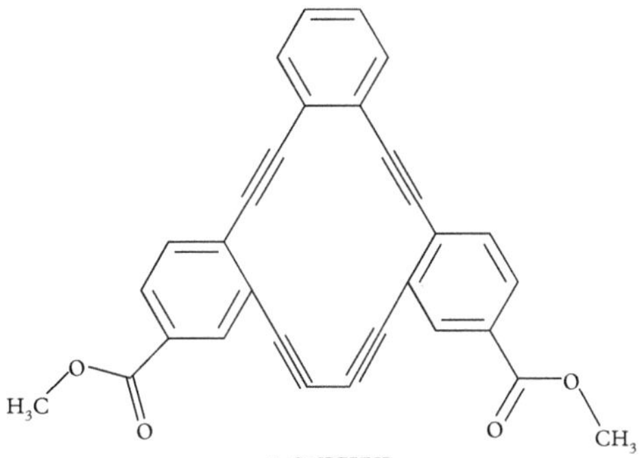

6-LXXVI

The connection between fluorescence and J-aggregation (see Section 6.5) is described in a study of polymorphic substituted diphenylbutadienes **6-LXXVII** (Davis et al. 2004).

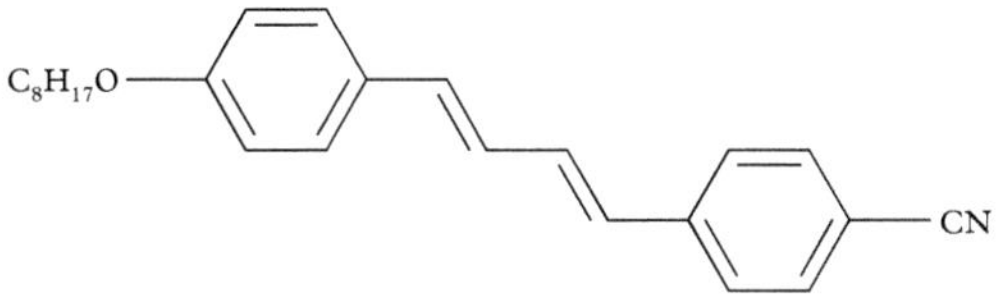

6-LXXVII

Another emission phenomenon that has been studied by utilizing polymorphic systems is ASE which is produced when a medium which exhibits laser gain is pumped to produce a population inversion. Some molecular polymorphic systems that have been studied include derivatives of di(*p*-methoxyphenyl)-dibenzofulvene **6-LXXVIII** (Gu et al. 2012), tetramorphic 4,4′-(thiazolo[5,4-*d*] thiazole-2,5-diyl)bis(*N*,*N*-diphenylaniline) **6-LXXIX** (Wang et al. 2014), and salicylidene(4-dimethylamino)aniline **6-LXXX** (Zhang et al. 2016) in which one of the polymorphs also displays size-dependent fluorescent properties.

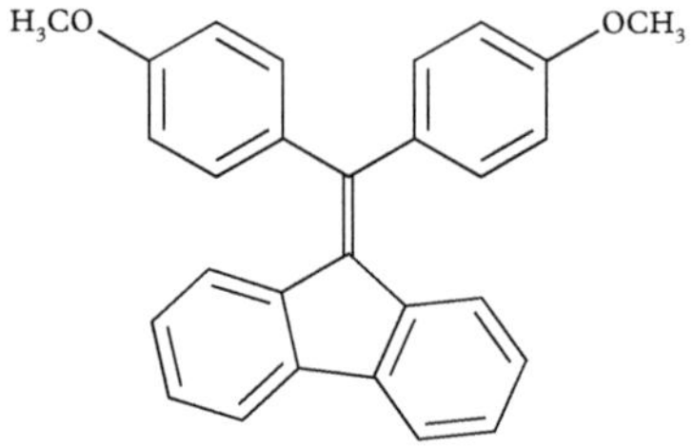

6-LXXVIII

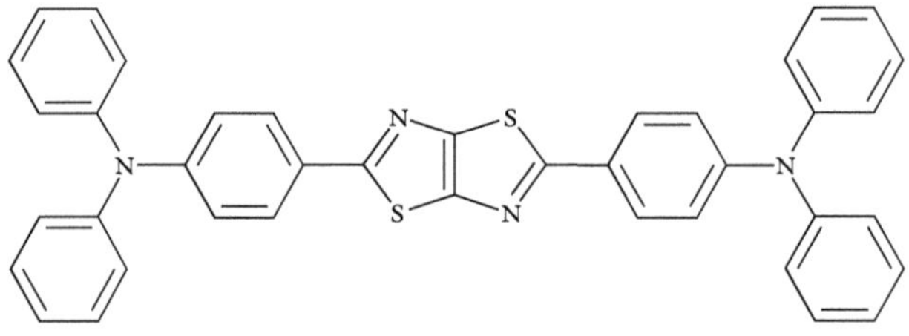

6-LXXIX

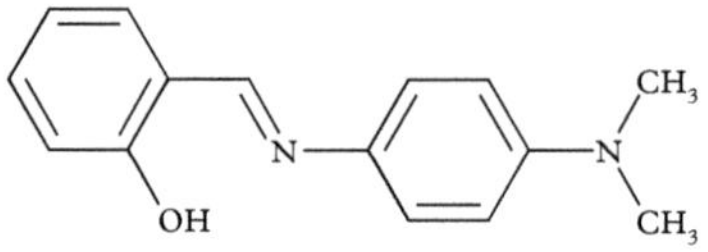

6-LXXX

The (as yet apparently) unique system of vapor-sensitive luminescent poly-morphic system of two gold complexes **6-LXXXI** has been reported by Malwitz et al. (2012). Both compounds exhibit yellow polymorphs that convert to color-less forms upon exposure to vapors that are not incorporated into the transformed structures. The initial yellow forms exhibit a green fluorescence, while the color-less forms fluoresce in the blue. A number of structural similarities and differences are discussed, but this is another rather dramatic example of the subtleties of the connection between structural and spectral differences.

6-LXXXI

6.8 Concluding remarks

Polymorphic systems present unique opportunities for the study of structure–property relations. Just as virtually any property of a material may vary from polymorph to polymorph, so any structure–property relationship may be studied by utilizing polymorphic systems. In this chapter I have attempted to present some representative examples, with perhaps some prejudice resulting from my own experience and interests. But for the limitation on space many more examples could have been cited, and it is hoped that the selection made demonstrates the advantages to be gained by investigating such systems and provides the impetus to give polymorphic systems a high priority in choosing materials for structure–property studies.

7

Polymorphism of pharmaceuticals

After discovery of the first cases of polymorphism with dramatic differences in biological activity between two forms of the same drug...no pharmaceutical manufacturer could neglect the problem.

Borka (1991)

There are many mysteries of nature that we have not yet solved. Hurricanes, for example continue to occur and often cause massive devastation. Meteorologists cannot predict months in advance when and with what velocity a hurricane will strike a specific community. Polymorphism is a parallel phenomenon. We know that it will probably happen. But not why or when. Unfortunately, there is nothing we can do today to prevent a hurricane from striking any community or polymorphism from striking any drug.

Eugene Sun (1998)

7.1   Introduction

As noted earlier (Section 1.1, graphs of citations) the increasing awareness and importance of polymorphism in the past forty years or so is perhaps nowhere more evident than in the field of pharmaceuticals. The early landmark chapters of Buerger and Bloom (1937) and McCrone (1965) did not place any special emphasis on pharmaceutical materials. One noteworthy exception was the Kofler and Kofler book of 1954 whose authors were Professors of the Institute of Pharmacognosy at the University of Innsbruck (Austria) (Webster: *Pharmacognosy* is the branch of pharmacology that treats or considers the natural and chemical history of unprepared medicines). During the early twentieth century the main analytical tool for the investigation of substances isolated from plants was thermomicroscopy. In the course of their studies they paid particular attention to changes of the thermal analytical behavior of the substances investigated at different temperatures and published a wealth of information on the polymorphic behavior of organic compounds, much of it summarized in Kofler and Kofler's (1954) text. In spite of that pioneering work, the seminal paper on the subject of polymorphism of pharmaceuticals was the review by Haleblian and McCrone (1969), which set the scope and the standards for many subsequent works.

Polymorphism in Molecular Crystals. Second Edition. Joel Bernstein. © Joel Bernstein 2020. Published in 2020 by Oxford University Press. DOI: 10.1093/oso/ 9780199655441.001.0001

The literature on the polymorphism of pharmaceuticals is now best described as vast. A quick Google search using the words "polymorphism *and* pharmaceuticals" led to 453 000 results. Even prior to the first edition there was a growing number of reviews (e.g., Haleblian 1975; Kuhnert-Brandstätter 1975; Bouche and Draguet-Brughmans 1977; Giron 1981; Burger 1983; Threlfall 1995; Streng 1997; Caira 1998; Yu et al. 1998; Winter 1999) covering various aspects of polymorphism as related directly to problems in the pharmaceutical field and/or pharmaceutical compounds, and several monographs appeared covering various aspects of the subject (e.g., Byrn 1982; Brittain 1999b; Byrn et al. 1999), including some of the economic and intellectual property implications (Henck et al. 1997).

Reflecting the interest and activity in the subject, especially with regard to the pharmaceutical industry, in the interim since the first edition a number of monographs and significant reviews have appeared.[1] In toto they present an excellent picture of the current state of the art. Therefore, it would be foolhardy to attempt to present another comprehensive review of the subject here. Rather, in keeping with the general philosophy of this book, I provide here references to the review publications and in the remainder of the chapter cite and describe some developments and trends that I view as important components of this increasingly important scientific and technological discipline.

A major contribution to the field and a very useful reference is the Hilfiker-edited (2006) tome on *Polymorphism in the Pharmaceutical Industry*. This was complemented by the second edition of the Brittain-edited (2009b) *Polymorphism in Pharmaceutical Solids*. During the same period Brittain has published very useful regular literature surveys on polymorphism and solvatomorphism in the pharmaceutical industry (Brittian 2007, 2008, 2009a, 2010, 2011, 2012), followed by a ten year perspective on developments, problems, and challenges (Bernstein 2011).

A review on the materials science aspects of pharmaceutical solids (Cui 2007) was followed by a summary of the proliferation, preparation, and use of many new types of solid forms and the relationships among them (Braga et al. 2010; Sarma et al. 2011). Chen et al. (2011) reviewed advances in understanding of the fundamentals of nucleation, the production and scale-up of novel crystal forms and the development and application of continuous processing to pharmaceutical crystallizations. Analytical techniques for studying and characterizing those forms have been reviewed by Chieng et al. (2011), while the importance of recognizing and dealing with polymorphism and solvate formation in chemical processing development has been reviewed by Lee et al. (2011a).

Eighteen years following the initial report on manufacturing problems with the AIDS drug ritonavir, a detailed update on that now iconic disappearing polymorph (see Section 7.12) provides new insights on the nature of that particular

---

[1] As this is being written the second edition of *Polymorphism in the Pharmaceutical Industry*, edited by R. Hilfiker is scheduled for publication in the middle of 2018. This contains essentially all new chapters on many aspects of the same and updated subject matter of the first 2006 edition.

problem and the efforts both to solve it and to understand it (Chakraborty et al. 2016).

7.2   Occurrence of polymorphism in pharmaceuticals

The development of a new drug from a promising lead compound to a marketed product is a long and expensive process. A survey (Herper 2012) covering the years 1997–2011 of research and development spending by twelve innovator companies provides some useful information on the cost of developing a new drug. Over that fourteen year period one company developed five new drugs while the maximum number developed by a single company was twenty one. The research and development costs per new drug ranged from a minimum of $3.69 billion to a maximum of $11.8 billion.

The strict quality control requirements and the intellectual property implications of the drug industry lead to thorough and intensive investigations of the formation and properties of solid substances intended for use in pharmaceutical formulations, both active ingredients and excipients. These efforts, often extending over long periods of time and with many potential experimental and environmental variables, can create conditions that may lead to the appearance of polymorphic forms, intentionally or serendipitously. While it may not be surprising that many pharmaceutically important materials have been found to be polymorphic, or that any particular compound may turn out to be polymorphic, every compound is essentially a new situation (Price and Reutzel-Edens 2016), and the current state of our knowledge and understanding of the phenomenon of polymorphism is still such that we cannot predict with any degree of confidence if a compound will be polymorphic, prescribe how to make possible (unknown) polymorphs, or predict what their properties might be.

Historically there have been attempts to compile instances of polymorphism in pharmaceutically important materials. Since even the definition of what comprises "pharmaceutically important materials" is itself subject to debate (Lipinski 2004) it is difficult to judge how comprehensive such compilations might be. However, generally they serve as useful points of reference and are given here. One of the first organized attempts at such a compilation for steroids, sulfonamides, and barbiturates was by Kuhnert-Brandstätter (1965; see also Kuhnert-Brandstätter and Martinek 1965). Much of those data may be found in her subsequent book (Kuhnert-Brandstätter 1971), which also contains a compilation of many of the thermal studies of pharmaceutical compounds that revealed polymorphic behavior. Numerous additional reports of studies of materials by the Innsbruck school of polymorphic pharmaceutical compounds (using thermomicroscopy and IR spectroscopy) appeared in the literature starting in the middle 1960s; many of those have been listed by Byrn et al. (1999). Borka and Haleblian (1990) compiled a list of over 500 references to the reports of polymorphism in over 470 pharmaceutically important compounds.[2] This was shortly followed (Borka 1991) by a review of polymorphic substances included in Fasciculae 1 to 12 of the European Pharmacopoeia (EP), including a comparison of melting points in the EP and the original literature. The latter review was subsequently updated in 1995 by including EP entries for Fasciculae 13 to 19 (Borka 1995).

Griesser and Burger (1999) compiled the information regarding 559 polymorphic forms, solvates (including hydrates) of drug solids at 25 °C in the 1997 edition of the EP. They also noted that of the 10 330 compounds in the 1997 edition Merck Index only 140 (1.4%) are specifically noted as polymorphic, 540 (5%) are noted as hydrates and 55 (0.5%) have been specified as solvates. These numbers reflect a failure to report these phenomena rather than representative statistics suggesting the contemporary state of awareness of polymorphism on the part of compilers of such compendia and reference works. More recent statistics from a number of sources were summarized by Bernstein (2011). References to drugs discovered prior to 1971 that form solvates have been compiled, along with a separate summary of the thermomicroscopic behavior of drug hydrates by Byrn et al. (1999). It was recently shown that within the family of molecular compounds pharmaceutically important molecules do not exhibit any exceptional tendency to exhibit polymorphic behavior (Cruz-Cabeza et al. 2015). The October 2017 release of the CSD contains the following number of hits for descriptors of solid forms: *form* (481), *phase* (5606), *polymorph* (27 542), *solvate* (178 161), *hydrate* (115 971).

7.3   Importance of polymorphism in pharmaceuticals

Polymorphism can influence every aspect of the solid-state properties of a drug. Many of the examples given in preceding chapters on the preparation of different crystal modifications, on analytical methods to determine the existence of polymorphs and to characterize them and to study structure/property relations, were taken from the pharmaceutical industry, in part because there is a vast and growing body of literature to provide examples. In this section I will present some additional examples of the variation of properties relevant to the use, efficacy, stability, etc. of pharmaceutically important compounds that have been shown to vary among different crystal modifications. It will be clear from the context of what follows that the exploration of the solid form landscape can be a complex process (e.g., Braun et al. 2009a, 2009b; Chekal et al. 2009; Campeta et al. 2010) and one that continues over the entire lifetime of the drug, often leading to the discovery of new forms with considerable beneficiary properties to both developer and patient (e.g., Domingos et al. 2015; Pogoda et al. 2016; Covaci et al. 2017; Higashi et al. 2017).

---

[2] Dr. Borka communicated with this author that he and Dr. Haleblian did not receive galley proofs of this paper, which unfortunately contains "numerous printing errors." The list of errata actually contains 48 of them. Even with that cautionary note, it is a useful compilation.

7.3.1   Dissolution rate and solubility

The dissolution properties and solubility are often crucial factors in the choice of the crystalline form for formulation of a drug product (Carstensen 1977). In general, these two factors play a major, if not an overriding, role in determining the bioavailability of the drug substance (see also Section 7.3.2). The physiological absorption of a solid dosage form usually involves the dissolution of the solid in the stomach, and the rate and extent of that dissolution is often the rate determining step in the overall process. Since different crystalline forms can exhibit different dissolution kinetics and limits, these properties are routinely studied in great detail for any drug substance, whether polymorphic or not; clearly, characterization for polymorphic substances is even more critical. As a result, there is a vast array of literature covering such studies. An early review contained a compilation of references to many of the earlier studied materials (Kuhnert-Brandstätter 1973).

The fundamentals of the measurements of solubility and dissolution properties of pharmaceuticals are given elsewhere in considerable detail (Vachon and Grant, 1987; Grant and Brittain 1995; Byrn et al. 1999, esp. Chapter 6). Reviews of the physical principles of dissolution and solubility behavior of organic materials (Grant and Higuchi 1990) and pharmaceuticals (Grant and Brittain 1995) have been given and, particularly relevant to this chapter, typical examples of the variation of these processes among crystalline forms of pharmaceuticals are comprehensively described by Brittain and Grant (1999).

7.3.2   Bioavailability

The rate and extent of the physiological absorption of an active drug substance are crucial factors in its overall efficacy. These can vary among different crystal modifications, and they have become an important scientific and regulatory issue (Ahr et al. 2000).

While a number of studies of the connection between the crystal modification and bioavailability have been published, it is reasonable to assume that many more remain the intellectual property of pharmaceutical companies or in confidential documentation submitted to regulatory agencies. Also many in vitro studies, especially those of extent or rate of dissolution, are used to extrapolate to expected bioavailabilities (e.g., Chikaraishi et al. 1995; Shah et al. 1999), which is indeed proven in some cases, for example, the tetramorphic tolbutamide (Kimura et al. 1999). I cite here a few additional examples from the open literature, limiting myself (except for one example) to cases in which the bioavailability does differ among crystal modifications. Much of the following in this section has been retained from the first edition, since the examples suitably demonstrate the principles of the connection between polymorphism and bioavailability. Perhaps the most notorious case of the connection between polymorphism and bioavailability is that of ritonavir (Chemburkar et al. 2000; Bauer et al. 2001; see Section 3.7). Others, especially those involving poorly soluble drugs, have been reviewed by Censi and Di Martino (2015). Among the examples cited by Censi and Di Martino are atorvastatin calcium and axitinib, both of which have been reported to have over sixty solid forms (Chekal et al. 2009; Jin 2012).

The two polymorphs of barbiturate pentobarbital exhibit significantly different rates of absorption and area under the curve for oral administration (Figure 7.1) (Draguet-Brughmans et al. 1979). Comparison of rectal absorption of suppositories containing the two polymorphs of indomethacin showed higher levels for the α form than for the γ form (Yokoyama et al. 1979). The trimorphic antileukemic mercaptopurine form III was found to be 6–7 times more soluble than form I, but the bioavailability (in rabbits) was approximately 1.5 times greater for form III than for form I (Yokoyama et al., 1980). The anti-anxiety agent nabilone has at least four polymorphs, designated A, B, C, and D, which appear to be equally hydrophobic and insoluble. Forms B and D are bioavailable in dogs, while A and C are not. On prolonged storage, heating, or grinding, all convert to the non-bioavailable form A (Thakar et al. 1977).

The bioavailability can also vary between crystalline and amorphous modifications (Section 7.7), as well as among polymorphic forms. The amorphous form of the antibacterial agent azlocillin sodium has more activity than the crystalline form (Kalinkova and Stoeva 1996). Similarly, an anti-nematode drug, code-named PF1022A, exists in four modifications, designated α (amorphous) and crystalline I, II, and III; α and III have higher solubility and are more effective than I and II against tissue-dwelling nematodes (Kachi et al. 1998). An amorphous form of another code-named drug (YM022), generated by spray drying, has enhanced bioavailability over two crystalline forms (Yano et al. 1996).

The degree of hydration of different modifications also plays a role in the bioavailability. One of the most studied systems in this regard is the anhydrate/

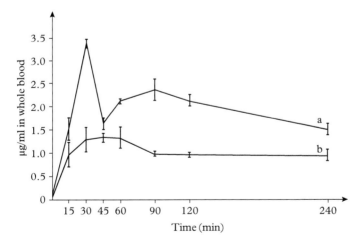

Figure 7.1 *Rate of absorption of the two polymorphs of orally administered pentobarbital (from Draguet-Brughmans et al. 1979, with permission).*

trihydrate of the antibiotic ampicillin, although the results have not always led to consistent conclusions. Early in vitro solubilities (Poole et al. 1968) and rates of dissolution (Poole and Bahal 1968) were shown to differ. In vivo the anhydrous form reaches a maximum. In vivo bioavailability studies by Ali and Farouk (1981) (Figure 7.2) clearly indicated differences between the two. Other studies, however, indicated similar bioavailabilities (Cabana et al. 1969; Mayersohn and Endrenyi 1973; Hill et al. 1975), and Brittain and Grant (1999) have suggested that the differences reported by various groups suggest that the bioavailability in this case is strongly influenced by the nature of the formulation of the dosage form.

There are also some examples of solvates (as opposed to hydrates) that have been reported to exhibit differing bioavailability. For instance, for implants of the methanol solvate of prednisolone *tert*-butyl acetate the mean absorption rate was found to be 4.7 times that of the anhydrous form, which in turn was similar to the hemiacetone solvate (Ballard and Biles 1964). The same authors also reported that the mean absorption rate for all solvates of cortisol *tert*-butyl acetate were significantly different from that of the anhydrous form.

While the *rate* of absorption may differ, the *extent* of absorption may still be equivalent. Such is the case for sulfameter (sulfamethoxydiazine). The different polymorphic forms have been shown to exhibit different equilibrium solubilities and dissolution rates (Moustafa et al. 1971). Form II is thermodynamically less stable and has an absorption rate about 1.4 times that of Form III, which is more stable in water (Khalil 1972). This group determined that the two forms have

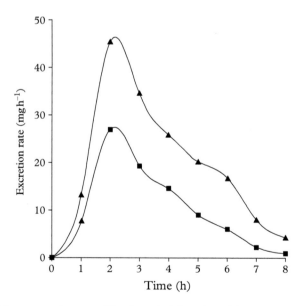

Figure 7.2 *Urinary excretion rates following administration of the anhydrate (▲) and trihydrate (■) forms of ampicillin (from Brittain and Grant 1999, with permission).*

different absorption, but using 72-hour excretion data showed that the extent of the absorption of the two forms was equivalent (Khalafallah et al. 1974).

Perhaps the classic example of the dependence of bioavailability on polymorphic form is chloramphenicol palmitate. Chloramphenicol is a broad spectrum antibiotic and antirickettsial which was developed in the 1960s and had a significant portion of the market until the appearance of side effects limited its use to topical application. The exceedingly bitter taste of the active chloramphenicol led to its formulation as an oral suspension of the tasteless 3-palmitate (CAPP). The early physical and physiological studies on the material were summarized by Aguiar et al. (1967) (see also Mitra et al. 1993). There are three polymorphic forms (A, B, and C) in addition to an amorphous form. The characterization of the various forms by melting point and IR analyses proved problematic, even inconclusive, due to polymorphic transitions during grinding for sample preparation (Borka and Backe-Hansen 1968).

The A form is the most stable, but only the B and amorphous forms are biologically active. Aguiar et al. (1967) determined the physiological absorption rate as a function of the A and B polymorphs, as shown in Figure 7.3. The suspension containing only the metastable B form gives higher blood levels following oral doses than those containing only Form A, by nearly an order of magnitude. Since particle size was shown to have little effect on blood levels, it was concluded that the structure of the solid plays an intimate role in determining the physiological absorption rate. As a result of this finding the mechanism of this absorption and its connection with the polymorphism were investigated in considerable detail.

CAPP is nearly insoluble in water; hence it must be hydrolyzed by enzymes in the small intestine before absorption can take place. According to one possible

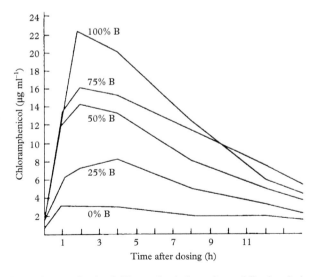

Figure 7.3 *Peak blood serum levels of chloramphenicol two hours following dosing for pure polymorphs A and B and various mixtures following a single oral dose equivalent to 1.5 g of chloramphenicol palmitate. (From Haleblian and McCrone 1969 (after Aguiar et al. 1967), with permission).*

proposed mechanism (Aguiar et al. 1967) the first and rate determining step in the total process is a dissolution of the ester followed by enzymatic hydrolysis of CAPP. Such a mechanism is consistent with the generally accepted dissolution/absorption mechanism for most solid drugs. However, Andersgaard et al. (1974) proposed a second mechanism, in which *solid* CAPP is enzymatically attacked in the small intestine. If dissolution is the first and rate determining step of the total process, then there should be a close relationship between the rates of dissolution and the rates of enzymatic hydrolysis of polymorphs A and B. On the other hand, no relationship of this sort is expected if CAPP is attacked in the undissolved state.

The rates of dissolution on the one hand and in vitro hydrolysis of the solid by the enzyme pancreatic lipase on the other hand, are given in Figure 7.4. If dissolution is the first step in the total hydrolysis process, the reaction scheme may be written as:

$$\text{undissolved CAPP} \Rightarrow \text{dissolved CAPP} \Rightarrow \text{hydrolyzed CAPP}$$

Since the rate of the second step of this process must be the same for forms A and B of CAPP, this model for the absorption process leads to the conclusion that any differences in the rate of formation of hydrolyzed CAPP must be due to a difference in the rate of dissolution of the two polymorphs. The data presented in Figure 7.4 are not compatible with the assumption that dissolution is the first and rate determining step, since the slopes at time zero are significantly different from those expected for such a mechanism. Rather, Andersgaard et al. (1974) claimed that it is more reasonable to assume that CAPP is attacked in the undissolved state, probably by pancreatic lipase, which is known to act on substances insoluble in water (Waki 1970). Further studies on CAPP and the analogous stearate (Cameroni et al. 1976) apparently corroborate this assumption, indicating that

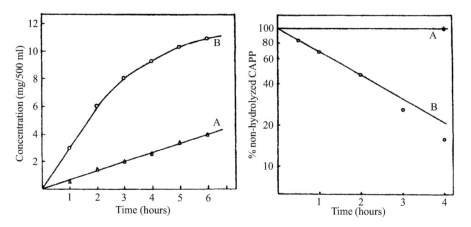

Figure 7.4 *(a) Rates of dissolution of polymorphs A and B of CAPP. (b) Rates of in vitro enzymatic hydrolysis by pancreatic lipase of polymorphs A and B of CAPP (after Andersgaard et al. 1974, with permission).*

there is a direct connection between the solid-state structure and the enzymatic hydrolysis of chloramphenicol palmitate. It has been found that storage of the B form at the relatively high temperature of 75 °C leads to conversion to the inactive A form in a matter of hours, suggesting that extended storage at higher temperatures could lead to reduced efficacy of a formulated product (Devilliers et al. 1991).

Another rather dramatic manifestation of the importance of the crystal habit as well as the solid form in the activity of an active ingredient was demonstrated by Yang et al. (2017a, 2017b) for the well-known—but heavily restricted—insecticide DDT (1,1,1-trichloro-2,2-bis(4-chlorophenyl)ethane). Following the 1962 publication of Rachel Carson's book *Silent Spring* describing the effect of DDT on the biological food chain, the previously widely used insecticide was essentially removed from the market. Its mode of action is based on contact between the solid form of the compound and the hydrophobic footpads of insects. Yang et al. (2017a) recently reported the discovery of a second polymorph of DDT which exhibited considerably enhanced lethality in the solid state over the only previously known crystal form.

The effect of the transformation of crystal form after ingestion has been of increasing interest in the past decade (see also Chapter 10). In a study of the two anhydrous forms (I and III) and a dihydrate of carbamazepine (Kobayashi et al. 2000), it was shown that the initial in vitro dissolution rate was of the order III > I > dihydrate, with form III being transformed more rapidly to dihydrate than form I. The solubilities of the two anhydrous forms were 1.5–1.6 times that of the hydrated form. When dosed to dogs at 40 mg/body weight there were no significant differences between forms for the area under the curve. However, at doses of 200 mg/body weight significant differences in plasma concentration versus time were observed. It was suggested that the difference was due to rapid transformation from form III to dihydrate in the gastrointestinal fluids.

Three brief examples involving commercially successful drugs indicate the diversity in bioavailability among different polymorphic systems. At one extreme the two polymorphic forms of ranitidine hydrochloride (GlaxoSmithKline's H$_2$-antagonist Zantac®) have been shown to be bioequivalent, which is one of the reasons that ethical and generic companies were involved in litigations over the two forms (see Section 10.2). At the other end of the spectrum is Abbot's protease inhibitor Norvir®, generically ritonavir described earlier (Section 3.7). The apparently greater stability of a newly prepared crystal form of glibenclamide has also been attributed to reduction in dissolution and bioavailability of its tablets (Panagopoulou-Kaplani and Malamataris 2000).

7.3.3 Following phase changes and mixtures of forms

One of the important aspects of polymorphism in pharmaceuticals is the possibility of interconversion among polymorphic forms, whether by design or happenstance.

This topic was earlier reviewed by (Byrn et al. 1999, especially Chapter 13). The ability to follow and understand phase changes adds an additional degree of control over the choice and preparation of the solid form chosen for eventual development and formulation. The methods and techniques described here comprise a (perhaps biased) selection from a wide variety of continually developing measures for tuning that control.

One of the first requirements for controlling a crystallization procedure is determining which forms are present and to what extent. Giron et al. (2007) have provided guidelines for comparing quantitative analysis of those mixtures.

Boldyreva (2007, 2016) has demonstrated a number of experimental techniques employing non-ambient conditions that can lead to the formation of new phases or phase changes to new forms: mechanical, ultrasonic treatment, hydrostatic compression, high temperature or cryogenic spray-drying, micronizing, and crystallization from supercritical solvents. It has also been shown that mechanochemistry should be considered an alternate strategy for carrying out an effective solid form screen (Hasa and Jones 2017).

In situ monitoring by Raman (Nanubolu and Burley 2015) by SSNMR (Pindelska et al. 2015) and by atomic force microscopy (Thakuria et al. 2013) can be employed to follow the transient phase changes in solids. The in situ melting temperatures of solid samples in a 96-well plate crystallization of an active pharmaceutical ingredient (API) were determined within ca. 5 minutes using IR thermal camera technology, and were shown to agree well with melting data obtained from DSC measurements (Kawakami 2010).

Raman spectroscopy has also been employed for the real-time monitoring of phase changes in solution (Bras and Loureiro 2013), and magnetic levitation has been shown to be useful for the separation of polymorphs from a mixture (Atkinson, et al. 2013). Slow evaporation from aqueous microdroplets yielded the normally difficult to crystallize metastable polymorphs of D-mannitol (δ form) and glycine (β form) (Poornachary et al. 2013). In another instance, variation of the pH led to one polymorph of an amino acid in the unstable zwitterion form, while a polymorph of the stable non-zwitterionic molecule was obtained at a different pH (Orola et al. 2012).

The phase changes in ROY have been studied in situ using Raman spectroscopy in tandem with time-resolved X-ray diffraction (Gnutzmann, et al. 2014). P-tentially interfering factors such as temperature, humidity, and contact with solid surfaces were eliminated by containing the sample in a specially designed acoustic levitator. The technique allows for the selection of pure individual phases in this highly polymorphic system.

It was pointed out earlier (Section 2.5) that the order of thermodynamic stabilities may differ in the nano regime from that in the macro regime. The nanocrystallitization can be monitored in situ using automated electron diffraction tomography, as demonstrated on the caffeine system by Gorelik et al. (2012).

7.4   Screening for crystal forms

The past two decades have witnessed a general recognition in the pharmaceutical industry of the absolute necessity for exploring and mapping the crystal form landscape as an integral component of the research and development program for any potential new drug product (Hilfiker et al. 2006; Kratochvil 2007; Florence 2010). That exploration and mapping has often been termed a "polymorph screen" or a "solid form screen," commonly with the adjectival precedent "routine." While the entire research and development program of bringing a new drug to market may involve a routine set of steps (formulation, toxicology, stability, clinical trials, etc.), none of those steps is in fact routine, least of all perhaps, as I demonstrate in this section, that of exploring and mapping the solid form landscape.

The often mistaken impression that the polymorph screen is a "cookbook routine" stems at least in part from the 1995 publication of a "decision tree" or flowchart presumably outlining or summarizing the search and characterization of polymorphs (Byrn et al. 1995; Figure 7.5).

While the figure does contain a number of the features of a polymorph screen, it is deceivingly brief. Many, if not most, of those aspects of the polymorph screen involve considerable variability in the choice of materials and conditions leading to an almost infinite number of choices for every experiment proposed or carried out (Aaltonen et al. 2009; Malamatari et al. 2017; Vioglio et al. 2017). That

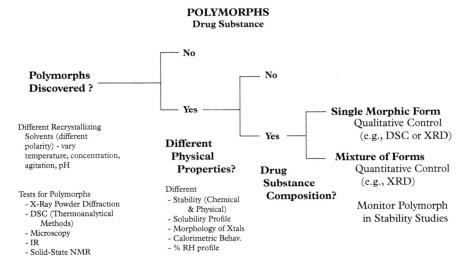

Figure 7.5 *Proposed flowchart or "decision tree" for carrying out a search for polymorphs of solid drug substances. (From Byrn, S. R., Pfeiffer, R., Ganey, M., Hoiberg, C., and Poochikian, G. (1995). Pharmaceutical solids: a strategic approach to regulatory considerations.* Pharm. Res., *12, 945–54. With permission of Springer Nature.)*

variability may be lost on the untrained or inexperienced hand (Newman, 2013). The following discussion will touch on many of the aspects of the flowchart in Figure 7.5 and additional aspects of the solid form screen. Some examples of the complexity of that solid form landscape—totally unknown at the start if its exploration—have been given by many others including Bolla et al. (2014), Braun et al. (2017b), Price and Reutzel-Edens (2016), Price et al. (2016), and Zilka et al (2017). But before moving on to the details of this decision tree it is important to emphasize the very important role of serendipity in the discovery of new crystal forms even in this somewhat formalized framework (Rafilovich and Bernstein 2006; Braga et al. 2011b).

7.4.1 Solvent selection

As described in other sections of this book, there are many, and an increasing number, of experimental techniques for searching for solid forms. The traditional and most common starting point is recrystallization from solution, which requires the choice of solvent and/or solvent mixtures. Practicing chemists maintain a battery of solvents for such a purpose, and a number of suggested somewhat limited lists of common solvents have been given, for instance, by Guillory (1999) and Newman (2013). Since the number of experiments is often limited by the amount of material (and time) available judicious choice of solvents and solvent combinations is essential. Often the choice is based on chemical experience and intuition combined with some early preliminary observations and familiarity with the compound. However, detailed consideration of the challenge reveals a vast array of choices and still no proven method for making those choices. An excellent detailed summary of that situation and guidelines to help make that selection have been presented by Allesø et al. (2008), covering a database of 218 organic solvents with 24 property descriptors, including log P, vapor pressure, hydrogen bond formation capabilities, polarity, number of π bonds, and descriptors derived from molecular interaction field calculations (e.g., size/shape parameters and hydrophilic/hydrophobic regions) and others. The authors note that for pharmaceutical applications solvent toxicity is also a crucial parameter.

Following this rather comprehensive study, a combined cheminformatics approach was applied to ninety-four solvents in an attempt to provide a rational approach for choosing suitable solvents for physical form screening (Johnston et al. 2017).

In spite of the fact that the choice of solvent together with solvent combinations is almost infinite one solvent can yield more than one polymorph, even concomitantly as noted earlier (Section 2.5) (Bernstein et al. 1999). For example, Getsoian et al. (2008) have demonstrated with carbamazepine that with controlled variation of the crystallization temperature and level of supersaturation it is possible to obtain three of the four polymorphs (known at that time) with the same cumene (isopropyl benzene) solvent.

7.4.2   Screening specifically for hydrates and solvates

The formation of hydrates and solvates of solid APIs was once considered circumstantial at best, and often an annoyance to be avoided (Tian and Rantanen 2011; Loschen and Klamt 2016). However, increased understanding of the solid-state chemistry of those solid forms, how they may be recognized, understood and ultimately be utilized in drug development and production has been demonstrated in a number of specific cases (e.g., Braun et al. 2015), increasingly employing a combination of computational and a variety of experimental techniques (Braun et al. 2017b). Earlier general and very useful reviews of this topic have been provided by Griesser (2006) and Morris (1999). I cite a number of the more recent developments in this area.

A general approach for the screening of an API for solvate formation has been given by Douillet et al. (2012), while Tian et al. (2010a, 2010b) have reported a specific study that attempts to identify the factors determining formation of hydrates and anhydrates.

One example of a screen designed specifically for solvates was reported by Zvoníček et al. (2017) on the anticancer drug ibrutinib. The eleven solvents were chosen on the basis of a tailor-made design and included a variety of experimental methods, leading to eight solvates. Subsequent characterization showed that in this case solvate formation could improve the dissolution rate by a factor of 8.5 over the most stable unsolvated form.

A CSD survey of 124 molecules containing five- and six-membered heterocyclic aromatic moieties attempted to determine the propensity to form hydrates (Bajpai et al. 2016). On the one hand 18.5% did form hydrates; on the other hand, an experimental screen of eleven N-heterocyclic aromatic compounds containing at least two hydrogen bond acceptors and no competing hydrogen bond donors resulted in the formation of hydrates in 70% of the cases. Such studies again demonstrate the care that must be exercised in choosing a data base for statistical studies of solid form occurrence.

Crystal structure prediction studies (see Section 3.6) had indicated that the possible existence of a high-density monohydrate form of 4-aminoquinaldine (Braun et al. 2016). Combined computational and experimental studies on the anhydrous and kinetically preferred monohydrate, with a higher nucleation rate than the calculated more stable form, led to a successful strategy employing impurities for producing the computationally predicted more stable form.

Similarly, computational efforts to generate possible hydrates of the poorly water-soluble creatine, suggested that these were considerably higher in energy than the predicted anhydrous forms, thus providing a rationale for the lack of formation of hydrate (Braun et al. 2014). Crystallization experiments yielded anhydrous forms in the appropriate order of thermodynamic stability, including the stable room-temperature form. Braun et al. (2011b) also demonstrated the utility of computational methods to predict possible hydrates. For a dihydroxybenzoic acid a novel hydrate predicted on the basis of those calculations was obtained.

The role of solvates and hydrates in preparing unsolvated forms by desolvation was demonstrated for phenobarbital, an important drug compound with a particularly complex solid form behavior (Zencirci et al. 2014). The desolvation processes can be considerably complex, but in the end may be utilized to produce an unsolvated material that is not accessible by other means.

The importance of the consideration of solvate and hydrate formation in screening for salt forms has been reviewed by Shevchenko et al. (2011). These authors actually carried out the screen by systematically characterizing successive batches during salt synthesis optimization. In the process they found two anhydrous and one hemihydrate form, the last being the most stable and chosen for further development.

7.4.3   Use of gels

The technique of gel crystallization is described in the classic text by Henisch (1996), originally published in 1970. There is considerable discussion of inorganic systems, but this technique has been successfully applied to molecular systems as well (Desiraju et al. 1977). The gel environment facilitates slow diffusion of the crystallizing substance(s), and hence is one approach to near equilibrium (i.e., thermodynamic) crystallization conditions. However, the composition of the gel may be varied in many ways, leading to a variety of possibilities for obtaining crystal forms, and hence is an appealing option in performing solid form screens.

Overcoming nucleation difficulties of hydrophobic APIs can be facilitated by creating either a calixarene-based tetrahydrazide gelator or a co-gel of a calixarene-containing building block together with a bis-crown ether (Kaufmann et al. 2016). These substrates provide potential anchor sites for nucleation and the subsequent growth of new crystal forms.

The principles were elegantly demonstrated in a report by Foster et al. (2017) on the iconic polymorphic molecule ROY (see Section 5.8). The synthesis of a special template designed gelator based on bis urea led to the crystallization of a metastable red form of ROY rather than the thermodynamically stable yellow form obtained from all other gels and solution control experiments.

An example of the variety of crystal forms in gel crystallizations that may be achieved was demonstrated by Braga et al. (2011a) using a newly synthesized gelator of silver(I) complexed with 1-phenyl-3-(quinolin-5-yl)urea. In sealed vials four different polymorphs were obtained depending on the gelling solvent.

7.4.4   Use of ionic liquids

Another recently utilized significant variation in the crystallization medium has been the use of ionic liquids (ILs) as solvent (Egorova et al. 2017) with specific applications to polymorphic systems (Shamshina et al. 2013; Domingos et al. 2015; Zeng et al. 2018). As the authors note, in common with normal solvents, mixtures of ILs over a full range of composition may be prepared, providing additional and virtually unlimited possibilities for creation of new polymorphic forms. Moreover, the fact that ILs are comprised of at least one ionic pair means that an ionized API can be one of the components to facilitate the preparation of liquid dosage forms. An additional advantage for the use of ILs is the green chemistry aspect, including the avoidance of traditional solvents and the need to recover or properly to dispose of them.

Two recent examples testify to the potential for utilizing ILs in drug development. Martins et al. (2017) employed ILs at room temperature in a successful search for new forms of the neuroleptic drug gabapentin, less stable than those already known. The previously known, but very difficult to prepare and maintain, Form IV was isolated and prepared in bulk, with stability being achieved by storing soaked in the imidazolium based IL or mixtures thereof. A complementary molecular dynamics study provided insight into the reasons for the appearance and stability of this form in the IL.

Adefovir dipivoxil, the diester prodrug of adefovir for the treatment of hepatitis B, was crystallized with ILs using antisolvent crystallization (An and Kim 2013). Variation of the IL composition and the temperature produced a new form, or the conventional form-1, or no crystallization. The authors attempted to rationalize the mechanism of the nucleation of the forms with the temperature and concentration conditions of the crystallizing medium.

There is no question that ILs open vast new opportunities for the preparation of new solid forms of APIs, and that they will become an integral component of the chemical toolbox for exploring the solid form landscape.

7.4.5 Obtaining a difficult stable or unstable new form

The general recognition of the importance of gaining detailed knowledge of, and control over the solid form landscape has fostered the continuing search for solid forms or developing new and creative techniques for those that may have disappeared (Section 5.8). Many examples are cited throughout the text; a few more should encourage future workers in the field to be imaginative and creative in carrying out those searches.

Boldyreva (2016) has reviewed a number of non-ambient conditions that have been employed in the search for new forms: mechanical and ultrasonic treatment, hydrostatic compression, high temperature or cryogenic spray-drying, crystallization from super-critical solvents, and utilization of solid-state reactions. Epitaxial growth was described in Section 5.8. The technique has been employed to retrieve a disappearing polymorph (Tidey et al. 2015). A polymorph of a palladium complex, obtained routinely following its first appearance, could no longer be obtained for over a year following the appearance of an additional form. Seeding with an isomorphous form of an analogous compound of the disappeared polymorph led to its recovery.

Although griseofulvin has been used as an antifungal drug for over fifty years no polymorphism had been observed until the report by Mahieu et al. (2013).

Crystallization by quenching the melt, followed by carefully detailed DSC studies and variable temperature XRPD revealed two additional enantiotropically related polymorphs both metastable with respect to the long-known form. Consistent with this metastable relationship the authors point out that all memory of the original form must be eliminated from the melt in order to obtain the two new forms.

The nucleation behavior of paracetamol in aqueous solutions over a range of concentrations was investigated using rapid cooling techniques (Sudha and Srinivasin 2013). The nucleation regimes for the three different polymorphs were thus defined, providing strategies for selective nucleation of the three polymorphs from the same solvent without the addition of additives. In a similar approach employing controlled kinetic conditions fast evaporation has been employed to selectively crystallize Form II of aspirin and to obtain a new polymorph of niflumic acid (Bag and Reddy, 2012).

In a review, Brandel et al. (2016) demonstrated that the relatively rare occurrence of solid solutions of two enantiomers can provide insights into chiral discrimination mechanisms at the crystal lattice scale. They attempted to define the molecular and crystallographic criteria for stereoselectivity in the formation of solid solutions to further the understanding and technique of chiral separations.

O'Mahoney et al. (2013) developed a technique for measuring the solubility of a quickly transforming metastable polymorph of carbamazepine in a solution-mediated transformation, a phenomenon similar to that observed for benzamide (Bernstein et al. 1999).

7.4.6 New techniques and conditions

In addition to the novel techniques for producing new forms previously cited, there are a number of additional impressive developments for exploring the solid form landscape. The fact that the number of experimental parameters for obtaining new solid forms is virtually unlimited has encouraged the development of new techniques for both varying those conditions and for analyzing the results of those experiments. The following survey touches on a number of those developments that deserve to be closely followed both in the near and longer term future.

A method employing solid-state solvent exchange has been developed to transform undesirable solvates to more suitable hydrates or co-solvates with water (Ramakrishnan et al. 2017). Another innovative experimental technique is the application of a Taylor–Couette system, which consists of a crystallizer containing two rotating cylinders between which a viscous fluid is confined. Park et al. (2015) employed this technique to study the polymorphic nucleation and transformation of sulfamerazine.

Interest and activity in mechanochemistry has been rapidly increasing in the past decade (e.g., Naumov et al. 2015) leading also to the application of mechanochemical techniques to the screening of solid forms of pharmaceuticals (Delori et al. 2012).

Since the exploration of the solid form landscape is an expensive and time-consuming activity, and design of experiment principles have been applied to increase the efficiency and productivity of the process (McCabe 2010), Nordquist et al. (2017) have demonstrated that a significant reduction in time (in some cases ~50%) can be achieved by inducing nucleation using surface-energy-modified glass substrates, which have been applied to the 96- and 384-well plates used in high throughput crystallizations. A method that allows screening of milligram quantities of material in which the characterization of the solid forms is based on using in situ phono-mode Raman spectroscopy has been reported (Al-Dulami et al, 2010). Raman microscopy was also used in another significant reduction in the scale of the crystallization experiment by Lee et al. (2008) which was carried out on patterned self-assembled monolayers. These studies also led to the production of metastable forms.

Since many pharmaceutical compounds are chiral molecules for which the enantiopure crystals lack an inversion center, Simon et al. (2015) suggested the wider use of second harmonic generation on powders, including temperature-resolved and microscopic techniques for the characterization of powder samples. Raman spectroscopy has also been integrated with second harmonic generation to increase the sensitivity in the detection of polymorphic transitions (Chowdhury et al. 2017). In another manifestation of the new application of sophisticated analytical techniques to the identity and characterization of polymorphic forms Schmitt et al. (2016) demonstrated that nonlinear optical Stokes ellipsometric microscopy can be applied to polymorphic systems with a very high degree of polymorphic discrimination.

A great deal of data may be generated in the course of exploring the solid form landscape. Uniquely identifying and distinguishing the various forms and the composition of mixtures represents a considerable challenge. That challenge can be met in part, at least, by the application of chemometrics, as demonstrated by Calvo et al. (2018).

A continuing series of state-of-the-art status papers on the increasingly symbiotic relationship between crystal energy calculations and experimental exploration is extremely useful in maintaining perspective on the field and recognizing both successes in this relationship as well as the considerable challenges that remain. Some examples of these are the papers by Price (2013, 2014), Braun et al. (2014a), and Iuzzolino et al. (2017).

The connection between polymorphism and structures with varying Z', particularly those with $Z' > 1$ has recently been reviewed in detail by Steed and Steed (2015).

7.5 Polymorphism in pharmaceutical co-crystals

Co-crystals (*aka* molecular complexes) have been known since Wöhler (1844) first prepared quinhydrone (quinone + hydroquinone). They were studied intensively by Pfeiffer (1927), Briegleb (1961), and Mulliken and Person (1962), and the structural chemistry was comprehensively reviewed by Herbstein (2005). Interest in the solid-state chemistry of these materials was renewed by the work of Etter

(1991), and the possible application to pharmaceuticals was outlined by Almarsson and Zaworotko (2004). In fact, two APIs had been co-crystallized and patented as early as 1937 (Hoffmann-La Roche 1937; Lemmerer et al. 2011).

Etter (1991) demonstrated that co-crystals could be obtained by simply grinding the two components together, sometimes assisted by the addition of small amounts of liquid. This approach has since been adopted by many groups, and in a notable development Hasa et al. (2016) have shown that different amounts of the same added liquid can result in different polymorphic forms of the same multicomponent material.

Subsequently there has been considerable activity in this area, and the first (modern) co-crystal pharmaceutical product, Novartis' Enresto® was recently launched on the market (Gao et al. 2017). In the present context it was earlier thought that co-crystals might exhibit a lower propensity for polymorphism than single entity solids (Vishweshwar et al. 2005) but the accumulated data to date do not warrant such a conclusion (Cruz-Cabeza et al. 2015) and by taking advantage of the possibility of altering the geometry of the fundamental interacting unit between co-crystal components polymorphic structures have been generated (Lemmerer et al. 2013). There have been a number of reviews of progress in this area (e.g., Brittain 2013; Steed 2013; Gadade and Pekamwar 2016; Ross et al. 2016; Healy et al. 2017; Malamatari et al. 2017) with two that specifically address the issue of polymorphism in pharmaceutical co-crystals (Aitipamula et al. 2014; Duggirala et al. 2016).

One of the early successes at preparing a pharmaceutical co-crystal in an attempt to overcome the low solubility of the native material involved the antifungal drug itraconazole (Remanar et al. 2003). The co-formers were 1,4-dicarboxylic acids, and in a number of instances the solubility of the co-crystal obtained was approximately that of the amorphous material.

7.6   Chemical similarity is not a predictor of similar polymorphic behavior

Chemical logic often involves analogies among functional groups, molecular structure, etc. to gain some understanding of the relationship between structure and expected or observed properties. With respect to polymorphic structures such analogies more often fail than succeed, as the statistics indicating the lack of connection between molecular properties and the propensity for polymorphic behavior demonstrate (Cruz-Cabeza et al. 2015), although as in much of chemistry, there may be some exceptions. A few examples of this lack of parallel behavior are given here.

Kitaigorodskii's (1961) early observation of the similarity of chlorine and methyl substituents (group volume 20 and 24 Å³, respectively) led to the so-called "chloro–methyl exchange rule" (Desiraju 1987), which suggested that exchanging these two substituents might be expected to lead to the same crystal structure, however, Desiraju pointed to a number of cases where similar crystal structures were not observed, even for simple molecules.

In terms of polymorphic behavior, the benzylideneanilines **7-I**, X,Y = Cl, Me (see also Section 6.3.2) are illustrative (Bar and Bernstein 1987, and references therein).

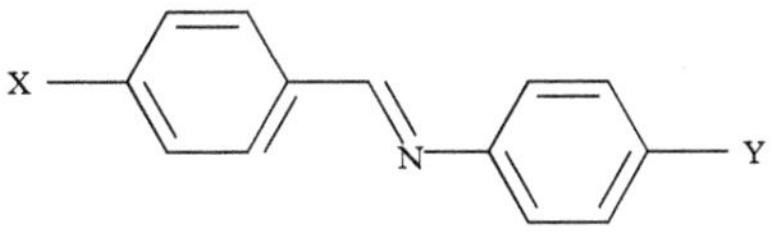

7-I

The 4,4′-dichloro derivative is dimorphic, exhibiting a metastable triclinic form and a stable orthorhombic form. The analogous dimethyl analogue is trimorphic with two structures in centrosymmetric monoclinic space group $P2_1/c$ while the third form is in the chiral monoclinic space group $P2_1$. While there is no isostructurality in these first two "hybrid" compounds, the two compounds with a methyl group in one para position and a chlorine group in the other para position are isostructural, and the near equivalency of the chloro and methyl groups is demonstrated by the statistical orientation disorder in the two structures.

The two agricultural fungicides **7-II** with identical hydrogen bonding capabilities differ in methyl (TM) or ethyl (TE) substitution at the molecular extremities (Nauha et al. 2011). The former exhibited two polymorphs and fourteen solvates, while for the latter, four polymorphs and seven solvates were obtained, with essentially no similarity in the observed structures. The identity of the hydrogen bonding functionality is not manifested in the hydrogen bonding patterns and/or the packing of the structures. However, Hirshfeld fingerprint plots (Section 5.10.3) indicated that the structure of the hydrogen bonded sheets in Form I of the ethyl derivative is identical to that of the methyl derivative in the 1,2-dichlorobenzene solvate, which again highlights the subtleties of the sum total of intermolecular interactions in the formation of a molecular crystal.

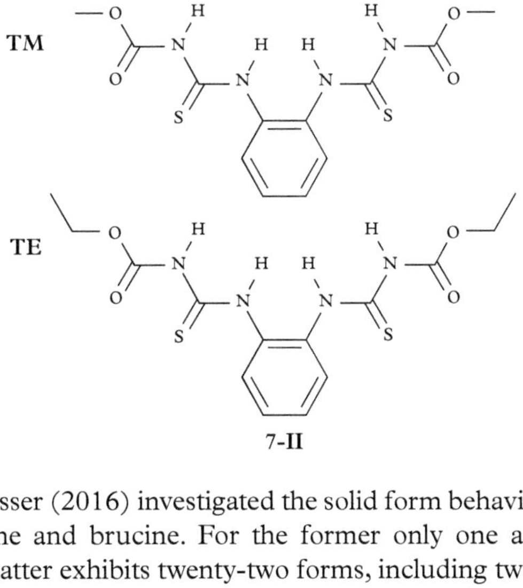

7-II

Braun and Griesser (2016) investigated the solid form behavior for two alkaloid steroids, strychnine and brucine. For the former only one anhydrous form is known, while the latter exhibits twenty-two forms, including two anhydrates, four hydrates, twelve solvates, and four solvates containing more than one type of solvent. The computational and experimental study confirmed that the differences in essentially all aspects of the solid form behavior, as the authors note, "are not intuitive from the molecular structure alone, as both molecules have hydrogen bond acceptor groups but lack hydrogen bond donor groups."

Two very similar Lilly drug candidate molecules, internally designated B5 and DB7, were screened for solid form behavior (Price and Reutzel-Edens 2016). Only one solid form was obtained for the former, while the latter exhibited a rich solid form landscape (Figure 7.6).

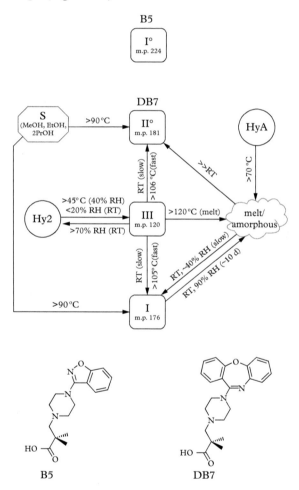

Figure 7.6 *Contrasting summaries of experimental solid-form screening and crystal structure prediction (CSP) for the similar molecular structures of Lilly compounds LY2806920 "B5" and LY2624803 "DB7." The clearly different solid form landscapes of the two molecules and their interrelations are shown. (In part from Price, S. L. and Reutzel-Edens, S. M. (2016). The potential of computed crystal energy landscapes to aid solid form development.* Drug Disc. Today, **21**, *912–23. With permission of Elsevier.)*

7.7 Excipients

Pharmaceutical formulations contain the active drug ingredient(s) as well as excipients that serve a variety of purposes: fillers, stabilizers, coatings, drying agents, etc. (Sheskey et al. 2017). As solid materials, excipients exhibit varying degrees of crystallinity, from the highly crystalline calcium hydrogen phosphate to nearly amorphous derivatives of cellulose. These materials can also exhibit polymorphism which may influence their performance in the formulation. Giron (1995, 1997) has listed many of the excipients that are known to exhibit a number of crystal forms (polymorphs, solvates, amorphous). They include many of those that are widely used: lactose, sorbitol, glucose, sucrose, magnesium stearate, various calcium phosphates, and mannitol (Burger et al. 2000).

The crystal form of the excipient may influence the final physical form of the tablet, such as a tendency to stick (Schmid et al. 2000) or may induce a polymorphic conversion of the active ingredient (Kitamura et al. 1994). Hence, there have been attempts to develop protocols for the selection of compatible active ingredient–excipient compositions (Serajuddin et al. 1999).

NMR spectroscopy has been employed to study the structural changes in epichlorohydrin crosslinked high amylose starch excipient (Shiftan et al. 2000) and has also been used to discriminate between two polymorphs of prednisolone present in tablets with excipients, even at low concentrations (5% w/w) of the active ingredient (Saindon et al. 1993). The characterization of excipients by thermal methods has also been reviewed by Giron (1997).

In spite of the fact that excipients are almost universally incorporated in solid pharmaceutical formulations, the reported incidence of polymorphism is relatively low. This is not surprising, since polymorphic behavior could compromise the end performance of the formulation no less than polymorphism of the API, so polymorphic excipients would tend to be avoided. Nevertheless, there are some examples of polymorphism among fairly commonly used excipients.

The naturally produced disaccharide lactose is employed as a diluent or filler in solid formulations, since it has a low hygroscopicity, bland taste, good physical and chemical stability, is generally compatible with APIs, and is water soluble. Considerable confusion surrounded the solid form landscape until it was described and characterized in detail by Kirk et al. (2007), resulting in the definition of four single phase polymorphs.

α,α-Trehalose is another widely used disaccharide excipient. The four known solid forms are a crystalline dihydrate, α- and β-anhydrate, and an amorphous form, and a fifth form obtained by dehydration and identified by SSNMR. The rather complex solid-state chemistry of these forms (Sussich and Cesaro 2000) has been studied and clarified in considerable detail by Pyszczynski and Munson (2013). Mannitol is a preferred excipient for the formulation of tablets due to its non-hygroscopic character, compatibility with primary amines, high sweetness, cooling mouth sensation, high solubility, and fast disintegration. Three polymorphs of mannitol have been described: α-, β-, and δ-mannitol. However, most commercially available mannitol grades consist of α or β forms or mixtures thereof; it is noteworthy that the melting points of the α and β forms differ by less than one degree. The energy–temperature diagram has been reported by Burger et al. (2000). A thorough study of the physical properties of the polymorphs that influence its use as an excipient has been given by Vanhoorne et al. (2016).

A widely used lubricant excipient is magnesium stearate, which prevents particles from sticking to each other or to equipment during processing and compression. It is also used in dry coating processes. Its polymorphic behavior and that of the analogous magnesium palmitate were described by Sharpe et al. (1997) and have recently been studied and characterized in detail (Delaney et al. 2017).

Glyceryl monostearate is another lubricant excipient that may also be used in the formulation of sustained-release solid dosage forms (Sheskey et al. 2017). It has two polymorphic forms with different properties: the α-form is dispersible and foamy, and thus is employed as a preservative and emulsifying agent, while the more dense β form is utilized in the preparation of wax matrices.

7.8   Microscopy and thermomicroscopy of pharmaceuticals

In Chapter 4, I noted the efficiency, simplicity, and utility of microscopy and thermomicroscopy in the characterization and study of polymorphs. Perhaps the most widespread and systematic application of these techniques has been in the field of pharmaceutical materials, as promoted particularly by the Institute of Pharmacognosy at the University of Innsbruck and summarized in the (now out of print) book by Kuhnert-Brandstätter (1971). In addition to introductory chapters on techniques of hot stage microscopy the book contains a compilation of the results of hot stage studies on approximately 1000 pharmaceutically important compounds. Many of these results were reported in more detail in a large series of papers beginning in the 1960s (Kuhnert-Brandstätter 1962) and continuing through the 1980s by Kuhnert-Brandstätter and her late successor, A. Burger. Later contributions often dealt with compounds actually listed in a pharmacopoeia (e.g., Burger et al. 1986).

Recent descriptions of microscopy and thermomicroscopy applied to pharmaceutically important compounds, along with quite comprehensive references to standard texts on the fundamentals and apparatus of the technique may be found in Nichols (2006), Byrn et al. (1999), and Brittain (1999b).

Prior to the development of routine and increasingly sophisticated analytical methods, hot stage microscopy competed as one of the principal tools for polymorph characterization and classification. As noted in Chapter 4, successful strategies for the investigation of polymorphs require the application of as many analytical techniques as possible, and hot stage microscopy should be considered as one of the first, if not *the* first to be employed in a comprehensive characterization of a compound (Morris et al. 1998). In spite of our current ability to generate a great deal of precise analytical data, there is often no substitute for physically observing solid materials and their behavior as a function of temperature, preferably on the polarizing microscope. Moreover, the ease of obtaining and publishing digital photos of crystal forms and videos of thermal events (even in supplementary material) makes such a practice essentially obligatory for polymorphic studies. They provide incontrovertible evidence for reported phenomena and the examples to which future workers may refer for verification or replication of earlier studies. The studies by Hean et al. (2015) and Lemmerer et al. (2011) are typical examples of the mélange of information that may be obtained from microscopic hot stage investigations.

In some cases, new phases that may not be detectable by other methods may be detected optically (Chang et al. 1995). Solid-state conversions and their monotropic (Burger et al. 1997) or enantiotropic nature (Henck et al. 2000), or the products of desolvations may be easily recognized (Schinzer et al. 1997). Intimate processes of polymorphic behavior such as nucleation, crystal growth, habit transformation, sublimation, and properties of the melt (e.g., degradation) may be readily observed and video recorded (De Wet et al. 1998).

As noted at the conclusion of Chapter 4, the amalgamation of a number of analytical techniques into a single instrument considerably expands the possibilities for detecting and characterizing polymorphs. This is a particularly powerful combination when optical/thermomicroscopy (preferably with video recording) is combined with other analytical methods. Thermomicroscopy combined with DTA led to the verification of a monotropic transformation between two forms of ibopamin (Laine et al. 1995). Combining hot stage microscopy with Raman spectroscopy led to the direct characterization and correlation of the thermal and spectroscopic information on three polymorphs of paracetamol, as well as the first report of lufenuron, a chitin synthesis inhibitor used in pest control and crop protection, and the identity of the polymorphic form in the marketed tablets (Szelagiewicz et al. 1999). FTIR spectroscopy was incorporated into a hot stage microscope to simultaneously obtain the visible and spectroscopic characterization of the three polymorphic forms of carbamazepine (Rustichelli et al. 2000). Finally, microspectroscopic FTIR and FT-Raman were combined with hot stage microscopy to study the polymorphism in (*R,S*)-proxyphylline, including the production and characterization of a new, kinetically very unstable form that most likely could not be detected or analyzed by any other technique (Griesser et al. 2000).

7.9   Thermal analysis of pharmaceuticals

The fundamentals of the application of thermal analysis in the study and characterization of polymorphs are given in Section 4.3, and many of the examples presented there are on pharmaceutically important compounds. The use of thermal analysis in the pharmaceutical area, including specific applications to polymorphic materials was reviewed in the early 1980s (Wollmann and Braun 1983; Giron-Forest

1984), the former reference containing a specific list (with citations) of pharmaceutically important compounds (both actives and excipients) that had been studied using thermal methods. The rapid technological developments in this area led to increasingly wider use in the pharmaceutical industry in general (Giron and Goldbronn 1997; Clas et al. 1999; Thompson 2000; Craig 2006; Bhattacharya et al. 2009), with a number of other reviews covering various aspects of polymorphism (Giron, 1990, 1995, 1997, 1998; Kuhnert-Brandstätter, 1996).

Many of the developments in thermal methods as applied to pharmaceuticals in general and polymorphic systems in particular were reviewed by Giron (1999). A study combining the use of thermal methods with a variety of additional techniques to characterize the polymorphic behavior of a purine derivative (designated MKS 492 by Novartis) again demonstrates the complementarity of these methods and the advantages to be obtained by applying as many of these as possible (Giron et al. 1999). The study revealed an amorphous form in addition to six crystalline modifications, of which four are pure crystalline (i.e., unsolvated) forms. Temperature resolved X-ray diffraction was used in conjunction with DSC to aid in interpretation of thermal events, leading to the identity of the most stable form at room temperature, which was chosen for further development. Using microcalorimetry the authors developed a protocol for quantitative determination of the amorphous material as a contaminant in the desired polymorphic form. The method was particularly useful at amorphous content < 10% with a limit of detection of about 1%.

The development of new thermal analytical techniques has naturally led to the reinvestigation of previously studied systems. The thermal behavior of sulfapyridine was used to demonstrate many aspects of "traditional" DSC described in Chapter 4, and a thorough study leading to the energy–temperature diagram was published by Burger et al. (1980). Nevertheless, there was still considerable confusion among various authors as to the naming and identity of the five polymorphs (Bar and Bernstein 1985). The compound was reinvestigated by Bottom (1999) combining the traditional DSC with the recently developed modulated temperature DSC (MTDSC) monitored by thermomicroscopy, resulting in considerably more understanding and insight into the nature of the solid–solid transitions, in particular the glass transition, all of which were more easily interpretable using the MTDSC technique.

Tetracaine hydrochloride was also revisited using thermal methods, revealing a complicated system containing six anhydrous crystalline forms, an amorphous form, a hemihydrate, a monohydrate and a tetrahydrate. As with the purine derivative noted above, the identification and classification of these modifications was considerably aided by the use of temperature resolved X-ray diffraction (Giron et al. 1997).

I mention briefly here some additional developments in the application of thermal methods to polymorphism in pharmaceuticals. Subambient DSC has been used to determine the melting behavior of dosage forms which may vary depending on the polymorph that might crystallize out below room temperature (Schwarz and Pfeffer 1997). Pressure differential scanning calorimetry (PDSC) has been used in conjunction with variable temperature X-ray diffraction to quantify the relative amounts of the anhydrate and trihydrate of ampicillin in mixtures of the two. PDSC could also be used to detect changes in crystallinity due to milling on a quantitative basis with a high degree of precision (Han et al. 1998).

So-called "hyphenated techniques," incorporating thermal methods as one of the combined analytical tools, are sure to play an increasing role in the identification and characterization of crystalline forms of pharmaceutical substances. The combination of TGA with FTIR allows the simultaneous quantitative analysis of weight changes during thermal processes with the IR identification of the decomposition products (e.g., solvent) resulting from those processes (Materazzi 1997). For substances with low volatility the FTIR analysis may be replaced with mass spectroscopy (Materazzi 1998).

7.10   The importance of metastable forms

As the energy–temperature phase diagram (Section 2.2.3) indicates, at any particular temperature only one polymorphic form is the thermodynamically stable one (except, of course at the temperature of a transition point). The stable form is also the least soluble form at a given temperature (Byrn et al. 1999). All other phases are higher in energy and metastable with respect to the most stable phase. A metastable phase may be fleeting in nature (e.g., benzamide, Bernstein et al. 1999) or may coexist indefinitely with its more stable counterpart in the absence of any perturbations (e.g., ranitidine hydrochloride; Cholerton et al. 1984; see also Giron 1988; Giron et al. 1990).

The lower solubility of stable forms may limit their pharmacological utility (e.g., ritonavir; Chemburkar et al. 2000), so that it may be advantageous to selectively obtain and maintain a metastable form in a formulation (e.g., Shah et al. 1999). In such cases, crystallization strategies may be designed on the basis of the principles derived from the energy–temperature or pressure–temperature diagrams (Toscani 1998), as described in Chapter 3. It will be recalled that even if qualitative in many aspects, such diagrams serve to summarize a great deal of information in a very compact manner. For instance, characterization of the two polymorphs of taltirelin, a central nervous system activating agent, indicated that they were enantiotropic, but the α form, metastable its crystallization temperature of 10 °C, was preferred for formulation. Critical evaluation of the crystallization parameters isolated the factors that led to conversion to the stable form, and these were controlled to prevent conversion (Maruyama et al. 1999). For a two-component system generation of the phase diagram can also prove very useful in developing strategies for obtaining a number of crystal modifications, including a metastable one (Henck et al. 2001).

Together with knowledge of the phase diagram an increasing variety of techniques have been designed and employed to generate metastable modifications. Seeding, of course, is one of those strategies, and Beckmann et al. (1998) developed a seeding strategy for a batch cooling crystallization to obtain quantitatively and reproducibly a metastable form of abecarnil, regardless of the purity of the material. In another approach, after thorough characterization of three polymorphic modifications by a variety of analytical methods, a desired metastable form of (R,S)-proxyphylline was crystallized in gram quantities from the supercooled melt, and proved to have considerable kinetic stability under dry atmospheric conditions (Griesser et al. 2000). A variation on that same theme was the successful high temperature crystallization from the amorphous material of the metastable α form of indomethacin, whereas the low temperature crystallization yielded the stable γ form (Andronis and Zografi 2000).

One traditional strategy for screening a compound for polymorphic behavior involves the trial of a variety of solvents and solvent mixtures. Our understanding of the role and choice of solvent has improved considerably and this information, combined with a knowledge of zones of stability can aid in determining crystallization conditions for obtaining metastable forms (Threlfall 2000). In addition, there has also been considerable progress in understanding and utilizing the interactions of solvent with the growing crystal. Combining the detailed structural information available from the single crystal structure determinations of polymorphs with crystal morphological data (i.e., crystal habit and the orientation of molecules projecting from the particular faces exposed) and with known intermolecular interactions between solute molecules and solvent functional groups allows the rational choice of solvent to select a particular polymorphic form. An analysis of this nature was carried out and experimentally confirmed by Blagden et al. (1998a) for polymorphic modifications of sulfathiazole.

Another manifestation of the strategy of using detailed structural information to steer a crystallization involves the design of tailor-made additives (Weissbuch et al. 1995) which can be used to inhibit the growth of the stable form in order to favor the growth of a metastable form. In the classic examples employing this strategy the additives are chosen to favorably interact with the molecules or parts of molecules projecting from particular crystal faces. Upon interacting they create a crystal surface which is no longer suitable for successive growth, thus leading to the inhibition. In a variation on this theme, when the crystal forms involved are conformational polymorphs, the choice of additive can be aided by considering the conformational differences between the polymorphic forms, a strategy that was successfully followed to inhibit the growth of the stable (β) form of L-glutamic acid to obtain the desired metastable α form (Davey et al. 1997). Another group, faced with the same problem at the batch level, found that seeding with the α form (seeds that do not contain any other solid phase) and increasing the cooling rate could lead to the selective crystallization of the desired α phase (Yokota et al. 1999).

This is clearly an area where the combination of thermodynamic, kinetic, and structural information potentially can lead to successful strategies for controlling the polymorphic form obtained, in specific instances even a metastable form, and developments since the first edition have in many ways realized that expectation (Anwar and Zahn 2017). Llinàs and Goodman (2007) provide a number of examples of the techniques to generate metastable modifications and to inhibit their conversion to more stable forms.

A few examples serve to demonstrate some of the additional developments since the publication of the first edition. He et al. (2001) employed an approach similar to that of Davey et al. (1997) to show that the heterogeneous nucleation mediated by isomorphous tailor-made additives could lead to new forms. A detailed analysis of the amounts of additives incorporated into the resulting structures led to an understanding of the role of the additive in determining the composition of the mixture of resulting forms.

Capillary crystallization (see also Section 7.4.6) was employed to generate a new, metastable polymorph of metformin hydrochloride (Childs et al. 2004), and subsequent thermal microscopy experiments were employed to generate single crystals suitable for structure analysis. The authors note that entire screens may be carried out with capillary crystallizations using less than 50 mg of material and concluded that metastable phases can be obtained with the very high supersaturation levels and slow evaporation rate attainable in the capillaries. The tendency for the formation of metastable forms at high levels of supersaturation has been confirmed for other systems (e.g., Lu and Rohani 2009a, 2009b; He et al. 2011) suggesting that this should be a preferred strategy for searching for those forms.

Detailed examination of the speciation diagram and application of pH modulated crystallization resulted in the preparation of anhydrous and hydrate forms of the lung-cancer drug ceritinib (Chennuru et al 2017).

I noted earlier (Section 7.3.3) the generation of metastable forms of D-mannitol at the solution–substrate contact line, where the solution–air interface meets the glass substrate surface (Poornachary et al. 2013). Some useful statistics on the appearance of stable versus metastable forms in solid forms screens have been given by Stahly (2007).

The epilepsy drug carbamazepine often serves as a subject for polymorphism studies of pharmaceutically important materials (see Sections 7.2 and 7.3), with five identified polymorphs. Metastable Form IV has proven elusive, although first prepared by Lang et al. (2002) on a polymorph template. Halliwell et al. (2017) developed a reproducible and robust method to produce this form by spray drying of a methanolic solution, which overcomes experimental limitations with solution-based crystallization methods.

Lenalidomide (LDM) is a thalidomide analogue known for its immunomodulation, antiangiogenic, and antineoplastic properties. A rather thorough solid form screen led to seven well characterized crystalline forms (Chennuru et al. 2017). Upon desolvation all solvates convert to the thermodynamically stable anhydrous Form-1, while all hydrates upon dehydration convert to a novel anhydrous and metastable Form-4; heating of the latter leads to the stable Form-1. This distinct different conversion pathway for solvates and hydrates was interpreted in terms of the isostructurality of the initial solvated/hydrated forms.

Historically, the concept of metastable forms originated with Ostwald, since he noted that later developing forms tended to be thermodynamically more stable than their predecessors. The first-formed metastable polymorphs have thus often been viewed as kinetically preferred, while the later forming ones are thermodynamically preferred. This competition has been visited repeatedly in the last 120 years, with the latest significant contribution coming from Black et al. (2018).

Finally, Price (2013) has provided a stimulating review on the computational and experimental search for polymorphs, a state of the art summary that should provide both perspective, stimulation and a benchmark for further progress (Woollam et al. 2018).

7.11   The importance of amorphous forms

Although polymorphism generally refers to different *crystalline* forms, no real crystal is perfect. The lack of perfection is manifested in disorder, and in the extreme when the entire material lacks long range order (although it may maintain some short-range order), the result is an amorphous material (Klug and Alexander 1974). Amorphous materials are generally more energetic than crystalline materials; hence they tend to have higher solubilities and rates of dissolution, properties which may make an amorphous form advantageous over a crystalline one in a pharmaceutical formulation (Hancock and Parks 2000). By the same token the presence of some amorphous material in a crystalline sample may profoundly influence the properties of the material. Because of their metastable state, amorphous materials are often difficult to prepare and handle; moreover, there is always the possibility that a stable crystalline state will crystallize from the amorphous material, leading to changes in properties and possibly rendering the resulting form unsuitable for pharmaceutical use (Craig et al. 1999). Nevertheless, quite a few pharmaceutical products are marketed with amorphous material as the active ingredient (e.g., Accolate® (zafirlukast), Ceftin® (cefuroxime axetil), Accupril® (quinapril hydrochloride), Viracept® (nelfinir mesylate)). One increasingly used method to stabilize the amorphous form is to disperse it in a polymer matrix (Harmon et al. 2009) (e.g., Cesamet® (nabilone), Gris-PEG® (griseofulvin), Kaletra® (Lopinavir/ritonavir), Isoptin® (verapamil)), although there is still the possibility that crystalline material will crystallize out (Kimura et al. 2000). Thus, increasing attention is being paid towards the preparation, detection, characterization, and stabilization of amorphous forms (Byrn et al. 1995, 1999; Giron 1997; Hancock and Zografi 1997; Guillory 1999; Yu 2001; Petit and Coquerel 2006; Taylor and Shamblin 2009; Talaczynska et al. 201).

A number of combined computational and experimental studies have attempted to provide information on the propensity for glass formation as opposed to crystallization.

In one study (Alhalaweh et al. 2014) glass forming ability versus crystallization was investigated experimentally by melt quenching 131 compounds with DSC and modeled computationally employing a number of molecular descriptors. A topological molecular descriptor based on size and shape was found to be a better descriptor for glass forming ability than molecular weight in 86% of the cases. Similar studies have aimed at developing a computational protocol for predicting glass transition temperatures for drug-like molecules (Alzghoul et al. 2014).

Amorphous pharmaceutical materials are typically obtained by employing crystallization procedures far from equilibrium such as (rapid) solidification from the melt, lyophilization (freeze drying) or spray drying, removal of solvent from a solvate, precipitation by changing pH, or by mechanical processing such as granulation, grinding, or milling (Guillory 1999), but "in-process methods" to produce amorphous API in the course of production have also been demonstrated (Jojart-Laczkovich and Szabo-Revesz 2011). Supercritical fluid technology is certain to play a role in preparation of amorphous and metastable forms (Shekunov and York 2000). The interest in understanding the processes involved in preparing amorphous materials or those that often result from the inherent tendency of amorphous materials to undergo changes to energetically lower states has led to some comprehensive studies (Surana et al. 2004; Bhugra and Pikal 2008). Many of these involve the same analytical techniques used to characterize crystal modifications. For example, the *formation* of amorphous anhydrous carbamazepine from the crystalline dihydrate was studied in situ by DSC, TGA, and variable temperature X-ray diffraction (Li et al. 2000). As demonstrated by Ceolin et al. (1995) in another study, the amorphous state may also be obtained by quenching the melt. On the other hand, the *degradation* of amorphous quinapril hydrochloride, prepared by two methods was followed by DSC, TGA, X-ray powder diffraction, polarized optical microscopy, scanning electron microscopy, and IR spectroscopy (Guo et al. 2000).

Since amorphous material may act as a contaminant, it is often necessary to quantitatively determine its concentration in a crystalline sample. A variety of techniques of analysis for the quantitative detection of an amorphous phase in a crystalline phase have been developed, and these are undergoing rapid change and improvement (Buckton and Darcy 1999). As always, the choice of method will depend on the material and the circumstances, but a few examples are noted here to demonstrate some recent developments. Taylor and Zografi (1998) employed FT-Raman spectroscopy to quantify the degree of crystallinity of indomethacin. They found a linear correlation over a 0–100% range of crystallinity with a limit of detection of 1% crystalline or amorphous content. Diffuse near-infrared reflectance spectroscopy (NIRS) was employed on the same compound together with mixtures of amorphous/crystalline sucrose (a potential excipient) (Seyer et al. 2000). The method was found to be slightly more precise than analyses carried out by the more traditional X-ray (Klug and Alexander 1974; Zevin and Kimmel 1995) and DSC techniques. The tendency of amorphous materials to absorb water has led to the application of dynamic vapor sorption to determine the amorphous content in a predominantly crystalline material (Sheokand et al. 2014).

The evolution in calorimetry technology has also led to the development of protocols for quantitative analysis (Buckton and Darcy 1999). Fiebich and Mutz

(1999) determined the amorphous content of desferal using both isothermal microcalorimetry and water vapor sorption gravimetry, with a level of detection of less than 1% amorphous material. Using the heat capacity jump associated with the glass transition of amorphous materials MTDSC was used to quantitate the amorphous content of a micronized drug substance with a limit of detection of 3% w/w of amorphous substance (Guinot and Leveiller 1999). Significant variations in the chemical stability profiles of an experimental drug could be attributed to differences in amorphous content of around 5%, which was also quantified by MTDSC (Saklatvala et al. 1999).

Finally, Zhu and Yu (2017) have recently reported and supported with considerable experimental evidence a case of *polyamorphism* for D-mannitol. This may at first appear to be a non-sequitur—different structures of a substance that lacks structure. But the general working definition of amorphous solids is that they lack *long range* structure. One could imagine, say, a pair of structures that both lack long range structure but differ in their short-range structure. A hypothetical possibility might result from the manifestation of the two hydrogen bonding motifs for crystalline tetrolic acid: dimers and chains. If two amorphous solids could be formed in which a different one of those motifs dominated the short-range order then in principle, at least, the structural condition for polyamorphism could be satisfied.

Zhu and Yu demonstrate that similar to the case of water with low and high density amorphous ices the supercooled liquid of D-mannitol spontaneously transforms to another amorphous lower energy phase with larger volume and stronger hydrogen bonding. The thermal evidence is also present, since that transition exhibits an anomalous heat loss concomitant with the volume change. The authors correlate those physical phenomena with a structural change that involves competition between the loose packing resulting from hydrogen bonds and the closer packing dominated more by van der Waals interactions. There are certainly many candidate systems in which this competition is possible and it remains to be seen if more examples of the phenomenon will be documented.

7.12   Obtaining a difficult stable or unstable new form

Many of the polymorphic systems presented in this chapter contain examples of the preparation or recovery of difficult-to-obtain forms. In the interest of conserving space the description details for preparing those forms are not detailed here. However, there are additional individual instances that warrant recognition of the efforts and experimental creativity and sophistication in obtaining crystal forms and to provide the impetus to future workers to continue to seek new ways to influence crystallizations. The list is not intended to be comprehensive but rather to provide a glimpse into the potential for the synergistic development of new physical and analytical techniques on the one hand and polymorphism research on pharmaceutically important compounds on the other hand. In addition, it will be noted that a number of these examples deal with substances that may be considered pharmacological icons.

As noted earlier, the preparation of unstable forms may be important as part of understanding the solid form landscape, and even for potential use, if more stable forms prove unsuitable. Although the solid form landscape of phenobarbital has been quite thoroughly investigated (Zencirci et al. 2009), Roy et al. (2016) prepared a new form and the elusive Form V employing polymer-induced heteronucleation and determined its crystal structure, in addition to elucidating additional aspects of the phase behavior.

Bobrovs et al. (2015) investigated the enigmatic crystallization behavior of the stable form IV of theophylline, which had been obtained only by the slurrying in specific solvent. NMR and FTIR methods were employed to monitor prenucleation aggregates, indicating that the molecule self-associates in solvents providing good hydrogen-bond donors, which hinders nucleation and the phase transformation. The slurrying from Form II to Form IV was determined to be a transformation involving nucleation and growth, with the nucleation being homogeneous.

In spite of considerable research on the solid form behavior of griseofulvin, as noted earlier, only one form had been identified and characterized (e.g., Willart et al. 2012). A combination of a detailed DSC study combined with variable temperature XRPD revealed two new enantiotropically related polymorphs, both not surprisingly metastable with respect to the known form. The authors pointed out that neither of these two new forms could be obtained if any traces of the original, stable form were present in the melt, which explains why they were not discovered for many years.

The question of the propensity for the formation of polymorphic hydrate forms and how to attempt to obtain them was addressed by Tian et al. (2010a, 2010b) for nitrofurantoin monohydrate and niclosamide monohydrate.

Molecule VI in the second CCDC blind test, an aminosulfonamide, defied structure prediction by the participants. It was known to be dimorphic when Roy and Matzger (2009) carried out polymer-induced heteronucleation crystallizations in ethanol and obtained the two known forms as well as a third, new form, which similarly had not been predicted.

A metastable, and therefore more highly soluble form of an HIV/AIDS drug was obtained using supercritical CO_2 as an antisolvent (Bettini et al. 2010). The new metastable form also exhibited lower stability to mechanical stress.

The special requirements for the physical properties of solid form inhalants, including melting point $\geq$ 150 °C, achieving a specific particle size distribution with a small range, chemical and thermal stability, lack of hygroscopicity, superior flow properties, etc. make this search of appropriate crystal forms particularly important, and absolutely required in the early development stages of the drug. The questions, challenges, and potential solutions have been reviewed and Weers and Miller (2015) and by Selby et al. (2011).

Of course, in any search, even if successful in preparing and characterizing many solid forms, there is always the question of when to cease looking. This question was addressed by Braun et al. (2011a) in the study of β-resorcylic acid. A considerable amount of time and effort exploring the solid form landscape led to nine crystalline forms: two concomitantly crystallizing polymorphs, five novel solvates, in addition to the previously reported hemihydrate and a new monohydrate. In addition, a new enantiotropically related polymorph was prepared by desolvation of some of the solvates, sublimation, and a thermally induced solid transformation. Following this extensive investigation, the authors concluded that "the most practically important features of β-resorcylic acid crystallization under ambient conditions have been established; however, it appears impractical to guarantee that no additional metastable solid-state form could be found."

7.13 Concluding remarks

The development of a pharmaceutical product is a long, arduous, and expensive process. For those that are marketed in solid dosage forms that process deals with many matters of solid-state chemistry, among them the detection, characterization, and preparation of various solid modifications. I have attempted here to touch on many of those issues, and as noted at the beginning of the chapter, examples of these various aspects of polymorphism and their manifestations in the pharmaceutical industry may be found through this book.

For scientific, regulatory, and intellectual property reasons recent years have witnessed increasing efforts to discover and characterize as many different crystal modifications as possible of substances associated with a pharmaceutical product. This effort has been aided and enhanced by the rapid developments in instrumentation and analytical techniques, so that often old compounds have been revisited for study to the same extent that new ones have been subject to investigation. In light of the new and often more sensitive techniques available many questions and problems noted twenty or more years ago warrant reinvestigation. One of these involves the large numbers of crystal modifications reported for some substances, for instance, in addition to those noted previously in this chapter, phenobarbitone (thirteen modifications) (Cleverly and Williams 1959; Mesley et al. 1968; Stanley-Wood and Riley 1972) or cimetidine (seven modifications) (Hegedus and Görög 1985; Sudo et al. 1991). The competition for discovery and characterization of the system with the most known polymorphs continues unabated (e.g., Braun et al. 2009a; Yu 2010; Lopez-Mejias et al. 2012; Zeidan et al. 2016). In a re-examination of a number of the sulfa drugs Threlfall reported discovering a new polymorph of sulfathiazole, two new polymorphs of sulfapyridine, 120 solvates of sulfathiazole, and thirty solvates of sulfapyridine (Threlfall 1999, 2000). As the techniques for exploring the solid form landscape and characterizing the newly found forms become increasingly sophisticated it is clear that many more of these multipolymorphic systems will be described. These developments will lead to a better understanding of polymorphism and better control over the production and utilization of polymorphic systems for improved pharmaceutical products and the intellectual property that protects them.

Another issue that I have not touched on is that of the difficulties that are often encountered in the process of scaling up from laboratory quantities and procedures, through the pilot plant and into full production (Reutzel-Edens 2006; Variankaval et al. 2008). Equipment changes, differences between the quality of laboratory grade and bulk chemicals, variations in heating/cooling rates, stirring procedures (Genck 2000), seeding (Brittain and Fiese 1999; Beckmann 2000), etc. can all influence the result of a crystallization procedure and the polymorph obtained. Little of this is documented in the literature (although for an exception see Wirth and Stephenson 1997), for it is often a matter of empirical testing and development maintained as trade secrets or incorporated into the collective memory of an industrial concern, although passing mention may appear in the literature (Yazawa and Momonaga 1994; Giron 1995; Giron et al. 1999; Rodriguez-Hornedo and Murphy 1999). Scale up and subsequent formulation also involve more complex transfer and processing procedures, which are affected by the physical and mechanical properties of the solids, likewise a function of crystal modification. Some of these have been described by Hulliger (1994), including tensile strength (Summers et al. 1977), compression (e.g., Di Martino et al. 1996, 2001; Suihko et al. 2000), flow properties (Beach et al. 1999), and filtration and drying characteristics (Crookes 1985; Crosby et al. 1999).

I have mentioned in passing the regulatory aspects of polymorphism in the pharmaceutical industry. As noted in one of the quotations at the start of this chapter, increasing attention is being paid by regulatory agencies to the preparation, identity, characterization, purity, and properties of the crystal form used in pharmaceutical products (Byrn et al. 1999) such information is now required by the U.S. Food and Drug Administration in a New Drug Application and Byrn et al. (1995) have presented a set of decision trees (see, for example, Section 7.4) to aid in presenting the data on different crystal forms (polymorphs, solvates, desolvated solvates, and amorphous forms) to regulatory agencies. The International Committee on Harmonization has set up guidelines (Q6A (2)) for addressing the issue of polymorphism in pharmaceuticals (Federal Register 2018).

8

Polymorphism of pigments and dyes

We never really perceive what color is physically.

Josef Albers

Color which, like music is a matter of vibrations, reaches what is most general and therefore most indefinable in nature; its inner power...

Paul Gaugin

The representatives of organic chemistry are coming more and more to the conclusion that the formulae of limiting states which we have been using so far fail to reflect the real conditions prevailing in nature, for there exist such subtle differences in the state of matter compared to which our methods of description that still are very simple sometimes appear wholly inefficient.

Ismailsky (1913)

8.1 Introduction

Humans have always been fascinated and charmed with color. The dye indigo has been in use for over four millennia, and frequently is the source of the color of the ubiquitous blue jeans around the world. It is probably not an exaggeration to state that dyes and pigments formed the basis of modern industrial organic chemistry. The synthesis of "aniline purple" (mauvein) by the 18-year-old Perkin in 1856 catalyzed a revolution in the chemical industry, with widespread commercial and even cultural ramifications. The dye and pigment industry quickly became the cornerstone of the chemical industry in both England and Germany and generated many additional new technologies and new industries. As of 2014 the worldwide consumption of organic pigments was valued at approximately 22.9 billion dollars (IHS Markit 2017), although reliable data are notoriously difficult to obtain (Herbst and Hunger 1997). Globally, printing inks account for 55–60% of total demand, coatings about 20%, plastics about 15%, and other industries such as textiles for the remainder. The next-largest market is coatings for automotive production. By pigment class (*vide infra*) 50–55% of the world value share is azo pigments and phthalocyanines have a 20% share.

Polymorphism in Molecular Crystals. Second Edition. Joel Bernstein. © Joel Bernstein 2020. Published in 2020 by Oxford University Press. DOI: 10.1093/oso/ 9780199655441.001.0001

The basis of the theory of all aspects of the sources and perception of color, including an excellent discussion on the physics and chemistry of dyes and pigments has been given by Nassau (1983) (see also Tilley 1999). Actually, the precise distinction between dyes and pigments is still a matter of discussion, even controversy, much the same as the definition of polymorphism (see Section 1.2.1), but as in the latter case the working definition is generally accepted by most workers in the field. Most dyes are generally soluble, while pigments are regarded as insoluble in the medium being considered. Hence, the physics of the color discerned by the viewer is different for the two classes (Evans 1974). As Nassau (1983) has noted, one feature common to dyes and pigments, indeed to all the substances discussed in this chapter, is that they involve electrons delocalized onto more than one atom in organic molecules, as opposed, say, to systems in which electrons are transferred from one molecular moiety to another (see also Dähne and Kulpe 1977). These are then termed "colorants," as opposed to charge transfer systems. Because they are used as solids, the solid-state chemistry and properties of pigments and the dependence of those properties on polymorphic structure are of particular importance, an observation made by Susich (1950) over seventy years ago in citing a number of early-known polymorphic colorant systems. However, in addition to their use as colorants, where the solubility and dissolution properties are important aspects of ultimate use, dyes once played a crucial role in the chemistry of photographic processes (Nassau 1983), and an increasingly important role in the development of digital printing and other new technologies (Section 6.2).

Regarding the polymorphism of colorants, another feature common to them and high energy materials (Chapter 9) is that much of the information may be found in sources of limited accessibility, in the case of colorants in company records, or in old or not widely circulated literature.

There are a number of properties that determine the ultimate use of a compound as a pigment (Erk 1999; Hao and Iqbal 1997), and these properties can vary among polymorphic forms. I review the properties briefly here; further details are given by Hunger and Schmidt (2018), Erk et al. (2004), Zollinger (2004), McKay et al. (1994), and Jaffe (1996). The *tinctorial strength* of a pigment is a measure of its ability to absorb and to scatter light, a property associated with the maximum molar extinction coefficient, ε_{max}. Other properties that influence the tinctorial strength are the particle size distribution, the morphology, and the structure of the pigment. The molar extinction coefficient of a solid is essentially derived from molecular properties which have undergone perturbations due to the incorporation of the molecules in the solid; hence different polymorphic solids can lead to different perturbations.

Solubility (or insolubility, which is also a property of the solid) is one of the most important properties of a pigment and plays a role in many of the other factors that follow. *Lightfastness* and *weatherfastness* are measures of the pigment's ability to resist degradation upon exposure to light and to weather. Fastness is also measured as a function of laundering, bleaching, ironing, exposure to water or perspiration, and abrasion. The *specific surface area*, measured as the surface area per unit weight, although difficult to measure precisely, depends in part on the habit of the individual pigment particles. The tendency to *agglomerate* (stick together) or aggregate (form bridges, edge-to-edge or edge-to-face, due to intermolecular forces) determines how well the pigment can be dispersed in the medium in which it is used. Finally, the *crystallinity* of a pigment (as opposed to amorphous characteristics of a solid) is important in determining the use of a pigment. All of these factors can and do vary with polymorphic form. Other, often directly related physical properties can directly influence the performance of a colorant (Erk 1999), for example, agglomeration, flocculation, wetting capability, dispersibility, dissolution rate, filtration properties, tendency to cake, dispersion properties, and powder flow properties. A number of reviews deal with the structural chemistry of pigments and the relationship between crystal structures and properties of pigments (Fryer 1997; Hunger 1999; Erk 2001). In spite of the variety of properties that must be considered in the choice of a pigment for use, in the end the overriding factor is the color (Erk 2000).

8.2   Occurrence of polymorphism among pigments

No specific compilation of polymorphic pigments is available. Hunger and Schmidt (2018) cite many instances of polymorphs throughout their text and show almost all published crystal structures of industrial organic pigments. Lomax (2010) has provided a collection of X-ray powder diffraction (XRPD) data for over 200 pigments, including some polymorphic ones. Whitaker's (1995) rather comprehensive compilation of the XRPD studies of organic colorants contains many references to polymorphic systems.[1] These citations have been compiled in Table 8.1 along with some additional references. As expected, consistent with McCrone's (1965) observation, those pigments that have been widely used and studied also tend to be those with the more extensive documented polymorphic behavior. Table 8.1 is not meant to be a complete compilation of polymorphic behavior of pigments, but rather to provide ready access to some of the literature on that behavior among recognized organic colorants. Amongst the high performance pigments the most widely sold polymorphic ones are three forms of copper phthalocyanine, two forms of quinacridone, and two forms of indanthrone. A number of these polymorphic classes of pigments will now be discussed in more detail.

---

[1] In spite of the wealth of information contained therein, as of this writing the only citations to this 1995 compilation of Whitaker is Lomax (2010), and combined both are useful resources for reference powder diffraction patterns of many organic colorants, polymorphic and non-polymorphic. The fact that Whitaker (1995) was published in a specialist monograph has perhaps rendered it less accessible. Whatever the reason, it appeared to be of some benefit and convenience to the reader to compile here those primary references to polymorphic materials that are designated as colorants by the Colour Index (C.I.), with the primary sources given by Whitaker. Whitaker and Hunger and Schmidt (2018) also present references for many other related dye materials that have not received such a designation.

Table 8.1   *Collection of references to polymorphic behavior of pigments[a]*

Pigment	Number of forms	CSD Refcodes	Reference or page number[b]
Quinacridone pigments	At least 3		41
Acid Blue 324	2		Sandefur and Thomas 1984
Acid Orange 156	2		Höhener and Smith 1987
Dioxazine violet	2		Hayashi and Sakaguchi 1981, 1982
Disperse Brown 1	α, β	VEJKAX	Kruse and Sommer 1976; Seo et al. 2007
Disperse Orange 5	2		Ghinescu et al. 1984
Disperse Orange 29	3		Hähnle and Optiz 1976
Disperse Red 65	α, β		Sommer and Kruse 1979; Lee et al. 2009
Disperse Red 73	3		Von Rambach et al. 1974; Wolf et al. 1986a; Lee et al., 2009
Disperse Yellow 23	2		Koch et al. 1987a, 1987b
Disperse Yellow 42	2	ZAMROU	Burkhard et al. 1968; Flores and Jones 1972; Koch et al. 1987a, 1987b; Glowka et al. 1995
Disperse Yellow 68	3		Sommer et al. 1974; Wolf et al. 1986b
Pigment Red 9	At least 2		285
Pigment Red 12	At least 2		285
Pigment Red 49	Several modifications		317
Benzimidazolone pigments	"polymorphism common"		351
"several azome-thine pigments"	?		411
"Polyester red A" $C_{25}H_{23}N_5O_4S$	3		Kuhnert-Brandstätter and Riedmann 1989
"Polyester red B" $C_{27}H_{25}N_5O_5S$	3		Kuhnert-Brandstätter and Riedmann 1989

continued

Table 8.1 Continued

Pigment	Number of forms	CSD Refcodes	Reference or page number[b]
Linear trans acridones	Multiple (α, β, γ)		461
"	β, two more γ modifications		462
Pigment Black 31	I and II	DICNIM DICNIM01	Hädicke and Graser (1986a); Mizuguchi (1998b)
Pigment Violet 19	5	QNACRD QNACRD01 QNACRD02 QNACRD03 γ QNACRD04 β QNACRD05 α' QNACRD06 β QNACRD07 γ QNACRD08 β QNACRD09 γ QNACRD10	465–6 Buchsbaum and Schmidt 2007; Chung and Scott 1971; Koyama et al. 1966; Mizuguchi et al. 2002a; Nishimura et al. 2006; Paulus et al. 2007; Potts et al. 1994; Senju et al. 2007; Panina et al. 2007
Pigment Violet 23	2		533; Sakaguchi and Hayashi 1981, 1982; Curry et al. 1982
Pigment Violet 122	4		465, 469
Pigment Orange 5		CICCUN	Schmidt et al. 2007a
Pigment Orange 36	2	β HOYVOJ	Dainippon 1982; van de Streek et al. 2009
Pigment Red 1	4?	NBZANO NBZANO01 NBZANO02 NBZANO11 NBZANO12	Grainger and McConnell 1969; Whitaker 1979, 1980a, 1980b, 1981, 1982; Bushuyev et al. 2013
Pigment Red 31	3		Griffiths and Monahan 1976
Pigment Red 53:1 (Ba^{2+} salt)	α and β		Hoechst 1982; Duebel et al. 1984; Schui et al. 1988;
Pigment Red 53:2 (Ca^{2+} salt)	15 polymorphic and solvated forms		Farbwerke vorm Meister Lucius & Bruening 1902; Schmidt 1999, 2000; Schmidt and Metz 1998a, 1998b, 1999a, 1999b

Pigment	Number of forms	CSD Refcodes	Reference or page number[b]
Pigment Red 53:3 (Sr^{2+} salt)	4		Dainippon 1996a, 1996b, 1996c, 1996d
Pigment Red 57 (Ba^{2+} salt)	α and β		Dainippon 1988
Pigment Red 57 (Sr^{2+} salt)	α and β		Dainippon 1986
Pigment Red 57:1 (Ca^{2+} salt)	α and β	FAWQIF	Chen 1990; Bekö et al. 2012
Pigment Red 57:2	4		Kobayashi and Ando 1988a, 1988b, 1988c
Pigment Red 122	4 (?)	tcACD02 MHQACD03	Eshkova et al. 1976; Kelly and Giambalvo 1966; Mizuguchi et al. 2002a; Senju et al. 2007
Pigment Red 149	α, β, and γ	AFOMOX	Bäbler 1983; Imahori and Hirako 1976; Spietschta and Tröster 1988a, 1988b; Mizuguchi and Tojo 2001
Pigment Red 168	α	KAHTEU	Schmidt et al. 2010
Pigment Red 170	3 α form of Me, F, Cl, Br, and NO_2 derivatives	α ECITIU γ ECITIU01	305, Ribka 1969, 1970; Schmidt et al. 2006
Pigment Red 177	3	DANTHQ	Koshelev et al. 1987; Ogawa and Scheel 1969
Pigment Red 187	At least 2		258, 308, Ribka 1961
Pigment Red 188		α KAHITEU	Schmidt et al. 2010
Pigment Red 194	2	ENOFAQ01 ENOFA102 ENOFA103	Shtanov et al. 1980, 1981; Pushkina and Shelyapin 1988; Shelyapin et al. 1987; Teteruk et al. 2016
Pigment Red 202	3	MAMGUD MAMGUD01	465, 470 Senju et al. 2005b
Pigment Red 207	4 (?)		Wagener and Meisters 1970a, 1970b

continued

Table 8.1 Continued

Pigment	Number of forms	CSD Refcodes	Reference or page number[b]
Pigment Red 209	3 or 4	SATTIR	Curry et al. 1982; Deuschel et al. 1964; Senju et al. 2006
Pigment Red 224	α and β	β SUWMIG α SUWMIG01 α SUWMIG02 β SUWMIG03 α SUWMIG04	Ogawa et al. 1999; Zykova-Timan et al. 2008; Forrest and Zhang 1994; Tojo and Mizuguchi 2002a
Pigment Red 247 (Ca²⁺ salt)	α and β		341, Hoechst 1988; Froelich 1989
Pigment Red 254	α and β	WEBKET	Hao and Iqbal 1997; Mizuguchi et al. 1993
Pigment Red 255	2	SAPDES SAPDES01	Ruch and Wallquist 1997; Iqbal et al. 1988; Mizuguchi et al. 1992
Pigment Yellow 5	α and β		Whitaker 1985a, 1985b
Pigment Yellow 10	4 or 5	VUYHEC	Momoi et al. 1976; Bäbler 1978; Fuji et al. 1980; Whitaker 1988
Pigment Yellow 12	3	MEWGAW MEWGAW01	Kawamura 1987; Tuck et al. 1997; Curry et al. 1982; Barrow et al. 2000; Schmidt et al. 2007b
Pigment Yellow 13		BARNUE BARNUE01	Barrow et al. 2002; Schmidt et al. 2007b
Pigment Yellow 14		BARPAM BARPAM01	Barrow et al. 2002; Schmidt et al. 2007b
Pigment Yellow 16	5		Rieper and Baier 1992
Pigment Yellow 17	α and β		Shenmin et al. 1992; van de Streek et al. 2009
Pigment Yellow 83		BARSOD BARSOD01	Barrow et al. 2002; Schmidt et al. 2007b
Pigment Yellow 154	α	HOYWEY	van de Streek et al. 2009
Pigment Yellow 181	β	GITWUC	Pidcock et al. 2009
Pigment Yellow 183	α, β	HOMMEC	Schmidt et al. 2009

Pigment	Number of forms	CSD Refcodes	Reference or page number[b]
Pigment Yellow 191	α, β	HOMMMIG	Schmidt et al. 2009; Ivashevskaya et al. 2009
Pigment Yellow 213	α	HOYVEX	Schmidt et al. 2009
Zapon Fast Yellow C2G	3		Susich 1950
Solvent Orange 63	2		Shimura et al. 1988
Solvent Yellow 18	3		Whitaker 1990, 1992
Sudan Orange R	4		Susich 1950
Vat Green 1	2		Popov et al. 1981
Pigment Blue 60 (Indanthrone)	4		516; FIAT 1948; Susich 1950; Bailey 1955; Dainippon 1994
Dioxazine	More than one		533
Indigo	A and B	INDIGO INDIGO01 INDIGO02 INDIGO03 INDIGO04	Von Eller 1955; Süsse and Wolf 1980; Süsse et al. 1988; Kettner et al. 2011
Thiazine indigo	α, β	SAJMOH SAJMOH1	Buchsbaum et al. 2011;
Cu phthalocyanine	α, β, γ, δ, ε As many as 9	CUPOCY10 CUPOCY11 CUPOCY12 CUPOCY17	41, 425, 434–443; Erk 1998; Xia et al. 2008
Metal phthalocyanines	2–4		Law 1993
Metal-free phthalocyanine	7	PMTHCY PMTHCY01 PMTHCY02 PMTHCY05 PMTHCY06 PMTHCY08 PMTHCY09 PMTHCY10 PMTHCY11	439 Whitaker 1995; Hammond et al. 1996; Zugenmaier et al. 1997; Janczak 2000

continued

Table 8.1 Continued

Pigment	Number of forms	CSD Refcodes	Reference or page number[b]
4-Formyl-3-hydroxy-2-naphthoic acid-4′-methoxyanilide ($C_{38}H_{30}N_4O_6$)	α, β	WAKCIW WAKCIW01	Brüning et al. 2009
Latent Pigment DPP-Boc	α, β, γ	EBIGUR03 EBIGUR04	MacLean et al. 2009; Mizuguchi 2003
Maize 1	I, II, III	UNEWUF UNEWUF01 UNEWUF02	Price et al. 2005
Dithiophene-DPP(Boc)-dithiophene	α, β	EPANEQ EPANEQ01 EPANEQ02	Salammal et al. 2014

[a] This table does not purport to be inclusive. It contains entries encountered in researching this book, with an attempt to update for the second edition. It is meant to serve as an entry into the literature on polymorphic colorants designated by the Colour Index. There are no doubt many additional instances that have not been noted (e.g., Whitaker 1995) for reasons described in earlier chapters. However, it is hoped that it might serve as a first reference for those seeking information on polymorphism of colorants, particularly pigments.
[b] Refers to location in Herbst and Hunger (1997).

8.3 Polymorphism in some specific groups of pigments

8.3.1 Quinacridones

The synthesis of quinacridone **8-I** was first claimed in a series of papers beginning in 1896 (Niementowski 1896, 1906; Ullman and Maag 1906; Baczynski and von Niementowski 1919), which all later proved to be an angular isomer of the linear compound first actually prepared by Liebermann (1935). Little was done with the compound until 1955 when Du Pont chemists recognized its polymorphic behavior and associated favorable photochemical stability

8-I

(Jaffe 1992). Jaffe also noted that many acridone derivatives are polymorphic. Many other relevant references may be found in the book by Hunger and Schmidt (2018).

Quinacridone itself has been believed to be tetramorphic for a long time (Manger and Struve 1958; Struve 1959; Mizuguchi et al. 2002a),[2] the four forms being designated as α, β, and γ, distinguished by their X-ray powder patterns. However, there is no unique "α phase," but two different phases that differ in their colors, powder patterns, and crystal structures, and were called α^I and α^{II} by Schmidt in 2007 (Paulus et al. 2007). The phases α^{II} and β are reddish violet, while α^I and γ are red. On the basis of some subtle variations in the powder pattern of the γ form there was also an earlier claim of the existence of an additional, γ' phase (Whitaker 1977a). The data from the full crystal structure analysis of the γ form (see also Mizuguchi et al. 2002a) were used to calculate the powder pattern, and it was demonstrated that those subtle variations could be simulated by varying the average crystallite size included in the calculation (Potts et al. 1994), thus quite conclusively proving the identity of the γ and γ' forms. As noted, the properties of pigments are functions of a variety of factors, one of them of course being the polymorphic form, but as discussed in Chapter 4 the existence and characterization of a new polymorphic form should be determined by the use of a number of analytical techniques, including preferably X-ray diffraction (XRD) as applied in this case.

The structures of the β and γ forms were determined by single crystal XRD (Paulus et al. 1989; Nishimura et al. 2006; Mizuguchi et al. 2002a). For the α^I and α^{II} phases, no single crystals can be grown. In 1996, Leusen predicted the possible crystal structures by global lattice-energy minimizations and reproduced the β and γ phases in the correct stability order. For the energetically less favorable predicted structures, powder diagrams were simulated and compared with the experimental diagrams, revealing the structure of the α^I phase (Leusen 1996; Paulus et al. 2007). The structure of the α^{II} was only recently determined by electron diffraction (Gorelik et al. 2016); it is strongly affected by stacking disorder. In the α^I, α^{II}, and β phases, the molecules form chains, with different arrangement of the chains. In the γ phase the molecules arrange in a hunter's fence packing.

A detailed account of the polymorphic structures of quinacridone has been given by Panina et al. (2007), combining experimental and computational techniques. The laboratory XRPD data were measured for the α^I form obtained by slow dilution of concentrated sulfuric acid solution of the compound. Possible structures were modeled by computational methods (see Chapter 5); the fifteen lowest

---

[2] At least fourteen polymorphic forms of quinacridone have been claimed, mostly in the patent literature. All these phases are claimed to differ in their powder patterns. A critical review revealed that actually only the four phases α^I, α^{II}, β, and γ exist; the differences in the other powder patterns are caused by impurities, particle size effects, measurement artefacts such as preferred orientation, etc. (Paulus et al. 2007).

Polymorphic behavior has also been described for the 2,9-dichloro- (Bohler and Kehrer, 1963; Deuschel et al. 1964; Nagai et al. 1965; Senju et al. 2005a, 2005b) and 2,9-dimethoxy- derivatives (Ciba 1965).

energy solutions were considered. The γ' form of Deuschel et al. (1963) was found to be identical to the γ form. The report contains a detailed and informative discussion of the methodology of using these computational techniques, including which solutions are more likely correct based on the behavior of the refinement, the force fields employed, the quality of the XRPD data, and the order of the stabilities. Those considerations combined with earlier sublimation studies indicated that the β form is less stable than γ over the entire temperature range. The structure of the β form was confirmed by a nearly contemporaneous report by Nishimura et al. (2006).

8.3.2   Perylenes

Perylene pigments, with the basic structure **8-II**, were reviewed by Mizuguchi and Tojo (2002). The anhydride (X = O) (Pigment Red 224) is an important intermediate in the preparation of other derivatives but has also been used as a pigment in a number of applications (Hunger and Schmidt 2018). The crystal structures of two polymorphs reported earlier (Lovinger et al. 1984; Möbus et al. 1992) were later carried out using electron crystallography on microcrystalline thin films (Ogawa et al. 1999) and by single crystal XRD (Tojo and Mizuguchi 2002a, 2002b, 2002c). The two forms have similar packing modes, with experimentally significant differences in cell constants and mutual orientation of neighboring molecules.

8-II

Perhaps the most widely used of the perylene pigments is Pigment Red 179 (**8-II**, X = N-CH$_3$), also known as perylene red. In spite of the intensive investigation and widespread use, this compound was long in the ranks of sucrose and naphthalene as very commonly crystallized compounds that as yet have shown no evidence of polymorphism. As noted in Section 3.2, a high pressure form of sucrose has been prepared and structurally characterized and the structures of the α and β forms of Pigment Red 179 have been reported, first by electron crystallography (Ogawa et al. 1999) and then by single crystal X-ray methods (Tojo and Mizuguchi, (2002a, 2002b, 2002c).

Pigment Red 149 (X = (3′,5′-dimethyl)phenylimido) is also widely used, especially in textile applications. There are three polymorphs (Mitsubishi Chemical Industries 1976; Bäbler, 1983; Mizuguchi and Tojo 2002), exhibiting different shades of red. The β form has a more yellow tint than the thermodynamically stable α form that is the polymorph used commercially. The structure determination of one form described simply as red has been reported by Mizuguchi and Tojo (2001).

Following the initial synthetic (Graser and Hädicke 1980, 1984) and structural (Hädicke and Graser 1986a, 1986b) investigations Klebe et al. (1989) reviewed the *crystallochromy*—effect of solid state on the color properties—in a series of perylene **8-II** (X = N–R) derivatives for a variety of R-substituents. The latter group found twenty-four different modes of overlap between neighboring perylene moieties, and correlated the packing with the absorption properties of the solid pigments. Kazmaier and Hoffmann (1994) also investigated this problem theoretically and proposed the concept of a quantum interference effect to account for spectral shifts, a model which contrasts with the exciton explanation given by Mizuguchi (1997) and Mizuguchi and Tojo (2002). Klebe et al. (1989) found that packing patterns could be correlated with molecular conformation and steric requirements, but this information could not be used for reliable predictions of crystal structure from the structure of an isolated molecule. In the past, attempts to computationally model and predict some polymorphic structures of perylene pigments were only partially successful (McKerrow et al. 1993) at that time (Section 5.10; Lommerse et al. 2000; Erk 2001; Motherwell 2001; Mizuguchi and Tojo 2002). With today's density functional theory methods, a crystal structure prediction of a perylene derivative may be feasible.

Mizuguchi (1997, 1998a) carried out combined structural and spectroscopic studies on evaporated films of two perylene pigments **8-III** and **8-II** (X = $NCH_2CH_2Phenyl$) [PDC] that indicate that the films can undergo polymorphic changes upon exposure to solvent vapors. In the case of **8-III** the original evaporated film is violet, while exposure to acetone leads to a color change to reddish-purple. The XRD patterns of the two films indicate that a structural change has indeed taken place.

8-III

The single crystal structure determination of PDC (Pigment Black 31) was reported by Hädicke and Graser (1986a), but Mizuguchi found that vapor phase growth initially led to a brilliant amorphous red film, rather than the black color usually assigned to this pigment. Exposure to acetone vapor or to temperatures above 100 °C led to a new crystalline phase that differs from that of Hädicke and Graser (Figure 8.1). The first structure is characterized by a parallel arrangement

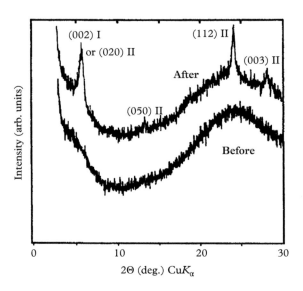

Figure 8.1 *X-ray diffraction pattern of films of PDC (Pigment Black 31) 8-II (X = NCH₂CH₂Phenyl) before and after exposure to acetone vapors. The material is clearly amorphous prior to exposure, and crystallinity is readily apparent following treatment. The diffraction peaks have been assigned Miller indices by the author based on correspondence with one of the two known crystal structures. Modification II dominates the transition to the crystalline phase (from Mizuguchi (1998a), with permission).*

of neighboring molecules in *C*2/*c*, while that determined by Mizuguchi from vapor-phase grown crystals (Mizuguchi 1998b) exhibits a herring-bond arrangement in *P*2₁/*c*. Mizuguchi also presented solid-state polarized reflection spectra (Section 6.3.2) and discussed how the amorphous (red) to crystalline Form II (black) transition (and vice versa), involving an optical absorption at about 635 nm, can be utilized in optical disk technology applications. Mizuguchi's interpretation of these effects on the basis of exciton theory differs from that given by Kazmaier and Hoffmann (1994).

8.3.3   Phthalocyanines

Phthalocyanines **8-IV** (Pcs) comprise one of the most important and widely used single class of pigments, used in printing inks, paints, plastics, and automotive finishes. Annual production of all Pcs currently amounts to about 80 000 tonnes/year with copper phthalocyanine (M = Cu) CuPc accounting for a little over 60% of the total. Activity in the field has spawned a number of books (Moser and Thomas 1963, 1983; Woehrle 1989; Leznoff and Lever 1996), reviews (Booth 1971; Fryer et al. 1981; Löbbert 2000), and a specialist journal entitled *Journal of Porphyrins and Phthalocyanines*.

8-IV

Historically, the first mention of Pcs appears to have been by Braun and Tcherniac (1907) who described a greenish residue on their filter, but provided no further characterization of the material. Twenty years later de Diesbach and von der Weid noted that the same preparation led to a blue which was recognized and characterized at Scottish Dyes (1929a, 1929b). In one of the historically significant single crystal structure determinations Robertson reported the first structure determination of the metal-free (Robertson, 1936) and Ni derivatives (Robertson and Woodward 1937), at that time the largest molecules to be determined by single crystal methods. Robertson (1935) also reported the cell constants of a number of other Pc derivatives. Other early reports on the polymorphic behavior include the XRPD patterns (FIAT 1948; Susich 1950) and IR spectra (Ebert and Gottlieb 1952).

8.3.3.1 *Copper phthalocyanine (CuPc)*

For commercial reasons, CuPc has received the most attention. In most cases new polymorphic forms resulted from changes in crystallization process conditions or in changes in synthetic procedures. At least five modifications—designated α, β, γ, δ, and ϵ—are known (Horn and Honigman 1974; Löbbert 2000, and references therein), and at least four more have been claimed in the literature (Gieren and Hoppe 1971): "R" (Pfeiffer 1962); "π" (Brach and Six 1975); "X" (Moser and Thomas 1983); and "ρ" (Komai et al. 1977). Additional patents (Byrne and Kurz 1967; Sharp et al. 1972; Brach and Lardon 1973; Brach and Six 1973) have led to some confusion, indeed controversy (Assour 1965), as to the existence of various forms. As in other cases in which there are clearly multiple polymorphic modifications, part of the problem lies in nomenclature (see Section 1.2.3), as well as in the determination of clear distinguishing characteristics for each of the claimed forms. For instance, in a patent Eastes (1956) claimed the β form, which all subsequent workers have referred to as the γ form (e.g., Brand 1964).

The diffraction patterns for the first five CuPc modifications are shown in Figure 8.2. In addition, Whitaker (1995) carried out a survey and critical evaluation of the XRPD patterns of the various forms of CuPc. He notes that in an earlier review (Whitaker 1977b) he had compiled claims for nine different crystalline modifications of CuPc. On the basis of the later review, in which he cited the sources of many of the powder patterns, he suggested the following sources as the preferred references for the XRPD patterns of the various forms:

α: preferred listing is given in the Powder Diffraction File (PDF, see Section 1.3.4), (PDF 36-1883).

β: preferred listing (PDF 37-1846) has been indexed from the single crystal structure determination (Brown 1968a).

γ: Whitaker (1977b); see also Komai et al. (1977) and Wheeler (1979).

δ: the powder patterns published subsequent to the Whitaker (1977b) review (Komai et al. 1977; Enokida and Hirohashi 1991) differ significantly from that in the review.

Whitaker's comparison of the patterns indicates that one of the Komai patterns claimed for the so-called δ_K form matches that for the ϵ form, while a second, purported new pattern for this δ_K form actually corresponds to the X form. The pattern given by Enokida and Hirohashi matches that in the Whitaker review except for an apparent typographical error for one d-spacing which should be 8.75 Å in the review rather than 7.85 Å.

ρ: this form has been reported by Reznichenko et al. (1984) and in a series of patents by Ninomiya et al. (1979). As Whitaker points out, most of these are from the same source, raising the question of reproducibility, but the pattern does seem to be unique.

σ: according to Whitaker, the pattern claimed by Enokida and Hirohashi (1991) appears to be unique.

In addition, Whitaker noted that the pattern claimed by Wheeler (1979) is identical to that of the α form, the one claimed by Suzuki et al. (1989a) is an improved crystalline specimen of the π form, and that claimed by Suzuki et al. (1989b) can be interpreted as a poorly crystalline sample of the α form.

Erk and Hengelsberg (2003) compiled a review on crystal structures and polymorphism of CuPc, describing ten forms. Furthermore, apparently an eleventh form exists (Erk 2012; Hunger and Schmidt 2018).

On the basis of solubilities in benzene Horn and Honigman (1974) estimated the relative thermodynamic stability of the five most commonly recognized polymorphs as $\alpha = \gamma < \delta < \epsilon < \beta$. Beynon and Humphreys (1955) determined the enthalpy difference between the two commercially most significant α and β forms as 10.75 kJ mol^{-1}, which is consistent with the results of the solvent-mediated phase transformation from α to β reported by Cardew and Davey (1985). Other modifications also transform to the β form upon heating in inert, high-boiling solvents (Löbbert 2000). Synthesis of CuPc generally leads to the β form, although

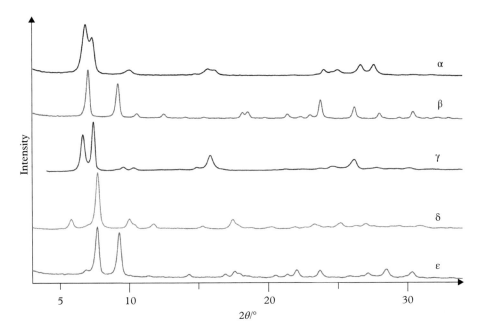

Figure 8.2 *Copper phthalocyanine blue. X-ray powder diffraction diagrams of different crystal modifications, measured in transmission mode. (Courtesy of Professor Martin Schmidt, University of Frankfurt.)*

Moynihan and Claudon (2010) have demonstrated a method for preparing the α form that they claim does not readily convert to β. Dissolution of this modification in concentrated H_2SO_4 or treatment with 55–90% H_2SO_4 leads to the greenish yellow sulfate (Honigmann 1964), which upon hydrolysis gives α-CuPc. Grinding in the presence or the absence of additives induces the transformation from β to α, which has also been studied by a variety of methods (Suito and Uyeda 1963, 1974). Investigations of crystal growth have been carried out using both electron microscopy (Ashida et al. 1971) and XRD (Haman and Wagner 1971), and many other aspects of the transitions between the α and β phases have been reviewed (Fryer et al. 1981).

The crystal structure of the α and β forms (Figure 8.3) have been determined from single crystal data (Brown 1968a; Erk 2002; Hoshino et al. 2003);[3] the crystal structures of the γ and ε phases were solved from powder diffraction data, and refined by combining lattice energy minimizations and Rietveld refinements (Erk and Hengelsberg 2003; Erk et al. 2004).

---

[3] The reference to the determination of the cell constants of α-CuPc is given as "C.J. Brown, to be published." The present author could find no evidence that those cell constants have been published.

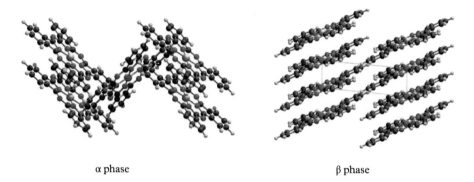

α phase   β phase

Figure 8.3 *Crystal structures of copper phthalocyanine blue polymorphs. The molecular columns run in the vertical direction. (Courtesy of Professor Martin Schmidt, University of Frankfurt.)*

The IR, UV-Vis, and XPS dark current and photocurrent have been compared for a number of the polymorphs of CuPc (Knudsen 1966; Enokida and Hirohashi 1991). The most important characteristic of a pigment is its color; the spectral response of five CuPc polymorphs, presented in Figure 6.9, shows considerable variability, which is manifested in different shades of blue. Hunger and Schmidt (2018) have listed six commercially available CuPc pigments, based on the α, β, and ϵ crystal modifications (including two with 0.5–1 Cl atom per molecule, but retaining the α structure) that vary from greenish to reddish blue.

Vergnat et al. (2011) investigated the growth mechanism of zinc phthalocyanine on oriented polymer substrates of bisphenol A, in which the polymorphic result depended on the size regime. For $33\,°C \le T(s) = 115\,°C$, ZnPc nanocrystals grew exclusively in a structure isomorphous with the α form of CuPc. At $115\,°C \le T(s) \le 200\,°C$, ZnPc films consist of both α and β (isomorphous with the β form of CuPc) nanocrystals with a gradual increase of the proportion of β-form with increasing $T(s)$. The onset of the β polymorph growth coincides with a marked change in the nanocrystal size. The β phase appeared when the ZnPc nanocrystals reach some critical dimensions which were estimated from the $T(s)$ dependence of the nanocrystal size. The relative stability of the α and β polymorphs was explained by the balance of the bulk and the surface energy of the nanocrystals (see also Sections 2.5 and 3.5 for additional discussion of the effect of the size regime on polymorphic crystallization).

8.3.3.2   *Metal-free phthalocyanine (MfPc)*

Metal-free phthalocyanine (MfPc) is another important class of polymorphic phthalocyanines.[4] Many of the aspects of the polymorphic behavior were covered

---

[4] The second most important Pcs are $Cl_{16}CuPc$ (Pigment Green 7) and $Cl_{12}Br_4CuPc$ (Pigment Green 36), which have not yet been reported to be polymorphic; there are suggestions in two archived XRPD patterns (PDF 36-1870 and PDF 36-1871) that these may be two similar but different structures.

in the previous section, but because of its historic importance (Robertson 1936, 1953), and current potential use as a photoconductor for laser printers (Loutfy et al. 1988), additional information is provided here.

The number of known and characterized polymorphs of MfPc appears to be still a matter of some question. Herbst and Hunger (1997) indicate that there are five different crystal modifications, designated α, β, γ, κ, and τ. Whitaker (1995) notes that in his earlier (Whitaker, 1977b) review there were X-ray powder patterns for three (α, β, and X) forms, but that in the intervening years the situation had become more complicated, with the claimed discovery of several more forms, named ϵ, τ, modified, and γ. He also notes the difficulties that may arise in obtaining polymorph-defining powder diffraction patterns due to uncertainty about polymorphic purity in the samples.[5] The original crystal structure of the β form, carried out in two-dimensional projection (Robertson 1936), was redone by Matsumoto et al. (1999). On the basis of the cell constants and (equivalent) space group this structure determination is identical with that reported by Kubiak and Janczak (1992).[6] They have also determined the structure of α-MfPc (Janczak 2000). The optical properties of the α, β, and X forms have been studied by Loutfy (1981). Only the α form is marketed, as Pigment Blue 16.

8.3.4 Some other pigments—old and new

Indanthrone **8-V** (Pigment Blue 60) is one of the oldest vat dyes known, having been originally synthesized by René Bohn at BASF in 1901. The polymorphic behavior was well characterized by Susich well before World War II (FIAT 1948; Susich 1950). The most stable of the four known polymorphs is α, appearing as greenish needles, and for which the crystal structure is known (Bailey 1955), while the δ form appears as plates. The β, γ, and δ forms convert to α upon heating to 250 °C. It is also possible to interconvert the various forms by dissolving them in sulfuric acid or by transforming them to the leuco compound (FIAT 1948). The β form has a reddish-blue shade, the γ form has a red shade (Hunger and

---

[5] From his own studies and a careful evaluation of the available literature, Erk (2009) concluded that there are seven modifications of MfPc, for which α, β, and γ correspond to the respective congeners for CuPc, and four additional forms, which he designated τ, τ', and τ''.

[6] However, although Kubiak and Janczak located and reported the hydrogen atoms on the periphery of the molecule in both of these structures, they did not report any information regarding the two hydrogens required in the core for the molecule to be aromatic. Indeed, they indicate that the two hydrogens are absent in the β-MfPc. Matsumoto et al. (1999) did crystallographically locate the hydrogens in the core, thereby confirming their presence. In their determination of the crystal structure of α-MfPc (Janczak and Kubiak 1992), which does appear to be isostructural with α-PtPc (itself used as the model for α-CuPc—see text), the hydrogen atoms were treated in a similar way as in their β-MfPc, although some residual electron density in the core was viewed as partial occupancy (0.0096 per site) of bismuth, rather than interpreted as disordered hydrogen. The structure of the X-form was determined by Hammond et al. (1996) using a combination of XRPD data and computational modeling. While in the earlier structures there was some question about the chemistry of these reports regarding the treatment of the core hydrogens, it appears that the molecular frameworks and crystal structures are essentially correct.

Schmidt 2018), and the δ form has a greenish-grey color (FIAT 1948). In the same publication Susich noted the difficulty of distinguishing among the forms, although Jelinek et al. (1964) later showed that the δ form could be distinguished from the α using the electron microscope.

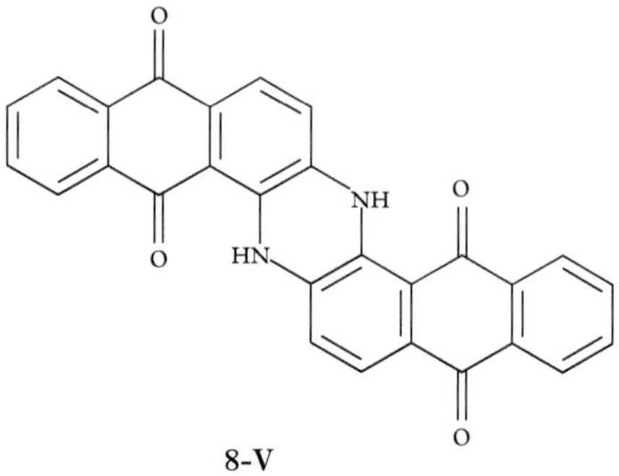

8-V

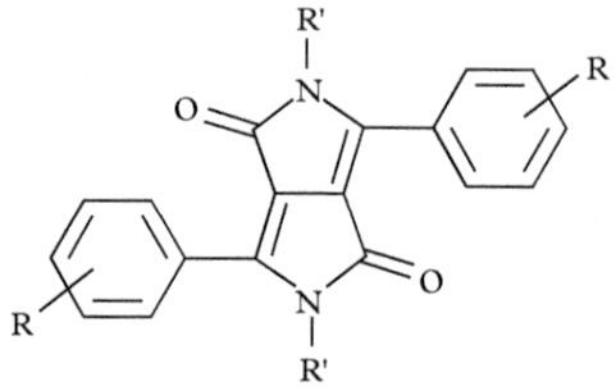

8-VI

On the other hand diketopyrrolopyrrole, prototypically **8-VI** (R, R′ = H, DPP), is a relatively new family of red pigments (Hunger and Schmidt 2018), having been discovered serendipitously by Farnum et al. (1974). The most widely used is Pigment Red 254 (R = 4-Cl, R′ = H), which is dimorphic (Mizuguchi et al. 1993; Hao et al. 1999c). The *m*-chloro (R = 3-Cl, R′ = H) and *m*-methyl (R = 3-CH₃, R′ = H) derivatives have also recently been shown to be polymorphic (Mizuguchi et al. 1993; Hao et al. 1999a, 1999b). The parent compound (R, R′ = H) has been shown to be trimorphic.

MacLean et al. (2009) studied the dimorphic behavior of the pigment precursor ("latent" pigment) derivative of **8-VI** (R = COO-*t*-butyl, R′ = H), abbreviated DPP-Boc, subsequently reporting the crystal structures of three forms. The "latency" is due to the thermal decomposition reaction of the compound resulting in the commercially important pigment DPP. The α form of DPP-Boc contains three half molecules in the asymmetric unit (see also Ellern et al. 1994) while the β form contains one half molecule per asymmetric unit. Hence they are easily distinguishable by solid-state NMR as well as by XRPD. The crystal structure

VIsolution from powder data and Rietveld refinement of both polymorphs is an exemplary study demonstrating the potential of these methods in determining the detailed crystal structure of these compounds which are not often difficult to crystallize. Similarly combining these crystallographic techniques Salammal et al. (2014) found that the two polymorphs of the dithiophene derivatives of DPP-Boc crystallize differently in bulk or in thin films.

8.4   Isomorphism of pigments

Little reference has been made in this volume to the subject of *isomorphism*, the appearance of essentially identical crystal structures of different compounds, a phenomenon also recognized by Mitscherlich (1820). The two structures might be more correctly described as isostructural, but the former term has gained wide acceptance and use in the literature (e.g., Kálmán et al. 1993a, 1993b). The phenomenon is not uncommon among pigments, by virtue of the nature of the way compounds for potential use as pigments have been developed. As implicitly noted previously, pigments are usually classified by families, and synthetic efforts at preparing new pigments and modifying the properties of known compounds have been directed at changing substituents in a rather systematic manner, for example, H, $-CH_3$, $-OCH_3$, Cl, Br, etc. The "core" of many of the prototypical dye families (or chromophore class) are quite large and often approximately planar molecules which essentially determine the limited number of possibilities for the molecules to pack in a crystal structure, even if polymorphism is possible. That, combined with the similarity of size (i.e., van der Waals radius) and polarity of some of these substituents can lead to very similar structures for different compounds, which is manifested in XRPD patterns that are also very similar.

   For example the monobromo and dibromo analogues of the monoazo C.I. Pigment Yellow 3 are isomorphous (Chapman and Whitaker 1971). In the isomorphous C.I. Pigment Yellow 14 **8-VII**, R = CH_3, and C.I. Pigment Yellow 63 **8-VII**, R = Cl, two methyl groups in the former are replaced by Cl in the latter (Shenmin et al. 1992). The powder patterns (Figure 8.4) clearly indicate the isostructurality. Substitution at the same position by the larger and more polar methoxy group **8-VII**, R = OCH_3, is sufficient to overcome the tendency for isostructurality, and in fact apparently leads to a dimorphic system, for which both powder patterns are shown in Figure 8.5. Similarly, the 2,9-dimethyl- and 2,9-dichloroacridones are isomorphous (Paulus et al. 1989). In another variation on this theme, a 50:50 mixture of **8-VIII** (R = n-C_3H_7 and R = n-C_4H_9) has been reported to crystallize in four polymorphic forms (Brandt et al. 1982). In the copper phthalocyanines, Pigment Blue 15, containing no Cl, is isomorphous with Pigment Blue 15:1, which on average contains 0.5–1.0 atoms of chlorine per molecule (Hao and Iqbal 1997). Recognition of this phenomenon can be useful, for instance, in attempting to work out the crystal structures of unknown polymorphic forms.

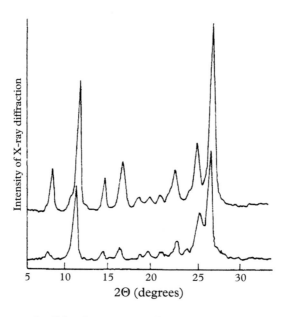

Figure 8.4 *X-ray powder diffraction patterns of Pigment Yellow 14 8-VII, R = CH₃ (upper) and Pigment Yellow 63 8-VII, R = Cl (lower) (from Shenmin et al. 1992, with permission).*

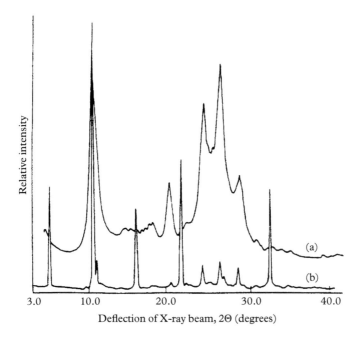

Relative intensity

Deflection of X-ray beam, 2Θ (degrees)

Figure 8.5 *X-ray powder diffraction patterns for the two polymorphs of Pigment Yellow 17 8-VII, R = OCH₃. (a) Modification obtained by heating sample at 90 °C for one hour after synthesis; (b) modification obtained by crystallization from nitrobenzene. The lower pattern shows evidence of virtually every peak that appears in the upper pattern, indicating that it may be a pure modification. Nevertheless, there are diffraction maxima that do not appear to be unique to a second form. (From Shenmin et al. 1992, with permission.)*

9

Polymorphism of high energy materials

Alfred Nobel's discoveries are characteristic; powerful explosives can help men perform admirable tasks. They are also a means to terrible destruction in the hands of the great criminals who lead peoples to war.

Pierre Curie

Although personally I am quite content with existing explosives, I feel we must not stand in the path of improvement…

Winston Churchill

9.1 Introduction

High energy materials include a variety of substances that react rapidly to produce light, heat, and gas. Some of these materials are used for nonexplosive purposes, while those that are explosive (i.e., cause a noisy, large-scale rapid expansion of matter into a volume much greater than its original volume) can be classified into three general categories: pyrotechnics, propellants, and high explosives. The last, in turn, are subdivided into primary, or initiating, explosives and secondary explosives (Cooper and Kurowski 1996). Some of the factors that influence the performance of explosives include sensitivity to detonation (by impact, friction, shock, or electrostatic charge), the rate of deflagration to detonation transition (i.e., the rate at which a burning reaction transforms to a much faster reaction), the detonation velocity, the detonation pressure, the crystal density, thermal and shock stability, and crystal morphology. Many of these factors ultimately depend on the solid-state structure of the solid high energy material (e.g., Rudel et al. 1990; Dick 1995; Klapötke 2017), and hence on the polymorphic modification, which historically was often ignored (e.g., Zeman 1980).

A number of comparative studies of different high energy materials have been carried out. Since it is crucial to assess the effectiveness and safety of these materials in their ultimate end form, a great deal of work is normally carried out to fully characterize them (e.g., Miller and Garroway 2001). Sorescu and colleagues have published a series of papers on the application of density functional theory (DFT)

Polymorphism in Molecular Crystals. Second Edition. Joel Bernstein. © Joel Bernstein 2020. Published in 2020 by Oxford University Press. DOI: 10.1093/oso/ 9780199655441.001.0001

methods to energetic molecular crystals, including a number of polymorphic systems (Sorescu and Rice 2016, and references therein). The role of metastable forms and controlling their presence in the production of cyclic nitramine energetic materials (e.g., RDX, HMX, and CL-20) has been studied by both DFT and experimental methods (Ghosh et al. 2016). Ma et al. (2014) have noted that for a group of polymorphic explosives with a same chemical component (e.g., HMX or CL-20), the most compacted form (in these cases β-HMX or ϵ-CL-20) is usually desired in practical applications, since the highest density form is generally consistent with the highest energy polymorph (see Sections 2.2.2 and 2.2.3). Zeman and Jungova (2016) have reviewed the sensitivity and performance of energetic materials, also noting the importance of the polymorphic form in determining these characteristics. A study of the kinetics of polymorphic transitions in a number of materials has been reported (Chukanov et al. 2016*)* showing that the general kinetic regularities of these processes (stepwise and continuous regimes) depend on their topotactic mode (frontal or quasi-homogeneous, respectively). A nucleation stage is not observed in the reverse phase transitions, which was explained by the persistence of nuclei of the low temperature (initial) polymorph in the preheated sample. The influence of mechanical effects on the kinetics of polymorphic transitions in these materials is also discussed. Moreover, a fundamental requirement of such a material is the reproducibility of its performance, and that reproducibility can be compromised by polymorphic variations and transformations. Hence, it is not surprising that polymorphism has been discovered among many compounds used as high energy materials.

For obvious reasons many of these compounds have been developed by government laboratories and military research establishments or programs funded externally but with broad restrictions on publication. Some of the reports emanating from such research are eventually declassified, but even then often they are not widely abstracted, referenced, or readily accessible. There are excellent sources of general information on high energy materials (e.g., Meyer 1987; Cooper and Kurowski 1996; Klapötke 2017, and general references compiled in these monographs) and the study of their properties (e.g., Brill 1992; Brill and James 1993, and references therein, a continuing series of papers by Brill and coworkers, e.g., Beal and Brill 2000, and reviews by Agrawal 1998, 2005). I have attempted here to compile some of the information specifically relating to the polymorphism of highly energetic molecular materials in the hope that some structural, physical, and chemical data will now be more readily accessible in one source. Where space considerations have limited the amount of information to be included I have attempted to provide details on the primary sources of that information. There are of course many high energy compounds for which polymorphism has not yet been reported; these are not covered here. Moreover, in keeping with the general theme of this monograph, I have not included here discussions of some well-known polymorphic inorganic high energy materials, such as ammonium nitrate and lead (and other) azides.

9.2 The "alphabet" of high energy molecular materials

Just as pigments are often specified by their Colour Index designation (Chapter 8), so high energy materials are usually designated by acronyms. These are defined here for reference throughout the chapter:

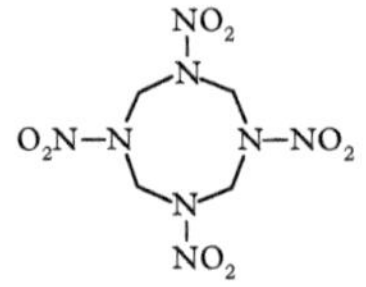

HMX

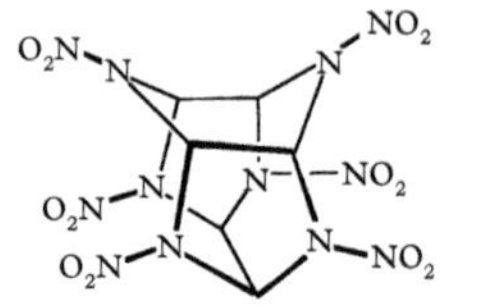

HNIW

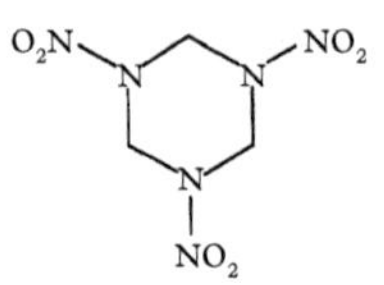

RDX

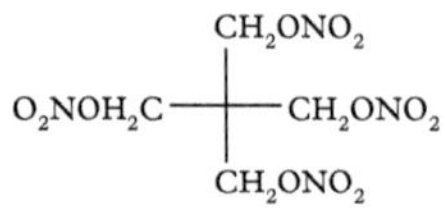

PETN

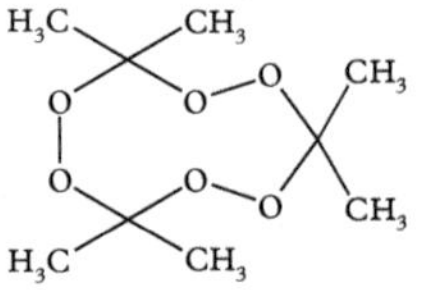

TATP

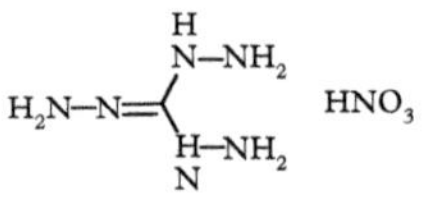

TAGN

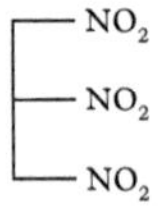

PTTN

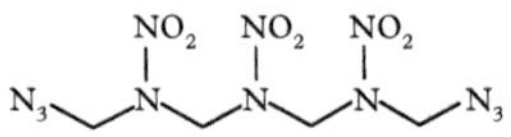

DATH

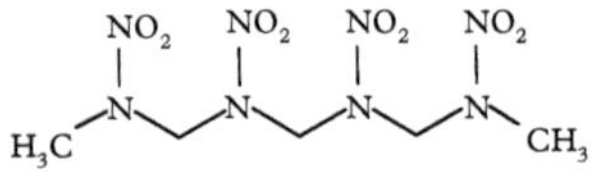

OHMX

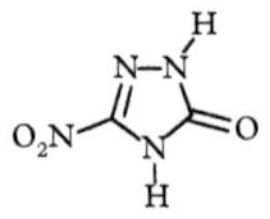

NTO

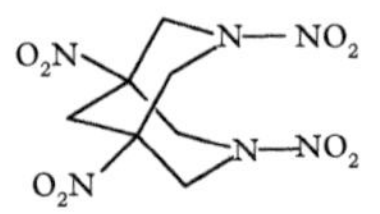

TNDBN

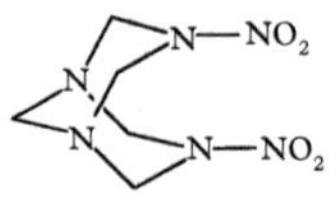

DPT

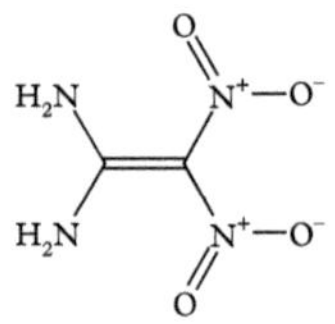

FOX-7/DADNE

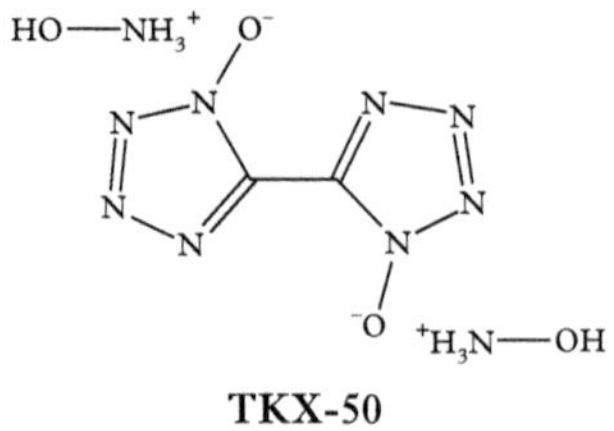

TKX-50

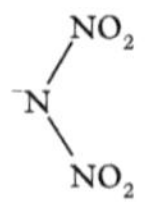

dinitramide anion

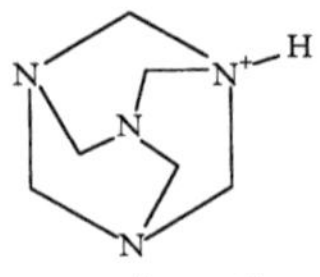

hexammonium cation

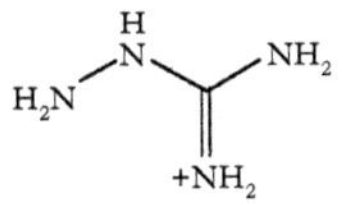

aminoguanidinium cation

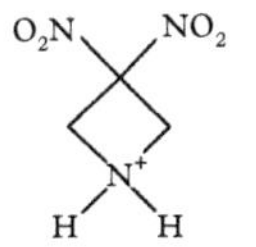

azetidinium cation

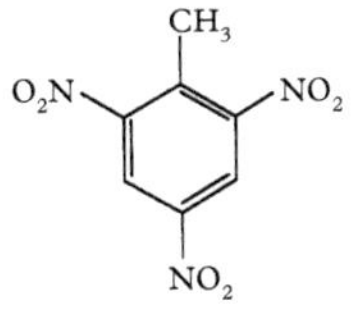

TNT

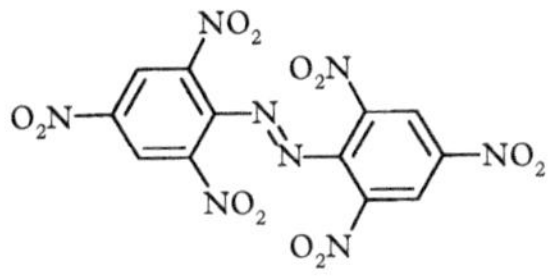

HNAB

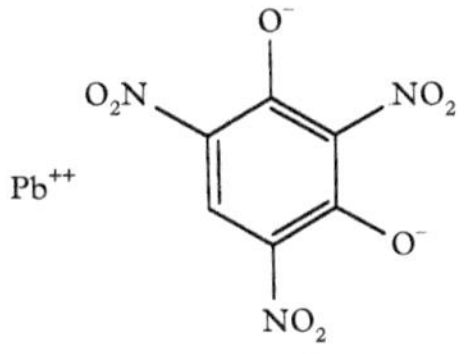

lead styphnate

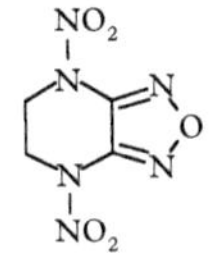

DNFP

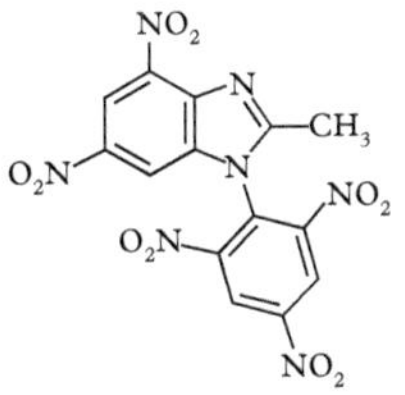

2-methyl-1-(2',4',6'-trinitrophenyl)-4,6-dinitrobenzimidazole

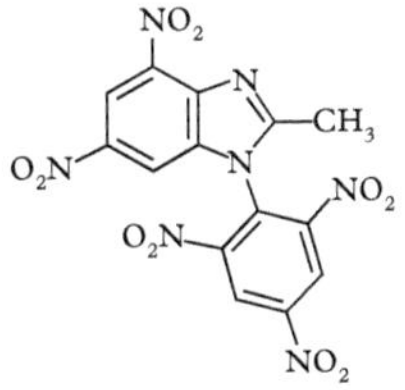

1-(2',4',6'-trinitrophenyl)-6-nitrobenzotriazole

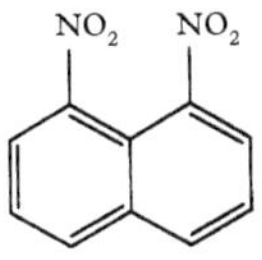

1,8-dinitronaphthalene

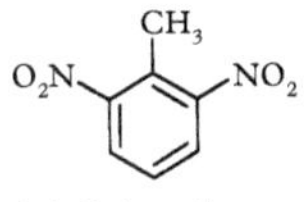

2,6-dinitrotoluene

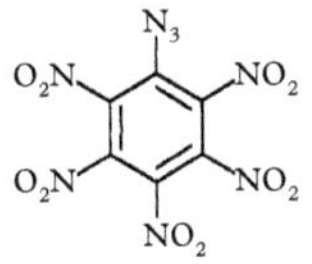

pentanitrophenylazide

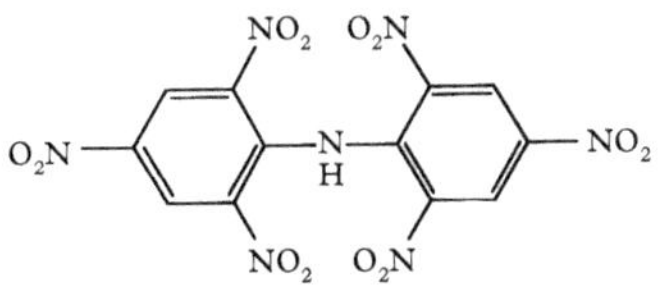

HND

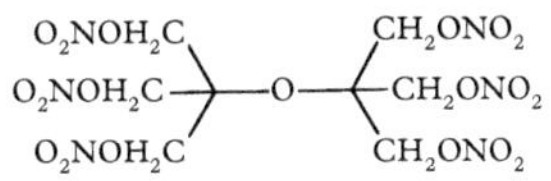

DiPEHN

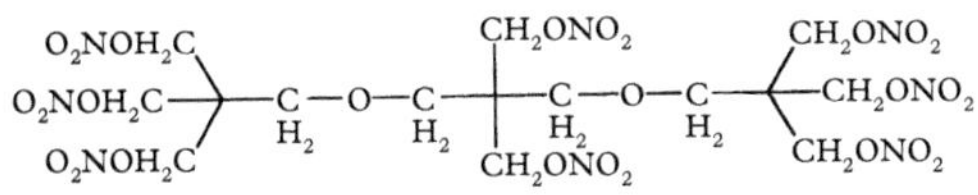

triPEON

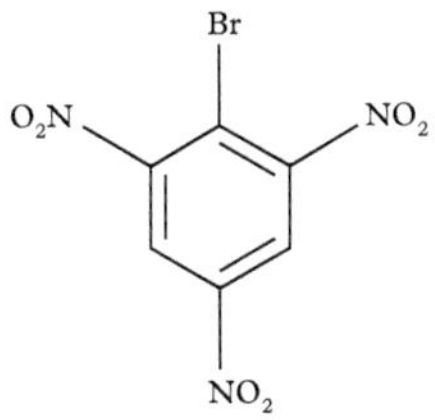

Picryl bromide

AND	ammonium dinitramide
CL-20	(see **HNIW)**
DADNE	(see **FOX-7**)
DATB	1,3-diamino-2,4,5-trinitrobenzene
DATH	1,7-diazido-2,4,6-trinitro-2,4,6-triazaheptane
DiPEHN	dipentaerythritol hexanitrate
DNAZ (DZ)	dinitroazetidinium
DPT	3,7-dinitro-1,3,5,7-tetrazabicyclo[3.3.1]nonane
FOX-7	1,1-diamino-2,2-dinitroethylene, DADNE
HMX	cyclotetramethylene tetranitramine
HNAB	hexanitroazobenzene
HND	hexanitrodiphenylamine
HNIW	hexanitrohexa-azaisowurtzitane, CL-20
NTO	5-nitro-2,4-dihydro-3H-1,2,4-triazol-3-one
OHMX	1,7-dimethyl-1,3,5,7-tetranitrotrimethylenetetramine
PETN	pentaerythritol tetranitrate
PTTN	1,2,3-propanetriol trinitrate
RDX	cyclo-1,3,5-trimethylene-2,4,6-trinitramine
TAGN	triaminoguanidinium nitrate
TATP	triacetone-triperoxide
TKX-50	dihydroxylammonium 5,5′-bistetrazole-1,1′-diolate
TNDBN	1,3,5,7-tetranitro-3,7-diazabicyclo[3.3.1]nonane
TNT	2,4,6-trinitrotoluene
triPEON	tripentaerythritol octanitrate

2-methyl-1-(2′,4′,6′-trinitrophenyl)-4,6-dinitrobenzimidazole
1,1′-dinitro-3,3′-azo-1,2,4-triazole
1,8-dinitronaphthalene
hexaammonium dinitramide
lead styphnate (2,4,6-trinitroresorcinol)
pentanitrophenylazide
2,6-dinitrotoluene
picryl bromide

9.3 Individual systems

In the following descriptive sections, I have attempted to collect the structural data, or when readily accessible, references to structural data on the polymorphic system. If a crystal structure has been reported and is entered in the Cambridge Structural Database (see Section 1.3.3), then the appropriate six-letter Refcode(s) are given (without the appended numerical designations for multiple entries) for direct access to that information. Less accessible information is quoted here, together with the appropriate references. Other descriptions of studies of polymorphic systems by a variety of increasingly sophisticated techniques are meant to be representative, rather than comprehensive. Two of the most widely studied systems are HMX and HNIW, the former having been developed in the USA during the Second World War (Blomquist and Ryan 1944), and the latter more recently developed as "the densest and most energetic explosive known" (Miller 1995), as quoted by Sorescu et al. (1998a). A 1:2 co-crystal of these widely used materials with an excellent spectrum of properties has recently been reported (Bolton et al. 2012). The reports on these systems provide excellent examples of the nature of investigations on polymorphic systems and hence will be covered in more detail than the others. Following a common practice, high energy materials are broadly classified as aliphatic or aromatic.

9.3.1   Aliphatic materials

9.3.1.1   HMX

Four polymorphs of **HMX** (OCHTET), currently designated α, β, γ, and δ, have been identified and quite well characterized in terms of many physical properties (Cady and Smith 1962; Holston Defense Corp. 1962). McCrone (1950a) also published a thorough optical and thermal study of the four polymorphs (including diagrams indicating morphology, form and habit, and interfacial angles) albeit with a different labelling system for the four modifications than that given here. A summary of some of the comparative data on HMX polymorphs is given in Table 9.1. Cyclotetramethylene tetranitramine has also been prepared as stoichiometric solids with over 100 organic molecules (George et al. 1965; Selig 1982) and among them the solvate with dimethylformamide has been claimed to be polymorphic (Cobbledick and Small 1975; Haller et al. 1983), although Marsh (1984) has shown that the latter structure (reported in space group $C2/c$) is most likely identical to the earlier one (reported in space group $R\bar{3}c$).

Some of the crystallization conditions reported by McCrone (1950a), reflect the relative stabilities and the need to use kinetic conditions to obtain the less stable forms. β is the stable form at room temperature, and can be prepared, for instance, by very slow cooling from solutions in acetic acid, acetone, nitric acid, or nitromethane. The α modification can be prepared from the same solvents, but with more rapid cooling, and even more rapid cooling can yield the γ form. δ may be obtained from solvents in which it is only slightly soluble, and even then by rapidly chilling, even to pouring the solution over ice. Teetsov and McCrone (1965) quantitatively determined the stability ranges and transformation temperatures using the method of eutectics later followed by Yu et al. (2000a) (Section 5.8); the results on HMX are also summarized in Table 9.1. Teetsov and McCrone (1965) have noted that some of these transition temperatures may be quite variable, in particular $\alpha \leftrightarrow \beta$, depending on the amount of strain in the β form. Also, a number of authors (e.g., Cady and Smith 1962) reported that trace amounts of residual RDX can affect the results from studies on HMX transformations; the two substances may be easily distinguished on the polarizing microscope by the difference in their indices of refraction (McCrone 1957).

Lee et al. (2011b) used two of the crystallization techniques of supercritical CO_2 (see Section 7.4) to crystallize HMX. With the aerosol solvent extraction system they obtained γ or δ with a well-controlled particle size distribution. The gas antisolvent process yielded micronized γ or β, but when the solvent for HMX

Table 9.1 *Data on polymorphs of HMX.*[a]

	β (I)	α (II)	δ (IV)	γ (III)
Crystal structure, CSD REFCODE	OCHTET01 OCHTET04 OCHTET12	OCHTET	OCHTET03	DEDBUJ[c]
Calculated density[b,d]	1.893, 1.902 (1.96)	1.839 (1.87)	1.759 (1.78)	1.78, 1.82 (1.82)
Melting point (°C)[e]	246–247	256–257	279–280	280–281.5
Transition temperatures (°C)[f]	$\beta \leftrightarrow \delta$ 167–183 $\beta \leftrightarrow \gamma$ 154	$\alpha \leftrightarrow \delta$ 193–201 $\alpha \leftrightarrow \beta$ 116		$\gamma \leftrightarrow \delta$ 167–182
Transition temperatures (°C)[g]	$\beta \leftrightarrow \alpha$ 102–104	$\alpha \leftrightarrow \gamma$ metastable $\alpha \leftrightarrow \delta$ 160–164		
Impact sensitivity (cm)[h]	31–32	5–50	6–25	6–12
Heat of sublimation (kcal mol^{-1})[i]	41.9	39.3		38.0
Heat of solution (kcal mol^{-1})[h]	4.4	3.8	3.7	1.5

[a] The polymorphic designation in parentheses is that of McCrone (1950a).
[b] Calculated from X-ray data as reported in crystal structure or presented here.
[c] The cell constants reported by Cady and Smith (1962) are actually those of a hemihydrate of HMX (Main et al. 1985) and not those of the γ form (as recognized also by a different REFCODE in the CSD). This misnomer has also been perpetuated in Kohno et al. (1996). Cady and Smith did cite a report by Krc (1955) that gave the following cell constants for the γ form (no space group indicated): $a = 16.80$, $b = 7.95$ $c = 10.97$ Å, $\beta = 130°$, $Z = 4$, and indicated that these could not be transformed to the cell that later turned out to be that of the monohydrate.
[d] Value in parenthesis given by Meyer (1987).
[e] Teetsov and McCrone (1965).
[f] Meyer (1987).
[g] Gibbs and Popolato (1980).
[h] Holston Defense Corp. (1962).
[i] Taylor and Crookes (1976).
OCHTET, OCHTET01—Cady et al. (1963).
OCHTET03—Cobbledick and Small (1974)
OCHTET04—Zhitomirskaya et al. (1987).
OCHTET12—Choi and Boutin (1970).

was acetone, this process could be fine-tuned via a number of operational experimental parameters to yield micronized β with a narrow particle size distribution (Kim et al. 2011).

The polymorphs of HMX have been studied by a wide variety of techniques. One of the crucial parameters for high energy materials is the impact sensitivity, which in this case was determined by determining the height from which a dropped 5 kg weight would initiate reaction. Low values therefore indicate increased sensitivity, which can also be influenced by the presence of grit and to some degree by crystal size and shape. In particular, the authors noted that crystal habit had a significant effect in the cases of α- and γ-HMX, but very little on the δ form. A rather thorough study of this property was carried out by Cady and Smith (1962), who concluded that (a) the impact sensitivity of β is reproducible and independent of particle size, and (b) in spite of variations in the results of different tests, the order of sensitiveness of HMX polymorphs is $\delta > \gamma > \alpha > \beta$.

Structurally, the polymorphs of HMX exhibit different conformations, as demonstrated in Figure 9.1. Those conformations have been studied in detail in the gas phase (Molt et al. 2013). Brill and Reese (1980) analyzed the relative stabilities of the α, β, and δ forms in terms of coulombic forces to interpret the thermophysical behavior of the three forms. They concluded that the chair conformation found in the β modification is more stable than the chair–chair conformation in the other two forms. The relative stability to pyrolysis could also be accounted for by the analysis of the coulombic attractions and repulsions around the molecule in each of the modifications. More recently, Henson et al. (1999) followed the change in second harmonic generation response during the $\beta \rightarrow \delta$ phase change, which has long been implicated in the thermal decomposition of HMX (Karpowicz and Brill 1982). A virtual-melting mechanism has been employed for HMX embedded in a polymer matrix to explain this reversible phase transformation (Levitas et al. 2009). The thermal expansion of the polymorphs has been studied by Zhou and Huang (2012).

Nanocrystals of HMX have been prepared by a number of methods. Qiu et al. (2012) employed spray drying of an acetone solution incorporating a polymeric

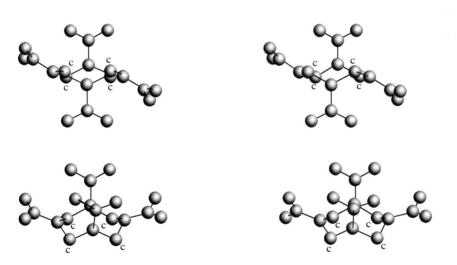

Figure 9.1 *Stereoviews of the HMX molecule in the β (top), α (bottom) forms. In both structures the view is on the plane of C–N–C of the ring (from Bernstein 1987, with permission).*

binder to obtain ~1 μm composite granules of γ and δ polymorphs dispersed in that polymer matrix. In another report, antisolvent methods were used to prepare the β form with an average particle size of 30–128 nm (Kumar et al. 2015).

The IR spectra of all four crystal modifications were reported by Cady and Smith (1962) and by Holston Defense Corp. (1962). The latter did point out some distinguishing features among the polymorphs, but Cady and Smith noted that problems with sample preparation and conversions among forms indicated that the optical properties described by McCrone (1950a) were the basis for the best rapid qualitative and even rough quantitative analysis. Raman spectroscopy, which requires less potentially destructive sample preparation, has been used to distinguish the polymorphs (Goetz and Brill 1979). It has been demonstrated that terahertz absorption spectroscopy reveals distinct spectra for the four forms of HMX, suggesting that this technique might be used in stand-off detection technology of explosives (Konek et al. 2010). The low resolution ^{1}H NMR spectra of the four crystal modifications were reported by Landers et al. (1985). The ^{14}N nuclear quadrupole resonance spectrum of the β form has been determined and has been used to suggest a mechanism for the $\beta \rightarrow \delta$ phase transition (Landers et al. 1981).

There have been many computational studies on the structures of polymorphs of high energy materials, in particular the nitramines, including HMX.

The molecular energetics were studied by Smith and Bharadwaj (1999), using high level (B3LYP/6-311G**) quantum mechanical geometry optimizations, with subsequent single-point energy calculations at the MP2/6-311G** level. Of the four low-energy conformers, two corresponded to the conformers found in the crystal structures, but the two lowest energy conformers have not been observed experimentally. This led the authors to suggest that the conformations in the observed structures are influenced by intermolecular interactions. Lattice energy calculations, based on the atom–atom potential method (see Section 5.3), including MNDO-derived partial charges led to sublimation energies similar in magnitude to those listed in Table 9.1, but did not correspond to the observed order of stability. While the authors attributed the discrepancy to the overestimation of the coulombic energy, this study illustrates some of the difficulties and limitations of computing lattice energetics of polymorphs, as discussed in Chapter 5. In a subsequent study employing a different computational strategy, the molecular structures and energetic stabilities of the α, β, and δ modifications were computed using a first-principles electronic structure method (Lewis et al. 2000). The computed results were compared with experimental molecular volumes, the bulk modulus and its pressure derivative "in reasonable accord with experiment;" the predicted energetic ordering of the three polymorphic phases does correspond with experiment.

Isothermal–isobaric molecular dynamics simulations of the α, β, and δ modifications have been carried out over the temperature range 4.2–553 K, using a force field developed for RDX, together with charges derived from *ab initio* calculations (Sorescu et al. 1998b). These gave results in close agreement with the experimentally determined crystal structures. Another molecular dynamics study (Kohno et al. 1996) indicated the importance of compressed N—N bonds in the initial decomposition process of nitramines in general, and the polymorphs of HMX in particular.

The phase transitions and mechanical properties of the α, β, and δ phases have also been studied theoretically (Cui et al. 2010). The lattice parameters and volumes show nonlinear dependence on the temperature and pressure. In the high-temperature regime, there exist two phase transitions, that is, $\beta \rightarrow \alpha$ at 360 K and $\alpha \rightarrow \delta$ at 440 K. Under high pressure, the $\beta \rightarrow \delta$ takes place at 27 GPa and 298 K. Within the range of temperature and pressure studies, it can be deduced that the β form exists at lower temperatures and higher pressures, the α form is preferred at higher pressures, and the δ form at lower temperatures have better malleability. The geometric aspects of the thermal expansion of the β phase have been studied in detail (Deschamps et al. 2011).

Given the differences in properties among the polymorphs, a number of attempts have been made to control or limit the change of phases. Starting with the β form, Peterson et al. (2007) heated a sample of HMX to 184 °C to convert it to the β form, which remained stable for days under a variety of conditions but converted to α at room temperature and 20–40% relative humidity in a few days. Soni et al. (2011) used hot water as a solvent to convert the γ phase to the β phase. Zhang et al. (2016) investigated the phase transition of HMX under static compression up to 50 GPa, noting a possible $\gamma \rightarrow \beta$ phase transition at ~2.10 GPa, while finding no evidence for previously reported $\beta \rightarrow \delta$ or $\epsilon \rightarrow \delta$ transitions at 27 GPa. Finally, seeding was used to preferentially obtain microcrystals of the β form over the γ form (Damiri et al. 2017).

9.3.1.2 *HNIW aka CL-20 (PUBMUU)*

While HMX is one of the oldest and widely used high energy materials, **HNIW** (also often referred to as CL-20) is a relatively more recent cyclic nitramine and has generated considerable interest, although the early work is in references of limited accessibility (see for instance, references 1–4 in both Russell et al. 1992, 1993, and Sorescu et al. 1998a, and references 1–7 therein). These same authors agree that five polymorphs are known, although Nedelko et al. (2000) indicate the existence of six crystal modifications.

Nielsen et al. (1998) have described the preparation of modifications α, β, γ, and ϵ, in considerable detail. They also reported the single crystal structure determinations of the β, γ, and ϵ forms. Subsequent reports (Bolotina et al. 2004; Golovina et al. 2004) included variable temperature studies and a report on the disorder of one of the nitro groups (Meents et al. 2008). Since the publication of the first edition of this volume HNIW has become a major player, if not *the* major player in the field of high energy materials, resulting in an explosion of literature. For instance, as of mid-2017 the Nielsen paper has been cited ~250 times. I attempt here to highlight the current literature of the polymorphic behavior of this material and some of the recent co-crystal chemistry associated with it, with the caveat that even as this is published there appear to be very rapid additional developments. Nevertheless, the early literature is still of value to demonstrate how knowledge and understanding of the polymorphic behavior of this important material have been recognized and dealt with.

The α modification crystallizes as a hemihydrate, from which the water may be removed to yield an anhydrate. The other three modifications are unsolvated. The same authors and another group (Jacob et al. 1999) have reported the crystal structures of all four modifications (under Refcodes PUBMIInn (α) and PUBMUUnn (β, γ, ϵ)). A compilation of the available literature on obtaining the apparently preferred ϵ-HNIW (Ghosh et al. 2012) has recently been published, with references to the newer forms of HNIW, including derivatives and co-crystals (Ghosh et al. 2016).

In a study of the crystallization behavior at room temperature, Kim et al. (1998) found that the initially appearing β form gradually converted to the ϵ form in a solution-mediated process, indicating that the latter is more stable in accord with Ostwald's rule (see Section 2.3). These authors also present comparative X-ray diffraction (XRD) patterns and qualitative and quantitative Fourier transform infrared (FTIR) analyses of mixtures of these two forms. Another study, ostensibly to determine the polymorphic purity of the material, in fact used a qualitative visual examination to estimate the maximum amount of the β form as an impurity in samples of the ϵ form from three different suppliers (Bunte et al. 1999).

The solution crystallization has been investigated more recently in considerable detail (Xu et al. 2012). Subsequent studies dealt with the propensity of the β form to initially precipitate in many crystallizations and the energetics of the transformation to the preferred ϵ form have been studied by Wei et al. (2016). An earlier study provides information on obtaining the ϵ form from the hydrated α form (Ghosh et al. 2014). A detailed study of the effect of additives on the formation and transformation of the polymorphic forms has been presented by Zhang et al. (2014). FTIR, in combination with thermal and microscopic methods has also been used to determine the thermally generated polymorphic forms among the modifications (Foltz et al. 1994b), with a specific application to a formulation based on the ϵ form as the active ingredient (Foltz 1994). Spectroscopic characterizations of the various forms, including FT Raman, NMR, CIMS, and UV, are also cited by Nielsen et al. (1998); SEM images of the various phases have been presented by Foltz et al. (1994a, 1994b). Goede et al. (2004) reported an FT Raman study on the four stable phases of HNIW in combination with quantum mechanical calculations to assign the bands. Those were used to develop a quantitative analysis protocol for detecting polymorphic impurities below 2% of the ϵ form.

At room temperature and atmospheric pressure, the relative thermal stability of the four modifications has been determined by solvent-mediated transformations and thermal analysis to be $\epsilon > \gamma > \alpha$-hemihydrate $> \beta$ (Foltz 1994; Foltz et al. 1994a, 1994b). Of the anhydrate materials ϵ does indeed have the highest density as expected from this ordering (see Section 2.3), but in contradiction to the expected correlation between density and relative stability β is more dense than γ (Nielsen et al. 1998). The polymorphic behavior has been reviewed in considerable detail through 2004 by Nair et al. (2005).[1] Li and Brill (2007) studied the ambient pressure isothermal solid–solid phase transitions $\epsilon \rightarrow \gamma$, $\alpha \rightarrow \gamma$, and $\beta \rightarrow \gamma$ by transmission FTIR spectroscopy followed by multivariate regression analysis. The connection between crystal structure and thermal behavior of some of these polymorphs and their derivatives has been studied in considerable detail by Yan et al. (2013) and by Liu et al. (2016) using in situ X-ray powder diffraction (XRPD). The phase purity of the various polymorphs as a function of crystal growth conditions was studied in detail by Urbelis and Swift (2014).

In a rather detailed study Russell et al. (1993) used a specially designed high-temperature/high pressure diamond anvil cell to determine the phase diagram and stability fields between −125 °C and 340 °C and atmospheric pressure to 14.0 GPa. They characterized the ζ phase which exists at room temperature above 0.7 GPa pressure and obtained FTIR spectra for all five polymorphs in various regions of P–T space. The data were used to characterize nine observed interphase transitions, among which four are reversible and five are unidirectional. The pressure and thermal ranges of decomposition were determined for the α, γ, and ζ polymorphs. A more recent high pressure study reported by Millar et al. (2010a) led to the preparation and the structure determination of the high pressure ζ-form.

The kinetics of thermal decomposition of three of the modifications were studied by thermogravimetry, IR spectroscopy, and optical and electron microscopy (Nedelko et al. 2000), with the conclusion that the rate increases in the series $\alpha > \gamma > \epsilon$. However, it was found that the results for a particular polymorph also depend upon the morphological features of the crystals as well as their size distribution and mean size.

The molecular conformations found in the crystal structures of the β, ϵ, and γ modifications differ (Sorescu et al. 1998a, 1998b). The conformation found in the α hemihydrate is similar to that in the γ form. The computed lattice energies, employing an intermolecular potential developed for RDX (Sorescu et al. 1997) including charges calculated at the HF/6-31G** level, gave an ordering of the relative stability of the ϵ (−50.35 kcal mol⁻¹), β (−49.62 kcal mol⁻¹), and γ (−48.19 kcal mol⁻¹) forms compatible with the experimental results of Russell et al. (1993); the differences in lattice energies between polymorphs are also in the range expected. More recently DFT studies on the four polymorphs of crystalline CL-20 and the influences of hydrostatic pressure on the ϵ form have been reported (Xu et al. 2007), leading to a prediction that the N–NO$_2$ bond is the trigger bond during thermolysis. The thermal expansion properties of the four polymorphs have also been studied and compared (Pu et al. 2016).

Crystal structure prediction methods (see Section 5.9) have been applied in a detailed study of a number of high energy materials, including HNIW, and the

---

[1] The CL-20 designation comes from China Lake, the United States Naval Weapons Center in California where this material was first synthesized and studied.

polymorphism therein (Pepe et al. 2009), although as noted elsewhere there is nothing intrinsically different between these and other molecular crystals.

There have been a number of studies of the mechanism of thermal decomposition of HNIW (Patil and Brill 1993) with a particular emphasis on the role of free radicals (Pace 1991, 1992). Ryzhkov and McBride (1996) compared the reactions at low temperature in the α and β modifications and found that the same cavities that contain water in the hemihydrate play an important role in differences in the solid-state chemistry between the two modifications.

9.3.1.3 RDX (CTMTNA)

RDX (cyclo-1,3,5-trimethylene-2,4,6-trinitramine) is one of the most important and widely used high energy materials and the stable α form is well characterized (Meyer 1987), including its crystal structure (Choi and Prince 1972; Hakey et al. 2008) and variable temperature studies thereof (Zhurov et al. 2011). Structures of the β, γ, and ϵ forms have been reported by Millar et al. (2009, 2010b) and Davidson et al. (2008). McCrone (1950b, 1957) was apparently the first to prepare the unstable β form on the microscope, where he determined some optical properties, and noted the difficulties in obtaining it in quantities suitable for further characterization. However, visually the β form is easily distinguishable from the α form, although apparently not detectable by thermal methods (Hall 1971). Subsequent efforts to obtain the β form (Sergio 1978) eventually led to the preparation, by crystallization from high-boiling solvents, of quantities sufficient to determine the FTIR spectra (Karpowicz et al. 1983). The study of the spectra of both solid forms of RDX with those in the vapor and solution phases and a comparison with those of HMX suggested that the former also exhibit different conformations in the different polymorphs (Karpowicz and Brill 1984). Miller et al. (1991) subsequently carried out a P–T study similar to that described for HNIW, and using FTIR, energy dispersive XRPD, and optical microscopy in a diamond anvil cell, further characterized these two phases in addition to a high pressure γ phase, which transforms to either α or β rather than decomposing.

Increasingly sophisticated computational and experimental techniques have been applied to RDX in the last 15 years. Dreger (2012) re-determined the phase diagram using both Raman spectroscopy and optical imaging and studied both the static and shock compression to investigate the decomposition of the γ phase. He reported the discovery of a new high pressure–high temperature (ϵ) phase. Hunter et al. (2013) also combined experimental and computational DFT-D methods in which they determined the heat of fusion for the β phase, as lattice energies and differences between forms. High pressure phonon calculations were shown to agree with Raman spectral frequencies, and also showed that the heat capacities for the α, γ, and ϵ forms are not strongly affected by pressure.

Goldberg and Swift (2012) had actually developed robust drop cast crystallization methods to prepare the previously "highly metastable" (as noted by Hunter et al. 2013) β phase, which then remained stable for extended periods. It was later demonstrated (Figueroa-Navedo et al. 2016) that the α and β phases could be formed by deposition on stainless steel substrates using spin coating methodology; conditions were developed to the latter phase readily and predominantly forming the β phase. Finally, it is not surprising that the presence of polymorphs of RDX (or any high energy material for that matter) may complicate the forensic detection of trace amounts, so that increasingly sophisticated techniques are being developed to deal with these situations (e.g., Emmons, et al. 2012).

9.3.1.4 PETN (PERTN)

The polymorphism of the secondary explosive **PETN** was first reported by Blomquist and Ryan (1944), and the crystal structure of the common Form I was originally reported by Booth and Llewellyn (1947) and later by Trotter (1963) and Zhurova et al. (2006). It crystallizes in a space group ($P\bar{4}2_1c$) that is quite rare for molecular substances. It is also noteworthy that the $\bar{4}$ crystallographic site available in this space group corresponds to the molecular point symmetry, since Kitaigorodskii (1961) indicated that the exigencies of packing efficiency generally lead to situations in which the crystallographic site symmetry for a molecule is lower than the full molecular point symmetry.

The early work on PETN suggested that Form I was stable to its melting point of 142.9 °C. Cady and Larson (1975) found that careful measurement of the melting points of the two polymorphs indicated that Form II actually melts 0.2 °C *higher* than Form I, suggesting that at that temperature Form II is the more stable form. Form II also forms spontaneously on a face of Form I growing from a supercooled melt. At lower temperatures (i.e., < 130 °C) Form II transforms rapidly to Form I. The details of these transformations were studied in considerable detail, but unfortunately are available in a document that is difficult to access (Cady 1974). Cady and Larson also determined the crystal structure of Form II in space group *Pcnb*, in which obviously there is no longer a crystallographic site symmetry of $\bar{4}$, but the molecular symmetry is nearly retained.

Sorescu et al. (1999) successfully modeled Form I up to pressures of about 5 GPa. The calculations were followed by elevated pressure (to 14 GPa) and temperature (to 550 K) measurements on single crystals in a diamond anvil cell (Dreger and Gupta, 2013). PETN Form III transforms to Form IV at high temperatures, while at room temperature Form I undergoes a phase change to Form III. These data were used to prepare the appropriate phase diagram.

9.3.1.5 TATP (HMOCN)

The preparation of **TATP** (triacetone-triperoxide) was first reported by Wolffenstein (1895). A single crystal structure was reported in by Groth (1969), followed by Dubnikova et al. (2005) at 140 °C, but its polymorphic behavior was not reported until recently (Reany et al. 2009), along with thermal analysis of a number of forms to determine relative thermodynamic stabilities. The results of further polymorphic studies by Parsons and Karaliota (2015) and Schollmeyer and Ravindran (2015) have been deposited at the CCDC. Because of its ease of preparation from commonly available materials, and relative difficulty in forensic-ally detecting it, TATP has become increasingly widely used in terrorist attacks since the turn of the twenty-first century. Six polymorphs are obtained concomi-tantly from the reaction mixture, but the components and the relative proportions differed with the reaction conditions. They also can be prepared from different solvents. Peterson et al. (2013) studied the crystallization conditions in greater detail to achieve considerable control over the polymorphic outcome of any par-ticular crystallization.

9.3.1.6 TAGN (TAGUDN)

The crystal structure of the room-temperature phase of **TAGN** was determined simultaneously by two groups (Bracuti 1979; Choi and Prince 1979). Oyumi and Brill (1985) reported endotherms in the DTA traces at 258, 270, and 407 K. In the course of a study of the rigid body motion of the nitrate ion in this material, a low temperature (−10 °C) polymorph was discovered (Bracuti 1988), as antici-pated in the data presented by Oyumi and Brill. The room-temperature structure crystallizes in space group *Pbcm*, while that at low temperature is *Pbca*. The *b* and *c* crystallographic axes are essentially identical for the two structures, but the *a* axis is quadrupled (from 8.366(2) to 33.47(1) Å) on going to the low temperature modification.

9.3.1.7 PTTN (nitroglycerine) (CORYIR)

Nitroglycerine **PTTN** is a liquid at room temperature with a solidification point of 13.2 °C for the stable modification. The compound has been known since 1914 to be polymorphic (Hibbert 1914), with a less stable modification solidifying at 2.2 °C. The structure of the stable orthorhombic modification has been reported twice (Espenbetov et al. 1984; Litvinov et al. 1985), while that of the triclinic metastable form has not yet been reported.

9.3.1.8 DATH (FIPZIN)

The crystallization behavior of **DATH** has been described by Oyumi et al. (1987b). An amorphous phase, unstable at room temperature, can be prepared under kinetic conditions by rapid removal of solvent (acetone or acetonitrile) or rapid cooling of the melt. It initially appears as a transparent waxy material that transforms into a crystalline material over about an hour. It can also be prepared on a DTA instrument by cooling from the melting point of the crystalline material (406 K) to the temperature range 333–290 K. On the other hand, crystals suitable for single crystal structure determination (carried out by the same authors) can be grown by slow evaporation from the same solvents.

9.3.1.9 OHMX (GEJXAU)

OHMX (1,7-dimethyl-1,3,5,7-tetranitrotrimethylenetetramine) is the acyclic analogue of HMX. Thermal analysis and IR spectroscopy were used to identify five polymorphs (Oyumi et al. 1987a). Rapid evaporation of solutions of a number of different solvents led to polycrystalline Form IV. Forms II and III were obtained by slow evaporation, the first from DMF, and the second from acetone. Form I could be obtained by cooling warm Form II to room temperature or by applying mechanical friction to Form II or Form IV on an NaCl IR sample cell. All four can be maintained at room temperature. Form V was described as a "pre-melt phase (which) is nearly thermally neutral with respect to Form IV..." The only crystals suitable for single crystal structure determination were of Form II; those of Form III exhibit twinning.

Differences and similarities in the IR spectra (that of Form V is indistinguishable from Form IV) were used to interpret structural changes and constancies resulting from transitions between the various modifications.

9.3.1.10   NTO (QOYJOD)

As an insensitive high explosive, **NTO** was once considered as the high energy material in the activation of auto air bags (Wardle et al. 1990) as well as replacement material for RDX in bomb fill (Lee and Gilardi 1993). (It has been replaced by sodium azide.) The compound was reported to be dimorphic (which may have led to some of the confusion in the thermal decomposition data (Williams and Brill 1995; Botcher et al. 1996), both forms being obtained from aqueous solutions. The more stable α form apparently crystallizes under thermodynamic conditions, while the β form is obtained (with some difficulty and in small amounts) under kinetic conditions as well as by recrystallization from methanol or ethanol/methylene chloride solvents. They can be distinguished by morphology and IR spectra. The crystal structures of both have been determined (Lee and Gilardi 1993) with α crystallizing in $P\bar{1}$ and β in $P2_1/c$. Both are characterized by (different) chains of hydrogen-bonded molecules along the a crystallographic axis. In the α modification this leads to polymer-type behavior with crystals shattering parallel to that axis and crystal ends fraying much like rope. They can even be bent without breaking (see also Yakobson et al. 1989). Subsequent structural studies were reported by Zhurova and Pinkerton (2001) and Bolotina et al. (2003, 2005). The influence of pressure and temperature on the thermodynamic and structural properties of NTO were investigated by *ab initio* methods (Rykounov 2015), who calculated the vibrational spectra and predicted a phase change at ~17.5 GPa.

9.3.1.11   TNDBN (NADHOP)

TNDBN (1,3,5,7-tetranitro-3,7-diazabicyclo[3.3.1]nonane) is another compound that is structurally and compositionally similar to HMX; it also exhibits conformational variation among the polymorphs. Oyumi et al. (1986a) identified five polymorphic modifications of TNDBN by DTA and variable temperature (above and below room temperature) mid-IR spectroscopy. Phases III and IV could be obtained at room temperature by crystallization from acetone and acetonitrile with IV appearing more often than III. The latter, crystallizing as orthorhombic crystals in space group *Pbca* with $Z' = 2$, is the only crystal structure reported to date. The two molecules in the asymmetric unit adopt different conformations. On the basis of this observation the authors considered the possibility that transitions among other phases might involve conformational changes; the analysis of the IR spectra indicated that the nitro groups are similarly disposed in the other four polymorphs, suggesting that the conformation is also constant. The structure of a monoclinic form was later published by Gilardi et al. (2001).

9.3.1.12   DPT (DNPMTA)

DPT (3,7-dinitro-1,3,5,7-tetrazabicyclo[3.3.1]nonane) is yet another analogue of HMX, for which the crystal structure of one form was reported by Choi and Bulusu (1974). Although it was the subject of thermal analysis (Hall 1971), prior to a study by Oyumi et al. (1986a) no polymorphism had been reported. By following the mid-IR spectrum as a function of temperature the latter authors detected a gradual, reversible phase transition beginning at 343 K and being completed at 440 K. A careful redetermination of the DSC indicated a phase transition at 339 K with ΔH_t = 1.2 kcal mol^{-1}. The changes in the IR spectra are consistent with a possible inversion in the amine lone pairs accompanied by some increase in the rotational freedom of the molecule.

9.3.1.13   FOX-7 (SEDTUQ)

1-Diamino-2,2-dinitroethene **FOX-7** (also known as DADNE), is a relatively late entry into the family of solid energetic materials. In spite of its molecular simplicity, it was first synthesized only in 1998 by the Swedish National Defense Research Institute (Ostmark et al. 1998; Latypov et al. 1998) with the report of the crystal structure (Bemm and Ostmark 1998), and has gained considerable attention due to its stability and insensitivity. The polymorphism was first reported by Evers et al. (2006), with a first order phase transformation from the room-temperatue α form to the β form at 389 K, which has been classified as displacive. The twists of the nitro groups and the hydrogen bonding differ in the two structures. It has been suggested that the apparent strengthening of the C–NO$_2$ bonds may be associated with the low sensitivity and high activation energy to impact of this material. A subsequent structure refinement based on synchrotron data described the electron densities (Meents et al. 2008). The kinetics of the reversible $\alpha \leftrightarrow \beta$ phase transition was studied by Zakharov et al. (2015). The forward reaction was determined to be adequately described by a first-order autocatalysis while the reverse was fitted to a kinetic model for two parallel first-order processes. Zakharov et al. (2016) also studied the structural transformations of FOX-7 over the temperature range 298–513 K and confirmed the existence of the δ form at $T > 480$ K, with the metastable γ form resulting from a transformation from the δ form.

An early high pressure, high temperature study of FOX-7 that purported the detection of the elusive γ form was reported by Pravica et al. (2010). A number of computational studies and experimental studies at elevated temperatures and pressures in the search for additional polymorphs and the definition of the phase diagram were reviewed by Bishop et al. (2014). Hunter et al. (2015) reported a combined high pressure and DFT-D computational study and determined that the high pressure form is different from the γ form, thereby establishing the presence of the ϵ form, which is characterized by a planar, layered structure. Subsequent high pressure studies (Bishop et al. 2015) determined further details on the phase diagram, including some phase transitions and the α–β–δ and β–γ–δ triple points.

These reports were followed in 2016 by a number of additional elevated temperature and pressure studies by Dreger et al. (2016a, 2016b, 2016c), in some instances augmented by crystal structure determinations of the ϵ phase and the so-called α' phase (the apparent result of a phase change but according to the authors indistinguishable from the α phase) and DFT-D2 calculations. These studies also led to the triple point for the α–β–γ phases, and some claimed corrections for the pressure–temperature phase diagram. The crystallographic structural parameters in the various phases were studied as a function of pressure up to ~13 GPa, indicating that the $\alpha \rightarrow \epsilon$ transition at 4.5 GPa significantly perturbs all molecular and crystal properties, while the $\alpha \rightarrow \alpha'$ transition at 2 GPa is apparently "associated only with subtle molecular and intermolecular changes."

9.3.1.14   TKX-50

TKX-50 is an early entry into a developing class of energetic ionic salts (Dreger et al. 2015). In a rather detailed study Lu et al. (2017) determined the existence of a metastable phase at ~180 °C as indicated by both Raman and TGA-DSC measurements combined with computational studies.

9.3.1.15   Dinitramide salts (DEKDAX)

The dinitramide anion was a relatively late entry in the field of energetic materials (Bottaro et al. 1991; Russell et al. 1997 and references therein) and a number of salts were prepared and structurally characterized (Gilardi and Butcher 1998b and references 7–12 therein; Sitzmann et al. 2000, and references 2b–i therein), including some polymorphic ones, described briefly here.

9.3.1.15.1 Hexaammonium dinitramide   The material is dimorphic: crystallization from water yields the monoclinic $P2_1/c$ modification, while crystallization from polar organic solvents leads to a triclinic $P\overline{1}$ form (Gilardi and Butcher 1998b). Both structures contain $R_2^2(9)$ hydrogen-bonded rings, with two nitro oxygens as acceptors and one —NH and one —CH as donors. The overall packing is quite similar with alternating layers of cations and anions.

9.3.1.15.2 Aminoguanidinium dinitramide   The material was originally prepared from an aqueous medium as white crystals, with a melting point of 70–79 °C. Several recrystallizations from ethyl acetate led to a melting point of 91–94 °C. The crystal structures of a triclinic $P\overline{1}$ modification (density = 1.639 g cm^{-3}) and a noncentrosymmetric monoclinic Pc modification (density = 1.650 g cm^{-3}) are reported, but it is not clear if the melting points are indeed different. The anions differ in conformation by about 11.5° in one torsion angle, suggesting that the pair of structures are conformational polymorphs.

9.3.1.15.3 Ammonium dinitramide (AND) and dinitroazetidinium dinitramide (DNAZ-DZ)

For both of these materials the pressure/temperature and reaction phase diagram have been determined using a high temperature–high pressure diamond anvil cell with FTIR spectroscopy, Raman spectroscopy and optical microscopy. For ammonium dinitramide energy dispersive XRD was also employed (Russell et al. 1996, 1997).

At ambient conditions ammonium dinitramide has not been found to be polymorphic; the crystal structure of the orthorhombic α form under those conditions has been reported by Gilardi et al. (1997). The pressure/temperature studies led to the discovery of a second monoclinic phase (β) formed from the α phase at 2.0 ± 0.2 GPa. As the pressure is further raised, a solid-state rearrangement, melting, and thermal decomposition take place.

The structure of the one known crystal (α) modification of dinitroazetidinium dinitramide at ambient conditions has been reported by Gilardi and Butcher (1998a). It crystallizes in space group $Cmc2_1$. The pressure studies led to a rapid and reversible phase transition at 1.05 ± 0.05 GPa to the β modification, which could not be maintained to ambient conditions. Additional studies led to the determination of the stability fields of both of these polymorphs and the liquidus phase, as well as a determination of the temperature/pressure conditions for decomposition (Russell et al. 1997). The FTIR spectra of both solid phases were also analyzed in detail.

In a number of these studies, Russell et al. showed that the exploration of temperature/pressure phase space with a relatively simple apparatus (Block and Piermarini 1976), combined with a number of rapidly developing analytical techniques, can lead to detailed understanding of polymorphic systems, the relationships among crystal modifications and the characterization of their properties. More studies of this type, especially to explore the pressure domain of polymorphic systems, should be encouraged.

9.3.2 Aromatic materials

9.3.2.1 *TNT (ZZZMUC)*

> *TNT proved to be a very difficult material to prepare in single crystalline form…there is still a great deal of confusion regarding the structure of the crystalline products and the conditions under which they are obtained.*
>
> Sherwood and Gallagher (1984)

> *How can a material so well known be so poorly understood?*
>
> Lowe-Ma (2000, quotation from 1990)

The history (perhaps one should say "saga") of the crystal chemistry and polymorphism of this compound is summarized in these two statements by two prominent workers in the chemical crystallography of high energy materials. 2,4,6-Trinitrotoluene **TNT** was among the most common aromatic explosive (Cooper and Kurowski 1996). In 1949 McCrone reported that there was "no evidence of polymorphism for TNT;" only a few years later, he twice mentioned an unstable polymorph of the material (McCrone 1957), although Grabar et al. (1969) subsequently doubted whether McCrone had observed what later became recognized as the monoclinic modification. Yet the confusion on the polymorphism of TNT had begun well before.

The first chemical crystallographic (goniometric) studies of TNT were apparently reported by Friedländer (1879), who claimed that the crystals exhibited rhombic symmetry. The next report by Artini (1915) indicated monoclinic symmetry, although Ito (1950) noted the similarity in the axial ratios in these two descriptions. The orthorhombic/monoclinic dilemma continued with the initial X-ray investigations: Hertel and Römer (1930; ZZZMUC02) reported a monoclinic cell with a β angle close to 90° (89°29′29″), whereas Hultgren (1936; ZZZMUC03) reported a rhombic cell with unit cell axes sufficiently similar to those of the monoclinic cell to raise doubts about the significance of the difference.

Ito (1950) carried out his own study, including an analysis of the twinning that he recognized from Weissenberg photographs, and concluded that TNT is trimorphic, with one rhombic form and two monoclinic forms. The three forms reported by Ito have identical b and c axial lengths, with the length of the a axis doubling and quadrupling from the rhombic to the two monoclinic forms.

Burkhardt and Bryden (1954) then reported for crystals grown by sublimation at 78 °C cell constants with estimated standard deviations (esds) for a $P2_1/c$ monoclinic cell with $\beta = 111.2°$ that could be transformed to a pseudo-orthorhombic cell with $\beta = 90.5°$. The transformed cell again had b and c axes identical to an orthorhombic form (grown from ether at −70 °C) in space group *Pmca* or *P2ca*, within experimental error, with the a axis length (20.07 Å) slightly less than half that of the transformed monoclinic cell (39.81 Å). However, they also reported three other monoclinic and two other orthorhombic modifications obtained under different crystallization conditions and some transformations that took place with ageing and grinding.

J.R.C. Duke (Lowe-Ma 2000; also reference 10 (dated 1981) in Gallagher and Sherwood 1996) determined the crystal structures of both modifications. He prepared the monoclinic form with $P2_1/c$ ($Z = 8$) (called Form A) by annealing cast TNT. Orthorhombic crystals were prepared by quenching the melt and were reported as $Pb2_1a$ ($Z = 8$) (called Form B). Again, the two have very similar, but nevertheless statistically significant different cell constants. The latter appears to be identical to the orthorhombic structure reported (in equivalent, but transformed space group *Pca2_1*) by Carper et al. (1982; ZZZMUC01), wherein the authors again noted the propensity for twinning of TNT single crystals. In both structures $Z' = 2$, the two independent molecules in the asymmetric unit exhibiting different degrees of twist of the nitro groups out of the plane of the benzene ring.

Much of the confusion surrounding the crystal chemistry of TNT as noted in the quotations at the start of this section has now been resolved, due to a large extent, to the availability of the crystal structures of the monoclinic and orthorhombic forms (Carper et al. 1982; Golovina et al. 1994) and the work of Gallagher and Sherwood (1996), who actually made use of the unpublished crystal structures of Duke (1981 *vide supra*). There is general agreement among the crystallographic constants reported by these various authors (Table 9.2), but some do differ by more than 3 esds of the reported values. Gallagher and Sherwood found that the orthorhombic form could be obtained from cyclohexanol and ethanol; the monoclinic form, stable from room temperature to the melting point of 82 °C, crystallizes from a number of other solvents, including methanol, toluene, acetone (almost always with twinning), and ethyl acetate (apparently free from twinning).

Some of the past confusion regarding the two polymorphs is obvious from the similarities in the cell constants and may be understood by examining the crystal structures (Figure 9.2). By visual inspection, the packing is very similar. Gallagher and Sherwood carefully analyzed the symmetry relations between the two independent molecules (A and B) in the asymmetric unit for both structures. In the monoclinic modification, centrosymmetrically related pairs of molecules are arranged in layers almost parallel to the *b–c* plane, with alternate layers related by a pseudo-glide plane. This leads to a molecular stacking sequence of …AA BB AA BB… For the orthorhombic structure, there is a true glide plane parallel to the

Table 9.2 *Comparison of some of the crystallographic constants reported for TNT[a]*

	Carper et al. (1982)	Golovina et al. (1994)	Duke (1981)[b]
Orthorhombic form			
a (Å)	14.991(1)	20.041(20)	15.005(2)
b (Å)	6.077(1)	15.013(8)	20.024(4)
c (Å)	20.017(2)	6.084(5)	6.107(3)
Space group	$Pca2_1$	$P2_1ab$	$Pb2_1a$
R-factor	0.057	0.055	0.043
Monoclinic form			
a (Å)		21.407(20)	21.275(2)
b (Å)		15.019(8)	6.093(3)
c (Å)		6.0932(5)	15.025(1)
Monoclinic angle (°)		(γ) 111.00(2)	(β) 110.14
Space group		$P2_1/b$	$P2_1/c$
R-factor		0.061	0.049

[a] The axis assignments of equivalent space groups are given here as reported by the authors. Appropriate transformations should be made for the purposes of comparison.
[b] See text.

(a)

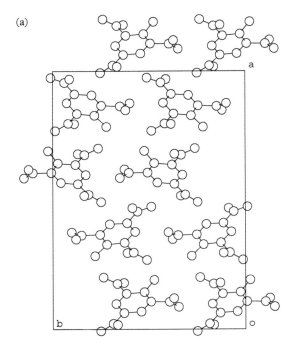

(b)

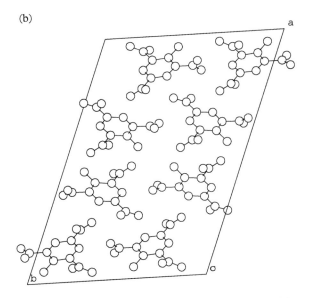

Figure 9.2 *Packing diagrams of (a) the monoclinic and (b) the orthorhombic structures of TNT, including the orientation of the unit cell. In both cases the view is on the best plane of the "A" molecule of the asymmetric unit as defined by Golovina et al. (1994) (after Golovina et al. 1994).*

b axis, so that the molecular stacking sequence is…AB AB AB AB…The sheet structure appears almost identical, but subsequent sheets are displaced relative to one another by *c*/2 with respect to the monoclinic axial system.

The similarity of the two crystal structures leads to very similar XRPD patterns (Figure 9.3), reminiscent of the situation in terephthalic acid (Section 4.4). Careful inspection reveals that the orthorhombic modification can be distinguished by a peak (the 511 reflection in the Golovina et al. (1994) cell) at $2\theta = 27.24°$, while the monoclinic modification can be distinguished by a peak (from 211) at $2\theta = 19.26°$.

Thermal (DSC) studies clearly confirm that the monoclinic form is stable from room temperature up to its melting point, while the orthorhombic modification goes through a small endotherm at 70 °C, corresponding to a solid–solid phase transition with a heat of transition of 0.22 kcal mol^{-1} (Connick et al. 1969; Gallagher and Sherwood 1996). Over a two-month period, the orthorhombic material transforms to the monoclinic.

Gallagher et al. (1997) also carried out lattice energy calculations on both forms. They obtained a value of −28.83 kcal mol^{-1} for the monoclinic form and −28.24 kcal mol^{-1} for the orthorhombic. The difference is that expected for polymorphs (see Chapter 5); moreover, the absolute values compare favorably with experimental sublimation enthalpies reported by Edwards (1950) (28.3 ± 1.0 kcal mol^{-1}, albeit with no specification of the polymorphic form), although values as low as 23.7 ± 0.5 kcal mol^{-1} (Pella 1976, 1977) have been reported (Chickos 1987). The Gallagher et al. (1997) calculations were also used to investigate and compare the specific interactions between molecules in the twinned monoclinic phase and the two untwinned phases.

TNT also provides a good historical example of Ostwald's rule and some other aspects of crystal growth and habit. Groth's entry (Groth 1917, p. 364) for the compound (taken from Friedländer 1879), shown in Figure 9.4, is clearly that of the orthorhombic form (also portrayed by McCrone 1949), now known to be less stable. A supplementary note on p. 766 of the same volume of Groth refers to the Rendic 1915 work describing the subsequently discovered more stable monoclinic form. In their studies of TNT, Gallagher et al. (1997) determined the habit and computationally modeled the expected morphology for the orthorhombic, monoclinic, and twinned monoclinic phases, using attachment energies. For the orthorhombic case, there is similarity in the prominent crystal faces to be expected, but a lack of correspondence between theory and experiment on the relative sizes of those faces.

That was the situation at the time of the publication of the first edition. Much of the confusion was clarified in a seminal paper by Vrcelj et al. (2003). They demonstrated that many of the problems encountered in the storied history of this polymorphic system, including determination and interpretation of the crystal structures arose from problems in crystallizing the two forms, as documented earlier by Gallagher and Sherwood (1996). They then showed that the orthorhombic form is less stable than the monoclinic form by ~0.6 kcal mol^{-1} with a monotropic relationship between the two forms.

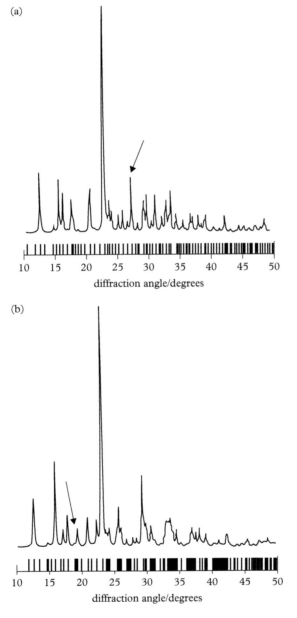

Figure 9.3 *X-ray powder diffraction patterns for the (a) orthorhombic and (b) monoclinic poly-morphs of TNT, calculated from the structures reported by Golovina et al. (1994). One of the most distinguishing reflections for each pattern is marked with an arrow (see text). The experimental powder patterns reported by Connick et al. (1969) are also listed in the PDF.*

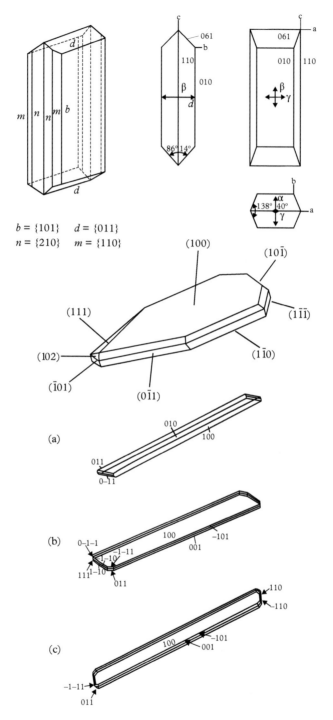

$b = \{101\}$ $d = \{011\}$
$n = \{210\}$ $m = \{110\}$

(a)

(b)

(c)

Figure 9.4 *Morphologies of TNT crystals. Upper: experimental morphology of the orthorhombic form (from Groth 1917 and McCrone 1949, with permission); middle: experimental morphology of the monoclinic form (from Gallagher and Sherwood 1996, with permission); lower: computationally predicted morphology of the (a) orthorhombic, (b) monoclinic, and (c) twinned monoclinic phases (from Gallagher et al. 1997, with permission).*

The carefully re-determined crystal structures of the two forms enabled a resolution and understanding of many of the earlier encountered problems and conundrums.

The very similar unit cell constants for the two forms are given in Table 9.3. It is seen that in addition to the difference in space group, and the ~20° difference in the β cell angle, the c axes differ by ~1.2 Å; note that both structures have $Z' = 2$. Although the bond lengths and bond angles are essentially identical the authors note subtle, but real differences in conformations. The unit cell contents are compared in Figure 9.5.

Note that this representation (of the original authors) differs from our earlier recommendation (see Section 2.4.2.1) because it demonstrates the effect of the above noted changes in the cell constants on the packing.

The authors also point out that the subtle differences in packing seen in Figure 9.5 lead to "large scale polytypism" usually observed in organic crystals of long chain molecules, although they did note some exceptions for small molecules. The phenomenon is demonstrated in the packing diagrams of Figures 9.6 and 9.7.

Figure 9.6a shows how the pairs form stacks, while Figure 9.6b shows stacks of two pairs of molecules. The unit cells are shown in Figure 9.7, demonstrating the difference in the packing between the two forms. It is noteworthy that this careful analysis reveals the rather subtle differences in the packing better than Figure 9.2, which has been retained in this edition specifically to demonstrate this point of the

Table 9.3 *Revised crystallographic data for the two forms of* **TNT** *(from Vrcelj et al. 2003). (Reprinted with permission from* Crystal Growth & Design *3, 6, 1027–32. Copyright 2003 American Chemical Society.)*

	monoclinic	orthorhombic
formula	$C_7H_5N_3O_6$	$C_7H_5N_3O_6$
molecular weight	227.14	227.14
system	monoclinic	orthorhombic
a (Å)	14.9113(1)	14.910(2)
b (Å)	6.0340(1)	6.034(2)
c (Å)	20.8815(3)	19.680(4)
β (deg)	110.365(1)	90.0
V (Å³)	1761.37(4)	1770.6(7)
T(K)	100	123
space group	$P2_1/a$	$Pca2_1$
Z	8	8

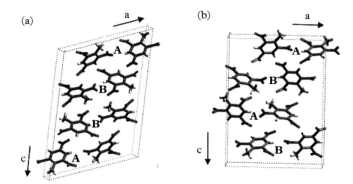

Figure 9.5 *The unit cells of (a) monoclinic and (b) orthorhombic TNT viewed along the b-axes. The letters A and B indicate the positions of the pairs of molecules of different conformations (from Vrcelj et al. 2003). (Reprinted with permission from* Crystal Growth & Design *3, 6, 1027–32. Copyright 2003 American Chemical Society.)*

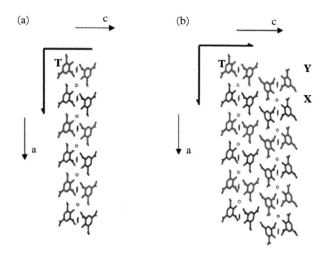

Figure 9.6 *(a) Stacks of pairs of TNT molecules and (b) stacks of two pairs of molecules. In both cases the stacks are viewed parallel to the b-axis (from Vrcelj et al. 2003). (Reprinted with permission from* Crystal Growth & Design *3, 6, 1027–32. Copyright 2003 American Chemical Society.)*

value of careful analysis and presentation of the structures of polymorphs in order to interpret and understand them. Vrcelj et al. (2001) then used these packing diagrams to propose mechanism of growth for the two forms, and to suggest that the previously reported unstable third form is due to the twinning of these two known forms.

A high pressure investigation of the phase (Stevens et al. 2008) revealed a sharp discontinuity at ~20 GPa, which was attributed to a phase transition

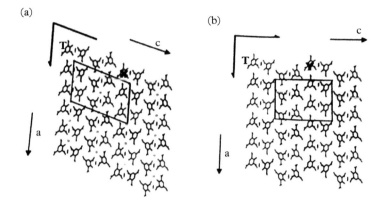

Figure 9.7 *Placement of the next cell to create (a) the monoclinic form and (b) the orthorhombic form (from Vrcelj et al. 2003). (Reprinted with permission from* Crystal Growth & Design *3, 6, 1027–32. Copyright 2003 American Chemical Society.)*

from the monoclinic to the orthorhombic form. A subsequent pressure study (Bowden et al., 2014), incorporating *in situ* synchrotron radiation combined with vibrational spectroscopy verified that the monoclinic phase persists at least up to 10 GPa.

These recent studies have transformed the previously enigmatic 2,4,6-TNT system into one which is now quite well understood and over which considerable control can be exercised. It should be noted here that 2,4,5-TNT, one of the main impurities in 2,4,6-TNT, has also been shown to be polymorphic (Chick and Thorpe 1970, 1971), and has been characterized by thermal and microscopic methods as well as by IR spectroscopy (Section 9.3.2.7).

9.3.2.2   HNAB (HNIABZ)

Hexanitroazobenzene **HNAB** is known to crystallize in five polymorphic forms (McCrone 1967). The structures of Form I, stable from room temperature to 185 °C and Form II, stable from room temperature to 205 °C have been published (Graeber and Morosin 1974), and were cited as early examples of conformational polymorphism (Figure 5.2) (Bernstein 1987). Form III has a stability similar to Form I and apparently has a strong tendency to form twinned crystals, while Forms IV and V have been obtained only upon supercooling of the melt. Thermal gradient sublimation has been used to grow clear single crystals of Form III (Firsich 1984), but the structure had not yet been determined as of the previous writing. The Form III structure determination was revisited by a group comprising two of the original authors (Rodriguez et al. 2005). It was found that the initially reported space group ($C222_1$) was due to pseudo-merohedral twinning in the space group $P2_1$, and a suitable refinement was carried out with the correct space group assignment. The effects of pressure on Form I have been studied computationally (Liu et al. 2012), while vapor deposition methods have been shown to yield an amorphous form as well as Form II and another unidentified form (Knepper et al. 2012).

9.3.2.3   Lead styphnate (lead trinitroresorcinate) (ZZZGSA)

The so-called "normal" material contains the dibasic resorcinate dianion. This material is used mainly as a primary or initiating explosive together with lead azide. Neither McCrone and Adams (1955) nor Meyer (1987) make any mention of polymorphic behavior; the trimorphic polymorphic behavior of monobasic lead styphnate has been known for quite some time (Böttger and Will 1846; Brün 1934; Tausen 1935; Hitchens and Garfield 1941).

The two normal lead styphnate structures are polymorphic monohydrates whose structures have been published by Pierce-Butler (1982, 1984).

9.3.2.4   DNFP (FEYLUQ)

Oyumi et al. (1986b) have carried out a rather thorough study on the tetramorphic system **DNFP**. Forms I and II may be obtained selectively from specific solvents; Form III is obtained by cooling Form IV; Form IV is obtained by heating any of the other three modifications. They have been characterized by thermal analysis, IR spectroscopy, and analysis of decomposition products. The crystal structure of Form I has been reported by two different groups (Oyumi et al. 1986b; Lowe-Ma et al. 1990).

9.3.2.5   1,1'-Dinitro-3,3'-azo-1,2,4-triazole (GICTIV)

This material has been considered as a potential candidate for high energy propellant applications. A yellow form may be obtained from ethanol, while a pale-orange modification crystallizes from acetone. They crystallize in (equivalent) $P2_1/c$ and $P2_1/a$ space groups respectively, both with $Z' = 0.5$, requiring that the molecule lies on a crystallographic inversion center and exhibits an essentially planar molecular conformation (Cromer et al. 1988).

9.3.2.6   2-Methyl-1-(2',4',6'-trinitrophenyl)-4,6-dinitrobenzimidazole

The compound crystallizes from an ethanol/acetone solvent as easily separable concomitant polymorphs (Section 3.5): Form I exhibits orange–brown triangular tabular platelets, while Form II appears as cream–yellow rods with hexagonal cross-section (Lowe-Ma et al. 1989; Freyer et al. 1992). The densities of the two forms differ significantly (1.658 and 1.712 g cm^{-3}) and the full crystal structure analysis indicated that they may also be considered conformational polymorphs, differing in the rotation of the phenyl ring about the exocyclic bond and in the rotations of the nitro groups on that phenyl ring. However, both forms have "unexceptional explosive properties…(and)…uninspiring detonation velocities… and detonation pressures…." (Freyer et al. 1992).

9.3.2.7 *Also worthy of note*

In this section, I list a number of other compounds, sometimes included among energetic materials, that have exhibited polymorphism.

1,8-Dinitronaphthalene (DNTNAP) is used in some explosive mixtures with ammonium nitrate (Meyer 1987). The detailed information on the preparation and characterization of the two dimorphs is given by McCrone (1951), who also studied the solution phase transformation (in thymol) between two polymorphs (McCrone 1957). The structure of the orthorhombic form was initially published by Akopyan et al. (1965), and subsequently refined at room temperature and 97 °C by Ciechanowicz-Rutkowska (1977). The CSD contains an entry (DNTNAP03) for the monoclinic form (Kozin 1964), but no coordinates are available.

2,6-Dinitrotoluene (ZZZQSC) is an important product in the manufacture of both powdery and gelatinous commercial explosives. It has been shown to be trimorphic (McCrone 1954). Well-formed modifications I (stable below 40 °C) and II (stable above 40 °C) can both be obtained from thymol. The structure of Form I was determined at room temperature by Nie et al. (2001), and at low temperature by Hanson et al. (2004). Form III is obtained as feathery dendrites only from the melt and is unstable at all temperatures. No crystal structures have been reported.

2,4,5-Trinitrotoluene (also known as γ-TNT) is one of the main impurities in military and commercial grades of TNT. Chick and Thorpe (1971) characterized two polymorphs. Form I (mp 376.2 K) may be obtained by recrystallization from alcohol or solidification of the melt. Form II (mp 347.2 K) is produced in small quantities with difficulty from an undercooled melt. It readily converts to Form I by mechanical perturbation or even spontaneously. Chick and Thorpe also determined latent heats of fusion, entropies of fusion, specific heats, and IR spectra. Due to the conversion induced by grinding no X-ray data were presented for either form. No crystal structures have been reported.

Pentanitrophenylazide is apparently at least dimorphic, the two forms exhibiting different impact sensitivity, H_{50} (α, 53 cm; β, 17 cm) (Lowe-Ma 2000).

Hexanitrodiphenylamine **HND** (HNIDPA), although toxic, has been employed in underwater explosives together with TNT and aluminum powder (Meyer 1987). Two polymorphs were characterized by McCrone (1952), including a listing of the XRPD lines for Form I, which he determined to be orthorhombic, with cell constants essentially matching those reported later by Dickinson and Holden (1977). Structure determinations of the orthorhombic form have been reported by Wozniak et al. (2002) and at low temperature by Huang et al. (2011).

Dipentaerythritol hexanitrate **DiPEHN** is formed as a byproduct in the synthesis of PETN, and hence can influence its performance. It is dimorphic, the stable modification is Form I (mp 76.0 °C microscope hot stage) (Cady 1974), which can be grown as dendrites from a supercooled melt. Form II starts growing from the supercooled melt, but large crystals can be grown from the melt above 70 °C. The melting point of Form II, also determined on a hot stage microscope is 74.4 °C. The similarity of the melting points and additional thermal data indicate that

Form I is more stable relative to Form II at room temperature than in the vicinity of the melting point.

Tripentaerythritol octanitrate **triPEON** is an additional impurity in the production of PETN. Using hot stage microscopic methods Cady (1974) identified four polymorphs, with melting points 83.3, 72.1, 74.6, and 69.0°C and heats of fusion (cal g^{-1}) of 18, 13, 13.1, and ~10.5 for Forms I–IV, respectively. This report also contains a rather detailed evaluation and discussion of previous work on the characterization of the polymorphic forms of this compound.

Picryl bromide (ZZZVXQ) is a common energetic precursor that has been shown to be pentamorphic (Parrish et al. 2008). The structures of all five forms have been determined, with all of them exhibiting high Z' values, ranging from 3 to 18, and being characterized by layered structures.

10

Polymorphism and patents

Modern science is not a solitary undertaking … Litigation is. Real science is the study of facts that are regular, but a courtroom trial is quintessentially singular. Science depends on placing facts in an orderly context, but a trial frames facts in isolation. Good science transcends the here and now, the individual and the idiosyncratic, the single laboratory, the single nations, the single planet, even the single galaxy, but a trial typically examines the single datum, and demands that scientific truths be rediscovered anew every time. Scientific facts emerge from many isolated observations, as data are accumulated, vetted for error, tested for significance, correlated, regressed, and reanalyzed, but trials are conducted retail. Good science is open, collegial, and cumulative, but the courtroom setting is discrete, insular, and closed—a one shot decision.

Huber (1991)

Scientists do not deal in certainties, only in likelihoods. In mathematical terms, we deal in "probabilities." Like most lawyers, we are not in the absolute *business of proving something is "true" and another thing is "false." Applied to our technological studies, we believe that this sort of "lawyers' language" would take us into the realm of metaphysics because such categorical statements are not relatable to opinions derived from observations.*

Smith (1993)

Courts of law are not the optimal fora for trying questions of scientific truth …

Zenith Laboratories, Inc. v. Bristol-Myers Squibb Co. (1994)

10.1 Introduction

A patent is a social contract, known since the Middle Ages.[1] According to the Oxford English Dictionary definition, it is

---

[1] One of the earliest patents was granted in 1421 to the Italian architect Brunelleschi for a barge with hoisting gear that carried marble along the Arno River for building the Duomo in Florence. The first known English patent was granted in 1449 to John of Utyman for a method for making stained glass for Eton College.

Polymorphism in Molecular Crystals. Second Edition. Joel Bernstein. © Joel Bernstein 2020. Published in 2020 by Oxford University Press. DOI: 10.1093/oso/ 9780199655441.001.0001

a license to manufacture, sell, or deal in an article or commodity, to the exclusion of other persons; in modern times, a grant from the government to a person or persons conferring for a certain definite time the exclusive privilege of making, using, or selling some new invention.

Historically many patents granted the right to sell, specifically tobacco, and were expanded to other commodities. Today's patents only give the right to exclude. As noted by Maynard and Peters (1991),

> patent systems reward the competitive, creative drive with a temporary, limited, exclusive right, in return for the cooperation of an inventor in teaching the rest of society how to use his or her findings for all time thereafter.

(See also Grubb 1986; Grubb and Thomsen 2010.)

Polymorphism presents a variety of challenging legal and scientific issues to patent systems. Since crystal modifications of a substance represent different crystal structures with potentially different properties, the discovery or preparation of a new crystal modification represents an opportunity to claim an invention that potentially can be recognized in the awarding of a patent. The rules and regulations of such patents differ from country to country, although some degree of standardization has resulted from World Trade Agreement and subsequent legislation, but there are still important differences in the nuances of the granting and enforcement of patents in various venues. As seen in earlier chapters, a particular crystalline modification can possess considerable chemical, physical, or biological advantages over its congeners, and the granting and maintenance of patent exclusivity over the rights to particular polymorphic form(s) may have considerable economic consequences. As a result, it is not surprising that the past thirty years or so have witnessed a number of patent litigations essentially involving different crystal modifications, many concerning the definitions, descriptions and analytical techniques presented in earlier chapters. Partly because of the size and economic impact of the pharmaceutical industry, some of the most visible cases involving crystal forms have involved some widely used drugs (Sodikoff et al. 2015). In addition, in the United States there are special patent provisions for pharmaceuticals (Engelberg 1999) that also may contribute to the frequency and nature of these legal battles (Barton 2000).

In this chapter, I will review some of the patent issues and cases that often serve as the meeting ground between the worlds of science and the law alluded to in the preambulary quotations of this chapter (see also Faigman 1999; Foster and Huber 1999). The intention is to demonstrate the scientific issues that were raised in the course of these litigations and how they were interpreted by the courts. Many of these cases also involved many principles and nuances of patent law, most of which also will not be covered here. However, two principles of patent law—novelty and obviousness—that are of particular relevance to polymorphism, indeed are almost a common denominator in many of these litigations, are discussed in the following sections.

Litigations involving patent issues can generate thousands of pages of documents and testimony, with the discovery of many facts and the expression of many (often opposing) scientific opinions on both sides of the issue (Bernstein 2017). I will very much limit the descriptions here to what is given in the official records of the specific cases considered, mainly from the patents, judicial decisions and the reports of them. Of course, even these are subject to controversy, since court rulings can be and are reversed, thus perhaps altering the way a scientific issue is subsequently viewed by a court of law and the society that is guided by that law. Moreover, both science and the law are dynamic, and the interpretations and ramifications of any particular case can and do change with time.[2]

10.2   Novelty and obviousness in crystal form patents

In general, two of the most important requirements for obtaining a patent are that the claimed invention must be (1) *novel*, that is, not inherently present or disclosed in the available prior art at the time that the invention was made or, under first to file systems, the time that the first patent application was filed, and (2) *non-obvious* to a hypothetical person of ordinary skill in the art in the relevant field. Accordingly, in challenging a patent or in defending against a claim of infringement, it is often argued that the patent in question is not valid since the invention was not novel or would have been obvious to a person of ordinary skill in the art.

These two issues are particularly common in the realm of prosecution and litigation of patents covering pharmaceutical solid forms. The issue of novelty is perhaps more straightforward than that of obviousness, although by no means is it always clear-cut. In the simplest sense, *novel* and *new* have the same etymology (Merriam-Webster). Therefore, a *new* crystal form, by definition, is *novel*: it has never previously been prepared. Although that would seem to settle the issue, the chemistry may be such that the prior art (including literature, conference proceedings, poster sessions, etc.) intentionally or unintentionally contained information sufficient to establish the inherent existence of the previously unknown novel crystal form. Considerable controversy and debate can surround such a situation.

An obviousness inquiry can be even more subject to controversy and debate. As noted above, in the first instance, the inquiry requires definition of a person of ordinary skill in the art before determination of whether or not the invention would have been obvious to that person. Then, to determine whether an invention is considered obvious, courts assess a variety of factual inquiries, including whether there was motivation to make the previously unknown crystalline form,

---

[2] With the passage of time some of these cases have acquired almost iconic status in the community of solid-state chemists. Since court records are somewhat difficult to access often the main source of information is secondary accounts, and the precise facts and legal decisions may become distorted over time. This unfortunately can lead to false impressions and understanding by a wider community unfamiliar with the nuances of the original legal proceedings and decisions.

whether there would have been a reasonable expectation of success in doing so, whether the selection and preparation of the crystalline form was the result of examining a finite number of identified, predictable solutions, and whether formation of the new crystalline form was unexpected or possessed properties that could not be predicted in advance of its creation. Each one of these issues requires determination of all of the facts at the relevant time, an understanding of the methodology of the discipline, and considerable judgment on any expectation of success.

I will deal here primarily with the issues of novelty and obviousness of patents covering polymorphs of pharmaceuticals, including the science, the current legal standards, and the case law as of this writing.

10.3　The scientific perspective

10.3.1　Novelty from a scientific perspective

The course of research and development of a drug starting from the identification of a lead compound to a final marketed product is currently by different estimations to cost anywhere from about $1.4 billion (DiMasi et al. 2014) or even up to $5 billion (Harper 2012) and to take an average of 7.5 years (Jarvis 2016). Since the vast majority of drugs are marketed as tablets or pills containing a solid form of the active pharmaceutical ingredient (API), that course of drug research and development may involve an investigation—often a continuing investigation—of the crystal form landscape of the API. In the course of that investigation, and not infrequently even serendipitously, new crystal forms (i.e., polymorphs, solvates, and hydrates) may be discovered. For a new chemical entity, by definition such crystal forms are novel. Even for a chemical entity known for many years, a new crystal form is novel. The common question that arises in assessing the novelty of a crystal form is whether that particular crystal form, with all of its identifying characteristics, previously existed even if it was not recognized.

10.3.2　Obviousness from a scientific perspective

From a scientific point of view, *obvious* is synonymous with *predictable*. Again turning to Webster "predict" is defined as "to be made known beforehand" (Merriam-Webster). Therefore, predictable means "can be known beforehand." In terms of current knowledge and understanding, no crystal form can be known beforehand. To put that statement into even more concrete terms, in spite of the fact that the community of chemists that deal with molecular crystals has made notable progress in understanding and working with those materials, as described earlier in this volume, it is still not possible to predict from a molecular formula the crystal form landscape, including the number of polymorphs, hydrates, and solvates that might be possible (if any), how one might make the as yet unknown crystal forms, or what their properties might be (Cruz-Cabeza et al. 2015). On that basis, under any definition of the person of ordinary skill in the art, a new crystal form can never be obvious.

As such, it seems that legally the result should be the same, as one cannot "predict" or "expect" the existence of a new crystalline form, including what its properties will be. Moreover, the exploration of the solid form landscape requires experimentation. And there is no textbook recipe for carrying out that experimentation. A plethora of techniques and options exist for chemists to pick and choose from in a search for new crystal forms. Furthermore, each compound presents a new situation with its own unique challenges. Lessons learned from one compound do not necessarily apply even to a structurally related compound or one with the same therapeutic activity. Some compounds exhibit a seemingly endless number of crystal forms (e.g., atorvastatin calcium, with at least seventy-eight reported; Jin 2012). Others, in spite of being common, and widely studied, to date have exhibited no tendency whatsoever for multiple crystal forms (e.g., naphthalene).

One of the enigmatic issues in addressing the question of obviousness on the basis of what appears in a granted patent is often the superficial simplicity with which the invention is described in the specification, and even more so in the claims. There is no need, indeed no requirement, to describe all of the steps, successes, and/or failures that occurred en route to the invention. A patent need only describe the invention in a manner that would allow a reader having ordinary skill in the art to identify, make, and use the invention. Hence, it is virtually impossible to extract from the patent any information about the time, effort, creativity, serendipity, unpredictability, and/or expense that lead to the claimed invention. The same is very often true for patents covering new crystal forms. As Milton (1667) perceptively pointed out approximately 350 years ago, "…[S]o easy it seemed, Once found, which yet unfound most would have thought impossible." Further, and equally as important, whether the new crystalline form appears as the result of years of toil or completely due to random chance is *irrelevant* in terms of obtaining a patent. Indeed, the patent statute itself recognizes this very point, by stating that "[p]atentability shall not be negated by the manner in which the invention was made" (35 U.S.C. § 103). Notwithstanding that the patent statute and controlling precedent of the Federal Circuit specifically mandate that the "spark of invention" standard is the wrong inquiry for determining obviousness, such challenges often grossly understate the time and effort often expended in inventing a new crystalline form.

It is of interest to contrast the patent status of obviousness *vis a vis* solid forms with that of chemical synthesis. It is not at all clear why patents drawn to new chemical entities obtained by chemical synthesis should be viewed any differently from patents drawn to new chemical forms obtained by experimentation with existing crystal forms or by serendipity, particularly where both inventions are *new materials*. When one synthesizes a new chemical compound, the picture of the compound one seeks to synthesize is known in advance even though it may take much effort and many years to achieve its creation and, of course, its properties remain unknown until prepared and characterized. In contrast, for a given specific molecule, there is *no* way to predict possible solid forms that might exist, methods of making any of those forms, or their properties. The lack of any a priori knowledge of the forms that might exist would seemingly suggest that crystalline forms are even less predictable and thus less obvious than the product of a specifically designed chemical synthesis.

10.4 Some representative crystal form patent litigations

Since the publication of the first edition there have been a number of additional significant litigations dealing with solid forms of pharmaceutical, including polymorphs. The legal aspects of many of the cases involving solid pharmaceutical forms have been reviewed by Sodikoff et al. (2015). Most of the specific cases I deal with here are consequences of the Hatch–Waxman 1984 law (Mossinghoff 1999) that restructured the generic drug market. Many deal with so-called "virtual" infringement—that is, whether a particular product would infringe still existing patents if and when it would go on the market upon the expiration of another patent covering the drug. The emphasis in what follows will be on the scientific aspects of the litigations, which cannot, of course, be totally isolated from the legal aspects. For a measure of completeness, I include here the cases from the first edition, which in some instances have achieved iconic nature (e.g., Thakur et al. 2015, Desiraju and Nangia 2016).

10.4.1 Cefadroxil

Cefadroxil (also sometimes spelled cephadroxil), an antibiotic originally developed by Bristol-Myers (BM) and marketed under the trade names Ultracef and Duricef, had combined sales in the United States of $100 million in 1988. A crystalline monohydrate (apparently discovered serendipitously), claimed in U.S. Patent 4 504 657 ("the '657 patent") issued in March 1985, is described in the patent in terms of its X-ray diffraction pattern (Figure 10.1). The improvement of this substance over the earlier known material included greater stability and high bulk density, properties which enabled the production of smaller pills.

Prosecution of this patent, the original application, dating from August 1979, was in fact quite lengthy and involved questions of similarities and differences in crystalline modifications (see Kalipharma, Inc. v. Bristol-Myers Co. 1989 for details of the history of the prosecution of the patent application). The prior art included a 1973 patent (U.S. Patent 3 781 282, "the '282 patent") in which a modification of cefadroxil had also been prepared. This was termed the "Micetich form" after one of the inventors. One of the questions surrounding the '657 patent was whether the crystalline material described therein (known as the "Bouzard

United States Patent [19]

Bouzard et al.

[11] Patent Number: 4,504,657

[45] Date of Patent: Mar. 12, 1985

[54] CEPHADROXIL MONOHYDRATE

[75] Inventors: Daniel Bouzard, Franconville; Abraham Weber; Jacques Stemer, both of Paris, all of France

[73] Assignee: Bristol-Myers Company, New York, N.Y.

[21] Appl. No.: 358,567

[22] Filed: Mar. 16, 1982

Related U.S. Application Data

[60] Continuation of Ser. No. 931,800, Aug. 7, 1978, abandoned, which is a continuation of Ser. No. 874,457, Feb. 2, 1978, Pat. No. 4,160,863, which is a division of Ser. No. 785,392, Apr. 7, 1977, abandoned.

[30] Foreign Application Priority Data

Apr. 7, 1977 [GB] United Kingdom 17028/76

[51] Int. Cl.³ C07D 501/22; C07D 501/12
[52] U.S. Cl. ....,................ 544/30
[58] Field of Search 544/30

[56] References Cited

U.S. PATENT DOCUMENTS

3,489,752 1/1970 Crast, Jr. 260/243

3,655,656 4/1972 Van Heyningen 260/243
3,781,282 12/1973 Garbrecht ....:................ 260/243
3,957,773 5/1976 Burton 260/243
3,985,741 10/1976 Crast, Jr. 260/243
4,091,215 5/1978 Bouzard 544/30
4,160,863 7/1979 Bouzard et al. 544/30
4,162,314 7/1979 Gottschlich 544/30

FOREIGN PATENT DOCUMENTS

829758 12/1977 Belgium .
1240687 7/1971 United Kingdom .

OTHER PUBLICATIONS

Dunn et al., The Journal of Antibiotics, 29: 65–80, (Jan. 1976).

Primary Examiner—Mark L. Berch
Attorney, Agent, or Firm—Robert E. Carnahan; David M. Morse

[57] ABSTRACT

A novel crystalline monohydrate of 7-[D-α-amino-α-(p-hydroxyphenyl)acetamido]-3-methyl-3-cephem-4-carboxylic acid is prepared and found to be a stable useful form of the cephalosporin antibiotic especially advantageous for pharmaceutical formulations.

1 Claim, 1 Drawing Figure

We claim:

65   1. Crystalline   7-[D-α-amino-α-(p-hydroxyphenyl-)acetamido]-3-methyl-3-cephem-4-carboxylic   acid monohydrate exhibiting essentially the following x-ray diffraction properties:

19

Line	Spacing d(Å)	Relative Intensity
1	8.84	100
2	7.88	40
3	7.27	42
4	6.89	15
5	6.08	70
6	5.56	5
7	5.35	63
8	4.98	38
9	4.73	26
10	4.43	18
11	4.10	61
12	3.95	5
13	3.79	70
14	3.66	5
15	3.55	12
16	3.45	74
17	3.30	11
18	3.18	14

20
-continued

Line	Spacing d(Å)	Relative Intensity
19	3.09	16
20	3.03	29
21	2.93	8
22	2.85	26
23	2.76	19
24	2.67	9
25	2.59	28
26	2.51	12
27	2.46	13
28	2.41	2
29	2.35	12
30	2.30	2
31	2.20	15
32	2.17	11
33	2.12	7
34	2.05	4
35	1.99	4
36	1.95	14
37	1.90	10

• • • • •

Figure 10.1 *(a) The title page of U.S. Patent 4 504 657 for the "Bouzard form" of cefadroxil monohydrate; (b) the claim for the Bouzard form, giving the powder diffraction pattern in tabular form.*

form" or "Bouzard monohydrate") was the same as or different from the "Micetich form." In the end, the '657 patent was granted, in essence recognizing the difference between the two.

In anticipation of the expiration of the '282 patent, there were attempts to make the Micetich form according to Example 19 in that patent. Those attempts invariably led to the Bouzard form, leading the companies involved in carrying out the experiments to claim inherency of the Bouzard form in the '282 patent. Bristol-Myers' explanation of these results included the role of unintentional seeding that tended to favor the formation of the Bouzard form rather than the Micetich form. A number of litigations ensued, which involved these issues (e.g., Kalipharma, Inc. v. Bristol-Myers Co. 1989; Bristol-Myers Co. v. United States International Trade Commission et al. 1989). In the former case, there was also a question of the identity of the various crystal modifications, determined mainly by powder X-ray diffraction, with conflicting opinions by various experts. The court found that the "…plaintiff had established a prima facie case of invalidity [of the '657 patent]," and that "…its scientists have replicated the claimed cefadroxil monohydrate according to…Gambrecht Example 7…." The court also did not accept the defendant's "…only challenge, that seeds from its cefadroxil monohydrate acted as a template such that any attempt to repeat Gambrecht Example 7 anywhere in the world would yield Bristol-Myers cefadroxil monohydrate…."

There have been many subsequent litigations and appeals. Many experiments have been done and many issues arose in those cases. Some of them are noted here:

(1) The X-ray powder pattern of the Bouzard form was indexed to determine that every line in the pattern could be accounted for by crystal modification.

(2) Attempts were made to prepare the Micetich form in an unseeded environment, with the measure of success being a matter of contention.

(3) Experiments were run in a clean room equipped with filters that would eliminate particles the size of bacteria (which are several microns in size), but could not eliminate seeds that can be hundreds of times smaller (see Section 3.2).

(4) A series of side-by-side experiments were run, with one sample sealed from seed crystals and the other not sealed—the seed-free produced the Micetich form and the open flasks produced the Bouzard form (Tarling 2001).

The series of litigations continued over other issues, including the nature of crystal modifications (Zenith Laboratories, Inc. v. Bristol-Myers Squibb Co. 1992). Zenith prepared and formulated a hemihydrate of cefadroxil, for which it submitted an Abbreviated New Drug Application (ANDA) to the Food and Drug Administration (FDA). Bristol-Myers (BM, now Bristol-Myers Squibb) contended that the FDA should require a much more extensive New Drug Application (NDA) for the hemihydrate, rather than approve it within the framework of the

FDA monograph for the monohydrate. BM also alleged that the Zenith product converted to the monohydrate, thereby infringing the '657 patent. BM's theory of infringement was that the Zenith product converts to the Bouzard monohydrate after ingestion in the gullet and the stomach, as it is mixed with liquid. A federal court found in favor of BM. Upon appeal, Zenith contended that the '657 patent does not cover Bouzard form crystals which might form momentarily in a patient's stomach, and there were both legal and scientific grounds for the basis of that contention. One of the main points of contention was whether the X-ray pattern of the material found by BM in a patient's stomach matched the claim of the Bouzard patent, as shown in Figure 10.1. BM had identified fifteen of the thirty-seven diffraction lines cited in the claim. The court found that, "Although the term 'essentially' recited in the claim permits some leeway in the exactness of the comparison with the specified 37 lines of the claim, it does not permit ignoring a substantial number of lines altogether," and that "…there was a failure of proof as to whether any crystals, assumed to form in the stomach from ingested cefadroxil…, literally infringe the '657 claim." On this basis, the appeals court reversed the lower court's decision. One result of this decision is that many subsequent patent applications including claims based on substances characterized by X-ray diffraction patterns and/or IR spectra were framed in language that differs from that used in the '657 patent.

10.4.2 Terazosin hydrochloride

Terazosin hydrochloride is a drug developed by Abbott Laboratories for the treatment of hypertension and benign prostatic hyperplasia, and marketed as the dihydrate under the trade name Hytrin since 1987. The patent of interest here is U.S. Patent 5 504 207 (the '207 patent), for which the fourth claim is a particular anhydrous crystalline form of terazosin hydrochloride, designated by Abbott as "Form IV,"[3] a form not marketed by Abbott. In the '207 patent, Form IV is defined by reference to the X-ray powder diffraction (XRPD) pattern, listing the peak positions of several of the peaks. To be within the scope of claim 4 of the '207 patent, a product must be anhydrous terazosin hydrochloride and must exhibit a powder X-ray diffraction pattern having each of the principal peaks identified in claim 4.

The defendants were planning to market anhydrous terazosin hydrochloride rather than the dihydrate found in Abbott's Hytrin tablets, the latter being protected by a patent that expired in 2000. Plaintiffs sued defendants for infringing the '207 patent. Geneva argued that the fourth claim in the '207 patent was invalid because Form IV terazosin hydrochloride was "on sale" in the United States more than one year before the filing date for the '207 patent. One of the defendant's experts in X-ray crystallography analyzed two samples of anhydrous terazosin

---

[3] Much of the following is taken essentially verbatim from the opinion of the court in Abbott Laboratories v. Geneva Pharmaceuticals, Inc. (1998).

hydrochloride purchased prior to the 1993 date and found that the patterns for both contained all of the "principal peaks" of claim 4 of the '207 patent (note the difference from the cefadroxil case, *vide ante*). One batch was pure Form IV terazosin hydrochloride while the second was a mixture of Form IV and Form II. Abbott argued that "it [was] conceivable that the substances [in those samples] were initially manufactured as a less stable crystal form" of terazosin hydrochloride, and that these "substances converted over time to Form IV." The court viewed this as "little more than speculation," adding that, "Despite Abbott's extensive work with terazosin hydrochloride, Abbott was unable to produce any expert testimony that such a transformation occurs or is likely to occur with terazosin hydrochloride." The court, therefore granted summary judgment in favor of the three defendants, holding claim 4 invalid because Form IV terazosin hydrochloride was on sale more than one year before Abbott filed its application for the '207 patent.

10.4.3   Ranitidine hydrochloride

Ranitidine was developed in the 1970s by Allen & Hanburys Ltd. of the Glaxo Group (later Glaxo Wellcome and now GlaxoSmithKline (GSK)) in the flurry of activity following the identification of the histamine H_2 receptor (Black et al. 1972) and other H_2 antagonists (Bradshaw 1993) for the treatment of peptic ulcers. In June 1977, David Collin, a Glaxo chemist, first prepared ranitidine hydrochloride (RHCl), and within a month Glaxo filed a U.S. patent application, which resulted in the issue of U.S. Patent 4 128 658 (Price et al. 1978, the '658 patent). Example 32 of this patent gives the procedure for the preparation of the hydrochloride (Figure 10.2).

   In the course of subsequent scale up, Glaxo developed a pilot plant process called 3A, and then one called 3B. On April 15, 1980, for still unknown reasons,[4] the thirteenth batch of RHCl prepared using the latter process produced crystals that gave different IR spectra and XRPD patterns from previous batches. Glaxo concluded that a new polymorph, designated Form 2, had been produced, and the earlier form, described in the '658 patent, was designated Form 1.[5] Glaxo subsequently developed a process, referred to as 3C, to manufacture all the RHCl it has sold commercially as the active ingredient in Zantac.

   In October 1981, Glaxo filed a patent application on Form 2, from which two patents were eventually granted in June 1985 as U.S. Patent 4 521 431 (the '431 patent) and June 1987 as U.S. Patent 4 672 133. The abstract of the '431 patent states simply, "A novel form of ranitidine...hydrochloride, designated Form 2, and having favorable filtration and drying characteristics, is characterized by its

---

[4] In fact, many of the discoveries of new crystal modifications have been made serendipitously (e.g., Silvestri and Johnson v. Grant and Alburn 1974; Meyers, 2007; Morrison et al., 2015), as have many other important scientific discoveries (Roberts 1989; Merton and Barber 2004).

[5] The following description is taken from the court's opinion in the case of Glaxo, INC. and Glaxo Group Limited v. Novopharm, Ltd. No. 91–759-CIV-5-BO.

United States Patent [19]

Price et al.

[11] **4,128,658**

[45] **Dec. 5, 1978**

[54] **AMINOALKYL FURAN DERIVATIVES**

[75] Inventors: Barry J. Price, Hertford; John W. Clitherow, Sawbridgeworth; John Bradshaw, Ware, all of England

[73] Assignee: Allen & Hanburys Limited, London, England

[21] Appl. No.: **818,762**

[22] Filed: **Jul. 25, 1977**

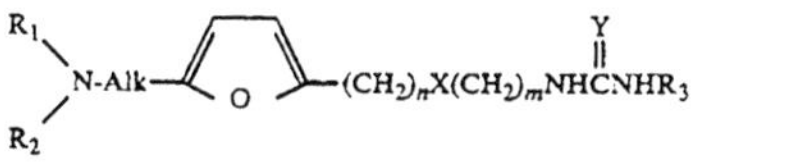

and physiologically acceptable salts thereof and N-oxides and hydrates, in which R₁ and R₂ which may be the same or different represent hydrogen, lower alkyl, cycloalkyl, lower alkenyl, aralkyl or lower alkyl interrupted by an oxygen atom or a group

EXAMPLE 32

N-[2-[[[5-(Dimethylamino)methyl-2-furanyl]methyl]thio]ethyl]-N'-methyl-2-nitro-1,1-ethenediamine hydrochloride

N-[2-[[[5-(Dimethylamino)methyl-2-furanyl]methyl]-thio]ethyl]-N'-methyl-2-nitro-1,1-ethenediamine (50 g, 0.16 mole) was dissolved in industrial methylated spirit 74° o.p. (200 ml) containing 0.16 of an equivalent of hydrogen chloride. Ethyl acetate (200 ml) was added slowly to the solution. The hydrochloride crystallised and was filtered off, washed with a mixture of industrial methylated spirit 74° o.p. (50 ml) and ethyl acetate (50 ml) and was dried at 50°. The product (50 g) was obtained as an off-white solid m.p. 133°–134°.

Figure 10.2 *Portion of the title page and Example 32 from the '658 patent, giving the procedure for the preparation of ranitidine hydrochloride, which subsequent to the discovery of a second polymorphic form became known as the procedure for the preparation of Form I.*

infrared spectrum and/or by its X-ray powder diffraction pattern." These advantageous filtering and drying characteristics are due, in part, at least, to the fact that Form 2 tends to crystallize as more needle-like crystals than the plate-like Form 1 (see Figure 4.42).

By 1991, Zantac sales had reached nearly $3.5 billion, nearly twice the sales of the next best-selling drug. A number of generic drug firms undertook efforts (under the provisions of the Hatch–Waxman Law; Engelberg 1999; Weiswasser and Danzis 2003) to prepare to go on the market with Form 1 in 1995, upon anticipated expiration of the '658 patent.[6] One of these generic companies was Novopharm, Ltd., the defendant in the first RHCl litigation that went to trial. Novopharm scientists unsuccessfully attempted to prepare Form 1 faithfully following the procedure of Example 32 of the '658 patent and therefore sought approval to market Form 2, claiming that the product is, and always has been, Form 2 RHCl. In November 1991, Novopharm filed an ANDA at the FDA to

---

[6] At the time a U.S. patent was valid for 17 years from the date of issue. Subsequent international trade agreements have led to changes in the period of enforcement, and during the changeover period there have been some extensions to the terms of some existing patents.

market Form 2 beginning in 1995. As required by the Hatch–Waxman Act, Novopharm notified Glaxo of its contention that the '431 patent was invalid. Glaxo sued Novopharm for infringement of the '431 patent. Novopharm admitted infringement of the '431 patent, but contended that it was invalid, claiming that the Form 2 was inherently anticipated in the '658 patent. Novopharm claimed that Glaxo *never* performed Example 32 precisely as written either before or after including it in the '658 patent (emphasis in the Court's original). Novopharm theorized that if Glaxo had performed Example 32, the result would have been Form 2, not Form 1; therefore the Form 2 patent was invalid.

Glaxo argued, *inter alia*, that Novopharm's experiments were contaminated with "seed" crystals of Form 2, and therefore, were not faithful replications of Example 32 (quotation marks in Court's original). To make its point on the inherency argument, Glaxo proved that Example 32 does not invariably lead to Form 2, and in fact had led to Form 1. In support of its position Glaxo compared David Collin's original experiments as described in his notebooks in minutiae with Example 32, and also presented evidence that Example 32 as performed in 1993 at Oxford University had led to Form 1. This was sufficient evidence to convince the court that Example 32 does lead to the Form 1 product and that the '431 patent was valid, preventing Novopharm from marketing Form 2.

Novopharm then examined the possibility of marketing Form 1, and developed a stable, reproducible process for the manufacture of Form I. In April 1994,[7] Novopharm filed a new ANDA, this time seeking approval to market Form 1 RHCl upon the expiration of the '658 patent. Shortly thereafter, Glaxo sued Novopharm again, alleging that Novopharm had sought permission to manufacture and market a product which would contain not pure Form 1, but rather a mixture of Form I and Form II, thereby infringing upon Glaxo's Form 2 patents. Novopharm's ANDA, as initially filed, specified that the marketed product be approximately 99% pure Form 1 RHCl (with impurities that may include Form 2 RHCl) as determined by IR. Amended ANDAs filed by Novopharm would have permitted the marketed product to have a Form 1 RHCl of polymorphic purity as low as 90%. Novopharm, however, submitted X-ray evidence at trial that demonstrated that its actual samples of RHCl did not contain detectable Form 2.

The court found that Novopharm had established that its product would not contain Form 2, and that if the product did contain Form 2, then it would be present as an independent component or impurity, not as the basis for some improvement or equivalent. The court thus allowed Novopharm to market mixtures of Form 1 and Form 2, even though Form 2 was still covered by the '431 patent.

The appeals court upheld the district court's decision holding that based on all the available evidence, Glaxo had not proven the Novopharm case likely to market

[7] Much of the following description is taken from the original district court opinion in the case of Glaxo, INC. and Glaxo Group Limited v. Novopharm, Ltd. No. 5:94-CV-527-BO(1), 931 F. Supp. 1280 and the appeals court's opinion in the case of Glaxo, Inc. and Glaxo Group Limited v. Novopharm, Ltd. 96-1466, DCT. 94-CV-527.

a product that contained Form 2. Notably the appeals court pointed to Glaxo's failure to test Novopharm's samples. In reviewing a later case (Glaxo, Inc. et al. v. Torpharm, Inc. et al. 1997), the same appeals court noted it had explicitly declined to address the question of whether small amounts of Form 2 RHCl in a mixture containing primarily Form 1 would infringe the '431 patent (Glaxo, Inc. et al. v. Torpharm, Inc. et al. 1998).

The two Glaxo v. Novopharm cases involved many aspects of the study and analysis of polymorphic materials described in earlier chapters: the (often seren-dipitous) discovery and recognition of polymorphic forms, the role of solvent, heating, stirring, and other experimental techniques in attempts to control the polymorph obtained, the use and development of analytical methods for the char-acterization of polymorphic forms, the relative stability of polymorphic forms, the phenomenon of disappearing polymorphs, the role of seeding, both intentional and unintentional, and the distinction between polymorphic identity and polymorphic purity. Many of these issues also arose in the other litigations involv-ing RHCl (e.g., Glaxo, Inc. v. Boehringer-Ingelheim Corp. 1996; Glaxo, Inc. et al. v. Torpharm, Inc. et al. 1997; Glaxo, Inc. v. Geneva Pharmaceuticals et al. 1994, 1996).[8]

10.4.4 Paroxetine hydrochloride

Paroxetine hydrochloride is a serotonin re-uptake inhibitor used for the treatment of depression, marketed as Paxil®, Seroxat® (and others). The chemical compound paroxetine was initially developed by the Danish company Ferrosan in the 1970s. Beecham (now part of GSK) purchased the rights to paroxetine in 1980 and undertook development of the hydrochloride salt of paroxetine as a drug product. Beecham developed a process that produced paroxetine hydrochloride in a crys-talline form that was later referred to as the "anhydrate" crystalline form. Late in 1984, however, in the course of pilot plant scale-up, the "hemihydrate" crystalline form suddenly appeared at two Beecham sites in the UK within a few weeks of

---

[8] An example of the misinterpretation of the court's decision appeared recently as sidebar entitled "Polymorphism in ranitidine hydrochloride (Zantac®)" in a review article by Thakur et al. (2015). The second Glaxo v. Novopharm case (*vide infra*) was briefly summarized as follows: "GSK now sued Novopharm for infringement of '431, stating that Novopharm's product was not pure Form 1 but a mixture of Forms 1 and 2. The court now ruled in favor of Novopharm, stating that 1% of Form 2 in Form 1 was no basis for improvement of the drug, and hence Novopharm should be allowed to sell its mixture of Forms 1 and 2." In fact as noted here the appeals court held that Glaxo had not demon-strated that all Form 2 peaks were present in Novopharm's intended product, although one Form 2 peak had clearly been detected. Since literal infringement had not been proven, Novopharm was per-mitted to market it, although the court did note that the amount of Form 2 present would not be "the basis for some improvement or equivalent."

In the subsequent Paxil case (*vide infra*), the court ruled at the district level that since the defendant had no intention of marketing the infringing hemihydrate of paroxetine hydrochloride and gained no commercial advantage by the presence of quantities below "high double digit amounts" defendant was not infringing. On appeal, however, the Federal Circuit ruled that any amount of the hemihydrate would infringe.

each other. The new hemihydrate form was designated Form 1 and the previously existing anhydrate was labeled Form 2.

The hemihydrate was not hygroscopic and exhibited handling properties superior to those of the anhydrate. In 1986 GSK applied in the United States for a patent on the hemihydrate, which was granted in 1988 (Barnes et al. 1988) Paroxetine hydrochloride was finally marketed as the hemihydrate form in 1993 under the name Paxil®.

During the 1980s in the course of the development of the compound for eventual launch, Beecham investigated the properties of both the anhydrate and hemihydrate. They determined that in the presence of water (or humidity) the anhydrate undergoes a conversion into the hemihydrate, a process that is accelerated by temperature, pressure, and the presence of seeds of the hemihydrate (Buxton et al. 1988). In Beecham's experience, it was difficult to avoid the conversion of the anhydrate to hemihydrate in the presence of water or humidity in a facility seeded with hemihydrate.

In 1998, Apotex, a Canadian generic drug manufacturer, filed an ANDA with the FDA to market a generic version of the off-patent anhydrous paroxetine hydrochloride (Form 2). The hemihydrate patent would expire in 2006. Again, under the terms of the Hatch–Waxman Act GSK opposed the Apotex request, arguing that the anhydrous form (Form 2) would convert into the hemihydrate (Form 1). GSK's argument was based, in part, on evidence that Apotex had begun development of its anhydrous product by bringing the hemihydrate into its manufacturing facility, thus providing the seeds that had been shown to be a factor in the conversion. In addition, there was contact with water in the manufacturing of the API, in the formulation, and in the production of the final water-based coating process of the pill, as well as the use of pressure in the last processing step.

A trial was held in February 2003. GSK's assertion that Apotex would infringe the hemihydrate patent was based on GSK and Apotex documents showing that many of Apotex's anhydrate batches had exhibited evidence of conversion: there were batches of anhydrate that converted almost entirely into hemihydrate when stored at 40 °C and 75% humidity within one month. This Apotex experience was bolstered by the results of GSK's testing of Apotex's API and its formulated tablets. Furthermore, Apotex's specification for release of bulk material was based on a visual method of comparing spectral data that could not detect less than 5–8% of infringing hemihydrate in the bulk API.

In its defense, Apotex argued that seeding is "junk science," not widely accepted in the scientific community and that the mechanism of the role of seeds and the conversion is not understood. Moreover, it claimed that the supplier of the bulk API had improved the process to avoid contact with water and was storing the material in improved storage bags, less permeable to water.

In response, GSK argued that Apotex produced the tablets under conditions of normal humidity and sprayed the tablets with an 88% water-based aqueous coating.

The court rejected Apotex's contentions concerning seeding, stating "that there is no scientific basis for believing that seeding occurs…is obviously wrong."

It found that Apotex's anhydrate converts into the hemihydrate (in accord with the earlier publication by SmithKline Beecham scientists) and that it "may continue until it reaches 100 per cent." The court found that Apotex's limit of detection of the hemihydrate in an allegedly anhydrous material was 5–8%, but it did not rely on this finding in determining whether Apotex was infringing the GSK patent.

Nevertheless, the court ultimately ruled in Apotex's favor, reasoning as follows. Claim 1 of GSK's hemihydrate patent recited "Crystalline paroxetine hydrochloride hemihydrates." The court ruled that this claim was valid, but that Apotex's product would not likely infringe the patent because Apotex would not be making it intentionally, and not "in any commercially significant quantity." The court interpreted Claim 1 as limited to only "commercially significant amounts of hemihydrate" and explained that the concentration would have to be in the "high double digits to contribute any commercial value." It further stated that GSK had not established that Apotex would be marketing material with high double-digit concentrations of hemihydrate and that therefore Apotex would not benefit monetarily from the hemihydrate.

The court thus found the patent valid, but also that Apotex would not be infringing it. The case was appealed to the U.S. Federal Circuit Court, which handles all patent appeals in the United States. The Federal Circuit ultimately ruled in favor of Apotex, but for different reasons than those invoked by the district court, and not connected with the crystal form issue. The Federal Circuit found the district court's claim construction to be incorrect, and that Claim 1 properly covered *any* amount of hemihydrate. In a ruling on the second appeal to the Federal Circuit, however, it reasoned that Claim 1 must, therefore, be invalid for inherency because if anhydrate converted into the hemihydrate now, it must have converted into hemihydrate in at least small amounts in the prior art. The U.S. Supreme Court refused to hear the case.

10.4.5   Armodafinil

The example of a litigation involving Cephalon's armodafinil (Nuvigil®) in the United States is particularly informative for demonstrating the nature of a number ex post facto scientific arguments for obviousness, but, of course, every litigation will be characterized by its own set of facts and contentions.

Nuvigil® is the levorotatory enantiomer of modafinil, used for treatment of sleepiness from narcolepsy, sleep apnea, or night shift work and disorders associated with Alzheimer's disease. A synthesis of the molecule is disclosed in U.S. Patent No. 4 927 855. The final step of that synthesis reads as follows:

> The methanol is evaporated off, the evaporation residue is taken up in ether and the product is filtered off and recrystallized from ethanol to give [the compound] with an overall yield of 32%. This product is in the form of white crystals which are soluble in alcohols and acetone and insoluble in water and ether. M.p.(inst.) = 153–154 °C.

U.S. Patent 7 132 570 ("the '570 patent") describes crystal forms of the optical enantiomers of modafinil, including Form 1 as claimed in independent claim 1:

> 1. The levorotatory enantiomer of modafinil in a polymorphic form that produces powder X-ray diffraction spectrum comprising intensity peaks at the interplanar spacings: 8.54, 4.27, 4.02, 3.98 Å.

Form 1 was found to be the most stable of the crystal forms described in the '570 patent, which discloses a number of methods for preparing that form as well as XRPD data for Form I.

In a U.S. patent litigation involving the '570 patent, opponents alleged that Form 1 was obvious.[9] Specifically, opponents argued that a person of ordinary skill would have been motivated to identify the most stable form of armodafinil— Form I for use in a pharmaceutical composition—and also would have had a reasonable expectation of obtaining that form using well-known, merely routine techniques and predictable steps. The patentee argued that the crystal structure and recrystallization would have been unpredictable to the extent that a skilled artisan would not have had a reasonable expectation of success in identifying or obtaining Form I specifically described and characterized in the patent.

The Court agreed with the patentee and concluded that "Form I would not have been obvious because there was no more than a general motivation to find new crystal forms of armodafinil with nothing directed to the unknown Form I itself." Since the existence, structure, and methods of making polymorphs were not predictable, a new form could only be prepared and identified by trial and error experimentation. That is not sufficient to establish legal obviousness. As discussed previously, legally, obviousness requires both motivation to make a new crystal form and a reasonable expectation of success in doing so.

The first step in such an inquiry would be whether or not one of ordinary skill in the art would have had a reasonable expectation that armodafinil was polymorphic. But whether armodafinil was polymorphic was not predictable in 2002. Indeed, the unpredictable nature of polymorphism in general had been discussed in numerous publications at the time:

> Unlike salts, which for the most part can be prophetically claimed based on an understanding of the chemical structure of the compound and its ionization constants, the existence and identity of...polymorphs have defied prediction (Morissette et al. 2004).

Also, quoting Moulton and Zaworotko (2001):

---

[9] The decision has been discussed in essentially the same format and terms in Bernstein and MacAlpine (2017). In that instance detailed citations of the court's findings are given. Here I cite simply the decision of the district court.

...[I]t remains in general impossible to predict the structure of even the simplest crystalline solids from a knowledge of their chemical composition,

and Peterson et al. (2002):

There are no failsafe methods to predict the extent of polymorphism of a given compound.

Even if the intended use of the armodafinil had been as an API, that would not have rendered it any more likely to have been polymorphic, as pharmaceutical compounds are no more polymorphic than other compounds (Bernstein 2011; Cruz-Cabeza et al. 2015). Similarly, the Court held that the fact that molecular structure of the armodafinil molecule was known would not have provided any basis on which polymorphism of the molecule could have been predicted (Desiraju 1997) explaining there were "major obstacles in routinely predicting crystal structures from molecular structure," including that "the crystal structures of many 'simple' organic compounds need not be simple at all" and "chemists seem unable to accurately foresee" how functional groups of molecules will interact to form crystals (Cruz-Cabeza et al. 2015). Thus, a person of ordinary skill in the art would not have had the requisite reasonable expectation that armodafinil was polymorphic.

The Court also disagreed with the defendants' assertion that a skilled artisan would have been motivated to find the most stable polymorphic form of armodafinil, Form I, and would have had a reasonable expectation of success in finding Form I because it would have been easy to obtain. The Court found that "the 'most stable' form of a crystal does not refer to a specific material, but, instead, is a relative term that refers to the lowest energy crystalline form known at a given time." Furthermore, "the motivation of a skilled artisan to find the 'most stable form' of armodafinil would be no different than the motivation to find an effective drug with the lowest toxicology profile, which likewise does not render obvious a specific drug that has the lowest toxicology profile." The Court also held that

considerations [] beyond thermodynamic stability are involved in the calculus that leads to the selection of a polymorph or solid form for use in a pharmaceutical product. Specifically, some drug products employ metastable or amorphous forms of the API because the 'most stable' form has undesirable characteristics.

Moreover, the Court concluded that, "[e]ven if there were a way of predicting that a compound would be polymorphic and what the crystal structures might be, the evidence presented shows that person of skill would not know how to make a specific polymorph or predict its properties." With respect to Form I, the Court held that defendants had presented insufficient evidence to demonstrate either that a skilled artisan would have had reason to select the route that produced Form I or that the prior art would have indicated to the skilled artisan which parameter(s) in that route were critical or likely to prove successful amongst the numerous possible conditions and variables.

The Court also concluded that defendants had not established legal obviousness on the basis that it would have been "obvious to try" to identify and prepare the most stable form of armodafinil. Rather, the Court held that "'[o]bvious to try' is not equivalent to obviousness in every case, particularly where, as here, the prior art provided at most general motivation to conduct trial and error experimentation in a decidedly unpredictable field." The Court held that, "even if the general idea of using crystallization experiments were obvious to try, such unpredictable trial and error experimentation fails to render Form I obvious because the testing required was more than simply routine."

The Court also disagreed with defendants' arguments that, for commercial success and regulatory reasons, a skilled artisan would have been motivated to conduct a polymorph screen of armodafinil and that it would have been "a simple and routine matter for [that artisan] to identify the most stable polymorph" using this technique. First, the Court concluded that, even if the requisite motivation had been present, because polymorphs are unpredictable, to try to make Form I, a skilled artisan would have had to resort to trial and error experimentation using a large number of conditions, and would not have had a defined, finite set of reasonably predictable experiments or variables. "Crystallizing new polymorphs often requires hundreds to thousands of experiments that analyze the effects of various parameters such as temperature, solvent and solvent mixtures, mixing time, cooling rates, stirring rates, and concentrations, as well as methods and processes for precipitation, cooling, evaporation, slurry, and thermo-cycling."

Second, the Court acknowledged what is described in considerable detail elsewhere (Cruz-Cabeza et al. 2015): there is no standard cookbook routine crystal form screen. As with every other aspect of the crystal form landscape, each compound is a new situation. There are no cookbook recipes, no limited number of conditions, for example for the choice of solvent(s), concentrations, cooling, heating, and/or stirring rates, time of crystallization, slow or rapid evaporation, etc. And these are only some of the variables for solution crystallizations. As noted in Chapter 5 a variety of other ways of searching for crystal forms have been, and continue to be developed: sublimation, hot stage studies, crystallization from the melt, capillary crystallizations, crystallization from super critical solvents, vapor diffusion, gel crystallizations, etc. Thus, "there was on [*sic*, no] way to reasonably predict the outcome of any vast number of possible conditions that could have been chosen, as the selection of certain sets of conditions, but not others, could have resulted exclusively in forms other than Form I or mixtures of forms" (Burger et al. 2000).

Accordingly, the Court found that it had not been demonstrated that a person of ordinary skill in the art in 2002 would have had a reasonable expectation of success of obtaining Form I armodafinil using routine techniques and methods and the use of high throughput screening did not make the results of the search for polymorphs any more predictable or obvious. As of this writing the situation has not significantly changed since 2002, which is the date of reference for the Court in the armodafinil case (see, for example, Braem et al. 2014).

10.4.6   Tapentadol hydrochloride

Tapentadol (sold under the brand names Nucynta®, Palexia®, and Tapal®) is a centrally acting opioid analgesic of the benzenoid class. It was originally developed at the German pharmaceutical company Grünenthal in the late 1980s. Marketing in many countries was licensed to Johnson & Johnson until early 2015, when the rights were sold to Depomed. The patent litigation was another Hatch–Waxman case with Grünenthal and Depomed as innovator plaintiffs and Actavis, Roxanne Laboratories, and Alkem Laboratories as the generic defendants. The applicable solid form patent in issue was U.S. Patent 7 994 364 ("the '364 patent").

Regarding the polymorphism issue plaintiffs asserted claims 1, 2, 3, and 25 of the '364 patent. Defendants contended that each of these claims is invalid. The relevant text for our discussion in the '364 patent is as follows:

The invention claimed is:

1. A crystalline Form A of $(-)$-$(1R,2R)$-3-(3-dimethylamino-1-ethyl-2-methylpropyl)-phenol hydrochloride exhibiting at least X-ray lines (2-theta values) in a powder diffraction pattern when measured using Cu K_a radiation at 15.1±0.2, 16.0±0.2, 18.9±0.2, 20.4±0.2, 22.5±0.2, 27.3±0.2, 29.3±0.2 and 30.4±0.2.

2. The crystalline Form A of $(-)$-$(1R,2R)$-3-(3-dimethylamino-1-ethyl-2-methylpropyl)-phenol hydrochloride according to claim 1 exhibiting at least X-ray lines (2-theta values) in a powder diffraction when measured using Cu K_a radiation at 14.5±0.2, 18.2±0.2, 20.4±0.2, 21.7±0.2 and 25.5±0.2.

3. The crystalline Form A of $(-)$-$(1R,2R)$-3-(3-dimethylamino-1-ethyl-2-methylpropyl)-phenol hydrochloride according to claim 1 exhibiting an X-ray pattern (2-theta values) in a powder diffraction when measured using Cu K_a radiation essentially the same as that provided in Fig. 1.

10.4.6.1   *Inherent anticipation*

Defendants contended that the three asserted claims of the '364 patent are invalid as inherently anticipated by an example in a prior art patent. To prove that the '364 patent is inherently anticipated, the defendants bore the burden of proving by clear and convincing evidence that the cited example "necessarily and inevitably" produces the Form A of tapentadol hydrochloride. Defendants argued that polymorphic Form A of tapentadol hydrochloride in the asserted claims is inherently anticipated because "[the e]xample…of the [prior art] describes a procedure for making tapentadol hydrochloride crystals that inevitably results in at least some Form A." On this issue, plaintiffs and defendants each presented testimony of two expert witnesses. It was undisputed that the prior art patent example does not explicitly mention polymorphs and it does not disclose the crystal form of any chemical compound, including tapentadol hydrochloride. Thus the dispute on this issue was whether defendants had met their burden of proving by clear and convincing evidence that the practice of the procedure set forth in [the example in the prior art] "necessarily and inevitably" results in the production of polymorphic Form A of tapentadol hydrochloride. For three reasons the court found that they did not meet that burden.

The invention claimed in the '364 patent is specific for Form A of tapentadol hydrochloride. In contrast, the prior art patent describes a procedure for synthesizing tapentadol hydrochloride without reference to the polymorphic form being synthesized.

On the issue of inherent anticipation, the parties presented evidence of samples of tapentadol hydrochloride that were purportedly synthesized in accordance with the procedure of the prior art patent. In one instance of attempting to reproduce the prior art the defendants performed only one step in a multiple step procedure in the earlier work, rather than reproducing the entire procedure provided in that patent, and they performed that only once.

The second instance involved a procedure carried out at Grünenthal that yielded Form B rather than Form A. There was considerable debate among the experts on the details of that experiment, but the court concluded that "[G]iven the dearth of evidence as to the execution of the process for preparing the [starting material], the Court does not find this sample to be sufficient evidence that performing [the example in the '364 patent] necessarily and inevitably produces Form A."

The third basis for rejecting inherent anticipation includes some important aspects of scientific and legal reasoning in this instance. The direct quotation from the court is particularly illustrative, particularly since it cites some other cases described in this chapter:[10]

> "[I]f the teaching of the prior art can be practiced in a way that yields a product lacking the allegedly inherent property, the prior art in question does not inherently anticipate." *Cephalon, Inc. v. Watson Labs., Inc.*, 939 F. Supp. 2d 456, 465 (D. Del. 2013). Thus it is insufficient for Defendants to demonstrate that performing [the example in the prior art] ***can*** produce Form A; rather, they must show by clear and convincing evidence that performing [the prior art example] ***necessarily and inevitably*** produces Form A—*i.e.*, that [the example in the prior art] cannot be performed without producing Form A. *See Glaxo Inc. v. Novopharm Ltd.*, 52 F.3d 1043, 1047-48 (Fed. Cir. 1995) (affirming the district court's finding of no inherent anticipation where testing evidence demonstrated that the prior art example "could yield crystals of either polymorph").
>
> Defendants have posited that [the prior art example] produced Form A crystals on two occasions...Defendants have failed to put forth sufficient evidence to meet their

---

[10] The cases referred to in the following text have already been described here and demonstrate the evolution of the court's interpretation of inherent anticipation regarding solid forms. These are specific references to the published court findings and decisions.

burden of showing that [the prior art example] necessarily and inevitably produces Form A crystals. In fact, [defendant's] synthesis generated a mixture of Form A crystals and Form B crystals, thus suggesting that the prior art example could yield crystals of either the claimed polymorph or a different polymorph. *See Cephalon*, 939 F. Supp. 2d at 465. Defendants have not clearly convinced this Court that [the prior art example] cannot be performed without producing Form A...

Under the law, even if Defendants have shown that following the procedure of [the prior art example] *can* produce Form A tapentadol hydrochloride, that showing would be insufficient to meet their burden on inherent anticipation. *Cephalon*, 939 F. Supp. 2d at 465. The evidence presented at trial by both parties leaves this Court unconvinced that practicing the procedure of [the prior art example] will *necessarily and inevitably* result in Form A. *See Glaxo*, 52 F.3d at 1047-48; *Cephalon*, 939 F. Supp. 2d at 470 ("[T]he court concludes that the defendants have not met their burden of demonstrating that [the prior art example] necessarily and inevitably results in [the claimed polymorph form] and, therefore, have not proved invalidity by inherent anticipation."). Accordingly, the Court finds that the '364 patent is not inherently anticipated by Example 25 of the '737 patent. Defendants' evidence does not carry their burden on inherent anticipation. (Emphasis in original)

10.4.6.2 Obviousness

Defendants alleged that the asserted claims of the '364 patent are invalid for obviousness under the appropriate provisions of the law.[11] To prove that the '364 patent is obvious, defendants bore the burden of proving by clear and convincing evidence that polymorphic Form A of tapentadol hydrochloride would have been obvious to a POSA in 2004 at the time of the invention.

Defendants contended that Form A of tapentadol hydrochloride would have been obvious in view of the [prior art] patent in combination with common knowledge in the field about polymorph screening such as found in [the] FDA Guidance [DTX-290] and a 1995 article about polymorph screening published by Dr. Stephen Byrn (the "Byrn article") (Byrn et al. 1995), [as presented by one of defendant's experts]. Plaintiffs argue that the claims of the '364 patent were not obvious because polymorphism and the claimed Form A are fundamentally unpredictable [as presented by one of plaintiff's expert witnesses].

1. Obviousness analysis

As discussed above, the obviousness inquiry requires analysis of four factors: (1) the scope and content of the prior art; (2) the level of ordinary skill in the art; (3) the differences between the claimed subject matter and the prior art; and (4) secondary considerations of non-obviousness. *See Graham*, 383 U.S. at 17-18.

a. The scope and content of the prior art

In analyzing the question of obviousness, the Court must first look at the scope and content of the prior art. *See Graham*, 383 U.S. at 17-18. In this case, the Court evaluates the disclosure of the '737 patent as well as the prior art knowledge about crystalline forms and polymorph screening.

---

[11] The subheadings in this section are from the court's decision.

i. The '737 patent

The '737 patent discloses many compounds, one of which is tapentadol hydrochloride. The '737 patent states that tapentadol hydrochloride is made in crystalline form, but it does not discuss the crystal structure. The word "polymorph" does not appear in the '737 patent.

ii. Knowledge about crystal forms

The crystal form landscape for a particular compound can include polymorphs, solvates, hydrates, and polymorphs of solvates and hydrates. [plaintiff's expert witness] testified that about 30–35% of all compounds are polymorphic. However, it is impossible to predict whether a compound will have multiple crystalline forms, how many it will have, and which forms it will have. "[N]o crystal form is predictable," "[T]here is no molecular feature which can be used as a predictor of polymorphism," and "[T]he state of our knowledge and understanding of the phenomenon of polymorphism is still such that we cannot predict with any degree of confidence if a compound will be polymorphic, prescribe how to make possible (unknown) polymorphs, or predict what their properties might be." The court then quoted from some of the literature cited by plaintiff's expert: "Unlike salts, which for the most part can be prophetically claimed based on an understanding of the chemical structure of the compound and its ionization constants, the existence and identity of hydrates, solvates, co-crystals and polymorphs have defied prediction" (Morissette et al. 2004). "One of the continuing scandals in the physical sciences is that it remains in general impossible to predict the structure of even the simplest crystalline solids from a knowledge of their chemical composition" (Maddox 1988).

iii. Polymorph screening

As early as 1987, the FDA had issued guidelines indicating that ["appropriate analytical procedures should be used to determine whether (or not) polymorphism occurs." Such appropriate measures include polymorph screening. ("Interest in the subject of pharmaceutical solids stems in part from the [FDA]'s drug substance guideline that states 'appropriate' analytical procedures should be used to detect polymorphic, hydrated or amorphous forms of the drug substance.")]

[One of defendant's experts] testified that screening for polymorphs is a routine part of pharmaceutical companies' drug development. [One of plaintiff's witnesses] testified that, regardless of regulatory guidelines, it was important to know if tapentadol hydrochloride was polymorphic "not only [for] the regulatory bodies, it was important for any further developments that, to have knowledge of solid form characteristics." Once tapentadol hydrochloride entered clinical development, Grünenthal considered it "important also to take consideration of the so-called solid phase of the compound intended for oral use," which included "investigat[ing] the hydrochloride salt if different polymorphs may exist." By 2000, Grünenthal generally understood that pharmaceutical companies "take consideration of polymorphism when they are developing pharmaceutical compounds."

A POSA would have understood as of June 2004 how to conduct a polymorph screen and there were publications that provided guidance on how to do such a screen. [One of plaintiff's fact witnesses] understood at that time that polymorph screening was a common technique that involved using different common solvents and temperatures to test for polymorphs. ("[P]olymorph investigation is to apply various parameters on the crystallization like temperature ranges, like various solvents to extend as broad as possible range of investigations in order to understand and characterize the compounds or the compound under consideration.") [One of plaintiff's experts] has written that "[c]rystallization from solution is one of the first laboratory skills that chemists acquire, and applying variations to the conventional methods has been the traditional strategy in the search for polymorphs." and that "[o]ne traditional strategy for screening a compound for polymorphic behavior involves the trial of a variety of solvents and solvent mixtures."

Defendants presented the Byrn et al. (1995) article as evidence of the knowledge in the art about common solvents and techniques that should be used in polymorph screens. The Byrn article explains steps to detect polymorphism. The Byrn article also states that "[t]he first step in the polymorphs decision tree is to crystallize the substance from a number of different solvents in order to attempt to answer the question: Are polymorphs possible?" The Byrn article specifically lists certain solvents to use in the recrystallization experiments, including "water, methanol, ethanol, propanol, isopropanol, acetone, acetonitrile, ethyl acetate, hexane, and mixtures if appropriate." [*Id.*] Figure 1 of the article is a "Flowchart/decision tree for polymorphs." The first question in the decision tree is "Polymorphs discovered?" [*Id.*] To answer this question, the paper states "Different recrystallizing solvents (different polarity)—vary temperature, concentration, agitation, pH." [*Id.*] Figure 1 also identifies techniques (including XRPD) that can be used to "Test for polymorphs." [*Id.*]

b. The level of ordinary skill in the art

The parties have proposed substantially similar definitions of a POSA for the '364 patent. All parties agree that a POSA would be a chemist, chemical engineer, or someone in a similar field who has experience in development, preparation, and characterization of polymorphic forms of compounds.

c. The differences between the claimed subject matter and the prior art

The claimed subject matter is a polymorph form of tapentadol hydrochloride, which is one of the many compounds disclosed in the '737 patent. There is nothing in particular in the '737 patent that would have motivated a POSA studying crystal structure to select tapentadol hydrochloride from amongst the many compounds disclosed in the '737 patent.

Had a POSA chosen to study the crystal form of tapentadol hydrochloride (from amongst the compounds in the '737 patent), its crystal structure could have taken many forms. Had a POSA specifically decided to screen for polymorphs of tapentadol hydrochloride, a POSA would not have been able to look at the structure of tapentadol hydrochloride and predict whether it is polymorphic.

Direct testimony of one of plaintiff's experts: "Q. Could the polymorphism of tapentadol have been predicted from the chemical structure? A. Not to my understanding it could have not been predicted at that point in time."

If it had been known that tapentadol hydrochloride was polymorphic, a POSA could not have predicted how many crystal forms it would have. Finally, if it had been known that tapentadol had two polymorph forms, a POSA could not have predicted the structure, properties, or relative stability of any of the forms.

Defendants contend that it was straightforward for Grünenthal to ask SSCI (a company founded by the same Dr. Byrn who authored Byrn et al. (1995)) to "produce as many new forms of [tapentadol hydrochloride] as possible." Defendants presented evidence suggesting that SSCI followed known procedures (including using eight of the nine solvents listed in the Byrn article and different cooling and evaporation conditions) to conduct the polymorph screen and arrive at the conclusion that tapentadol hydrochloride had two polymorphic forms—Form A and Form B. Defendants argued that "a routine polymorph screen, like the one described in the Byrn article, would have revealed Form A of tapentadol hydrochloride" as one of the two polymorphs.

Plaintiffs, however, presented evidence that the Byrn article merely summarized the initial question as whether a compound was polymorphic or not, condensing the many variables listed in other prior art references into a few lines. Plaintiffs also presented evidence that, while "solution crystallization" is a known technique that may be used to explore the polymorph landscape, it is a technique that includes a large variety of conditions that could be appropriate for a particular polymorph screen, including solvent, temperature, stirring, cooling rate, seeding, and whether an antisolvent is used. The variability of these conditions produces a huge number of possible choices that must be made during the course of a polymorph screen. Ultimately, there is no way to know how a screen will proceed before beginning to experiment and a POSA has no way of knowing whether more than one form will exist or what the physical properties of the forms are.

In sum, Plaintiffs demonstrated that, while each individual technique performed during a polymorph screen may be routine, polymorph screening consists of an unpredictable application of those routine techniques. Plaintiffs further showed that the results of the polymorph screen—that is, the determination of the structure and properties of the polymorph forms of tapentadol hydrochloride—would have been impossible to predict as "predictability is a touchstone of obviousness," *Depuy Spine, Inc. v. Medtronic Sofamor Danek, Inc.*, 567 F.3d 1314, 1326 (Fed. Cir. 2009), Defendants have failed to meet their burden of demonstrating that the specific polymorph Form A of tapentadol hydrochloride was obvious. *See Eisai Co. Ltd. v. Dr. Reddy's Labs., Ltd.*, 533 F.3d 1353, 1358 (Fed. Cir. 2008) ("To the extent an art is unpredictable, as the chemical arts often are, *KSR*'s focus on…'identified, predictable solutions' may present a difficult hurdle because potential solutions are less likely to be genuinely predictable.").

2. Obviousness in polymorph patent cases
The court then summarized the current state of obviousness in polymorph patent cases:

In making this obviousness challenge to the '364 patent, Defendants are the latest in a line of defendants who have argued that newly discovered polymorphs should not be patentable under the law of obviousness. *See, e.g., Cephalon*, 939 F. Supp. 2d at 490; *Merck & Cie v. Watson Labs., Inc.*, 125 F. Supp. 3d 503 (D. Del. 2015), *rev'd on other grounds*, 822 F.3d 1347 (Fed. Cir. 2016); *Takeda Pharm. Co. v. Handa Pharm., LLC*, Case No. C-11-00840, 2013 U.S. Dist. LEXIS 187604, at *232-33 (N.D. Cal. Oct. 17, 2013). In each case the defendants contended that a POSA would have been motivated to discover the claimed polymorph, and in each case the court rejected this argument. *Id.* In *Cephalon*, the court specifically noted that "for a patent challenger to establish obviousness, it is insufficient to allege a general motivation to discover an undefined solution that could take many possible forms." 939 F. Supp. 2d at 500; *see also Innogenetics, N.V. v. Abbott Labs.*, 512 F.3d 1363, 1373-74 (Fed. Cir. 2008) ("[K]nowledge of a problem and motivation to solve it are entirely different from motivation to combine particular references to reach the particular claimed method.") In *Takeda*, the court stated that "here the invention involved the selection from a broad range of available but unpredictable techniques to try to create a previously non-existent crystalline compound whose structure could not be predicted." 2013 U.S. Dist. LEXIS 187604, at *233.

Defendants attempt to distinguish the present case by arguing that "[e]ach court that has declined to find polymorph screening renders polymorph patent claims invalid has lacked the benefit of the Byrn et al. (1995) article, or any article that evidences how the patentee followed the specific and express teachings of the prior art...." However, the Byrn article was published in 1995, long before many of the cases discussed above. Moreover, while it is undisputed in this case that methods for conducting polymorph screening were known in the art, the relevant question is whether such knowledge renders the patent obvious in what courts have described as the "decidedly unpredictable field" of polymorphism. *See Cephalon*, 939 F. Supp. 2d at 501–02.

Defendants attempt to discredit [plaintiff's expert's] testimony by noting his opinion that "no polymorph patent can ever be obvious [b]ecause [polymorphism] is not predictable." While this Court does not speak on the hypothetical question of whether every polymorph patent is nonobvious, the Court notes that Defendants' argument appears to imply the opposite extreme—that every polymorph patent is obvious. This Court does not make such a finding, which would contravene the findings of other district courts and the Federal Circuit. Accordingly, the Court finds that the asserted claims of the '364 patent are not obvious.

10.4.7   Habit, not form: aspartame

Aspartame, the artificial sweetener marketed as NutraSweet®, is a dipeptide, which was discovered in 1965 by accident to be 100–200 times sweeter than sucrose. The discovery was originally made at G.D. Searle, which was later acquired by Monsanto.

In the original conventional manufacturing process, the crystallization from aqueous solutions tended to produce needles with diameter of 10 μm or less. The crystals were very fine, with large specific volume. These characteristics led to many problems in the filtration and drying processes, the formation of scale on reactor surfaces, and the high dustability and hygroscopicity of the final product, which made it difficult to handle and unsuitable for use as a direct (i.e., table top) sweetener (Ajinomoto Co. Inc. 1983; Kishimoto et al. 1989).

Five crystal modifications of Aspartame are known (Ajinomoto Co. Inc. 1983; Hatada et al. 1985; Nagashima et al. 1987; Kishimoto and Naruse 1988; Tsuboi et al. 1991; Furedi-Milhofer et al. 1999). A number of analytical methods have been used recently to characterize three of them: two hemihydrate polymorphs and a dihemihydrate (Leung et al. 1998a, 1998b; Zell et al. 1999). The crystal structure of one of the hemihydrate forms, known as Form I, was published earlier (Hatada et al. 1985). More recently, synchrotron radiation was used to determine the structure of the "low humidity" form (denoted Form Ib by the authors), with an asymmetric unit comprised of three aspartame molecules and two water molecules (Meguro et al. 2000).

Many attempts were made to improve the qualities of the crystals obtained from the conventional manufacturing process. The Japanese Ajinomoto Co. had a license to make aspartame and, as a result of "intensive investigations to improve the workability {of the crystallization} step in the production...," discovered that cooling aqueous solutions of aspartame without stirring led to "bundle-like" crystal aggregates of crystals (Kishimoto and Naruse 1987). These crystals had significantly improved handling characteristics compared to the conventional crystals. This process was subsequently refined and led to an Ajinomoto European patent, granted in 1985.

In April 1992, the Opposition Division of the European Patent Office concluded that the bundle-like crystals were the same as the conventional crystals. In response to that decision, Ajinomoto presented a detailed study of the correlation of the differences between the two (Tarling 1992). The two types of crystals were different habits rather than polymorphs. The bundle-like crystals, produced without stirring, had a higher degree of crystal perfection than the conventional crystals, produced with agitated stirring. The Tarling report identified some "significant and reproducible differences" in a number of properties determined with a variety of analytical techniques. The difference formed the basis of the Ajinomoto response to the Opposition, and included the following:

- specific volumes: the bundle-like crystals have smaller volumes and hence higher bulk density;
- rate of dissolution: because of their smaller surface to volume ratio the larger bundle-like crystals would be expected to dissolve more slowly than the conventional crystals; however, the opposite is true;
- water content: bundle-like crystals contain significantly less water than conventional crystals;
- density: a small but significant difference in the density was determined, with the conventional crystals having a lower density, consistent with the higher strain noted for those crystals;
- solid-state NMR spectra: consistent shifts in position and changes in the widths of peaks could be detected between the two modifications, indicating slightly different molecular environments;
- polarizing light microscopy: bundle-like crystals show very sharp extinction indicating a high degree of structural perfection, which is not exhibited by conventional crystals;

- XRPD: the powder patterns for the bundle-like and conventional crystals are shown in Figure 10.3. They are quite similar, as expected for materials that have essentially the same crystal structure. However, there are also some differences. There are some slight differences in the peak positions, which were attributed to difference in water content and degree of internal crystal strain. There are some more obvious differences in relative intensities, including the position of the strongest peak. There are clear differences in the peak widths, which also reflects differences in the degree of crystallinity;

- scanning electron microscopy, single crystal X-ray diffraction (Hatada et al. 1985; Kishimoto et al. 1989), and crystal habit modeling were used to determine the preferred orientation and to simulate the X-ray powder patterns of both forms to show a good comparison with the experimental data.

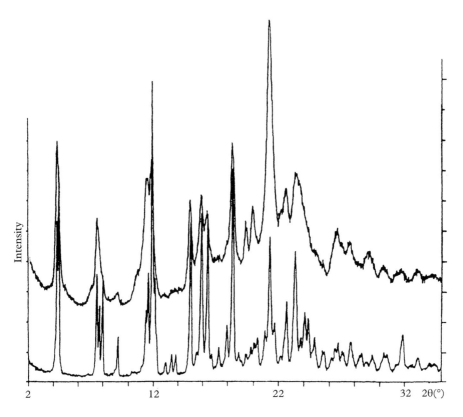

Figure 10.3 *Comparison of the measured X-ray diffraction patterns of aspartame. Upper, conventional crystals; lower, bundle-like crystals (from Tarling 2001, with permission).*

On May 27, 1997 the Boards of Appeal of the European Patent Office set aside the 1992 opposition to the Ajinomoto patent (European Patent Office 1997), and a new European patent specification was issued (Ajinomoto Co. Inc. 1998).

10.5 Concluding remarks

I have presented here but a sampling of a number of litigations that involved controversies over the nature, identity, quantity, and uniqueness of crystalline modifications. As in other aspects of the law each case has its own character and idiosyncrasies. As evidenced in numerous examples throughout this book, the same variability exists in the nature of the polymorphic behavior of a substance. That behavior differs because the substances are different, and to be properly understood each substance has to be investigated on its own. Nevertheless, the polymorphism of molecular systems can be understood and utilized on fundamental principles. In light of Buerger and Bloom's comments quoted at the opening of Chapter 1, I hope that this book serves to remove some of the mystery that they suggested surrounds the subject.

References

Aakeröy, C. B., Nieuwenhuyzen, M., and Price, S. L. (1998). Three polymorphs of 2-amino-5-nitropyrimidine: experimental structures and theoretical predictions. *J. Am. Chem. Soc.*, **120**, 8986–93. [239]

Aaltonen, J., Alleso, M., Mirza, S., Koradia, V., Gordon, K., and Rantanen, J. (2009). Solid form screening – a review. *Eur. J. Pharm. Biopharm.*, **71**, 23–37. [92, 96, 98f, 105, 353]

Abbott Laboratories v. Geneva Pharmaceuticals, Inc., Abbott Laboratories v. Novopharm Limited, Abbott Laboratories v. Invamed, Inc. (1998). Nos 96 C 3331, 96 C 5868, 97 C 7587; 1998 WL 566884 (N. D. Ill.), Sept. 1, 1998. [437]

Acree, W. and Chickos, J. S. (2016). Phase transitions, enthalpy measurements of organic and organometallic compounds. Sublimation, vaporization and fusion enthalpies of 1880 to 2015. Part 1. C_1–C_{10}. *J. Phys. Chem. Ref. Data*, **45**, 1–565. [218, 228]

Acree, W. and Chickos, J. S. (2017). Phase transitions, enthalpy measurements of organic and organometallic compounds. Sublimation, vaporization and fusion enthalpies of 1880 to 2015. Part 2. C_{11}–C_{192}. *J. Phys. Chem. Ref. Data*, **46**, 1–532. [228]

Addadi, L., Berkovitch-Yellin, Z., Weissbuch, I., van Mil, J., Shimon, L. J. W., Lahav, M., and Leiserowitz, L. (1985). Growth and dissolution of organic crystals with "tailor-made" inhibitors—implication in stereochemistry and materials science. *Angew. Chem. Int. Ed. Engl.*, **24**, 466–85. [61]

Adyeeye, C. M., Rowley, J., Madu, D., Javadi, M., and Sabnis, S. S. (1995). Evaluation of crystallinity and drug-release stability of directly compressed theophylline hydrophilic matrix tablets stored under varied moisture conditions. *Int. J. Pharm.*, **116**, 65–75. [9]

Aerts, J. (1996). Polymer crystal silverware: a fast method for the prediction of polymer crystal structures. *Polym. Bull.*, **36**, 645–52. [238]

Agatonovic-Kustrin, S., Wu, V., Rades, T., Saville, D., and Tucker, I. G. (1999). Powder diffractometric assay of two polymorphic forms of ranitidine hydrochloride. *Int. J. Pharmacol.*, **184**, 107–14. [158f, 169]

Agatonovic-Kustrin, S., Wu, V., Rades, T., Saville, D., and Tucker, I. G. (2000). Ranitidine hydrochloride X-ray assay using a neural network. *J. Pharm. Biomed. Sci.*, **22**, 985–92. [169]

Agrawal, J. P. (1998). Recent trends in high energy materials. *Prog. Energy Combust. Sci.*, **24**, 1–30. [399]

Agrawal, J. P. (2005). Some new high energy materials and their formulations for specialized applications. *Propellants, Explos., Pyrotech.*, **30**, 316–28. [399]

Agrawal, S. G. and Paterson, A. H. J. (2015). Secondary nucleation: mechanisms and models. *Chem. Eng. Commun.*, **202**, 698–706. [81]

Aguiar, A. J., Krc, J., Kinkel, A. W., and Samyn, J. C. (1967). Effect of polymorphism on the absorption of chloramphenicol from chloramphenicol palmitate. *J. Pharm. Sci.*, **56**, 847–53. [349, 349f, 350]

Ahn, S., Guo, F., Kariuki, B. M., and Harris, K. D. M. (2006). Abundant polymorphism in a system with multiple hydrogen-bonding opportunities: oxalyl dihydrazide. *J. Am. Chem. Soc.*, **128**, 8441–52. [19]

Ahr, G., Voith, B., and Kuhlmann, J. (2000). Guidances related to bioavailability and bio-equivalence: European industry perspective. *Eur. J. Drug Metab. Pharmacokinet.*, **25**, 25–7. [346]

Aitipamula, S., Chow, P. S., and Tan, R. B. H. (2014). Polymorphism in cocrystals: a review and assessment of its significance. *CrystEngComm*, **16**, 3451–65. [19, 360]

Ajinomoto Co. Inc. (1983). Process for crystallizing alpha-L-aspartyl-L-phenylalanine-methyl ester. European Patent Application 0 091 787 A1. Issued as European Patent Specification 0 091 787 B1 (04.09.85). [453]

Ajinomoto Co. Inc. (1998). Process for crystallizing alpha-L-aspartyl-L-phenylalanine-methyl ester. New European Patent Specification EP 0 091 787 B2. [456]

Akopyan, Z. A., Kitaigorodskii, A. I., and Struchkov, Yu. T. (1965). Steric hindrance and molecular conformation. XII. The crystal and molecular structure of 1,8-dinitronaph-thalene. *Zh. Strukt. Khim.*, **6**, 729–44. [427]

Al-Dulami, S., Aina, A., and Burley, J. (2010). Rapid polymorph screening on milligram quantities of pharmaceutical material using phonon-mode Raman spectroscopy. *CrystEngComm*, **12**, 1038–40. [359]

Aldoshin, S. M., Kozina, O. A., Gutsev, G. L., Atovmyan, E. G., Atovmyan, L. O., and Nedvetskii, V. S. (1988). Cocrystallization of two isomers of N'-(5-nitrofurfurylidene) benzhydrazide and features of the molecular, crystal, and electronic structure of their crystal hydrate. *Izv. Akad. Nauk SSSR, Ser. Khim.*, **10**, 2301–2308. [297]

Algra, R. E., Graswinckel, W. S., van Enckevort, W. J. P., and Vlieg, E. (2005). Alizarin crys-tals: an extreme case of solvent induced morphology change. *J. Cryst. Growth*, **285**, 168–77. [62]

Alhalaweh, A., Alzghoul, A., Kaialy, W., Mahlin, D., and Bergstron, C. A. S. (2014). Computational predictions of glass-forming ability and crystallization tendency of drug molecules. *Mol. Pharm.*, **11**, 3123–32. [370]

Ali, A. A. and Farouk, A. (1981). Comparative studies on the bioavailability of ampicillin anhydrate and trihydrate. *Int. J. Pharm.*, **9**, 239–43. [348]

Allemand, P. M., Fite, C., Srdanov, G., Keder, N., Wudl, F., and Canfield, P. (1991). On the complexities of short-range ferromagnetic exchange in a nitronyl nitroxide. *Synth. Met.*, **41–43**, 3291–5. [288]

Allen, F. A., Davies, J. E., Galloy, J. J., Johnson, O., Kennard, O., Macrae, C. F., Mitchell, E. M., Mitchell, G. F., Smith, J. M., and Watson, D. G. (1991). The development of Versions 3 and 4 of the Cambridge Structural Database system. *J. Chem. Inf. Comput. Sci.*, **31**, 187–204. [26]

Allen, F. H. and Kennard, O. (1993). 3D search and research using the Cambridge Structural Database. *Chem. Des. Autom. News*, **8**, 31–7. [26, 239]

Allen, F. H., Harris, S. E., and Taylor, R. (1996). Comparison of conformer distributions in the crystalline state with conformational energies calculated by *ab initio* techniques. *J. Comput. Aided Mol. Des.*, **10**, 247–54. [219]

Allen, F. H., Kennard, O., and Watson, D. G. (1994). Crystallographic databases: search and retrieval of information from the Cambridge Structural Database. In *Structure Correlation* (ed. H.-B. Bürgi and J. D. Dunitz), pp. 71–110. VCH, Weinheim. [16]

Allen, F. H., Kennard, O., Watson, D. G., Brammer, L., Orpen, A. G., and Taylor, R. (1987). Tables of bond lengths determined by X-ray and neutron diffraction. Part 1. Bond lengths in organic compounds. *J. Chem. Soc., Perkin Trans. 2*, S1–S19. [215]

Allesø, F. M., van den Berg, F., Cornett, C., Jørgenson, F. S., Halling-Sørenson, B., Lopez de Diego, H., Hovgaard, L., Aaltonen, J., and Rantanen, J. (2008). Solvent diversity in polymorph screening. *J. Pharm. Sci.*, **97**, 2145–59. [354]

Almarsson, Ö. and Zaworotko, M. (2004). Crystal engineering of pharmaceutical phases. Do pharmaceutical co-crystals represent a new path to improved medicines? *Chem. Commun.*, 1889–96. [5, 360]

Almeida, M. and Henriques, R. T. (1997). Perylene based conductors. In *Handbook of Organic Conductive Molecules and Polymers*, Vol. 1, pp. 87–149. John Wiley & Sons, Chichester. [276]

Alzghoul, A., Alhalaweh, A., Nahlin, D., and Begstrom, C. A. S. (2014). Experimental and computational prediction of glass transition temperature of drugs. *J. Chem. Inf. Model.*, **54**, 3396–403. [371]

Amici, G. B. (1844). Note sur un appareil de polarisation. *Ann. Chim. Phys., Ser. 3*, **12**, 114–20. [32]

Amorós, J. L. (1959). Notas sobre la historia de la cristalografia I. La controversia Haüy–Mitscherlich. *Bol. Real Soc. Espan. Hist. Nat.*, **57**, 5–30. [31]

Amorós, J. L. (1978). *Le gran aventura del cristal. Naturaleza y evolucion de la ciencia de los cristales*, pp. 205–9. Editorial de la Universidad Complutense, Madrid. [31]

Amundsen, S. J. (2013). Al U.S. Court Appeals Fed. Circuit Case 13–1360 Doc. 56 No. 2013–1360 -1361 -1364 -1365 -1366 -1367-1368 -1369 -1370 -1371. [91]

An, J.-H. and Kim, W.-S. (2013). Antisolvent crystallization using ionic liquids as solvent and antisolvent for polymorphic design of active pharmaceutical ingredient. *Cryst. Growth Des.*, **13**, 31–9. [357]

Andersgaard, H., Finholt, P., Gjermundsen, R., and Hoyland, T. (1974). Rate studies on dissolution and enzymatic hydrolysis of chloramphenicol palmitate. *Acta Pharm. Suec.*, **11**, 239–48. [350, 350f]

Anderton, C. (2004). A valuable technique for polymorph screening. *Eur. Pharm. Rev.*, **9**, 68–74. [102, 103f]

Andreev, G. A. and Hartmanová, M. (1989). Flotation method of precise density measurements. *Phys. Stat. Solidi. A: Appl. Res.*, **116**, 457–68. [207]

Andreev, Y. G., Lightfoot, P., and Bruce, P. G. (1997). A general Monte Carlo approach to structure solution from powder-diffraction data: application to poly(ethylene oxide)$_3$:LiN(SO$_2$CF$_3$)$_2$. *J. Appl. Crystallogr.*, **30**, 294–305. [155]

Andronis, V. and Zografi, G. (2000). Crystal nucleation and growth of indomethacin polymorphs from the amorphous state. *J. Non-Cryst. Solids*, **271**, 236–48. [368]

Anex, B. G. (1966). Optical properties of highly absorbing crystals. *Mol. Cryst.*, **1**, 1–36. [323]

Anex, B. G. and Fratini, A. V. (1964). Polarized single crystal reflection spectra of auramine perchlorate. *J. Mol. Spectrosc.*, **14**, 1–26. [323, 326]

Anex, B. G. and Simpson, W. T. (1960). Metallic reflection from molecular crystals. *Rev. Mod. Phys.*, **32**, 466–76. [323, 328]

Anthony, S. P. and Draper, S. M. (2010). Nano/microstructure fabrication of functional organic material: polymorphic structure and tunable luminescence. *J. Phys. Chem. C*, **114**, 11708–16. [318]

Anthony, S. P., Raghavalah, P., and Radhakrishnan, T. P. (2003). Extreme molecular orientations in a dimorphic system: polar/centric and polar/polar co-growth crystal architectures. *Cryst. Growth Des.*, **3**, 631–4. [303]

Antipin, M. Y., Timofeeva, T. V., Clark, R. D., Nesterov, V. N., Dolgushin, F. M., Wu, J., and Leyderman, A. (2001). Crystal structures and molecular mechanics calculation of nonlinear optical compounds: 2-cyclooctylamino-5-nitropyridine (COANP) and 2-adamantylamino-5-nitropyridine (AANP). New polymorphic modifcation of AANP and electrooptic effects. *J. Mater. Chem.*, **11**, 351–8. [303]

Anwar, J. and Zahn, D. (2011). Uncovering molecular processes in crystal nucleation and growth using molecular simulation. *Angew. Chem. Int. Ed. Engl.*, **50**, 1996–2013. [77]

Anwar, J. and Zahn, D., (2017). Polymorphic phase transitions: macroscopic theory and molecular simulation. *Adv. Drug. Deliv. Rev.*, **117**, 47–70. [368]

Apperley, D. C., Fletton, R. A., Harris, R. K., Lancaster, R. W., Tavener, S., and Threlfall, T. L. (1999). Sulfathiazole polymorphism studied by magic–angle spinning NMR. *J. Pharm. Sci.*, **88**, 1275–80. [189f]

Arnautova, E. A., Zakharova, M. V., Pivinia, T. S., Smolenskii, E. A., Sukhachev, D. V., and Shcherbukhin, V. V. (1996). Methods for calculating the enthalpies of sublimation of organic molecular crystals. *Russ. Chem. Bull.*, **45**, 2723–32. [228]

Arod, F., Pattison, P., Schenk, K. J., and Chapuis, G. (2007). Polymorphism in *N*-salicylideneaniline reconsidered. *Cryst. Growth Des.*, **7**, 1679–85. [310]

Artini, E. (1915). Sulla forma cristallina del trinitrotuluolo. *Rend. Fis. Acc. Lincei*, **21**, 274–9. [417]

Ashida, M., Uyeda, N., and Suito, E. (1971). Thermal transformation of vacuum condensed thin films of copper phthalocyanines. *J. Cryst. Growth*, **8**, 45–56. [391]

Ashizawa, K. (1989). Polymorphism and crystal structure of 2*R*,4*S*,6-fluoro-2-methyl-spiro[chroman-4,4′-imidazoline]-2′,5-dione (M79175). *J. Pharm. Sci.*, **78**, 256–60. [174]

Ashwell, G. J., Bahra, G. S., Brown, C. R., Hamilton, D. G., Kennard, C. H. L., and Lynch, D. E. (1996). 2,4-Bis[4-(*N*,*N*-dibutylamino)phenyl]squaraine: X-ray crystal structure of a centrosymmetric dye and the second-order non-linear optical properties of its non-centrosymmetric Langmuir–Blodgett films. *J. Mater. Chem.*, **6**, 23–6. [131, 294]

Ashwell, G. J., Wong, G. M. S., Bucknall, D. G., Bahra, G. S., and Brown, C. R. (1997). Neutron reflection from 2,4-bis(4-(*N*-methyl-*N*-octylamino)phenyl)squaraine at the air–water interface and the linear and nonlinear optical properties of its Langmuir–Blodgett and spin-coated films. *Langmuir*, **13**, 1629–33. [293]

Asiri, A. M., Heller, H. G., Hughes, D. S., Hursthouse, M. B., Kendrick, J., Leusen, F. J. J., and Montis, R. (2014). A mechanophysical phase transition provides a dramatic example of color polymorphism: the tribochromism of a substituted tri(methylene)-tetrahydrofurna-2-one. *Chem. Centr. J.*, **8**, 70. [317]

Assour, J. M. (1965). On the polymorphic modifications of phthalocyanines. *J. Phys. Chem.*, **69**, 2295–9. [389]

Atkinson, M. B. J., Bwambok, D. K., Chen, J., Chopade, P. D., Thuo, M. M., Mace, C. R., Mirica, K. A., Kumar, A. A., Meyerson, A. S., and Whitesides, G. M. (2013). Using magnetic levitation to separate mixtures of crystal polymorphs. *Angew. Chem. Int. Ed.*, **52**, 10208–11. [292, 352]

Augelli-Szafran, C. E., Barvian, M. R., Bigge, C. F., Glase, S. A., Hachiya, S., Kiely, J. S., Kimurfa, T., Lai, Y., Sajjab, A. T., Suto, M. J., Walker, L. C., Yasunaga, T., and Zhuang, N. (2005). Method of inhibiting amyloid protein aggregation and imaging amyloid deposits for the treatment of Alzheimer's disease. U.S. Patent 6 972 287. [245]

Authier, A. (2013). *Early Days of X-ray Crystallography*. Oxford University Press, Oxford. [31]

Azaroff, L. V. and Burger, M. J. (1958). *The Powder Method in X-ray Crystallography*. McGraw-Hill Book Co., New York. [156]

Bäbler, F. (1978). Stable gamma modification of an isoindolinone pigment—coloring PVC, lacquers, cellulose ether(s) and ester(s) etc., in greenish-yellow shades. Ciba-Geigy AG Patent DE 2 804 062; GB 1 568 198. [382t]

Bäbler, F. (1983). Perylene tetracarboxylic bisdimethylphenylimide—in gamma modification is useful for pigmenting plastics, paint, lacquer and ink. Ciba–Geigy AG EU Patent 0023191 B1. [381t, 386]

Babu, N. J., Cherukuvada, S., Thakuria, R., and Nangia, A. (2010). Conformational and synthon polymorphism in furosemide (Lasix). *Cryst. Growth Des.*, **10**, 1979–89. [6]

Baczynski, W. L. and von Niementowski, S. (1919). Structure of hydroxyquinacridone. *Chem. Ber.*, **52B**, 461–84. [384]

Badri, Z., Bouzkova, K., Foroutan-Nejad, C., and Marek, R. (2014). Origin of the thermodynamic stability of the polymorph IV of crystalline barbituric acid from solid-state NMR and electron density analyses. *Cryst. Growth Des.*, **14**, 2763–72. [193]

Bag, P. P. and Reddy, C. M. (2012). Screening and selective preparation of polymorphs by fast evaporation method: a case study of aspirin, anthranilic acid and niflumic acid. *Cryst. Growth Des.*, **12**, 2740–3. [358]

Bai, D.-R. and Wang, S. (2006). Organoplatinum polymorphs with varying molecular conformation, intermolecular interaction and luminescence. *Organometallics*, **25**, 1517–24. [338]

Baias, M., Dumez, J. N., Svenson, P. H., Schantz, S., Day, G. M., and Emsley, L. (2013). De novo determination of the crystal structure of a large drug molecule by crystal structure prediction-based powder NMR crystallography. *J. Am. Chem. Soc.*, **135**, 17501–7. [191]

Bailey, M. (1955). The crystal structure of indanthrone. *Acta Crystallogr.*, **8**, 182–5. [383t, 393]

Bailey, M. and Brown, C. J. (1967). The crystal structure of terephthalic acid. *Acta Crystallogr.*, **22**, 387–91. [65f, 159, 161t]

Bailey, M. and Brown, C. J. (1984). The crystal structure of terephthalic acid: errata. *Acta Crystallogr. C*, **40**, 1762. [65f, 159, 161t]

Bajpai, A., Scott, H. S., Pham, T., Chen, K.-J., Space, B., Lusi, M., Perry, M. L., and Zaworotko, M. J. (2016). Towards an understanding of the propensity for crystalline hydrate formation by molecular compounds. *IUCr J.*, **3**, 430–9. [355]

Ballard, B. E. and Biles, J. A. (1964). Effect of crystallizing solvent absorption rates of steroid implants. *Steroids*, **4**, 273–8. [348]

Ballester, L., Gil, A. M., Gutierrez, A., Perpinan, M. F., Azcondo, M. T., Sanchez, A. E., Amador, U., Campo, J., and Palacio, F. (1997). Polymorphism in [Cu(cyclam)(TCNQ)$_2$] (TCNQ) stacked systems (cyclam = 1,4,8,11-tetraazacyclotetradecane, TCNQ = 7,7,8,8-tetracyanoquinodimethane). *Inorg. Chem.*, **36**, 5291–8. [289]

Bally, T., Haselbach, E., Lanyiova, S., Mardcher, F., and Rossi, M. (1976). Electronic structure and physico-chemical properties of azo-compounds. Part XIX. Concerning the conformation of isolated benzylideneaniline. *Helv. Chim. Acta*, **59**, 486–98. [325]

Banister, A. J., Bricklebank, N., Clegg, W., Elsegood, M. R. J., Gregory, C. I., Lavender, I., Rawson, J. M., and Tanner, B. K. (1995). The first solid-state paramagnetic 1,2,3,5-dithidiazolyl radical; X-ray crystal structure of [p-NCC$_6$F$_4$NSSN]. *J. Chem. Soc., Chem. Commun.*, 679–80. [287, 291]

Banister, A. J., Bricklebank, N., Lavender, I., Rawson, J. M., Gregory, C. I., Tanner, B. K., Clegg, W., Elsegood, M. R. J., and Palacio, F. (1996). Spontaneous magnetization in a sulfur–nitrogen radical at 36 K. *Angew. Chem. Int. Ed. Engl.*, **35**, 2533–5. [287, 291]

Bannigan, P., Zeglinski, J., Lusi, M., O'Brien, J., and Hudson, S. P. (2016). Investigation into the solid and solution properties of known and novel polymorphs of the antimicrobial molecule clofazimine. *Cryst. Growth Des.*, **16**, 7240–50. [137]

Bar, I. and Bernstein, J. (1985). Conformational polymorphism. 6. The crystal and molecular structures of Form II, Form III and Form V of *4-amino-N-2-pyridinylbenze-nesulfonamide* (sulfapyridine). *J. Pharm. Sci.*, **74**, 255–63. [11, 366]

Bar, I. and Bernstein, J. (1987). Modification of crystal packing and molecular conformation via systematic substitution. *Tetrahedron*, **43**, 1299–305. [361]

Bardwell, D. A., Adjiman, C. S., Arnautova, Y. A., Bartashevich, E., Boerrigter, S. X., Braun, D. E., Cruz-Cabeza, A. J., Day, G. M., Della Valle, R. G., Desiraju, G. R., and Van Eijck, B. P. (2011). Towards crystal structure prediction of complex organic compounds—a report on the fifth blind test. *Acta Crystallogr. B*, **67**, 535–51. [241]

Barker, T. V. (1908). XII. Krystallographische Untersuchung der Dinitrobenzole und Nitrophenol. *Z. Kristallogr.*, **44**, 154–61. [302]

Barnes, A. F., Hardy, M. J., and Lever, T. J. (1993). A review of the applications of thermal methods within the pharmaceutical industry. *J. Therm. Anal.*, **40**, 499–509. [154]

Barnes, D., Wood-Kaczmar, M. W., Curzons, A. D., Lynch, I. R., Richardson, J. E., and Buxton, P. C. (1988). Anti-depressant crystalline paroxetine hydrochloride hemihydrate. U.S. Patent 4 721 723. [442]

Barr, G., Dong, W., and Gilmore, C. J. (2004). High-throughput powder diffraction. II. Applications of clustering methods and multivariate data analysis. *J. Appl. Crystallogr.*, **37**, 243–52. [271]

Barrow, M. J., Christie, R. M., and Monteith, J. E. (2002). The crystal and molecular structures of three diarylide yellow pigments, C.I. Pigments Yellow 13, 14 and 63. *Dyes Pigm.*, **55**, 79–89. [382t]

Barrow, M. J., Christie, R. M., Lough, A. J., Monteith, J. E., and Standring, P. N. (2000). The crystal structure of Pigment Yellow 12. *Dyes Pigm.*, **45**, 153–60. [382t]

Barry, S. J., Pham, T. N., Borman, P. J., Edwards, A. J., and Watson, S. A. (2012). A risk-based statistical investigation of the quantification of polymorphic purity of a pharmaceutical candidate by solid-state ^{19}F NMR. *Anal. Chim. Acta*, **712**, 30–6. [193]

Barton, J. H. (2000). Reforming the patent system. *Science*, **287**, 1933–4. [430]

Bassett, W. A. and Takahashi, T. (1974). X-ray diffraction studies up to 300 kbar. *Adv. High Pressure Res.*, **4**, 164–247. [336]

Bauer, J., Spanton, S., Henry, R., Quick, J., Dizki, W., Porter, W., and Morris, J. (2001). Ritonavir: an extraordinary example of conformational polymorphism. *Pharm. Res.*, **18**, 859–66. [346]

Bauer, M., Harris, R. K., Rao, R. C., Apperley, D. C., and Rodger, C. A. (1998). NMR study of desmotropy in irbesartan, a tetrazole-containing pharmaceutical compound. *J. Chem. Soc., Perkin Trans. 2*, 475–81. [186, 191f]

Bauer, N. and Lewin, S. Z. (1972). Determination of density. In *Physical Methods of Organic Chemistry*, Vol. 1, pt. I (ed. A. Weissberger and B. W. Rossiter), pp. 131–90. John Wiley & Sons, New York. [207]

Bauer, W. H. and Kassner, D. (1992). The perils of Cc: comparing the frequencies of falsely assigned space groups with their general population. *Acta Crystallogr. B*, **48**, 356–69. [238]

Bauer, W. H. and Tillmanns, E. (1986). How to avoid unnecessarily low symmetry in crystal structure determinations. *Acta Crystallogr. B*, **42**, 95–111. [159]

Bavin, M. (1989). Polymorphism in process development. *Chem. Ind.*, **21**, 527–9. [39, 240]

Bayard, F., Decoret, C., and Royer, J. (1990). Structural aspects of polymorphism and phase transition in organic molecular crystals. In *Structure and Properties of Molecular Crystals*. Studies in Physical and Theoretical Chemistry, Vol. 69, pp. 211–34. Elsevier, Amsterdam. [274]

Beach, S., Latham, D., Sidgwick, C., Hanna, M., and York, P. (1999). Control of the physical form of salmeterol xinofoate. *Org. Process Res. Dev.*, **3**, 370–6. [375]

Beal, R. W. and Brill, T. B. (2000). Thermal decomposition of energetic materials 77. Behavior of N–N bridged bifurazan compounds on slow and fast heating. *Propellants, Explos., Pyrotech.*, **25**, 241–6. [399]

Bechgaard, K., Jacobsen, C. S., Mortensen, K., Pedersen, H. J., and Thorup, N. (1980). The properties of five highly conducting salts:, $(TMTSF)_2 X, X = PF_6^-, AsF_6^-, SbF_6^-, BF_4^-$ and NO_3^-. *Solid State Commun.*, **33**, 1119–25. [276]

Bechgaard, K., Kistenmacher, T. J., Bloch, A. N., and Cowan, D. O. (1977). The crystal and molecular structure of an organic conductor from 4,4′,5,5′-tetramethyl-$\Delta^{2,2'}$-bis-1,3-diselenole and 7,7,8,8-tetracyano-*p*-quinodimethane. *Acta Crystallogr. B*, **33**, 417–22. [275]

Becker, R. and Döring, W. (1935). The kinetic treatment of nuclear formation in supersaturated vapors. *Ann. Phys.*, **24**, 719–52. [57]

Beckmann, W. (2000). Seeding the desired polymorph: background, possibilities, limitations and case studies. *Org. Process Res. Dev.*, **4**, 372–83. [100, 375]

Beckmann, W., Nickisch, K., and Budde, U. (1998). Development of a seeding technique for the crystallization of the metastable A modification of abecarnil. *Org. Process Res. Dev.*, **2**, 298–304. [367]

Begley, M. J., Crombie, L., Griffiths, G. L., and Jone, R. C. F. (1981). Charge-transfer and non-charge-transfer crystal forms of (E)–(5,5′-dimesitylbifuranylidenediones: an X-ray structural investigation. *J. Chem. Soc., Chem. Commun.*, 823–5. [308]

Beilstein, F. K. (1978). *How to Use Beilstein: Beilstein Handbook of Organic Chemistry*. Beilstein Institute for the Advancement of Chemical Sciences, Frankfurt am Main. [25]

Bekö, S. L., Hammer, S. M., and Schmidt, M. U. (2012). Crystal structures of the hydration states of Pigment Red 57:1. *Angew. Chem. Int. Ed. Engl.*, **51**, 4635–8. [381t]

Belenguer, A. M., Lampronti, G. I., Cruz-Cabeza, A. J., Hunter, C. A., and Sanders, J. K. M. (2016). Solvation and surface effects on polymorph stabilities at the nanoscale. *Chem. Sci.*, **7**, 6617–27. [77, 85, 100]

Bell, S. E. J., Burns, D. T., Dennis, A. C., and Speers, J. S. (2000). Rapid analysis of ecstasy and related phenethylamines in seized tablets by Raman spectroscopy. *Analyst*, **125**, 541–4. [177]

Bemm, U. and Ostmark, H. (1998). 1,1-Diamino-2,2-dinitroethylene: novel energetic material with infinite layers in two dimensions. *Acta Crystallogr. C*, **54**, 1997–9. [4114]

Bennema, P. and Hartman, P. (1980). The attachment energy as a habit controlling factor. *J. Cryst. Growth*, **49**, 145–56. [61]

Beran, G. J. O. (2016). Modeling polymorphic molecular crystals with electronic structure theory. *Chem. Rev.*, **116**, 5567–613. [240]

Berendt, R. T., Sperger, D. M., Munson, E. J., and Isbester, P. K. (2006). Solid-state NMR spectroscopy in pharmaceutical research and analysis. *Trends Anal. Chem.*, **25**, 977–84. [182]

Bergerhoff, G., Hundt, R., Sievers, R., and Brown, I. D. (1983). Inorganic Crystal Structure Database. *J. Chem. Inf. Comput. Sci.*, **23**, 66–9. [29]

Berkovitch-Yellin, Z. and Leiserowitz, L. (1982). Atom–atom potential analysis of the packing characteristics of carboxylic acids. A study based on experimental electron density distributions. *J. Am. Chem. Soc.*, **104**, 4052–64. [65f, 159, 160f]

Berman, H., Henrick, K., and Nakamura, H. (2003). Announcing the worldwide Protein Data Bank. *Nat. Struct. Biol.*, **10**, 980. [29]

Berman, H. M., Jeffrey, G. A. and Rosenstein, R. D. (1968). The crystal structures of the α and β forms of D-mannitol. *Acta Crystallogr. B*, **24**, 442–9. [88]

Bernstein, F. C., Koetzle, T. F., Williams, G. J. B., Meyer, A., Brice, M. D., Rodgers, J. R., Kennard, O., Shimanouchi, T., and Tasumi, M. (1977). The Protein Data Bank: a computer-based archival file for macromolecular structures. *J. Mol. Biol.*, **112**, 535–42. [16, 29]

Bernstein, H. I. and Quimby, W. C. (1943). The photochemical dimerization of *trans*-cinnamic acid. *J. Am. Chem. Soc.*, **65**, 1845–6. [332]

Bernstein, J. (1979). Conformational polymorphism. III. The crystal and molecular structures of the second and third forms of iminodiacetic acid. *Acta Crystallogr. B*, **35**, 360–6. [69, 222]

Bernstein, J. (1984). Crystal forces and molecular conformation. In *X-ray Crystallography and Drug Action: Current Perspectives* (ed. A. S. Horn and C. J. DeRanter), pp. 23–44. Oxford University Press, New York. [216t]

Bernstein, J. (1987). Conformational polymorphism. In *Organic Solid State Chemistry* (ed. G. R. Desiraju). Studies in Organic Chemistry, Vol. 32, Elsevier, Amsterdam. [4, 216, 220f, 221, 223f, 224f, 225f, 405f, 425]

Bernstein, J. (1991a). Polymorphism of L-glutamic acid: decoding the α–β phase relationship vis graph-set analysis. *Acta Crystallogr. B*, **47**, 1004–10. [66, 71]

Bernstein, J. (1991b). Polymorphism and the investigation of structure–property relations in organic solids. In *Organic Crystal Chemistry* (ed. J. B. Garbarczyk and D. W. Jones), pp. 6–26. International Union of Crystallography Book Series, Oxford University Press, Oxford. [275f, 276f]

Bernstein, J. (1992). Effect of crystal environment on molecular structure. In *Accurate Molecular Structures* (ed. A. Domenicano and I. Hargittai), pp. 469–97. Oxford University Press, Oxford. [216t]

Bernstein, J. (1993). Crystal growth, polymorphism and structure–property relationships in organic crystals. *J. Phys. D*, **26**, B66–76. [274]

Bernstein, J. (1999). Structural and crystallographic aspects of supramolecular engineering. In *Supramolecular Engineering of Synthetic Metallic Materials* (ed. J. Veciana, C. Rovira, and D. B. Amabilino), pp. 23–40. Kluwer Academic Publishers, Dordrecht. [279f, 288f, 354, 358]

Bernstein, J. (2005). And another comment on *pseudopolymorphism*. *Cryst. Growth Des.*, **5**, 1661–2. [8]

Bernstein, J. (2011). Polymorphism – a perspective. *Cryst. Growth Des.*, **11**, 632–50. [8, 254, 256f, 258f, 343, 345, 445]

Bernstein, J. (2017). Structural chemistry, fuzzy logic and the law. *Isr. J. Chem.*, **57**, 124–36. [255, 271, 431]

Bernstein, J. and Davis R. E. (1999). Graph Set Analysis of Hydrogen Bond Motifs. In *Implications of Molecular and Materials Structure for New Technologies* (eds J.A.K. Howard, F.H. Allen and G.P. Shields), NATO Science Series, Series E, Applied Sciences. **360**, 275–290, Kluwer Academic Publishers, Dordrecht. [71, 72]

Bernstein, J. and Goldstein, E. (1988). The polymorphic structures of a squarylium dye. The monoclinic (green) and triclinic (violet) forms of 2,4-bis(2-hydroxy-4-diethylaminophenyl)-1,3-cyclobutadienediylium 1,3-diolate. *Mol. Cryst. Liq. Cryst.*, **164**, 213–29. [128, 129f, 130f, 294]

Bernstein, J. and Hagler, A. T. (1978). Conformational polymorphism. The influence of crystal forces on molecular conformation. *J. Am. Chem. Soc.*, **100**, 673–81. [4, 219, 221, 257, 269]

Bernstein, J. and Henck, J.-O. (1998). Disappearing and reappearing polymorphs—an anathema to crystal engineering? *Cryst. Eng.*, **1**, 119–28. [134, 137]

Bernstein, J. and Izak, I. (1976). Molecular conformation and electronic structure. III. The crystal and molecular structure of the stable form of *N*-(*p*-chlorobenzylidene)-*p*-chloroaniline. *J. Chem. Soc., Perkin Trans.* 2, 429–34. [88, 325]

Bernstein, J. and MacAlpine, J. (2017). Pharmaceutical crystal forms and crystal form patents: novelty and obviousness. In *Polymorphism in Pharmaceutical Technology*, 2nd edn (ed. R. Hilfiker and M. von Raumer). Wiley-VCH, Weinheim. [444]

Bernstein, J. and Schmidt, G. M. J. (1972). Conformational studies. Part IV. The crystal and molecular structure of the metastable form of *N*-(*p*-chlorobenzylidene)-*p*-chloroaniline, a planar anil. *J. Chem. Soc., Perkin Trans.* 2, 951–5. [88, 325]

Bernstein, J., Anderson, T., and Eckhardt, C. J. (1979). Conformational influence on electronic spectra and structure. *J. Am. Chem. Soc.*, **101**, 541–5. [324f, 325, 325f, 326f, 327f]

Bernstein, J., Davey, R. J., and Henck, J.-O. (1999). Concomitant polymorphs. *Angew. Chem. Int. Ed.*, **38**, 3440–61. [55f, 56f, 58f, 114, 121, 122f, 155, 367]

Bernstein, J., Davis, R. E., Shimoni, L., and Chang, N.-L. (1995). Patterns in hydrogen bonding: functionality and graph set analysis in crystals. *Angew. Chem. Int. Ed. Engl.*, **34**, 1555–73. [69, 71, 72]

Bernstein, J., Dunitz, J. D., and Gavezzoti, A. (2008). Polymorphic perversity: crystal structures with many symmetry-independent molecules in the unit cell. *Cryst. Growth Des.*, **8**, 2011–18. [268f]

Bernstein, J., Etter, M. C., and MacDonald, J. C. (1990). Decoding hydrogen-bond patterns. The case of iminodiacetic acid. *J. Chem. Soc., Perkin Trans.* 2, 695–8. [71, 58]

Bernstein, J., Hagler, A. T., and Engel, M. (1981). An *ab initio* study of the conformational energetics of *N*-benzylideneaniline. *J. Chem. Phys.*, **75**, 2346–53. [324]

Berthold, C., Presser, V., Huber, N., and Nickel, K. G. (2011). 1 + 1 = 3 Coupling μ-XRD^2 and DTA. New insights in temperature-dependent phase transitions. *J. Therm. Anal. Calorim.*, **103**, 917–23. [209]

Berzelius, J. (1844). Verbindungen des phosphors mit schwefel. *Jahresbericht*, **23**, 44–55. [5, 29, 32]

Bettini, R., Menabeni, R., Tozzi, R., Pranzo, M. B., Pasquali, I., Chierotti, M. R., Gobetto, T., and Pellegrino, L. (2010). Didanosine polymorphism in a supercritical antisolvent process. *J. Pharm. Sci.*, **99**, 1855–70. [373]

Beynon, J. H. and Humphreys, A. R. (1955). The enthalpy difference between α- and β-copper phthalocyanine measured with an isothermal calorimeter. *Trans. Faraday Soc.*, **51**, 1065–71. [390]

Bhardwaj, R. M. (2016). Development and validation of high-throughput crystallization and analysis (HTCAA) methodology for physical form screening. In *Control and Prediction of Solid-State Pharmaceuticals*, pp. 39–75. Springer, Cham. [109]

Bhatt, P. M. and Desiraju, G. R. (2007). Tautomeric polymorphism in omeprazole. *Chem. Commun.*, 2057–9. [6, 125]

Bhattacharya, S., Brittain, H. G., and Suryanarayanan, R. (2009). Thermoanalytical and crystallographic methods. In *Polymorphism in Pharmaceutical Solids*, 2nd edn (ed. H. G. Brittain), pp. 318–46. Informa Healthcare, New York. [147, 366]

Bhugra, C. and Pikal, M. J. (2008). Role of thermodynamic, molecular, and kinetic factors in crystallization from the amorphous state. *J. Pharm. Sci.*, **97**, 1329–49. [371]

Biedermann, P.U., Stezowski, J. J., and Agranat, I. (2006). Polymorphism versus thermochromism: interrelation of color and conformation in overcrowded bistricyclic aromatic enes. *Chem. Eur. J.*, **12**, 3345–54. [313]

Biradha, K. and Zaworotko, M. (1998). Supramolecular isomerism and polymorphism in dianion salts of pyromellitic acid. *Cryst. Eng.*, **1**, 67–78. [121]

Bish, D. L. and Reynolds, R. C. (1989). Sample preparation for X-ray diffraction. In *Reviews in Mineralogy, Modern Powder Diffraction,* Vol. 20 (ed. D. L. Bish and J. E. Post), pp. 73–99. Mineralogy Society of America, Washington, DC. [156]

Bishop, M. M., Chellappa, R. S., Liu, Z., Preston, D. N., Sandstrom, M. M., Dattelbaum, D. M., Vohra, Y. K., and Velisavljevic, N. (2014). High pressure–temperature polymorphism of 1,1-diamino-2,2-dinitroethylene. *J. Phys.: Conf. Ser.*, **500**, 052005. [414]

Bishop, M. M., Velisavljevic, N., Chellappa, R. S., and Vohrai, Y. K. (2015). High pressure–temperature phase diagram of 1,1-diamino-2,2-dinitroethylene (FOX-7). *J. Phys. Chem. A*, **119**, 9739–47. [415]

Black, J. F. B., Cardew, P. T., Cruz-Cabeza, A. J., Davey, R. J., Gilks, S. E., and Sullivan, R. A. (2018). Crystal nucleation and growth in a polymorphic system: Ostwald's rule, *p*-aminobenzoic acid and nucleation transition states. *CrystEngComm*, **20**, 768–76. [370]

Black, J. W., Duncan, W. A. M., Durant, C. J., Ganellin, C. R., and Parsons, E. M. (1972). Definition and antagonism of histamine H_2-receptors. *Nature*, **236**, 385–90. [438]

Black, S. N. and Davey, R. J. (1988). Crystallization of amino acids. *J. Cryst. Growth*, **90**, 136–44. [86]

Black, S. N., Davey, R. J., Morley, P. R., Halfpenny, P., Shepherd, E. E. A., and Sherwood, J. N. (1993). Crystal-growth and characterization of the electrooptic material 3-(2,2-dicyanoethenyl)-1-phenyl-4,5-dihydro-1*H*-pyrazole. *J. Mater. Chem.*, **3**, 129–32. [302]

Black, S. N., Williams, L. J., Davey, R. J., Moffat, F., McEwan, D. M., Sadler, D. E., Docherty, R., and Williams, D. J. (1990). Crystal chemistry of 1-(4-chlorophenyl)-4,4-dimethyl-2-(1-*H*-1,2,4-triazol-1-yl)pentan-3-one, a paclobutrazol intermediate. *J. Phys. Chem.*, **94**, 3223–6. [61]

Blagden, N., Davey, R. J., Lieberman, H., Williams, L., Paynem, R., Roberts, R., Rowe, R., and Docherty, R. (1998a). Crystal chemistry and solvent effects in polymorphic systems: sulphathiazole. *J. Chem. Soc., Faraday Trans.*, **98**, 1035–45. [121, 368]

Blagden, N., Davey, R. J., Rowe, R., and Roberts, R. (1998b). Disappearing polymorphs and the role of reaction by-products: the case of sulphathiazole. *Int. J. Pharm.*, **172**, 169–77. [121]

Blake, A. J., Gould, R. O., Halcrow, M. A., and Schröder, M. (1993). Conformational studies on [16]aneS$_4$. Structures of α- and β-[16]aneS$_4$ ([16]aneS$_4$ = 1,5,9,13-tetrathiacyclohexadecane). *Acta Crystallogr. B*, **49**, 773–9. [120]

Block, S. and Piermarini, G. (1976). The diamond cell stimulates high pressure research. *Phys. Today*, **29**, 44–7. [416]

Block, S., Weir, C. E., and Piermarini, G. J. (1970). Polymorphism in benzene, naphthalene and anthracene at high pressure. *Science*, **169**, 586–7. [336]

Blomquist, A. T. and Ryan, J. F., Jr. (1944). *Studies related to the stability of PETN.* OSRD Report NDRC-B-3566. [411]

Bobrovs, R., Seton, L., and Dempster, N. (2015). The reluctant polymorph: investigation into the effect of self-association on the solvent mediated phase transformation and nucleation of theophylline. *CrystEngComm*, **17**, 5237–51. [373]

Bock, H., Rauschenbach, A., Näther, C., Havlas, Z., Gavezzotti, A., and Filippini, G. (1995). Orthorhombic and monoclinic 2,3,7,8-tetramethoxythianthrene—small structural difference—large lattice change. *Angew. Chem. Int. Ed. Engl.*, **34**, 76–8. [178]

Boese, R., Polk, M., and Bläser, D. (1987). Cooperative effects in the phase transformation of triethylcyclotriboroxane. *Angew. Chem. Int. Ed. Engl.*, **26**, 245–7. [335]

Bohler, H. and Kehrer, F. (1963). Pure crystal forms of 5,12-dihydroquino[2,3-b]acridine-7,14-diones. Sandoz Ltd. Belgian Patent 611 271; *Chem. Abstr.*, **58**, 4675. [385]

Boldyreva, E. (1999). Interplay between intra- and intermolecular interactions in solid-state reactions: general overview. In *Reactivity of Molecular Solids: The Molecular Solid State*, Vol. 3 (ed. E. Boldyreva and V. Boldyrev), pp. 1–50. John Wiley & Sons, Chichester. [336]

Boldyreva, E. (2007). High-pressure polymorphs of molecular solids: when are they formed, and when are they not? Some examples of the role of kinetic control. *Cryst. Growth Des.*, **7**, 1662–8. [352]

Boldyreva, E. (2016). Non-ambient conditions in the investigation and manufacturing of drug forms. *Curr. Pharm. Des.*, **22**, 4981–5000. [352, 357]

Boldyreva, E. V., Shakhtshieder, T. P., Vasilchenko, M. A., Ahsbahs, H., and Uchtmann, H. (2000). Anisotropic crystal structure distortion of the monoclinic polymorph of acetaminophen at high hydrostatic pressures. *Acta Crystallogr. B*, **56**, 299–309. [336]

Boldyreva, E. V., Sowa, H., Ahsbahs, H., Goryainov, S. V., Chernyshev, V. V., Dmitriev, V. P., Seryotkin, E. N., Kolesnik, E. N., Shakhtshneider, T. P., Ivashevskaya, S. N., and Drebushchak, T. N. (2008). Pressure-induced phase transitions in organic molecular crystals: a combination of single-crystal and powder diffraction, Raman and IR-spectroscopy. *J. Phys.: Conf. Ser.*, **121**, 022023. [336]

Bolla, G., Mittapalli, S., and Nangia, A. (2014). Pentamorphs of acedapsone. *Cryst. Growth Des.*, **14**, 5260–74. [354]

Bolotina, N. B., Hardie, M. J., Speer, R. L., and Pinkerton, A. A. (2004). Energetic materials: variable-temperature crystal structures of γ- and ϵ-HNIW polymorphs. *J. Appl. Crystallogr.*, **37**, 808–14. [407]

Bolotina, N. B., Kirschbaum, K., and Pinkerton, A. A. (2005). Energetic materials: alpha-NTO crystallizes as a four component triclinic twin. *Acta Crystallogr. B*, **61**, 577–84. [413]

Bolotina, N. B., Zhurova, E. A., and Pinkerton, A. A. (2003). Energetic materials: variable-temperature crystal structures of two biguianidium dinitramides. *J. Appl. Crystallogr.*, **36**, 280–5. [413]

Bolton, O., Simke, L. R., Pagoria, P. F., and Matzger, A. J. (2012). High power explosive with good sensitivity: a 2:1 cocrystal of CL-20:HMX. *Cryst. Growth Des.*, **12**, 4311–14. [403]

Boman, C.-E., Herbertsson, H., and Oskarsson, A. (1974). Crystal and molecular structure of a monoclinic phase of iminodiacetic acid. *Acta Crystallogr. A*, **30**, 378–82. [222]

Bombicz, P., Czugler, M., Tellgren, R., and Kalman, A. (2003). A classical example of a disappearing polymorph and the shortest intermolecular H$\cdots$H separation ever found in an organic crystal structure. *Angew. Chem. Int. Ed.*, **42**, 1957–60. [134]

Bonafede, S. J. and Ward, M. D. (1995). Selective nucleation and growth of an organic polymorph by ledge-directed epitaxy on a molecular crystal substrate. *J. Am. Chem. Soc.*, **117**, 7853–61. [100]

Bond, A. D. (2014). processPIXEL: a program to generate energy-vector models from Gavezzotti's *PIXEL* calculations. *J. Appl. Crystallogr.*, **47**, 1777–80. [269]

Bond, A. D., Boese, R., and Desiraju, G. R. (2007). On the polymorphism of aspirin: crystalline aspirin as intergrowths of two "polymorphic" domains. *Angew. Chem. Int. Ed.*, **46**, 618–22. [42, 119]

Bondi, A. (1963). Heat of sublimation of molecular crystals. A catalog of molecular structure increments. *J. Chem. Eng. Data*, **8**, 371–81. [228]

Boon, W. and Vangerven, L. (1992). Magnetization of electron-spin pairs in a free-radical at low-temperatures in a high magnetic field. *Physica B*, **177**, 527–30. [292]

Booth, A. D. and Llewellyn, F. J. (1947). The crystal structure of pentaerythritol tetranitrate. *J. Chem. Soc.*, 837–46. [411]

Booth, G. (1971). Phthalocyanines. In *The Chemistry of Synthetic Dyes*, Vol. V (ed. K. Venkataraman). Academic Press, New York. [388]

Borka, L. (1991). Review on crystal polymorphism of substances in the European Pharmacopoeia. *Pharm. Acta Helv.*, **66**, 16–22. [138, 342, 345]

Borka, L. (1995). Crystal polymorphism and related phenomena of substances in the European Pharmacopoeia. An updated review for fasciculae 13 to 19. *Pharmeuropa*, **7**, 586–93. [345]

Borka, L. and Backe-Hansen, K. (1968). IR spectroscopy of chloramphenicol palmitate. Polymorph alteration caused by the potassium bromide disk technique. *Acta Pharm. Suec.*, **5**, 271–8. [349]

Borka, L. and Haleblian, J. K. (1990). Crystal polymorphism of pharmaceuticals. *Acta Pharm. Jugoslav.*, **40**, 71–94. [344]

Borsenberger, P. M. and Weiss, D. S. (1993). *Organic Photoreceptors for Xerography*. Marcel Dekker, New York. [293]

Bos, M. and Weber, H. T. (1991). Comparison of the training of neural networks for quantitative X-ray fluorescence spectrometry by genetic algorithm, and backward error propagation. *Anal. Chim. Acta*, **247**, 97–105. [169]

Botcher, T. R., Berdall, D. J., Wight, C. A., Fan, L. M., and Burkey, T. J. (1996). Thermal decomposition mechanism of NTO. *J. Phys. Chem.*, **100**, 8802–6. [413]

Botha, S. A., Guillory, J. K., and Lotter, A. P. (1986). Physical characterization of solid forms of urapidil. *J. Pharm. Biomed. Anal.*, **4**, 573–87. [155]

Botsaris, G. D. Secondary Nucleation – A Review. In: Mullin, J. W. (ed), Industrial Crystallization, Springer, Boston, MA, pp. 3–22. [81]

Bottaro, J. C., Penwell, P. E., Ross, D. S., and Schmidt, R. J. (1991). Novel *N*, *N*-dinitroamide salts—useful as oxidizers in rocket fuels, exhibiting high temperature stability, high energy density and an absence of smoke generating halogen(s). World Intellectual Property Organization Application Number PCT.US91/04268. [415]

Böttger, R. and Will, H. (1846). Über eine neue, der Pikrinsaüer nahesteheude Saüer. *Annalen*, **58**, 275–300. [426]

Bottom, R. (1999). The role of modulated temperature differential scanning calorimetry in the characterisation of a drug molecule exhibiting polymorphic and glass forming tendencies. *Int. J. Pharm.*, **192**, 47–53. [366]

Bouas-Laurent, H. and Durr, H. (2001). Organic photochromism. *Pure Appl. Chem.*, **73**, 639–65. [307]

Bouchard, A., Jovanovic, N., Hofland, G. W., Mendes, E., Crommelin, D. J. A., Jisjoot, W., and Witkamp, G.-J. (2007). Selective production of polymorphs and pseudomorphs using supercritical fluid crystallization from aqueous solutions. *Cryst. Growth Des.*, **7**, 1432–40. [101]

Bouche, R. and Draguet-Brughmans, M. (1977). Polymorphism of organic drug substances. *J. Pharm. Belgique*, **32**, 23–51. [343]

Boucherle, J. X., Gillon, B., Maruani, J., and Schweizer, J. (1987). Crystal structure determination by neutron diffraction of 2,2-diphenyl-1-picrylhydrazyl (DPPH) benzene solvate (1/1). *Acta Crystallogr. C*, **43**, 1769–73. [291]

Bowden, P. R., Chellappa, R. S., Dattlemaum, D. M., Manner, V. W., Mack, N. H., and Liu, Z. (2014). The high-pressure phase stability of 2,4,6-trinitrotoluene (TNT). *J. Phys.: Conf. Ser.*, **500**, 052006. [425]

Boyd, P., Mitra, S., Raston, C. L., Rowbottom, G. L., and White, A. H. (1981). Magnetic and structural studies on copper(II) dialkyldithiocarbamates. *J. Chem. Soc., Dalton Trans.*, 13–22. [285]

Boyd, R. H. (1994). Prediction of polymer crystal-structures and properties. *At. Model. Phys. Prop. Adv. Polym. Sci.*, **116**, 1–25. [238]

Boyle, R. (1661). *Essay on the Unsuccessfulness of Experiments.* Henry Herrington, London. [78]

Brach, P. J. and Lardon, M. A. (1973). π-Form metal-free phthalocyanine. Xerox Corporation Patent US 3 708 293; DE-Os 2 218 788 (1972); GB 1 395 769 (1972); FR 2 138 865 (1972). [389]

Brach, P. J. and Six, H. A. (1973). π-Form metal phthalocyanine. Xerox Corporation Patent US 3 708 292; DE-Os 2 218 767 (1972); GB 1 395 615 (1972); FR 2 138 730 (1972). [389]

Brach, P. J. and Six, H. A. (1975). Process of making x-form metal phthalocyanine. Xerox Corporation Patent US 3 927 026; DE-Os 2 026 057 (1970); GB 1 312 946 (1975). [389]

Bracuti, A. J. (1979). 1,2,3-Triaminoguanidinium nitrate. *Acta Crystallogr.*, **35**, 760–1. [412]

Bracuti, A. J. (1988). *Discovery of a low temperature form of TAGN.* ARAED-TR-88009. Access No. E780-1782. U.S. Army Armament Research & Development Center, Picatinny Arsenal, NJ. [412]

Bradshaw, J. (1993). Ranitidine. In *Annals of Drug Discovery,* Vol. 3 (ed. D. Lednicer), pp. 45–81. American Chemical Society, Washington, DC. [438]

Braem, A., Deshpande, P. P., Ellsworth, B. A., and Washburn, W. N. (2014). Discovery and development of selective renal sodium-dependent glucose cotransporter 2 (SGLT2) dapagliflozin for the treatment of Type 2 diabetes. *Top. Med. Chem.*, **12**, 73–94. [446]

Braga, D. and Grepioni, F. (1991). Effect of molecular shape on crystal building and dynamic behavior in the solid state: from crystalline arenes to crystalline metal arene complexes. *Organometallics*, **10**, 2563–9. [126]

Braga, D. and Grepioni, F. (1992). Crystal structure and molecular interplay in solid ferrocene, nickelocene, and ruthenocene. *Organometallics*, **11**, 711–88. [239]

Braga, D. and Grepioni, F. (2003). Polymorphism, crystal transformations and gas-solid reactions. In *Crystal Design: Structure and Function,* Vol. 7 (ed. G. R. Desiraju), Chapter 8. John Wiley & Sons, Ltd, Chichester. [336]

Braga, D. and Grepioni, F. (2005). Making crystals from crystals: a green route to crystal engineering and polymorphism. *Chem. Commun.*, 3635–45. [336]

Braga, D., Cojazzi, G., Abati, A., Maini, L., Polito, M., Scaccianoce, L., and Grepioni, F. (2000). Making and converting organometallic pseudo-polymorphs via non-solution methods. *J. Chem. Soc., Dalton Trans.*, 3969–75. [134]

Braga, D., d'Agostino, S., D'Amen, E., and Grepioni, F. (2011a). Polymorphs from supramolecular gels: four crystal forms of the same silver(I) supergelator crystallized directly from its gels. *Chem. Commun.*, **47**, 5154–6. [356]

Braga, D., d'Agostino, S., Dichiarante, E., Maini, L., and Grepioni, F. (2011b). Dealing with crystal forms (The kingdom of Serendip?). *Chem. Asian J.*, **6**, 2214–23. [354]

Braga, D., Grepioni, F., and Maini, L. (2010). The growing world of crystal forms. *Chem. Commun.*, **46**, 6232–42. [343]

Braga, D., Grepioni, F., and Sabatino, P. (1990). On the factors controlling the crystal packing of first row transition metal binary carbonyls. *J. Chem. Soc., Dalton Trans.*, 3137–42. [218]

Braga, D., Grepioni, F., Polito, M., Chierotti, M. R., Ellena, S., and Gobetto, R. (2006). A solid–gas route to polymorph conversion in crystalline $[Fe^{II}(\eta^5\text{-}C_5H_4COOH)_2]$. A diffraction and solid-state NMR study. *Organometallics*, **25**, 4627–33. [336]

Brahadeeswaran, S., Venktaramanan, V., and Baht, H. L. (1999). Non-linear activity of anhydrous and hydrated sodium *p*-nitrophenolate. *J. Cryst. Growth*, **205**, 548–53. [302]

Brand, B. P. (1964). Preparation of delta copper phthalocyanine. Imperial Chemical Industries Patent US 3 150 150. [389]

Brand, J. C. D. and Speakman, J. C. (1960). *Molecular Structure, the Physical Approach*, pp. 248–50. E. Arnold, London. [216]

Brandel, C., Petit, S., Cartigny, Y., Coquerel, G. (2016). The structural aspects of solid solutions of enantiomers. *Curr. Pharm. Des.* **22**, 4929–41. [358]

Brandstätter, M. (1947). The isomorphous replacability of H, OH, NH_2, CH_3 and Cl in the *m*-dinitrobenzene series. *Monatsch. Chem.*, **77**, 7–17. [142]

Brandt, H., Hörnle, R., and Leverenz, K. (1982). Color stable modification of monoazo dyestuff—viz butyl cyano methyl nitrophenyl azo pyridone, used for dyeing polyester fibers. Bayer AG Patent DE 3 046 587. [395]

Bras, L. P. and Loureiro, R. M. S. (2013). Polymorphic conversion monitoring using real-time Raman spectroscopy. *Chim. Oggi*, **31**, 31–7. [352]

Braun, A. and Tscherniak, J. (1907). Über die Produkte er Einwirkung von Acetanhydrid auf Phthalamid. *Ber. Dtsch. Chem. Ges.*, **40**, 2709–14. [389]

Braun, D. E. (2008). *Crystal polymorphism and structure–property relationships of drug compounds: statistical aspects, analytical strategies and case studies.* PhD Thesis, University of Innsbruck, Switzerland. [17]

Braun, D. E. and Griesser, U. J. (2016). Why do hydrates (solvates) form in small neutral molecules? Exploring the crystal form landscapes of the alkaloids brucine and strychnine. *Cryst. Growth Des.*, **16**, 6405–18. [361]

Braun, D. E., Gelbrich, T., Kahlenberg, V., Laus, G., and Griesser, U. J. (2008). Packing polymorphism of a conformationally flexible molecule. *New J. Chem.*, **32**, 1677–85. [6]

Braun, D. E., Gelbrich, T., Kahlenberg, V., Tessadri, R., Wieser, J., and Griesser, U. J. (2009a). Stability of solvates and packing systematics of nine crystal forms of the antipsychotic drug aripiprazole. *Cryst. Growth Des.*, **9**, 1054–65. [14t, 152f, 345, 374]

Braun, D. E., Gelbrich, T., Kahlenberg, V., Tessadri, R. T., Wieser, J., and Griesser, U. J. (2009b). Conformational polymorphism in aripiprazole: preparation, stability and structure of five modifications. *J. Pharm. Sci.*, **98**, 2010–26. [14t, 150, 345]

Braun, D. E., Karamertzanis, P. G., and Price, S. L. (2011b). Which, if any, hydrates will crystallize? Predicting hydrate formation of two dihydroxybenzoic acids. *Chem. Commun.*, **47**, 5443–5. [355]

Braun, D. E., Karamertzanis, P. G., Arlin, J. B., Florence, A. J., Kahlenberg, V., Tocher, D. A., Griesser, U. J., and Price, S. J. (2011a). Solid-state forms of β-resorcylic acid: how exhaustive should polymorph screen be? *Cryst. Growth Des.*, **11**, 210–20. [374]

Braun, D. E., Koztecki, L. H., McMahon, J. A., Price, S. L., and Reutzel-Edens, S. M. (2015). Navigating the waters of unconventional crystalline hydrates. *Mol. Pharm.*, **12**, 3069–88. [355]

Braun, D. E., Lingreddy, S. R., Beidelschies, M. D., Guo, R., Muller, P., Price, S. L., and Reutzel-Edens, S. M. (2017b). Unraveling complexity in the solid form screening of a pharmaceutical salt: Why so many forms? Why so few? *Cryst. Growth Des.*, **17**, 5349–65. [354, 355]

Braun, D. E., McMahon, J. A., Koztecki, L. H., Price, S. L., and Reutzel-Edens, S. M. (2014a). Contrasting polymorphism of related small molecule drugs correlated and guided by the computed crystal energy landscape. *Cryst. Growth Des.*, **14**, 2056–72. [110f, 359]

Braun, D. E., Oberacher, H., Arnhard, K., Orlova, M., and Griesser, U. J. (2016). 4-Aminoquinaldine monohydrate polymorphism: prediction and impurity aided discovery of a difficult to access stable form. *CrystEngComm*, **18**, 4053–67. [355]

Braun, D. E., Orlova, M., and Griesser, U. J. (2014b). Creatine: polymorphs predicted and found. *Cryst. Growth Des.*, **14**, 4895–900. [355]

Brehmer, T. H., Weber, E., and Cano, F. H. (2000). Balance of forces between contacts in crystal structures of two isomeric benzo-condensed dibromodihydroxy-containing compounds. *J. Phys. Org. Chem.*, **13**, 63–74. [217]

Brescello, R., Cotarca, L., Smaniotto, A., Verzini, M., Polentarutti, M., Bais, G., and Plaisier, J. R. (2016). Method of detecting polymorphs using synchrotron radiation. European Patent Specification EP 2 710 355 B1. [169]

Briegleb, G. (1961). *Elektronen-Donator-Acceptor-Komplexen*. Springer-Verlag, Berlin. [359]

Brill, T. B. (1992). Connecting the chemical-composition of a material to its combustion characteristics. *Prog. Energy Combust. Sci.*, **18**, 91–116. [399]

Brill, T. B. and James, K. J. (1993). Kinetics and mechanisms of thermal-decomposition of nitroaromatic explosives. *Chem. Rev.*, **93**, 2667–92. [399]

Brill, T. B. and Reese, C. O. (1980). Analysis of intra- and intermolecular interactions relating to the thermophysical behavior of α, β, and δ octahydro-1,3,5,7-tetranitro-1,3,5,7-tetraazocine. *J. Phys. Chem.*, **84**, 1376–80. [405]

Bristol-Meyers Co. v. United States International Trade Commission, Gema, S.A., Kalipharma, Inc., Purepac Pharmaceutical Co., Istituto Biochimico Italiano Industria Giovanni Lorenzini, Institut Biochimique, S.A., and Biocraft Laboratories, Inc. (1989). No. 89–1530 [See also United States International Trade Commission, Investigation No. 337-TA-293]. [436]

Brittain, H. G. (1997). Spectral methods for the characterization of polymorphs and solvates. *J. Pharm. Sci.*, **86**, 405–12. [170, 182]

Brittain, H. G. (1999b). *Polymorphism in Pharmaceutical Solids*. Drugs and the Pharmaceutical Sciences, Vol. 95. Marcel Dekker, New York. [343, 364]

Brittain, H. G. (1999c). Application of the phase rule to the characterization of polymorphic systems. In *Polymorphism in Pharmaceutical Solids*, Drugs and the Pharmaceutical Sciences, Vol. 95 (ed. H. G. Brittain), pp. 35–72. Marcel Dekker, New York. [42]

Brittain, H. G. (2007). Polymorphism and solvatomorphism 2005. *J. Pharm. Sci.*, **96**, 705–28. [343]

Brittain, H. G. (2008). Polymorphism and solvatomorphism 2006, *J. Pharm. Sci.*, **97**, 3611–36.

Brittain, H. G. (2009a). Polymorphism and solvatomorphism 2007. *J. Pharm. Sci.*, **98**, 1617–42. [170]

Brittain, H. G. (ed.) (2009b). *Polymorphism in Pharmaceutical Solids*, 2nd edn. Informa Healthcare, New York. [136, 177, 343]

Brittain, H. G. (2010). Polymorphism and solvatomorphism 2008. *J. Pharm. Sci.*, **99**, 3648–64.

Brittain, H. G. (2011). Polymorphism and solvatomorphism 2009. *J. Pharm. Sci.*, **100**, 1260–79.

Brittain, H. G. (2012). Polymorphism and solvatomorphism 2010. *J. Pharm. Sci.*, **101**, 464–84.

Brittain, H. G. (2013). Pharmaceutical cocrystals: the coming wave of new drug substances. *J. Pharm. Sci.*, **102**, 311–17. [360]

Brittain, H. G. and Fiese, E. F. (1999). Effects of pharmaceutical processing on drug poly-morphs and solvates. In *Polymorphism in Pharmaceutical Solids*, Drugs and the Pharmaceutical Sciences, Vol. 95 (ed. H. G. Brittain), pp. 331–61. Marcel Dekker, New York. [375]

Brittain, H. G. and Grant, D. J. W. (1999). Effects of polymorphism and solid-state solvation on solubility and dissolution rate. In *Polymorphism in Pharmaceutical Solids*, Drugs and the Pharmaceutical Sciences, Vol. 95 (ed. H. G. Brittain), pp. 279–330. Marcel Dekker, New York. [155, 346, 348, 348f]

Brittain, H. G., Ranadive, S. A., and Serajuddin, A. T. M. (1995). Effect of humidity-dependent changes in crystal structure on the solid-state fluorescence properties of a new HMG-COA reductase inhibitor. *Pharm. Res.*, **12**, 556–9. [9]

Brock, C. P. (2016). High-Z' structures of organic molecules: their diversity and organizing principles. *Acta Crystallogr. B*, **72**, 807–21. [218]

Broderick, W. E., Eichorn, D. M., Liu, X., Toscano, P. J., Owens, S. M., and Hoffman, B. M. (1995). Three phases of $[Fe(C_5Me_5)_2]^+[TCNQ]^-$: ferromagnetism in a new structural phase. *J. Am. Chem. Soc.*, **117**, 3641–2. [287]

Brown, C. J. (1968a). Crystal structure of β-copper phthalocyanine. *J. Chem. Soc.*, 2488–93. [390, 391]

Bruce, A. D. and Cowley, R. A. (1981). *Structural Phase Transitions*. Taylor & Francis, London. [319]

Brückner, S. (1982). An unusual example of packing among molecular layers: the struc-tures of two crystalline forms of 2,2-aziridinecarboxamide, $C_4H_7N_3O_2$. *Acta Crystallogr. B*, **38**, 2405–8. [73, 74f]

Brün, W. (1934). Priming mixture. Remington Arms U.S. Patent 1 942 274. [426]

Brünger, A. T. (1991). Simulated annealing in crystallography. *Annu. Rev. Phys. Chem.*, **42**, 197–223. [243]

Brüning, J., Alig, E., Meents, A., van de Streek, J., and Schmidt, M. U. (2009). Structure determinations of three fluorescent organic pigments by powder diffraction and micro-crystal structure analysis. *Z. Kristallogr.*, **224**, 556–62. [384t]

Bruno, I. J., Cole, J. C., Kessler, M., Luo, J., Motherwell, W. D. S., Purkis, L. H., Smith, B. R., Taylor, R., Cooper, T. I., Harris, S. E., and Orpen, A. G. (2004). Retrieval of crystallographically-derived molecular geometry information. *J. Chem. Inf. Comput. Sci.*, **44**, 2133–44. [243]

Brus, J., Urbanova, M., Sedenkova, I., and Brusova, H. (2011). New perspectives of ^{19}F MAS NMR in the characterization of amorphous forms of atorvastatin in dosage for-mulations. *Int. J. Pharm.*, **409**, 62–74. [193]

Bučar, D.-K., Day, G. M., Halasz, I., Zhang, G. G. Z., Sander, J. R. G., Reid, D. G., MacGillivray, L. R., Duer, M. J., and Jones, W. (2013). The curious case of (caffeine)·(benzoic acid): how heteronuclear seeding allowed the formation of an elu-sive cocrystal. *Chem. Sci.*, **4**, 4417–25. [102]

Bučar, D.-K., Lancaster, R. W., and Bernstein, J. (2015). Disappearing polymorphs revisited. *Angew. Chem. Int. Ed.*, **54**, 6972–93. [116, 131]

Buchsbaum, C. and Schmidt, M. U. (2007). Rietveld refinement of a wrong crystal struc-ture. *J. Appl. Crystallogr.*, **40**, 105–14. [380t]

Buchsbaum, C., Paulus, E., and Schmidt, M. (2011). Crystal structures of thiazine-indigo pigments, determined from single-crystal and powder diffraction data. *Z. Kristallogr.*, **226**, 822–31. [383t]

Buckley, H. E. (1951). *Crystal Growth*. Wiley, New York. [61]

Buckton, G. and Darcy, P. (1999). Assessment of disorder in crystalline powders—a review of analytical techniques and their application. *Int. J. Pharm.*, **179**, 141–58. [371]

Budzianowski, A., Derzsi, M., Leszczynski, P. J., Cyranski, M. K., and Grochala, W. (2010). Structural polymorphism of pyrazinium hydrogen sulfate: extending chemistry of the pyrazinium salts with small anions. *Acta Crystallogr. B*, **66**, 451–7. [6]

Buerger, M. J. (1951). Crystallographic aspects of phase transformations. In *Phase Transformations in Solids* (ed. R. Smoluchowski, J. E. Mayer, and W. A. Weyl), pp. 183–211. John Wiley & Sons, New York. [43, 44, 44f, 47]

Buerger, M. J. (1956). *Elementary Crystallography: An Introduction to the Fundamental Geometrical Features of Crystals.* Wiley, New York. [59]

Buerger, M. J. and Bloom, M. C. (1937). Crystal polymorphism. *Z. Kristallogr.*, **96**, 182–200. [4, 16, 38, 342]

Bugay, D. E. (1993). Solid-state nuclear magnetic resonance spectroscopy: theory and pharmaceutical applications. *Pharm. Res.*, **10**, 317–27. [182, 183f, 190]

Bugay, D. E. (1999). Quantitative and regulatory aspects of polymorphism, Presented at the 1st International Symposium on Aspects of Polymorphism and Crystallization—Chemical Development Issues, Hinckley, UK, June 23–25, 1999. Scientific Update, Mayfield, UK. [174, 175f, 176f, 177, 190, 192f]

Bugay, D. E. and Brittain, H. G. (2006). Raman spectroscopy. In *Spectroscopy of Pharmaceutical Solids* (ed. H. G. Brittain), pp. 271–312. Taylor and Francis, New York. [177]

Bugay, D. E., Newman, A. W., and Findlay, W. P. (1996). Quantitation of cefepime.2HCl dihydrate in cefepime.2HCl monohydrate by diffuse reflectance IR and powder X-ray diffraction techniques. *J. Pharm. Biomed. Anal.*, **15**, 49–61. [174, 132]

Bunte, G., Pontius, H., and Kaiser, M. (1999). Analytical characterization of impurities in new energetic materials. *Propellants, Explos., Pyrotech.*, **24**, 149–55. [408]

Burg, J., Grobet, P., Van den Bosch, A., and Vansummeren, J. (1982). Magnetic interaction and spin diffusion in solvent-free DPPH. *Solid State Commun.*, **43**, 785–7. [292]

Burger, A. (1982a). Interpretation of polymorphism studies. *Acta Pharm. Technol.*, **28**, 1–20. [53, 106]

Burger, A. (1982b). Thermodynamic and other aspects of the polymorphism of drug substances. *Pharm. Int.*, **3**, 158–63. [53, 150]

Burger, A. (1983). The relevance of polymorphism. In *Topics in Pharmaceutical Science* (ed. D. D. Breimer and P. Speiser), pp. 347–58. Elsevier, Lausanne. [5, 343]

Burger, A. and Ramberger, R. (1979a). On the polymorphism of pharmaceuticals and other molecular crystals. I. Theory of thermodynamic rules. *Mikrochim. Acta*, **II**, 259–72. [37, 47, 52, 54, 207]

Burger, A. and Ramberger, R. (1979b). On the polymorphism of pharmaceuticals and other molecular crystals. II. *Mikrochim. Acta*, **II**, 273–316. [52, 54, 173, 207]

Burger, A., Henck, J.-O., Hetz, S., Rollinger, J. M., Weissnicht, A. A., and Stottner, H. (2000). Energy/temperature diagram and compression behavior of the polymorphs of D-mannitol. *J. Pharm. Sci.*, **89**, 457–68. [120, 138, 363, 364, 446]

Burger, A., Ratz, A. W., and Brox, W. (1986). Polymorphic pharmaceuticals in the European pharmacopoeia: oxytetracycline hydrochloride. *Pharm. Acta Helv.*, **61**, 98–106. [364]

Burger, A., Rollinger, J. M., and Bruggeller, P. (1997). Binary system of (*R*)- and (*S*)-nitrendipine—polymorphism and structure. *J. Pharm. Sci.*, **86**, 674–9. [365]

Burger, A., Schulte, K., and Ramberger, R. (1980). Aufklärung thermodynamischer Beziehungen zwischen fünf polymorphen Modifikationen von Sulfapyridin mittels DSC. *J. Therm. Anal.*, **19**, 475–84. [366]

Bürgi, H.-B. and Dunitz, J. D. (1994). Structure correlation; the chemical point of view. In *Structure Correlation*, Vol. 1 (ed. H.-B. Bürgi and J. D. Dunitz), pp. 163–204. VCH, Weinheim. [220]

Burkhard, H., Muller, C., and Senn, O. (1968). Process for the production of finely divided, heat stable 1-phenylamino-2-nitrobenzene-4-sulphonic acid, its production and uses. Sandoz Patents Ltd. GB Patent 1 040 607. [379t]

Burkhardt, L. A. and Bryden, J. H. (1954). X-ray studies of 2,4,6-trinitrotoluene. *Acta Crystallogr.*, 7, 135–6. [417]

Burns, G. R., Cunningham, C. W., and McKee, V. (1988). Photochromic formazans: Raman spectra, X-ray crystal structure and ^{13}C magnetic resonance spectra of the orange and red isomers of 3-ethyl-1,5-diphenylformazan. *J. Chem. Soc., Perkin Trans. 2*, 1275–80. [309]

Bushuyev, O. S., Singleton, T. A., and Barrett, C. J. (2013). Fast, reversible, and general photomechanical motion in single crystals of various azo compounds using visible light. *Adv. Mater.*, **25**, 1796–1800. [380t]

Busing, D., Jenau, M., Reuter, J., Wurflinger, A., and Tamarit, J. L. (1995). Differential thermal-analysis and dielectric studies on 2-methyl-2-nitro-propoane under high-pressure. *Z. Natur. A*, **50**, 502–4. [336]

Buttar, D., Charlton, M. H., Docherty, R., and Starbuck, J. (1998). Theoretical investigations of conformational aspects of polymorphism. Part 1. *o*-Acetamidobenzamide. *J. Chem. Soc., Perkin Trans. 2*, 763–72. [219]

Buxton, C., Lynch, I. R., and Roe, J. M. (1988). Solid-state forms of paroxetine hydrochloride. *Int. J. Pharm.*, **42**, 135–43. [442]

Byrn, S. R. (1982). *Solid-State Chemistry of Drugs*, pp. 7–10. Academic Press, New York. [7, 8, 343]

Byrn, S. R., Curtin, D. Y., and Paul, I. C. (1972). The X-ray crystal structures of the yellow and white forms of dimethyl 3,6-dichloro-2,5-dihydroxyterephthalate and a study of the conversion of the yellow form to the white form in the solid state. *J. Am. Chem. Soc.*, **94**, 890–8. [305]

Byrn, S. R., Pfeiffer, R., Ganey, M., Hoiberg, C., and Poochikian, G. (1995). Pharmaceutical solids: a strategic approach to regulatory considerations. *Pharm. Res.*, **12**, 945–54. [353, 353f, 370, 375, 449, 451, 453]

Byrn, S. R., Pfeiffer, R. R., and Stowell, J. G. (1999). *Solid State Chemistry of Drugs*, 2nd edn. SSCI, Inc., West Lafayette, IN. [8, 105, 182 ,183f, 343, 344, 345, 346, 352, 364, 367, 370, 375]

Byrn, S. R., Sutton, P. A., Tobias, B., Frye, J., and Main, P. (1988). The crystal structure solid state NMR spectra and oxygen reactivity of five crystal forms of prednisolone *tert*-butylacetate. *J. Am. Chem. Soc.*, **110**, 1609–14. [184]

Byrne, J. F. and Kurz, P. F. (1967). Metal free phthalocyanine in the new x-form. Patent US 3 357 989 (1967); DE 1 619 654 (1967); GB 1 169 901. [389]

Cabana, B. E., Willhite, L. E., and Bierwagen, M. E. (1969). Pharmacokinetic evaluation of the oral adsorption of different ampicillin preparations in beagle dogs. *Antimicrob. Agents Chemother.*, **9**, 35–41. [348]

Cady, H. H. (1974). *The PETN–DiPEHN–TriPEON system*. LA-4486-MS, Los Alamos Scientific Laboratory, U.S. Atomic Energy Commission Contract W-7405-ENG. 36. [411, 427, 428]

Cady, H. H. and Larson, A. C. (1975). Pentaerythritrol tetranitrate II: Its crystal structure and transformation to PETNI, an algorithm for refinement of crystal structures with poor data. *Acta Crystallogr. B*, **31**, 1864–9. [411]

Cady, H. H. and Smith, L. C. (1962). *Studies on the polymorphs of HMX*. Los Alamos Scientific Laboratory LAMS-2652, Chemistry (TID-4500 17th ed.) based on Contract W-7405-ENG 36 with the U.S. Atomic Energy Commission. [403, 404t, 405, 406]

Cady, H. H., Larson, A. C., and Cromer, D. T. (1963). The crystal of α-HMX and a refinement of the structure of β-HMX. *Acta Crystallogr.*, **16**, 617–23. [404t]

Cailleau, H., Luty, T., Le Cointe, M., and Lemée-Cailleau, M.-L. (1997). Cooperative mechanism at the neutral-to-ionic transition. *Biul. Inst. Chem. Fiz. I Teor. Politech. Wroclawskiej*, **5**, 19–34. [283f, 284f, 285]

Cains, P. W. (2009). Classical methods of preparation of polymorphs and alternative solid forms. In *Polymorphism in Pharmaceutical Solids*, 2nd edn. (ed. H. G. Brittain), pp. 76–138. Informa Healthcare, New York. [92, 95]

Caira, M. R. (1998). Crystalline polymorphism of organic compounds. *Top. Curr. Chem.*, **198**, 163–208. [343]

Caira, M. R., Peinaar, E. W., and Lotter, A. P. (1996). Polymorphism and pseudo-polymorphism of the antibacterial nitrofurantoin. *Mol. Cryst. Liq. Cryst., Sci. Technol. A*, **278**, 241–64. [9]

Calleja, F. J. B., Arche, A. G., Ezquerra, T. A., Cruz, C. S., Batallan, F., Frick, B., and Cabarcos, E. L. (1993). Structure and properties of ferroelectric copolymers of poly (vinylidene flouride). *Adv. Polym. Sci.*, **108**, 1–48. [40]

Calvo, N. L., Maggio, R. M., and Kaufman, T. S. (2018). Chemometrics assisted solid-state characterization of pharmaceutically relevant materials. *J. Pharm. Biomed. Anal.*, **147**, 518–37. [359]

Cameroni, R., Coppi, G., Gamberini, G., and Forni, F. (1976). Dissolution and enzymic hydrolysis of chloramphenicol palmitic and stearic esters. *Farmaco Ed. Prat.*, **31**, 615–24. [350]

Cametti, M., Llander, L., Valkonen, A., Nieger, M., Nissinen, M., Nauha, E., and Rissanen, K. (2010). Non-centrosymmetric tetrameric assemblies of tetramethylammonium halides with uranyl salophen complexes in the solid state. *Inorg. Chem.*, **49**, 11473–84. [297]

Cammenga, H. K. and Hemminger, W. F. (1990). Thermomikroskopie. *Labo*, **21**, 7–19. [209]

Campeta, A. M., Chekal, B. P., Abramov, Y. A., Meenan, P. A., Henson, M. J., Shi, B., Singer, R. A., and Horspool, K. R. (2010). Development of a targeted polymorph screening approach for a complex polymorphic and highly solvating API. *J. Pharm. Sci.*, **99**, 3874–86. [109, 237, 345]

Capes, J. S. and Cameron, R. E. (2007). Contact line crystallization to obtain metastable polymorphs. *Cryst. Growth Des.*, **7**, 108–12. [101]

Capiomont, A., Bordeaux, D., and Lajzérowicz-Bonneteau, J. (1972). Crystal structure of the tetragonal form of the nitroxide 2,2,6,6-tetramethylpiperidine 1-oxyl. *C. R. Acad. Sci., Ser. C*, **275**, 317–20. [291]

Capiomont, A., Lajzerowicz, J., Legrand, J.-F., and Zeyen, C. (1981). Structure of the ferroelectric–ferroplastic phase of tanane (neutron diffraction). Evaluation of the lattice energy. *Acta Crystallogr. B*, **37**, 1557–60. [291]

Cardew, P. T. and Davey, R. (1982). *Tailoring of Crystal Growth*. Institute of Chemical Engineers, North Western Branch, Symposium Papers, Number 2, pp. 1.1–1.8. [57]

Cardew, P. T. and Davey, R. J. (1985). The kinetics of solvent-mediated phase transformations. *Proc. R. Soc. London, A*, **398**, 415–28. [88, 114, 390]

Cardew, P. T., Davey, R. J., and Ruddic, A. J. (1984). Kinetics of polymorphic solid-state transformations. *J. Chem. Soc., Faraday Trans.*, **80**, 659–68. [114]

Cardozo, R. L. (1991). Enthalpies of combustion, formation, vaporization and sublimation of organics. *AIChE J.*, **37**, 290–8. [228]

Caridi, A., Di Profio, G., Caliandro, R., Guagliardo, A., Curcio, E., and Drioli, E. (2012). Selecting the desired crystal form by membrane crystallizers: crystals or cocrystals. *Cryst. Growth Des.*, **12**, 4349–56. [101]

Carles, M., Eloy, D., Pujol, L., and Bodot, H. (1987). Photochromic and thermochromic salicylideneamines: isomerization in the crystal, infrared identities and conformational influence. *J. Mol. Struct.*, **156**, 43–58. [310]

Carletta, A., Buol, X., Leyssens, T., Champagne, B., and Wouters, J. (2016). Polymorphic and isomorphic cocrystals of a *N*-salicylidene-3-aminopyridine with dicarboxylic acids: tuning of solid-state photo- and thermochromism. *J. Phys. Chem. C*, **120**, 10001–8.

Carlin, R. L. (1989). *Magnetochemistry*. Springer, Berlin. [291, 313]

Carlson, K. D., Wang, H. H., Beno, M. A., Kini, A. M., and Williams, J. M. (1990). Ubiquitous superconductivity near 4K in salts of the BEDT-TTF/I system: is there a common source? *Mol. Cryst. Liq. Cryst.*, **181**, 91–104. [118, 119]

Carlton, R. A., Difeo, T. J., Powner, T. H., Santos, I., and Thomson, M. D. (1996). Preparation and characterization of polymorphs for an LTD4 antagonist, RG 12525. *J. Pharm. Sci.*, **85**, 461–7. [153]

Carper, W. R., Davis, L. P., and Extine, M. W. (1982). Molecular structure of 2,4,6-trinitrotoluene. *J. Phys. Chem.*, **86**, 459–62. [417, 418, 418t]

Carson, Rachel, 1907–1964. Silent Spring. Boston: Houghton Mifflin, 2002. [351]

Carson, J. F., Waisbrot, S. W., and Jones, F. T. (1943). A new form of crystalline xylitol. *J. Am. Chem. Soc.*, **65**, 1777–8. [86, 103]

Carstensen, J. T. (1977). *Pharmaceutics of Solids and Solid Dosage Forms*. John Wiley & Sons, New York. [346]

Cash, D. J. (1981). Exclusion chromatography of anionic dyes. Anomalous elusion peaks due to reversible aggregation. *J. Chromatogr.*, **209**, 405–12. [122]

Cassoux, P., Valade, L., Kobayashi, H., Kobayashi, A., Clark, R. A., and Underhill, A. E. (1991). Molecular metals and superconductors derived from metal complexes of 1,3-dithiol-2-thione-1,5-dithiotale (dmit). *Coord. Chem. Rev.*, **110**, 115–60. [119]

Catala, L., Wurst, K., Amabilino, D. B., and Veciana, J. (2006). Polymorphs of a pyrazole nitronyl nitroxide and its complexes with metal(II) hexafluoroacetylacetonates. *J. Mater. Chem.*, **16**, 2736–45. [288]

Censi, R. and Di Martino, P. (2015). Polymorph impact on the bioavailability and stability of poorly soluble drugs. *Molecules*, **20**, 18759–76. [347]

Ceolin, R., Agafonov, V., Gonthier-Vassal, A., Swarc, H., Cense, J. M., and Ladure, P. (1995). Solid-state studies on crystalline and glassy flutamide—thermodynamic evidence for dimorphism. *J. Therm. Anal.*, **45**, 1277–84. [371]

Chakraborty, D., Sengupta, N., and Wales, D. (2016). Conformational energy landscape of the ritonavir molecule. *J. Phys. Chem. B*, **120**, 4331–40. [344]

Chalmers, J. M. and Dent, G. (2006). Vibrational spectroscopic methods in pharmaceutical solid-state characterization. In *Polymorphism in the Pharmaceutical Industry* (ed. R. Hilfiker), pp. 95–138. Wiley-VCH Verlag, Weinheim. [170]

Chamot, E. M. and Mason, C. W. (1973). *Handbook of Chemical Microscopy*, Vol. 1, 3rd edn. Wiley-Interscience, New York. [33, 59, 61f, 137]

Chan, F. C., Anwar, J., Cernik, R., Barnes, R., and Wilson, R. M. (1999). *Ab initio* structure determination of sulfathiazole polymorph V from synchrotron X-ray powder diffraction data. *J. Appl. Crystallogr.*, **32**, 436–41. [155]

Chandra, A. K., Lim, E. C., and Ferguson, J. (1958). Absorption and fluorescence spectra of crystalline pyrene. *J. Chem. Phys.*, **28**, 765–8. [330]

Chang, L. C., Caira, M. R., and Guillory, J. K. (1995). Solid-state characterization of dehydroepiandrosterone. *J. Pharm. Sci.*, **84**, 1169–79. [365]

Chapman, S. J. and Whitaker, A. (1971). X-ray powder diffraction data for some azo pigments. *J. Soc. Dyers Colour.*, **87**, 120–1. [395]

Chekal, B. P., Campeta, A. M., Abramov, Y. A., Feeder, N., Glynn, P. P., McLaughlin, R. W., Meenan, P. A., and Singer, R. A. (2009). The challenges of developing an API crystallization process for a complex polymorphic and highly solvating system. Part I. *Org. Process Res. Dev.*, **13**, 1327–37. [109, 345, 347]

Chemburkar, S. R., Bauer, J., Deming, K., Spiwek, H., Patel, K., Morris, J., Henry, R., Spanton, S., Dziki, W., Porter, W., Quick, J., Bauer, P., Donaubauer, J., Narayanan, B. A., Soldani, M., Riley, D., and McFarland, K. (2000). Dealing with the impact of ritonavir polymorphs on the late stages of bulk drug process development. *Org. Process Res. Dev.*, **4**, 413–17. [210, 346, 367]

Chen, J., Sarma, B., Evans, J. M. B., and Myerson, A. S. (2011). Pharmaceutical crystallization. *Cryst. Growth Des.*, **11**, 887–95. [318, 343]

Chen, S., Guzei, I. A., and Yu, L. (2005). New polymorphs of ROY and a new record for coexisting polymorphs of solid structures. *J. Am. Chem. Soc.*, **127**, 9881–5. [229]

Chen, X. F., Liu, S. H., Duan, C. Y., Xu, Y. H., You, X. Z., Ma, J., and Min, N. B. (1998). Synthesis, crystal structure and triboluminescence of 1,4-dimethylpyridinium tetrakis(2-thenoyltrifluoroacetonato)europate. *Polyhedron*, **17**, 1883–9. [315]

Chen, X., Zhuang, P., and Ren, S. (1990). Study on crystal transformation of C.I. Pigment Red 57:1. *Huadong Huagong Xueyuan Xuebao*, **16**, 434–9. (*Chem. Abstr.*, **115**, 10900a) [381t]

Cheng, S. Z. D., Li, C. Y., Calhoun, B. H., Zhu, L., and Zhou, W. W. (2000). Thermal analysis: the next two decades. *Thermochim. Acta*, **355**, 59–68. [209]

Cheng, X., Zhang, Y., Han, S., Li, F., Zhang, H., and Wang, Y. (2016). Multicolor amplified spontaneous emission based on organic polymorphs that undergo excited-state intramolecular proton transfer. *Chem. Eur. J.*, **22**, 4899–903. [338]

Chennuru, R., Koya, R. T., Kommavarapu, P., Narasayya, S. V., Prakash, M., Vishweshar, P., Babu, R. R., and Mahapatra, S. (2017). In situ metastable form: a route for the generation of hydrate and anhydrous forms of ceritinib. *Cryst. Growth Des.*, **17**, 6341–52. [369]

Chiang, L.-C., Caira, M. R., and Guillory, J. K. (1995). Solid state characterization of dehydroepiandrosterone. *J. Pharm. Sci.*, **84**, 1169–79. [210]

Chiarelli, R. and Rassat, A. (1991). Magnetic properties of some biradicals of D_{2d} symmetry. In *Magnetic Molecular Materials* (ed. D. Gatteschi et al.), pp. 191–202. Kluwer Academic Publishers, Dordrecht. [286]

Chick, M. C. and Thorpe, B. W. (1970). *Polymorphism in 2,4,5-trinitrotoluene*. Report 382. Department of Supply, Australian Defence Scientific Service, Defence Standards Laboratories, Maribyrnong, Victoria. [425]

Chick, M. C. and Thorpe, B. W. (1971). Polymorphism in 2,4,5-trinitrotoluene. *Aust. J. Chem.*, **24**, 191–5. [425, 427]

Chickos, J. (1987). Heats of sublimation. In *Molecular Structure and Energetics. Volume 2. Physical Measurements* (ed. J. F. Liebman and A. Greenberg), pp. 67–150. VCH Publishers, New York. [50, 420]

Chickos, J. S., Braton, C. M., Hesse, D. G., and Liebman, J. F. (1991). Estimating entropies and enthalpies of organic compounds. *J. Org. Chem.*, **56**, 927–38. [45, 50]

Chieng, N., Rades, T., and Aaltonen, J. (2011). An overview of recent studies on the analysis of pharmaceutical polymorphs. *J. Pharm. Biomed. Anal.*, **55**, 618–44. [105, 137, 211, 212t, 343]

Chierotti, M. R., Ferrero, L., Garino, N., Gobetto, R., Pellegrino, L., Braga, D., Grepioni, F., and Maini, L. (2010). The richest collection of tautomeric polymorphs: the case of 2-thiobarbituric acid. *Chem. Eur. J.*, **16**, 4347–58. [125]

Chikaraishi, Y., Otsuka, M., and Matsuda, Y. (1995). Dissolution behavior of piretanide polymorphs at various temperatures and pHs. *Chem. Pharm. Bull.*, **43**, 1966–9. [346]

Childs, S. L., Chyall, L. J., Dunlap, J. T., Coates, D. A., Stahly, B. C., and Stahly, G. P. (2004). A metastable polymorph of metformin hydrochloride: Isolation and characterization using capillary crystallization and thermal microscopy techniques. *Cryst. Growth Des.*, **4**, 441–9. [99, 369]

Childs, S. L., Rodriguez-Hornedo, N., Reddy, L. S., Jayasankar, A., Maheshwari, C., McCausland, L., Shipplett, R., and Stahly, B. C. (2008). Screening strategies based on solubility composition generate pharmaceutically acceptable cocrystals of carbamazepine. *CrystEngComm*, **10**, 856–64. [271]

Chisolm, J. A. and Motherwell, S. (2005). COMPACK: a program for identifying crystal structure similarity using distances. *J. Appl. Crystallogr.*, **38**, 228–31. [242, 247]

Choi, C. S. and Boutin, H. P. (1970). Study of the crystal structure of β-cyclotetramethylene-tetranitramine by neutron diffraction. *Acta Crystallogr. B*, **26**, 1235–40. [404]

Choi, C. S. and Bulusu, S. (1974). The crystal structure of dinitropentamethylenetetramine. *Acta Crystallogr. B*, **30**, 1576–80. [414]

Choi, C. S. and Prince, E. (1972). The crystal structure of cyclotrimethylene-trinitramine. *Acta Crystallogr. B*, **28**, 2857–62. [410]

Choi, C. S. and Prince, E. (1979). 1,2,3-Triaminoguanidinium nitrate by neutron diffraction. *Acta Crystallogr. B*, **35**, 761–3. [412]

Cholerton, T. J., Hunt, J. H., Klinkert, G., and Martin-Smith, M. (1984). Spectroscopic studies on ranitidine: its structure and the influence of temperature and pH. *J. Chem. Soc., Perkin Trans. 2*, 1761–6. [172f, 209, 367]

Choquesillo-Lazarte, D. and Garcia-Ruiz, J. M. (2011). Poly(ethylene) oxide for small-molecule crystal growth in gelled organic solvents. *J. Appl. Crystallogr.*, **44**, 172–6. [101]

Chowdhury, A. U., Ye, D. H., Song, Z. T., Zhang, S. J., Hedderich, H. G., Mallick, B., Thirunahari, S., Ramakrishnan, S., Sengupta, A., Gualtieri, E. J., Bouman, C. A., and Simpson, G. J. (2017). Second harmonic generation guided Raman spectroscopy for sensitive detection of polymorphic transitions. *Anal. Chem.*, **89**, 5959–66. [359]

Chukanov, N. V., Zakharov, V. V., Korsunskiy, B. L., Chervonnyi, A. D., and Vozchikova, S. A. (2016). Kinetics of polymorphic transitions in energetic compounds. *Cent. Eur. J. Energ. Mater.*, **13**, 483–504. [399]

Chung, F. H. and Scott, R. W. (1971). Vacuum sublimation and crystallography of quinacridones. *J. Appl. Crystallogr.*, **4**, 506–11. [380]

Chung, F. H. and Smith, D. K. (eds.) (2000). *Industrial Uses of X-ray Diffraction*. Marcel Dekker, New York. [156]

Chung, H. and Diao, Y. (2016). Polymorphism as an emerging design strategy for high performance organic electronics. *J. Mater. Chem. C*, **4**, 3915–33. [274, 285]

Chunwachirasiri, W., West, R., and Winokur, M. J. (2000). Polymorphism, structure, and chromism in poly(di-*n*-octylsilane) and poly(di-*n*-decylsilane). *Macromolecules*, **33**, 9720–31. [40, 304]

Churchill, W. S. (1950). *The Second World War*, Vol. III, p. 814. Houghton Mifflin Co. Boston.

Ciba Ltd. (1965). Quinacridone pigments. Dutch Patent Application 6,405,130. *Chem. Abstr.*, **63**, 15103. [385]

Cicoria, F., Santato, C., and Rosei, F. (2008). Two-dimensional nanotemplates as surface cues for the controlled assembly of organic molecules. *Top. Curr. Chem.*, **285**, 203–67. [207]

Ciechanowicz-Rutkowska, M. (1977). An independent investigation of the crystal structure of 1,8-dinitronaphthalene (orthorhombic form) at 22 and 97°C. *J. Solid State Chem.*, **22**, 185–92. [427]

Ciechanowicz, M., Skapski, A. C., and Troughton, P. G. H. (1976). The crystal structure of the orthorhombic form of hydridodicarbonylbis(triphenylphosphine)iridium(I): successful location of the hydride hydrogen atom from X-ray data. *Acta Crystallogr. B*, **32**, 1673–80. [57]

Clark, G. R., Waters, J. M., and Waters, T. N. (1975). Crystal and molecular structure of brown form of bis(*N*-methyl-2-hydroxy-1-naphthaldiminato)copper(II). *J. Inorg. Nucl. Chem.*, **37**, 2455–8. [225]

Clark, G. R., Waters, J. M., Waters, T. N., and Williams, G. J. (1977). Polymorphism in Schiff base complexes of copper(II): the crystal and molecular structure of a second brown form of bis(*N*-methyl-2-hydroxy-1-naphthaldiminato)copper(II). *J. Inorg. Nucl. Chem.*, **39**, 1971–5. [225]

Clas, S. D., Dalton, C. R., and Hancock, B. C. (1999). Differential scanning calorimetry: application in drug development. *Pharm. Sci. Technol. Today*, **2**, 311–20. [366]

Cleverly, B. and Williams, P. P. (1959). Polymorphism in substituted barbituric acid. *Tetrahedron*, **7**, 277–88. [374]

Clydesdale, G., Roberts, K. J., and Docherty, R. (1994a). Computational studies of the morphology of molecular crystals through solid-state intermolecular force calculations using the atom–atom method. In *Colloid and Surface Engineering: Controlled Particle, Droplet and Bubble Formation* (ed. D. J. Wedlock), pp. 95–135. Butterworth Heinemann, London. [61]

Clydesdale, G., Roberts, K. J., and Docherty, R. (1994b). Modeling the morphology of molecular crystals in the presence of disruptive tailor-made additives. *J. Cryst. Growth*, **135**, 331–40. [61]

Clydesdale, G., Roberts, K. J., and Docherty, R. (1996). HABIT95—a program for predicting the morphology of molecular crystals as a function of the growth environment. *J. Cryst. Growth*, **166**, 78–83. [61, 93, 181]

Clydesdale, G., Roberts, K. J., and Walker, E. M. (1997). The crystal habit of molecular materials: a structural perspective. In *Theoretical Aspects and Computer Modeling of the Molecular Solid State* (ed. A. Gavezzotti), pp. 203–32. John Wiley & Sons, Chichester. [61]

Cobbledick, R. E. and Small, R. W. H. (1974). Crystal structure of the δ-form of 1,3,5,7-tetranitro-1,3,5,7-tetraazacyclooctane (δ-HMX). *Acta Crystallogr. B*, **30**, 1918–22. [404t]

Cobbledick, R. E. and Small, R. W. H. (1975). Crystal structure of the complex formed between 1,3,5,7-tetranitro-1,3,5,7-tetraazacyclooctane (HMX) and *N,N*-dimethylformamide (DMF). *Acta Crystallogr. B*, **31**, 2805–8. [403]

Cohen, M. D. and Schmidt, G. M. J. (1964). Topochemistry. Part I. A survey. *J. Chem. Soc.*, 1966–2000. [34, 273]

Cohen, M. D., Hirschberg, Y., and Schmidt, G. M. J. (1964a). Topochemistry. Part VII. The photoactivity of anils of salicylaldehydes in rigid solutions. *J. Chem. Soc.*, 2051–9. [310]

Cohen, M. D., Hirschberg. Y., and Schmidt, G. M. J. (1964b). Topochemistry. Part VIII. The effect of solvent, temperature and light on the structure of anils by hydroxynaphthaldehydes. *J. Chem. Soc.*, 2060–7. [310]

Cohen, M. D., Schmidt, G. M. J., and Flavian, S. (1964c). Topochemistry. Part VI. Experiments on photochromy and thermochromy of crystalline anils of salicylaldehydes. *J. Chem. Soc.*, 2041–51. [218]

Cohen, M. D., Schmidt, G. M. J., and Sonntag, F. I. (1964d). Topochemistry. Part II. The photochemistry of *trans*-cinnamic acids. *J. Chem. Soc.*, 2000–13. [34, 332]

Cohen, R., Ludmer, Z., and Yakhot, V. (1975). Structural influence on the excimer emission from a dimorphic crystalline stilbene. *Chem. Phys. Lett.*, 34, 271–4. [330, 331f]

Colapietro, M., Domenicano, A., Marciante, C., and Portalone, G. (1984a). Angular ring distortions in benzene derivatives. *Acta Crystallogr. A*, **40**, C98–9. [161t, 218]

Colapietro, M., Domenicano, A., Portalone, G., Schulz, G., and Hargittai, I. (1984b). Molecular structure and ring distortion of *p*-dicyanobenzene in the gas phase and in the crystal. *J. Mol. Struct.*, **112**, 141–57. [218]

Colapietro, M., Domenicano, A., Portalone, G., Torrini, I., Hargittai, I., and Schultz, G. (1984c). Molecular structure and ring distortions of *p*-diisocyanobenzene in the gaseous phase and in the crystal. *J. Mol. Struct.*, **125**, 19–32. [218]

Coleman, L. B., Cohen, M. J., Sandman, D. J., Yamagishi, F. G., Garito, A. F., and Heeger, A. J. (1973). Superconducting fluctuations and the Peierls instability in an organic solid. *Solid State Commun.*, **12**, 1125–32. [275]

Collins, A., Wilson, C. C., and Gilmore, C. J. (2010). Comparing entire crystal structures using cluster analysis and fingerprint plots. *CrystEngComm*, **12**, 801–9. [255]

Colthup, N., Daly, L. H., and Wiberly, S. E. (1990). *Introduction to Infrared and Raman Spectroscopy*. Academic Press, London. [176]

Commins, P., Desta, I. T., Karothu, D. P., Panda, M. K., and Naumov, P. (2016). Crystals on the move: mechanical effects in dynamic solids. *Chem. Commun.*, **52**, 13941–54. [320]

Connick, W., May, F. G. J., and Thorpe, B. W. (1969). Polymorphism in 2,4,6-trinitrotoluene. *Aust. J. Chem.*, **22**, 2685–8. [420, 421f]

Cooke, P. M. (1998). Chemical microscopy. *Anal. Chem.*, **70**, 385R–423R. [137]

Coombes, D. S., Nagi, G. K., and Price, S. L. (1997). On the lack of hydrogen bonds in the crystal structure of alloxan. *Chem. Phys. Lett.*, **265**, 532–7. [239]

Cooper, P. W. and Kurowski, S. R. (1996). *Introduction to the Technology of Explosives*. VCH, New York. [398, 399, 417]

Cooper, W., Edmonds, J. W., Wudl, F., and Coppens, P. (1974). The 2-2'-bi-1,3-dithiole. *Cryst. Struct. Commun.*, **3**, 23–6. [65f]

Coppens, P. (2011). Molecular excited-state structure by time-resolved pump-probe X-ray diffraction. What is new and what are the prospects for further progress? *J. Phys. Chem. Lett.*, **2**, 816–21. [332]

Coppens, P. and Fournier, B. (2015). New methods in time-resolved Laue pump-probe crystallography at synchrotron sources. *J. Synchrotron Radiat.*, **22**, 280–7. [319]

Coppens, P., Fomitchev, D. V., Carducci, M. D., and Culp, K. (1998). Crystallography of molecular excited states. Transition-metal nitrosyl complexes and the study of transient species. *J. Chem. Soc., Dalton Trans.*, 865–72. [42, 332]

Coppens, P., Maly, K., and Petricek, V. (1990). Composite crystals: what are they and why are they so common in the organic solid state? *Mol. Cryst. Liq. Cryst.*, **181**, 81–90. [42, 119]

Cordes, A. W., Haddon, R. C., Hicks, R. G., Oakley, R. T., and Palstra, T. T. M. (1992a). Preparation and solid-state structures of (cyanophenyl)dithia- and (cyanophenyl)diselenadiazolyl. *Inorg. Chem.*, **31**, 1802–8. [117]

Cordes, A. W., Haddon, R. C., Hicks, R. G., Oakley, R. T., Palstra, T. T. M., Schneemeyer, L. F., and Waszczak, J. V. (1992b). Polymorphism of 1,3-phenyl bis(diselenadiazolyl). Solid-state structure and electronic properties of β-1,3-[(Se$_2$N$_2$C)C$_6$H$_4$(CN$_2$Se$_2$)]. *J. Am. Chem. Soc.*, **114**, 1729–32. [118]

Cornelissen, J. P., Haasnoot, J. G., Reedijk, J., Faulmann, C., Legros, J.-P., Cassoux, P., and Negrey, P. J. (1992). Crystal structures and electrochemical properties of two phases of tetrabutylammonium bis(1,3-dithiole-2-thione-4,5-diselenolato)-nickelate (III). *Inorg. Chim. Acta*, **202**, 131–9. [276]

Cornelissen, J. P., Pomarede, B., Spek, A. L., Reefman, D., Haasnoot, J. G., and Reedijk, J. (1993). Two phases of [Me$_4$N][Ni(dmise)$_2$]$_2$: synthesis, crystal structures, electrical conductivities and intermolecular overlap calculations of α and β tetramethylammonium bis[bis(2-selenoxo-1,3-dithiole-4,5-dithiolato)nickelate], the first conductors based on the M(C$_3$S$_4$Se) system. *Inorg. Chem.*, **32**, 3720–6. [276]

Cornell, W. D., Cieplak, P., Bayly, C. I., Gould, I. R., Merz, K. M., Jr., Ferguson, D. M., Spellmeyer, D. C., Fox, T., Caldwell, J. W., and Kollman, P. A. (1995). A second generation force field for the simulation of proteins, nucleic acids and organic molecules. *J. Am. Chem. Soc.*, **117**, 5179–97. [227]

Corpinot, M. K., Stratford, S. A., Arhangelskis, M., Anka-Lufford, J., Halasz, I., Judaš, N., and Bučar, D.-K. (2016). On the predictability of supramolecular interactions in molecular cocrystals – the view from the bench. *CrystEngComm*, **18**, 5434–9. [102]

Corradini, P. (1973). X-ray studies of conformation: observation of different geometries of the same molecule. *Chem. Ind. (Milan)*, **55**, 122–9. [4, 218, 220f, 221, 257]

Covaci, O.-I., Mitran, R. A., Buhalteanu, L., Dumitrescu, D. G., Shova, S., and Manta, C. M. (2017). Bringing new life into old drugs: a case study on nifuroxazide. *CrystEngComm*, **19**, 3584–91. [345]

Cox, S. R. and Williams, D. E. (1981). Representation of the molecular electrostatic potential by a net atomic charge model. *J. Comput. Chem.*, **2**, 304–23. [218]

Craig, D. Q. M. (2006). Characterization of polymorphic systems using thermal analysis. In *Polymorphism in the Pharmaceutical Industry* (ed. R. Hilfiker), pp. 43–79. Wiley-VCH, Weinheim. [147, 366]

Craig, D. Q. M., Royall, P. G., Kett, V. L., and Hopton, M. L. (1999). The relevance of the amorphous state to pharmaceutical dosage forms: glassy drugs and freeze-dried systems. *Int. J. Pharm.*, **179**, 179–207. [370]

Cramer, F., Sprinzl, R., Furgac, N., Freist, W., Saenger, W., Manor, P., Sprinzll, L., and Sternback, H. (1974). Crystallization of yeast phenylalanine transfer RNA: polymorphism and studies of sulfur-substituted mercury binding derivatives. *Biochim. Biophys. Acta*, **349**, 351–65. [30]

Craven, B. M., Vizzini, E. A., and Rodriguez, M. M. (1969). The crystal structures of two polymorphs of 5,5'-diethylbarbituric acid. *Acta Crystallogr. B*, **25**, 1978–92. [240]

Crawford, S., Kirchner, M. T., Bläser, D., Boese, R., David, W. I., Dawson, A., Gehrke, A., Ibberson, R. M., Marshall, W. G., Parsons, S., and Yamamuro, O. (2009). Isotopic polymorphism in pyridine. *Angew. Chem. Int. Ed.*, **48**, 755–7. [6]

Cromer, D. T., Lee, K.-Y., and Ryan, R. R. (1988). Structures of two polymorphs of 1,1'-dinitro-3,3'-azo-1,2,4-triazole. *Acta Crystallogr. C*, **44**, 1673–4. [426]

Crookes, D. L. (1985). Aminoalkyl furan derivative. U.S. Patent 4 521 431. [195, 375]

Crosby, J., Pittam, J. D., Wright, N. C. A., Ohashi, M., and Murkami, K. (1999). Impact of crystallization behavior on the development of the azole antifungal ZD0870. Presented at the 1st International Symposium on Aspects of Polymorphism and Crystallisation—Chemical Development Issues, Hinckley, UK, June 23–25, 1999. Scientific Update, Mayfield, UK. [375]

Crottaz, O., Kubel, F., and Schmid, H. (1997). Jumping crystals of the spinels of $NiCr_2O_4$ and $CuCr_2O_4$. *J. Mater. Chem.*, 7, 143–6. [320]

Cruz-Cabeza, A. and Bernstein, J. (2014). Conformational polymorphism. *Chem. Rev.*, 114, 2170–91. [4, 27, 84, 219, 221, 259, 260f, 261f, 262f, 263f, 264f, 265f, 267f, 345]

Cruz-Cabeza, A., Liebeschuetz, J. N., and Allen, F. H. (2012). Systematic conformational bias in small molecule crystal structures is rare and explicable. *CrystEngComm*, 14, 6797–811. [218]

Cruz-Cabeza, A., Reutzel-Edens, S. M., and Bernstein, J. (2015). Facts and fictions of polymorphism. *Chem. Soc. Rev.*, 44, 8619–35. [2f, 17, 19, 19t, 69, 108, 360, 433, 445, 446]

Cui, H.-L., Ji, G.-F., Chen, X.-R., Zhang, Q.-M., Wei, D.-Q., and Zhao, F. (2010). Phase transitions and mechanical properties of octahydro-1,3,5,7-tetranitro-1,3,5,7-tetrazocine in different crystal phases by molecular dynamics simulation. *J. Chem. Eng. Data*, 55, 3121–9. [407]

Cui, Y. (2007). The material science perspective of pharmaceutical solids. *Int. J. Pharm.*, 339, 3–18. [343]

Curcio, E., Di Profio, G., and Drioli, E. (2003). A new membrane-based crystallization technique: tests on lysozyme. *J. Cryst. Growth*, 247, 166–76. [100]

Curry, C. J., Rendle, D. F., and Rogers, A. (1982). Pigment analysis in the forensic examination of paints. I. Pigment analysis by X-ray powder diffraction. *J. Forensic Sci.*, 22, 173–7. [380t, 382t]

Curtin, D. Y. and Engelmann, J. H. (1972). Intramolecular oxygen–nitrogen benzoyl migration of 6-aroyloxyphenanthridines. *J. Org. Chem.*, 37, 3439–43. [10]

Curtin, D. Y. and Paul, I. C. (1981). Chemical consequences of the polar axis in organic solid-state chemistry. *Chem. Rev.*, 81, 525–41. [296]

Curtin, D. Y., Paul, I. C., Duesler, E. N., Lewis, T. W., Mann, B. J., and Shiau, W.-I. (1979). Studies of thermal reactions in the solid state. *Mol. Cryst. Liq. Cryst.*, 50, 25–42. [335]

Dähne, L. and Biller, E. (1998a). Excitonic interactions in dye arrays. *J. Inf. Recording*, 24, 171–7. [330]

Dähne, L. and Biller, E. (1998b). Color variation in highly oriented dye layers by polymorphism of dye aggregates. *Adv. Mater.*, 10, 241–5. [330]

Dähne, S. and Kulpe, S. (1977). *Structural Principles of Unsaturated Organic Compounds*. Abhandlungen der Akademie der Wissenschaften der DDR, Akademie-Verlag, Berlin. [377]

Dai, X., Moffat, J. G., Wood, J., and Reading, M. (2012). Thermal scanning probe microscopy in the development of pharmaceuticals. *Adv. Drug. Deliv. Rev.*, 64, 449–60. [154]

Dainippon (1982). Preparation of imidazolone series orange color pigment—by heating aqueous suspension of α-type diazonitrochloroaniline coupled with acetoacetylaminobenzoimidazolone. Japanese Patent 57–141 457; *Chem. Abstr.*, 98, 55548p. [380t]

Dainippon (1986). Japanese Patent 272 697-A. [381t]

Dainippon (1988). Japanese Patent 358 810-A. [381t]

Dainippon (1994). δ-Type and anthrone blue pigment—is a mixture containing titanium dioxide and is useful as car paint having good weatherability, crystal stability and dispersibility. Patent WO 9605255-A1. [383t]

Dainippon (1996a). New crystal type monoazo lake pigment. Japanese Patent 09194752-A. [381t]

Dainippon (1996b). New crystal type monoazo lake pigment. Japanese Patent 09227791-A. [381t]

Dainippon (1996c). Novel crystalline monoazo lake pigment. Japanese Patent 092411524-A. [381t]

Dainippon (1996d). New crystal type monoazo lake pigment. Japanese Patent 09268529-A. [381t]

Damiri, S., Namvar, S., and Pahahi, H. (2017). Micro-seeding and soft template effects on the control of polymorph and morphology of HMX micro particles in solvent-anti-solvent process. *Defense Technol.*, **13**, 1–5. [407]

Daniels, F., Williams, J. W., Bender, P., Alberty, R. A., Cornwell, C. D., and Harriman, J. E. (1970). *Experimental Physical Chemistry*, 7th edn, p. 53. McGraw-Hill, New York. [228]

DaSilva, J. G. and Miller, J. S. (2013). Pressure-dependent enhanced T_c and magnetic behavior of the metamagnetic and ferromagnetic polymorphs of $[Fe^{III}Cp^\star_2]^{.+}[TCNQ]^{.-}$ ($Cp^\star$ = pentamethylcyclopentadienide; $TCNQ$ = 7,7,8,8-tetracyano-*p*-quinodimethane). *Inorg. Chem.*, **52**, 1108–12. [287]

Davey, R. J. (1993). General discussion. *Faraday Discuss.*, **95**, 160–2. (See also Cardew and Davey, 1982.) [57, 134]

Davey, R. J., Back, K. R., and Sullivan, R. (2015). Crystal nucleation from solutions – transition states, rate determining steps and complexity. *Faraday Discuss.*, **179**, 9–26. [84]

Davey, R. J., Blagden, N., Potts, G. D., and Docherty, R. (1997). Polymorphism in molecular crystals: Stabilization of a metastable form by conformational mimicry. *J. Am. Chem. Soc.*, **119**, 1767–72. [101, 368, 369]

Davey, R. J., Blagden, N., Righini, S., Allison, H., Quayle, M. J., and Fuller, S. (2000). Crystal polymorphism as a probe for molecular assembly during nucleation from solutions: the case of 2,6-dihydroxybenzoic acid. *Cryst. Growth Des.*, **1**, 59–65. [87]

Davey, R. J., Maginn, S. J., Andrews, S. J., Black, S. N., Buckley, A. M., Cottier, D., Dempsey, P., Plowman, R., Rout, J. E., Stanley, D. R., and Taylor, A. (1994). Morphology and polymorphism in molecular crystals: terephthalic acid. *J. Chem. Soc., Faraday Trans.*, **90**, 1003–9. [61, 65f, 73, 159, 160f]

Davey, R. J., Maginn, S. J., Andrews, S. J., Buckley, A. M., Cottier, D., Dempsey, P., Rout, J. E., Stanley, D. R., and Taylor, A. (1993). Stabilization of a metastable crystalline phase by twinning. *Nature*, **366**, 248–50. [92]

Davey, R. J., Schroeder, S. L. M., and ter Horst, J. H. (2013). Nucleation of organic crystals—a molecular perspective. *Angew. Chem. Int. Ed.*, **52**, 2166–79. [34, 80, 83, 83f, 84]

David, R. and Giron, D. (1994). Crystallization. *Handb. Powder Technol.*, **9**, 193–241. [8]

David, W. I. F., Shankland, K., and Shankland, N. (1998). Routine determination of molecular crystal structures from powder diffraction data. *J. Chem. Soc., Chem. Commun.*, 931–2. [155]

David, W. I. F., Shankland, K., Pulham, C. R., Blagden, N., Davey, R. J., and Song, M. (2005). Polymorphism in benzamide. *Angew. Chem. Int. Ed.*, **43**, 7032–5. [156]

Davidson, A. J., Oswald, I. D. H., Francis, D. J., Lennie, A. R., Marshall, W. G., Millar, D. I. A., Pulham, C. R., Warren, J. E., and Cumming, A. E. (2008). Explosives under pressure – the crystal structure of gamma-RDX as determined by high-pressure X-ray and neutron diffraction. *CrystEngComm*, **10**, 162–5. [410]

Davies, E. S. and Hartshorne, N. H. (1934). Identification of some aromatic nitro-compounds by optical crystallographic methods. *J. Chem. Soc.*, 1830–6. [302]

Davis, R., Rath, N. P., and Das, S. (2004). Thermally reversible fluorescent polymorphs of alkoxy-cyano-substituted diphenylbutadienes: role of crystal packing in solid state fluorescence. *Chem. Commun.*, 74–5. [339]

Day, G. M., Cooper, T. G., Cruz-Cabeza, A. J., Hejczyk, K. E., Ammon, H. L., Boerrigter, S. X., Tan, J. S., Della Valle, R. G., Venuti, E., Jose, J., and Gadre, S. R. (2009). Significant progress in predicting the crystal structures of small organic molecules—a report on the fourth blind test. *Acta Crystallogr. B*, **65**, 107–25. [241]

Day, G. M., Motherwell, W. D. S., Ammon, H. L., Boerrigter, S. X. M., Della Valle, R. G., Venuti, E., Dzyabchenko, A., Dunitz, J. D., Schweizer, B., Van Eijck, B. P., and Erk, P. (2005). A third blind test of crystal structure prediction. *Acta Crystallogr B*, **61**, 511–27. [241]

Day, G. W., Hamilton, C. A., Peterson, R. L., Phelan, R. J., and Mullen, L. O. (1974). Effects of poling conditions on responsivity and uniformity of polarization of PVF2 pyroelectric detectors. *Appl. Phys. Lett.*, **24**, 456–8. [292]

Day, J. and McPherson, A. (1991). Characterization of two crystal forms of cytochrome c from *Valida membranaefaciens*. *Acta Crystallogr. B*, **47**, 1020–2. [127f, 128]

de Gelder, R., Wehrens, R., and Hageman, J. A. (2001). A generalized expression for the similarity of spectra: Application to powder diffraction patterns classification. *J. Comput. Chem.*, **22**, 273–89. [271]

De Ilarduya, M. C. T., Martin, C., Goni, M. M., and Martinez-Oharriz, M. C. (1997). Polymorphism of sulindac: isolation and characterization of a new polymorph and three new solvates. *J. Pharm. Sci.*, **86**, 248–51. [9]

de Jong, A. W. K. (1922a). Über die Einwirkung des Lichtes auf die zimtsaüren und über die Kontitution der Truxillsaüren. *Chem. Ber.*, **55**, 463–74. [332]

de Jong, A. W. K. (1922b). Über die konstitution der Truxill- und Truxinsaüren und über die Einwirkung des Sonnenlichtes auf die Zimtsaüren und Zimtsaüre-Salze. *Chem. Ber.*, **56**, 818–32. [332]

de Matas, M., Edwards, H. G. M., Lawson, E. E., Shields, L., and York, P. (1998). FT-Raman spectroscopic investigation of a pseudopolymorphic transition in caffeine hydrate. *J. Mol. Struct.*, **440**, 97–104. [9, 177]

de Wet, F. N., Gerber, J. J., Lotter, A. P., van der Watt, J. G., and Dekker, T. G. (1998). A study of the changes during heating of paracetamol. *Drug Dev. Ind. Pharm.*, **24**, 447–53. [249]

Decurtins, S., Shoemaker, C. B., and Wockman, H. H. (1983). Structure of a new modification of bromobis(N,N-diethyldithiocarbamato)iron(III), $Fe[S_2CN(C_2H_5)_2]_2Br$. *Acta Crystallogr. C*, **39**, 1218–21. [285, 365]

Deeley, C. M., Spragg, R. A., and Threlfall, T. L. (1991). A comparison of Fourier transform infrared and near-infrared Fourier transform Raman spectroscopy for quantitative measurements: an application in polymorphism. *Spectrochim. Acta A*, **47**, 1217–23. [177]

Deene, W. A. and Small, R. W. H. (1971). Refinement of the structure of rhombohedral acetamide. *Acta Crystallogr. B*, **27**, 1094–8. [240]

Deffet, L. (1942). *Repertoire des Composés Organique Polymorphes*. Editions Desoer, Liége. [21]

Deij, M. A., Vissers, T., Meekes, H., and Vlieg, E. (2007). Toward rational design of tailor-made additives using growth site statistics. *Cryst. Growth Des.*, **7**, 778–86. [62]

DeJong, E. J. (1979). Nucleation: a review. In *Industrial Crystallization 78* (7th Symposium, Warsaw) (ed. E. J. DeJong and S. J. Jančić), pp. 3–17. North-Holland, Amsterdam. [81]

Delaney, S. P., Nethercott, M. J., and Mays, C. J. (2017). Characterization of synthesized and commercial forms of magnesium stearate using differential scanning calorimetry, thermogravimetric analysis, powder X-ray diffraction, and solid-state NMR spectroscopy. *J. Pharm. Sci.*, **106**, 338–47. [364]

Delori, A., Friščić, T., and Jones, W. (2012). The role of mechanochemistry and supramolecular design in the development of pharmaceutical materials. *CrystEngComm*, **14**, 2350–62. [358]

Deschamps, J. R., Frisch, N., and Parrish, D. (2011). Thermal expansion of HMX. *J. Chem. Crystallogr.*, **41**, 366–70. [407]

Desiraju, G. R. (1983). Intermolecular proton transfers in the solid state. Conversion of the hydroxyazo into the quinone hydrazone tautomer of 2-amino-3-hydroxy-6-phenylazo-pyridine. X-ray crystal structures of the two forms. *J. Chem. Soc., Perkin Trans. 2*, 1025–30. [125]

Desiraju, G. R. (1987). Crystal engineering at 4 Å – short axis structure for planar chloro aromatics. In *Organic Solid State Chemistry* (ed. G. R. Desiraju), pp. 519–46. Studies in Organic Chemistry, Vol. 32, Elsevier, Amsterdam. [360]

Desiraju, G. R. (1989). *Crystal Engineering—the Design of Organic Solids*. Material Science Monographs, Vol. 54, Elsevier, Amsterdam. [218, 287]

Desiraju, G. R. (1997). Crystal gazing: structure prediction and polymorphism. *Science*, **278**, 404–7. [445]

Desiraju, G. R. (2004). Counterpoint: what's in a name? *Cryst. Growth Des.*, **4**, 1089–90. [8]

Desiraju, G. R. and Nangia, A. (2016). Use of the term "Crystal Engineering" in the regulatory and patent literature of pharmaceutical solid forms. *Cryst. Growth Des.*, **16**, 5585–6. [131, 434]

Desiraju, G. R., Curtin, D. Y., and Paul, I. C. (1977). Crystal growth by nonaqueous gel diffusion. *J. Am. Chem. Soc.*, **99**, 6148. [356]

Desiraju, G. R. and Steiner, T. (1999). *The Weak Hydrogen Bond*. Oxford University Press, Oxford. [218]

Desiraju, G. R., Paul, I. C., and Curtin, D. Y. (1977). Conversion in the solid state of the yellow to the red form of 2-(4′-methoxyphenyl)-1,4-benzoquinone. X-ray crystal structures and anisotropy of the rearrangement. *J. Am. Chem. Soc.*, **99**, 1594–601. [305]

Desiraju, G. R., Vittal, J. J., and Ramanan, A. (2011). *Crystal Engineering: A Textbook*. World Scientific, Singapore. [19]

Deuschel, W., Gundel, F., Wuest, H., and Daubach, E. (1963). Process for producing a modification of gamma 7,14-dioxo-5,7,12,14-tetrahydroquinolino-(2,3-b) acridine. U.S. Patent 3 074 950. [386]

Deuschel, W., Honigmann, B., Jettmar, W., and Schroeder, H. (1964). Quinacridone pigments. British Patent 923 069. *Chem. Abstr.*, **61**, 8447. [382t, 385]

Devilliers, M. M., van der Watt, J. G., and Lotter, A. P. (1991). The interconversion of the polymorphic forms of chloramphenicol palmitate (CAP) as a function of environmental temperature. *Drug Dev. Ind. Pharm.*, **17**, 1295–303. [351]

Dey, A. and Desiraju, G.R. (2006). Dimorphs of 4′-amino-4-hydroxy-2′-methylbiphenyl: Assessment of likelihood of polymorphism in flexible molecules. *CrystEngComm*, **8**, 477–81. [19]

Dhananjay, D., Thomas, S. P., Spackman, M. A., and Chopra, D. (2016). "Quasi-isostructural polymorphism" in molecular crystals: inputs from interaction hierarchy and energy frameworks. *Chem. Commun.*, **52**, 2141–4. [269]

Di Martino, P., Guyot-Hermann, A. M., Conflant, P., Drach, M., and Guyot, J. C. (1996). A new pure paracetamol for direct compression: the orthorhombic form. *Int. J. Pharm.*, **128**, 1–8. [375]

Di Martino, P., Scoppa, M., Joivís, E., Palmieri, G. F., Andres, C., Pourcelot, Y., and Martelli, S. (2001). The spray drying of acetazolamide as a method to modify crystal properties and to improve compression behavior. *Int. J. Pharm.*, **213**, 209–21. [375]

Di Profio, G., Tucci, S., Curcio, E., and Drioli, E. (2007). Selective glycine polymorph crystallization by using microporous membranes. *Cryst. Growth Des.*, **7**, 526–30. [101]

Diao, Y., Helgeson, M. E., Siam, Z. A., Doyle, P. S., Myerson, A. S., and Hatton, T. A. (2012). Nucleation under soft confinement: role of polymer–solute interactions. *Cryst. Growth Des.*, **12**, 508–17. [100]

Diao, Y., Lenn, K. M., Lee, W.-Y., Blood-Forsythe, M. A., Xu, J., Mao, Y., Kim, Y., Reinspacj, J. A., Park, S., Aspuru-Guzik, A., Xue, G., Clancy, P., Zhenan, B., and Mannsfeld, S. C. B. (2014). Understanding polymorphism in organic semiconductor thin films through nanoconfinement. *J. Am. Chem. Soc.*, **136**, 17046–57. [281]

Dick, J. J. (1995). Shock-wave behavior in explosive monocrystals. *J. Phys. IV*, **5**, 103–6. [398]

Dickinson, C. and Holden, J. R. (1977). Crystal structures of hexanitrodiphenol amine and its potassium salt. *American Crystallographic Association Series 2*, 5, 55 (Abstracts, American Crystallographic Association Summer Meeting, August 1977, East Lansing, MI, Abstract PA7). [427]

DiMasi, J. D., Grabowski, H. G., and Hansen, R. W. (2014). Cost to develop and win marketing approval for a new drug is $2.6 billion. Available at http://csdd.tufts.edu/news/complete_story/pr_tufts_csdd_2014_cost_study [432]

Ding, J., Herbst, R., Praefcke, K., Kohne, B., and Saenger, W. (1991). A crystal that hops in phase transition, the structure of *trans,trans,anti,trans,trans*-perhydropyrene. *Acta Crystallogr. B*, **47**, 739–42. [319]

Dinnebier, R. and Billinge, S. (eds.) (2009). *Powder Diffraction – Theory and Practice*. RSC Publishing, Boca Raton, FL. [156]

Dobler, M. (1984). 18-Crown-6: only a simple molecule? *Chimia*, **38**, 415–21. [218]

Dollimore, D. and Phang, P. (2000). Thermal analysis. *Anal. Chem.*, **72**, 27R–36R. [154]

Domenicano, A. and Hargittai, I. (1993). Gas/crystal structural differences in aromatic molecules. *Acta Chim. Hung. Models Chem.*, **130**, 347–62. [218]

Domenicano, A., Schultz, G., Hargittai, I., Colapietro, M., Portalone, G., George, P., and Bock, C. W. (1990). Molecular structure in nitrobenzene in the planar and orthogonal conformations. A concerted study by electron diffraction, X-ray crystallography and molecular orbital calculations. *Struct. Chem.*, **1**, 107–22. [161t]

Domingos, S., Andre, V., Quaresma, S., Maltins, I. C. B., da Piedade, M. F. M., and Duarte, M. T. (2015). New forms of old drugs: improving without changing. *J. Pharm. Pharmacol.*, **67**, 830–46. [345, 356]

Donohue, J. (1974). *The Structure of the Elements*. Wiley, New York. [29]

Donohue, J. (1985). Revised space group frequencies for organic compounds. *Acta Crystallogr. A*, **41**, 203–4. [238]

Donahue, M., Botonjuc-Sehic, E., Wells, D., and Brown, C. W. (2011). Understanding infrared and Raman spectra of pharmaceutical polymorphs. *Am. Pharm. Rev.*, **14**(2).

Dorset, D. L. (1996). Electron crystallography. *Acta Crystallogr. B*, **52**, 753–69. [156, 178f]

Dorset, D., McCourt, M. P., Gao, L., and Voigt-Martin, I. G. (1998). Electron crystallography of small molecules: criteria for data collection and strategies for structure solution. *J. Appl. Crystallogr.*, **31**, 544–53. [156]

Douillet, J., Stevenson, N., Lee, M., Mallet, F., Ward, R., Aspin, P., Dennethy, D. R., and Camus, L. (2012). Development of a solvate as an active pharmaceutical ingredient: development, crystallization and isolation challenges. *J. Cryst. Growth*, **342**, 2–8. [355]

Dračínský, M., Procházková, E., Kessler, J., Šebestík, J., Matějka, P., and Bouř, P. (2013). Resolution of organic polymorphic crystals by Raman spectroscopy. *J. Phys. Chem. B*, **117**, 7297–307. [177]

Draguet-Brughmans, M., Bouche, R., Flandre, J. P., and van den Bulcke, A. (1979). Polymorphism and bioavailability of pentobarbital. *Pharm. Acta Helv.*, **54**, 140–5. [347, 347f]

Dreger, Z. A. (2012). Energetic materials under high pressures and temperatures: stability, polymorphism and decomposition of RDX. *J. Phys.: Conf. Ser.*, **377**, 012047. [410]

Dreger, Z. A. and Gupta, Y. M. (2013). High pressure–high temperature polymorphism and decomposition of penterythritol tetranitrate (PETN). *J. Phys. Chem. A*, **117**, 5306–13. [411]

Dreger, Z. A., Stash, A. I., Yu, Z.-G., Chen, Y.-S., Tao, Y., and Gupta, Y. M. (2016a). High-pressure structural response of an insensitive energetic crystal: 1,1-diamino-2,2-dinitroethene (FOX-7). *J. Phys. Chem. C*, **120**, 1218–24. [415]

Dreger, Z. A., Stash, A. I., Yu, Z.-G., Chen, Y.-S., Tao, Y., and Gupta, Y. M. (2016b). High-pressure structural response of an insensitive energetic crystal: 1,1-diamino-2,2-dinitroethene (FOX-7). *J. Phys. Chem. C*, **120**, 27600–7. [415]

Dreger, Z. A., Tao, Y., and Gupta, Y. M. (2016c). Phase diagram and decomposition of 1,1-diamino-2,2-dinitroethene single crystals at high pressures and temperatures. *J. Phys. Chem. C*, **120**, 11092–8. [415]

Dreger, Z. A., Tao, Y., Averkiev, B. B., Gupta, Y. M., and Klapötke, T. M. (2015). High-pressure stability of energetic crystal of dihydroxylammonium 5,5′-bistetrazole-1,1′-diolate: Raman spectroscopy and DFT calculations. *J. Phys. Chem. B*, **119**, 6836–47. [415]

Dromzee, Y., Chiarelli, R., Gambarelli, S., and Rassat, A. (1996). Dupeyredioxyl (1,3,5,7-tetramethyl-2,6-diazaadamantane-*N*,*N*′-dioxyl). *Acta Crystallogr. C*, **52**, 474–7. [285, 286f]

Drucker, C. (1925). *Hand- und Hiflsbuch zur Ausführung physikochemischer Messungen*, pp. 228–33. Akademische Verlagsgesellschaft mbH, Leipzig. [208]

Du, W., Cruz-Cabeza, A. J., Woutersen, S., Davey, R. J., and Yina, Q. (2015). Can the study of self-assembly in solution lead to a good model for the nucleation pathway? The case of tolfenamic acid. *Chem. Sci.*, **6**, 3515–24. [84]

Dubey, R. and Desiraju, G. R. (2015). Exploring the crystal structure landscape with a heterosynthon module: fluorobenzoic acid:1,2-bis(4-pyridyl)thylene 2:1 cocrystals. *Cryst. Growth Des.*, **15**, 489–96. [102]

Dubnikova, F., Kosloff, R., Almog, J., Zeiri, Y., Boese, R., Itzhaky, H., Alt, A., and Keinan E. (2005). Decomposition of triacetone triperoxide is an entropic explosion. *J. Am. Chem. Soc.*, **127**, 1146–59. [411]

Duebel, R., Schui, F., and Wester, N. (1984). Crystal β-form of Pigment Red 53:1—with more yellow tone, for coloring paints, plastics and printing inks. Hoechst Patent EP 97 913. [380t]

Duggirala, N. K., Perry, M. L., Almarsson, O., and Zaworotko, M. J Z. (2016). Pharmaceutical co-crystals: along the path to improved medicines. *Chem. Commun.*, **52**, 640–55. [360]

Dunitz, J. D. (1979). *X-ray Analysis and the Structure of Organic Molecules*. Cornell University Press, Ithaca, NY. [59, 159, 215, 218]

Dunitz, J. D. (1991). Phase transitions in molecular-crystals from a chemical standpoint. *Pure Appl. Chem.*, **63**, 177–85. [5, 7, 41, 42, 319. 336]

Dunitz, J. D. (1995). Phase changes and chemical reactions in molecular crystals. *Acta Crystallogr. B*, **51**, 619–31. [4, 24, 126]

Dunitz, J. D. (1996). Thoughts on crystals as supermolecules. In *The Crystal as a Supramolecular Entity. Perspectives in Supramolecular Chemistry*, Vol. 2 (ed. G. R. Desiraju), pp. 1–30. Wiley, Chichester. [5, 227, 239]

Dunitz, J. D. (2016). Phase transitions in molecular crystals: looking backwards, glancing sideways. *Phys. Scripta*, **91**, 112501. [132]

Dunitz, J. D. and Bernstein, J. (1995). Disappearing polymorphs. *Acc. Chem. Res.*, **28**, 193–200. [34, 86, 116, 131, 132]

Dunitz, J. D. and Gavezzotti, A. (1999). Attractions and repulsions in molecular crystals: what can be learned from the crystal structures of condensed ring aromatic hydrocarbons? *Acc. Chem. Res.*, **32**, 677–84. [217]

Dunitz, J. D. and Gavezzotti, A. (2005a). Molecular recognition in organic crystals: directed intermolecular bonds or nonlocalized bonding? *Angew. Chem. Int. Ed.*, **44**, 1766–87. [267, 269]

Dunitz, J. D. and Gavezzotti, A. (2012). Supramolecular synthons: validation and ranking intermolecular interaction energies. *Cryst. Growth Des.*, **12**, 5873–7. [269]

Dunitz, J. D., Filippini, G., and Gavezzotti, A. (2000). Molecular shape and crystal packing: a study of $C_{12}H_{12}$ isomers, real and imaginary. *Helv. Chim. Acta*, **83**, 2317–35. [240]

Dunning, W. J. and Shipman, A. J. (1954). *Progress in Agricultural Industries 10th International Conference*, Madrid, pp. 1448–56. [84]

Earl, D. and Deem, M. W. (2005). Parallel tempering: theory, applications and new perspectives. *Phys. Chem. Chem. Phys.*, **7**, 3910–16. [243]

Eastes, J. W. (1956). Preparation of phthalocyanine pigments. American Cyanamid Patent US 2 770 629. [389]

Ebert, A. A., Jr. and Gottlieb, H. B. (1952). Infrared spectra of organic compounds exhibiting polymorphism. *J. Am. Chem. Soc.*, **74**, 2806–10. [169, 389]

Eddleston, M. D., Bithell, E. G., and Jones, W. (2010). Transmission electron microscopy of pharmaceutical materials. *J. Pharm. Sci.*, **99**, 4072–83. [195, 197f]

Eddleston, M. D., Hejczyk, K. E., Bithell, E. G., Day, G. M., and Jones, W. (2013). Polymorph identification and crystal structure determination by a combined crystal structure prediction and transmission electron microscopy approach. *Chem. Eur. J.*, **19**, 7874–82. [195]

Edwards, G. (1950). Vapor pressure of 2,4,6-trinitrotoluene. *Trans. Faraday Soc.*, **46**, 423–7. [420]

Egorova, K. S., Gordeev, E. G., and Ananikov, V. P. (2017). Biological activity of ionic liquids and their application in pharmaceutics and medicine. *Chem. Rev.*, **117**, 7132–89. [356]

Elacqua, E., Laird, R. C., and MacGillivray, L. R. (2012). Templated [2+2] photodimerizations in the solid state. In *Supramolecular Chemistry: From Molecules to Nanomaterials* (ed. J. W. Steed and P. A. Gale), pp. 3153–66. Wiley-VCH, Weinheim. [334]

Ellern, A., Bernstein, J., Becker, J. Y., Zamir, S., Shahal, L., and Cohen, S. A. (1994). New polymorphic modification of tetrathiafulvalene. Crystal structure, lattice energy and intermolecular interactions. *Chem. Mater.*, **6**, 1378–85. [65f, 394]

Ellison, R. D. and Holmberg, R. W. (1960). Cell dimensions and space group of 1,1-diphenyl-2-picrylhydrazine. *Acta Crystallgr.*, **13**, 446–7. [292]

Emmons, E. D., Farrell, M. E., Holtoff, E. L., Tripathi, A., Green, N., Moon, R. P., Guicheteau, J. A., Christensen, S. D., Pellegrino, P. M., and Fountain, A. W., 3rd. (2012). Characterization of polymorphic states in energetic samples of 1,3,5-trinitro-1,3,5-triazine (RDX) fabricated using drop-on-demand inkjet technology. *Appl. Spectrosc.*, **66**, 628–35. [411]

Encyclopaedia Britannica. (1798). [31]

Engelberg, A. B. (1999). Special patent provisions for pharmaceuticals: have they outlived their usefulness? A political legislative and legal history of U.S. law and observations for the future. *IDEA: J. Law Technol.*, **39**, 389. [430, 439]

Enokida, T. and Hirohashi, R. (1991). A new crystal of copper phthalocyanine synthesized with 1,8-diaza-bicyclo-(5,4,0)undecane-7. *Mol. Cryst. Liq. Cryst.*, **195**, 265–79. [390, 392]

Enokida, T. and Hirohashi, R. (1992). Electrophotographic dual-layered photoreceptors incorporating copper phthalocyanines. *J. Imag. Sci. Technol.*, **36**, 135–41. [296]

Ephraim, F. (1923). Über die Löslichkeit von Kobaltiaken. *Ber. Dtsch. Chem. Ges.*, **56**, 1530–42. [337]

Erdemir, D., Lee, A. Y., and Myerson, A. (2009). Nucleation of crystals from solution: classical and two-step models. *Acc. Chem. Res.*, **42**, 621–9. [80, 82]

Erk, P. (1998). *Proceedings of the 2nd International Symposium on Phthalocyanines*, Edinburgh. [383t]

Erk, P. (1999). Crystal design from molecular to application properties. In *From Molecules and Crystals to Materials* (ed. D. Braga, F. Grepioni and A. G. Orpen), NATO ASI Series, Vol. C538, pp. 143–61. Kluwer Academic Publishers, Dordrecht. [377, 378]

Erk, P. (2000). Private communication. [378]

Erk, P. (2001). Crystal design of organic pigments—a prototype discipline of materials science. *Curr. Opin. Solid State Mater. Sci.*, **5**, 155–60. [378, 387]

Erk, P. (2002). Communication to the Cambridge Structural Database. [391]

Erk, P. (2009). Crystal design of high performance pigments. In *High Performance Pigments*, 2nd edn (ed. E. B. Faulkner and R. J. Schwarz), pp. 105–27. Wiley, New York. [393]

Erk, P. (2012). Personal communication to M. U. Schmidt. [390]

Erk, P. and Hengelsberg, H. (2003). Phthalocyanine dyes and pigments. *Porphyrin Handb.*, **19**, 105–49. [390, 391]

Erk, P., Hengelsberg, H., Haddow, M. F., and van Gelder, R. (2004). The innovative momentum of crystal engineering. *CrystEngComm*, **6**, 475–83. [377, 391]

Erlenmeyer, E., Jr., Brakow, C., and Herz, O. (1907). Isomeric cinnamic acids. *Chem. Ber.*, **40**, 653–63. [332]

Eshkova, Z. I., Moiseeva, S. S., Konysheva, L. I., Bir, E. Sh., and Demina, L. V. (1976). X-ray study of binary systems on the basis of linear quinacridone. *FATIPEC Cong.*, **13**, 270–4. [381t]

Espenbetov, A. A., Antipin, M. Yu., Struchkov, Yu. T., Philippov, V. A., Trirel'son, V. G., Ozerov, R. P., and Svetlov, B. S. (1984). Structure of 1,3-propanethiol trinitrate (β-modification), $C_3H_5N_3O_9$. *Acta Crystallogr. C*, **40**, 2096–8. [412]

Etter, M. C. (1985). Aggregate structures of carboxylic acids and amides. *Isr. J. Chem.*, **25**, 312–19. [69]

Etter, M. C. (1990). Encoding and decoding hydrogen-bond patterns of organic compounds. *Acc. Chem. Res.*, **23**, 120–6. [68, 69]

Etter, M. C. (1991). Hydrogen bonds as design elements in organic chemistry. *J. Phys. Chem.*, **95**, 4601–10. [68, 69, 78, 81, 360]

Etter, M. C. and Siedle, A. R. (1983). Solid state rearrangement of (phenylazophenyl)palladium hexafluoroacetylacetonate. *J. Am. Chem. Soc.*, **105**, 641–3. [311]

Etter, M. C., Huang, K. S., Frankenbach, G. M., and Adsmond, D. (1991). Control of symmetry and asymmetry in hydrogen-bonded nitroaniline materials. *ACS Symp. Ser.*, **455**, 446–55. [296]

Etter, M. C., Kress, R. B., Bernstein, J., and Cash, D. J. (1984). Solid-state chemistry and structures of a new class of mixed dyes. Cyanine-oxonol. *J. Am. Chem. Soc.*, **106**, 6921–7. [121, 123f]

Etter, M. C., MacDonald, J. C. and Bernstein, J. (1990a). Graph-set analysis of hydrogen-bond patterns in organic crystals. *Acta Crystallgr. B*, **46**, 256–62. [69, 71]

Etter, M. C., Urbanczyk-Lipkowska, Z., Zia-Ebrahimi, M., and Panunto, T. W. (1990b). Hydrogen bond-directed cocrystallization and molecular recognition properties of diarylureas. *J. Am. Chem. Soc.*, **112**, 8415–26. [116]

European Patent Office (1997). Decision of 27 May 1997, Case Number T 0475/92—3.3.4, Application Number 83301951.6, Publication Number 0091787, IPC C07K 3/12. [456]

Evans, R. M. (1974). *The Perception of Color.* Wiley, New York. [377]

Even, J. and Bertault, M. (2000). Transformations in molecular crystals and chemical reactions: a physico-chemical approach. *Condens. Matter News*, **8**, 9–21. [319, 335]

Everall, N. J. (2000). Confocal Raman spectroscopy: why the depth resolution and spatial accuracy can be much worse than you think. *Appl. Spectrosc.*, **54**, 1515–20. [209]

Evers, J., Klapötke, T. M., Mayer, P., Oehlinger, G., and Welch, J. (2006). α- And β-FOX-7, polymorphs of a high energy density material, studied by X-ray single crystal and powder investigations in the temperature range 200 to 423 K. *Inorg. Chem.*, **45**, 4996–5007. [414]

Eyring, H., Gershinowitz, H., and Sun, C. E. (1935). The absolute rate of homogeneous atomic reactions. *J. Chem. Phys.*, **3**, 786–95. [84]

Fabbiani, P. A., Allan, D. R., David, W. I. F., Moggach, S. A., Parsons, S., and Pulham, C. R. (2004). High-pressure recrystallization: a route to new polymorphs and solvates. *CrystEngComm*, **6**, 505–11. [336]

Faber, J. and Blanton, J. (2008). Full pattern comparison of experimental and calculated powder patterns using the Integral Index method in PDF-4+. *Powder Diff.*, **23**, 141–5. [272]

Faigman, D. L. (1999). *Legal Alchemy. The Use and Misuse of Science and the Law.* W. H. Freeman and Co., New York. [430]

Fairrie, G. (1925). *Sugar.* Fairrie and Co. Ltd., Liverpool. [86f]

Falini, G., Albeck, S., Weiner, S., and Addadi, L. (1996). Control of aragonite or calcite polymorphism by mollusk shell macromolecules. *Science*, **271**, 67–9. [30]

Fanconi, B. M., Gerhold, G. A., and Simpson, W. T. (1969). Influence of exciton phonon interaction on metallic reflection from molecular crystals. *Mol. Cryst. Liq. Cryst.*, **6**, 41–81. [329]

Faraday Discussions No. 136. (2007). *Crystal Growth and Nucleation.* Royal Society of Chemistry, London. [81]

Faraday Discussions No. 179. (2015). *Nucleation: A Transition State to the Directed Assembly of Materials.* Royal Society of Chemistry, London. [80, 81]

Farbwerke vorm Meister Lucius & Bruening (1902). Verfahren zur Herstellung eines roten, besonders zur Berreitung von Farblacken geeigneten Monoazofarbstoffes aus *o*-chlor-*m*-toluidin-*p*-sulfosaeure und *beta*-naphthol. Deutsches Reichspatent DRP 145 908. [380t]

Farmer, V. C. (1957). Effects of grinding during the preparation of alkali-halide disks on the infrared spectra of hydroxylic compounds. *Spectrochim. Acta*, **8**, 374–89. [170]

Farnum D. G., Mehta, G., Moore, G. I., and Siegal, F. P. (1974). Attempted Reformatskii of benzonitrile 1,4-dioxo-3,6-diphenylpyrrolo[3,4-C]pyrrole, a lactam analog of pentalene. *Tetrahedron Lett.*, **29**, 2549–52. [394]

Fawcett, T. G. (1987). Greater than the sum of its parts: a new instrument. *Chemtech*, 564–9. [209]

Fawcett, T. G., Crowder, C., and Needham, F. (2009). Effective use of the powder diffraction pattern file. ICDD presentation at PPXRD-8, May 2009. Glasgow, UK. Available at http://www.icdd.com/ppxrd/08/presentation/fawcett-effectivepowderdiffraction-ppxrd-8.pdf [158f]

Fawcett, T. G., Harris, W. C., Newman, R. A., Whiting, L. F., and Knoll, F. J. (1989). Combined thermal analyzer and X-ray diffractometer. U.S. Patent 4 821 303. [209]

Fawcett, T. G., Martin, E. J., Crowder, C. E., Kincaid, P. J., Strandjord, A. J., Blazy, J. A., Newman, R. A., and Armentrout, D. N. (1986). Analyses of multi-phase pharmaceuticals using simultaneous differential scanning calorimetry and X-ray diffraction. *Adv. X-ray Anal.*, **29**, 323–32. [209]

Federal Register (2018). Available at https://www.federalregistger.gov/dE7-13171/page-37245 [375]

Feher, R., Wurst, K., Amabilino, D., and Veciana, J. (2008). Polymorphic and hydrate supramolecular solid state structures of a uracil derived nitronyl nitroxide. *Inorg. Chim. Acta*, **361**, 4094–9. [288]

Fernández-Mato, A., Garcia, M. D., Peinador, C., Quinteia, J. M., Sánachez-Andújar, M. S., Pato-Doldán, B., Señaris-Rodriguez, M. A., Tordera, D., and Bolink, H. J. (2013). Polymorphism-triggered reversible thermochromic fluorescence of a simple 1,8-naphthyridine. *Cryst. Growth Des.*, **13**, 460–4. [318]

Ferraris, J., Cowan, D. O., Walatka, V., and Perlstein, J. H. (1973). Electron transfer in a new, highly conducting donor–acceptor compound. *J. Am. Chem. Soc.*, **95**, 948–9. [275]

Ferraro, J. R. and Nakamoto, K. (1994). *Introductory Raman Spectroscopy*. Academic Press, London. [176]

Ferro, D. R., Bruckner, S., Meille, S. V., and Ragazzi, M. (1992). Energy calculations for isotactic polypropylene—a comparison between models of the alpha-crystalline and gamma-crystalline structures. *Macromolecules*, **25**, 5231–5. [238]

FIAT (1948). *Field Information Agency Technical: German Dyestuffs and Intermediates*, Final Report 1313, Vol. III. Dyestuff Research, Technical Industrial Intelligence Division, U.S. Department of Commerce, Washington, DC, pp. 434–63. [383t, 389, 393, 394]

Fiebich, K. and Mutz, M. (1999). Evaluation of calorimetric and gravimetric methods to quantify the amorphous content of desferal. *J. Therm. Anal. Calorim.*, **57**, 75–85. [371]

Figueroa-Navedo, A. M., Ruiz-Caballero, J. L., Pacheco-Londoño, L. C., and Hernández-Rivera, S. P. (2016). Characterization of α- and β-RDX polymorphs in crystalline deposits on stainless steel substrates. *Cryst. Growth Des.*, **16**, 3631–8. [411]

Filippini, G., Gavezzotti, A., and Novoa, J. J. (1999). Modeling the crystal structure of the 2-hydronitronylnitroxide radical (HNN): observed and computer-generated polymorphs. *Acta Crystallogr. B*, **55**, 543–55. [238]

Findlay, A. F. (1951). *The Phase Rule and its Applications*, 9th edn (ed. A. N. Campbell and N. O. Smith), pp. 7–19. Dover Publications, New York. [5, 16, 34, 35f, 42, 49]

Findlay, W. P. and Bugay, D. E. (1998). Utilization of Fourier transform Raman spectroscopy for the study of pharmaceutical crystal forms. *J. Pharm. Biomed. Anal.*, **16**, 921–30. [177]

Firsich, D. W. (1984). Energetic material separations and specific polymorph preparation via thermal gradient sublimation. *J. Hazard. Mater.*, **9**, 133–7. [425]

Fischer, P., Zolliker, P., Meier, B. H., Ernst, R. R., Hewat, A. W., Jorgensen, J. D., and Rotella, F. J. (1986). Structure and dynamics of terephthalic acid from 2 to 300 K. I. High resolution neutron diffraction evidence for a temperature dependent order-disorder transition—a comparison of reactor and pulsed neutron source powder techniques. *J. Solid State Chem.*, **61**, 109–25. [161t]

FIZ (2001). *Inorganic Crystal Structure Database (ICSD)*. Fachinformationszentrum Karlsruhe, Gesellschaft für wissenschaft lichtechische Information mbH, Eggenstein-Leopoldschafen, Germany. [29]

Fletton, R. A., Lancaster, R. W., Harris, R. K., Kenwright, A. M., Packer, K. J., Waters, D. N., and Yeadon, A. (1986). A comparative spectroscopic investigation of two polymorphs of 4′-methyl-2′-nitroacetanilide using solid-state infrared and high-resolution solid-state nuclear magnetic resonance spectroscopy. *J. Chem. Soc., Perkin Trans. 2*, 1705–9. [321]

Florence, A. J. (2009). Approaches to high-throughput physical form screening and discovery. In *Polymorphism in Pharmaceutical Solids*, 2nd edn (ed. H. G. Brittain), pp. 139–84, Informa Healthcare, New York. [92, 96, 97f, 105, 107f, 108, 108t, 135]

Florence, A. J. (2010). Polymorph screening in pharmaceutical development. *Eur. Pharm. Rev.*, Issue 4, 19 August. [353]

Flores, L. and Jones, F. (1972). Physicothermal stabilities of dye solids. II. Phase transitions and melting behavior of some disperse and vat dyes. *J. Soc. Dyers Colour.*, **88**, 101–6. [379t]

Foltz, M. F. (1994). Thermal stability of ε-hexanitrohexaazaisowurtzitane in an estane formulation. *Propellants, Explos., Pyrotech.*, **19**, 63–9. [408]

Foltz, M. F., Coon, C. L., Garcia, F., and Nichols III, A. L. (1994a). The thermal stability of the polymorphs of hexanitrohexaazaisowurtzitane. Part I. *Propellants, Explos., Pyrotech.*, **19**, 19–25. [408]

Foltz, M. F., Coon, C. L., Garcia, F., and Nichols III, A. L. (1994b). The thermal stability of the polymorphs of hexanitrohexaazaisowurtzitane. Part II. *Propellants, Explos., Pyrotech.*, **19**, 133–44. [408]

Fomitchev, D. V., Bagley, K. A., and Coppens, P. (2000). The first crystallographic evidence for side-on coordination of N_2 to a single metal center in a photoinduced metastable state. *J. Am. Chem. Soc.*, **122**, 532–3. [42]

Forrest, S. R. and Zhang, Y. (1994). Ultrahigh-vacuum quasiepitaxial growth of model van der Waals thin films. I. Theory. *Phys. Rev. Condens. Matter*, **49**, 11297–308. [382t]

Forster, A., Gordon, K., Schmierer, D., Soper, N., Wu, V., and Rades, T. (1998). Characterization of two polymorphic forms of ranitidine-HCl. *Internet J. Vib. Spectrosc.*, **2**, 1–12. [172f]

Förster, T. and Kasper, K. (1954). Ein Konzentrationsumschlag der Fluoreszenz. *Z. Phys. Chem. Neue Folge*, **1**, 275–7. [330]

Foster, J. A., Damodaran, K. K., Maurin, A., Day, G. M., Thomson, H. P. G., Cameron, G. J., Bernal, J. C., and Steed, J. W. (2017). Pharmaceutical polymorph control in a drug-mimetic supramolecular gel. *Chem. Sci.*, **8**, 78–84. [356]

Foster, K. R. and Huber, P. W. (1999). *Judging Science. Scientific Knowledge and the Federal Courts*. MIT Press, Cambridge, MA. [430]

Foster, R. (1969). *Organic Charge Transfer Molecules*. Academic Press, London. [218]

Francis, C. V. and Tiers, G. V. D. (1992). Straight-chain carbamyl compounds for second harmonic generation. *Chem. Mater.*, **4**, 353–8. [302]

Francis, F. and Piper, S. H. (1939). The higher *n*-aliphatic acids and their methyl and ethyl esters. *J. Am. Chem. Soc.*, **61**, 577–81. [142]

Frankenheim, M. L. (1835). *Die Lehre von der Cohäsion*. August Schulz, Breslau. [32]

Frankenheim, M. L. (1839). Ueber die Isomerie. *Praktische Chem.*, **16**, 1–14. [31]

Fraxedas, J., Caro, J., Santiso, J., Figueras, A., Gorostiza, P., and Sanz, F. (1999a). Molecular organic thin films of *p*-nitrophenyl nitronyl nitroxide: surface morphology and polymorphism. *Phys. Status Solidi B*, **215**, 859–63. [207, 208]

Fraxedas, J., Caro, J., Santiso, J., Figueras, A., Gorostiza, P., and Sanz, F. (1999b). Polymorphic transformation observed on molecular organic thin films: *p*-nitrophenyl nitronyl nitroxide radical. *Europhys. Lett.*, **48**, 461–7. [288]

Free, M. L. and Miller, J. D. (1994). Effect of sample and incident beam areas on quantitative spectroscopy. *Appl. Spectrosc.*, **48**, 891–3. [173]

Freer, S. T. and Kraut, J. (1965). Crystal structures of D,L-homocysteine thiolactone hydrochloride: two polymorphic forms and a hybrid. *Acta Crystallogr.*, **19**, 992–1002. [42, 119]

Freyer, A. J., Lowe-Ma, C. K., Nissan, R. A., and Wilson, W. S. (1992). Synthesis and explosive properties of dinitropicrylbenzimidazoles and the "trigger linkage" in dinitropicrylbenzotriazoles. *Aust. J. Chem.*, **45**, 525–39. [426]

Friedländer, P. (1879). IX. Krystallographische Untersuchung einiger organischen Verbindungen. *Z. Kristallogr.*, **3**, 168–79. [417, 420]

Froelich, H. (1989). Beta-crystal modification of an azo pigment. Hoechst EP 320774 A2. [382t]

Frommer, J. (1992). Scanning tunneling microscopy and atomic force microscopy in organic chemistry. *Angew. Chem. Int. Ed. Engl.*, **31**, 1298–328. [207]

Frydman, L., Oliviery, A. C., Diaz, L. E., Frydman, B., Schmidt, A., and Vega, S. (1990). A ^{13}C solid-state NMR study of the structure and dynamics of the polymorphs of sulfanilamide. *Mol. Phys.*, **70**, 563–79. [189]

Fryer, J. R. (1997). Pigments: myth shape and structure. *Surf. Coat. Int.*, **80**, 421–6. [42, 119, 378]

Fryer, J. R., McKay, R. B., Mather, R. R., and Sing, K. S. (1981). The technological importance of the crystallographic and surface properties of copper phthalocyanine pigments. *J. Chem. Technol. Biotechnol.*, **31**, 371–87. [388, 391]

Fu, J.-H., Rose, J., Tam, M. F., and Wang, B.-C. (1994). New crystal forms of a μ-class glutathione 5-transferase from rat liver. *Acta Crystallogr. D*, **50**, 219–22. [127f, 127]

Fuji, O., Takano, M., Sakatami, T., and Iwamoto, E. (1980). Yellow pigment with good heat and light resistance – is bistetrachloroisoindoline-1-one-3-ylidine phenylene-1,4-diamine cdp. Toyo Soda Mfg. Co. Ltd. Patent JP 55-12106; *Chem. Abstr.*, **93**, 27823c. [382t]

Furedi-Milhofer, H., Garti, N., and Kamyshny, A. (1999). Crystallization from microemulsions—a novel method for the preparation of new crystal forms of aspartame. *J. Cryst. Growth*, **199**, 1365–70. [453]

Fyfe, C. A. (1983). *Solid State NMR for Chemists*. CFC Press, Guelph, ON. [182]

Gadade, D. D. and Pekamwar, S. S. (2016). Pharmaceutical cocrystals: regulatory and strategic aspects, design and development. *Adv. Pharm. Bull.*, **6**, 479–94. [360]

Galek, P. T. A., Allen, F. H., Fabian, L., and Feeder, N. (2009). Knowledge-based H-bond prediction to aid experimental polymorph screening. *CrystEngComm*, **11**, 2634–9. [140]

Galindo, S., Tamayo, A., Leonardi, F., and Mas-Torent, M. (2017). Control of polymorphism and morphology in solution sheared organic field-effect transistors. *Adv. Funct. Mater.*, **27**, 1700526. [282]

Gallagher, H. G. and Sherwood, J. N. (1996). Polymorphism, twinning and morphology of crystals of 2,4,6-trinitrotoluene grown from solution. *J. Chem. Soc., Faraday Trans.*, **92**, 2107–16. [417, 418, 420, 422f]

Gallagher, H. G., Roberts, K. J., Sherwood, J. N., and Smith, L. A. (1997). A theoretical examination of the molecular packing, intermolecular bonding and crystal morphology of 2,4,6-trinitrotoluene in relation to polymorphic structural stability. *J. Mater. Chem.*, **7**, 229–35. [420, 422f]

Gallier, J., Toudic, B., Delugeard, Y., Cailleau, H., Gourdji, M., Peneau, A., and Guibe, L. (1993). Chlorine-nuclear-quadrupole-resonance study of the neutral-to-ionic transition tetrathiafulvalene-chloranil. *Phys. Rev. C*, **47**, 11688–95. [284]

Gao, P. (1996). Determination of the composition of delaviridine mesylate polymorph and pseudopolymorph mixtures using C-13 CP/MAS NMR. *Pharm. Res.*, **13**, 1095–104. [9]

Gao, Z., Rohani, S., Gong, J., and Wang, J. (2017). Recent developments in the crystallization process: toward the pharmaceutical industry. *Engineering*, **3**, 343–53. [360]

Garg, R. K. and Sarkar, D. (2016). Polymorphism control of *p*-aminobenzoic acid by iso-thermal anti-solvent crystallization. *J. Cryst. Growth*, **454**, 180–5. [197, 198f]

Garside, J. (1985). Industrial crystallization from solution. *Chem. Eng. Sci.*, **40**, 3–26. [81]

Garside, J. and Davey, R. J. (1980). Secondary contact nucleation: kinetics, growth and scale-up. *Chem. Eng. Commun.*, **4**, 393–424. [81]

Garti, N. and Sato, K. (1988). *Crystallization and Polymorphism of Fats and Fatty Acids.* Marcel Dekker, New York. [40]

Garti, N. and Zour, H. (1997). The effect of surfactants on the crystallization and polymorphic transformation of glutamic acid. *J. Cryst. Growth*, **172**, 486–98. [40]

Gavezzotti, A. (1991). Generation of possible crystal structures from molecular structure for low-polarity organic compounds. *J. Am. Chem. Soc.*, **113**, 4622–9. [219, 238]

Gavezzotti, A. (1994a). Are crystal structures predictable? *Acc. Chem. Res.*, **27**, 309–14. [215, 238]

Gavezzotti, A. (1996). Polymorphism of 7-dimethylaminocyclopenta[*c*]coumarine: Packing analysis and generation of trial crystal structures. *Acta Crystallogr. B*, **52**, 201–8. [238]

Gavezzotti, A. (1997). Computer simulations of organic solids and their liquid-state pre-cursors. *Faraday Discuss.*, **106**, 63–77. [238]

Gavezzotti, A. (2002). Calculation of intermolecular interaction by direct numerical inte-gration over electron densities. I. Electrostatic and polarization energies in molecular crystals. *J. Phys. Chem. B*, **106**, 4145–54. [218, 267]

Gavezzotti, A. (2003). Calculation of intermolecular interaction energies by direct numeri-cal integration over electron densities. 2. An improved polarization model and the evalu-ation of dispersion and repulsion energies. *J. Phys. Chem. B*, **107**, 2344–53. [267]

Gavezzotti, A. (2007a). A solid-state chemist's view of crystal polymorphism of organic compounds. *J. Pharm. Sci.*, **96**, 2232–41. [267]

Gavezzotti, A. (2007b). *Molecular Aggregation: Structure Analysis and Molecular Simulation of Crystals and Liquids.* Oxford University Press, Oxford.

Gavezzotti, A. and Desiraju, G. R. (1988). A synthetic analysis of packing energies and other packing parameters for fused-ring aromatic hydrocarbons. *Acta Crystallogr. B*, **44**, 427–34. [126]

Gavezzotti, A. and Filippini, G. (1995). Polymorphic forms of organic crystals at room conditions: thermodynamic and structural implications. *J. Am. Chem. Soc.*, **117**, 12299–305. [27, 45, 63, 227, 229, 238]

Gavezzotti, A., Filippini, G., Kroon, J., van Eijck, B. P., and Klewinghaus, P. (1997). The crystal polymorphism of tetrolic acid ($CH_3C{\equiv}CCOOH$): a molecular dynamics study of precursors in solution, and in crystal structure generation. *Chem. Eur. J.*, **3**, 893–9. [238]

Gdanitz, R. J. (1997). *Ab initio* prediction of possible molecular crystal structures. In *Theoretical Aspects and Computer Modeling of the Molecular Solid State* (ed. A. Gavezzotti), pp. 185–99. Wiley, Chichester. [238]

Gdanitz, R. J. (1998). *Ab initio* prediction of molecular crystal structure. *Curr. Opin. Solid State Mater. Sci.*, **3**, 414–18. [238]

Gebauer, D. and Cölfen, H. (2011). Prenucleation clusters and non-classical nucleation. *Nano Today*, **6**, 564–84. [82]

Gebauer, D., Kellermeier, M., Gale, J. D., Bergström, L., and Cölfen, H. (2014). Pre-nucleation clusters as solute precursors in crystallization. *Chem. Soc. Rev.*, **43**, 2348–71. [83]

Geladi, P. and Kowalski, B. R. (1986). Partial least-squares regression: a tutorial. *Anal. Chim. Acta*, **185**, 1–17. [176]

Gelbrich, T., Hughes, D. S., Hursthouse, M. B., and Threlfall, T. L. (2008). Packing similarity in polymorphs of sulfathiazole. *CrystEngComm*, **10**, 1328–34. [121]

Genck, W. J. (2000). The effects of mixing on scale-up—how crystallization and precipitation react. *Chem. Process.*, **63**, 47. [375]

George, R. S., Cady, H. H., Rogers, R. N., and Rohwer, R. K. (1965). Solvates of octahydro-1,3,5,7-tetranitro-1,3,5,7-tetrazocaine (HMX). Relatively stable monosolvates. *Ind. Eng. Chem. Prod. Res. Dev.*, **4**, 209–14. [403]

Geppi, M., Mollica, G., Borsacchi, S., and Veracini, C. A. (2008). Solid state studies of pharmaceutical systems. *Appl. Spectrosc. Rev.*, **43**, 202–302. [182]

Gervais, C. and Hulliger, J. (2007). Impact of surface symmetry on growth-induced properties. *Cryst. Growth Des.*, **7**, 1925–35. [62]

Getsoian, A., Lodaya, R. M., and Blackburn, A. C. (2008). One-solvent polymorph screen for carbamazepine. *Int. J. Pharm.*, **348**, 3–9. [354]

Geuther, A. (1883). VII. Ueber die Constitution der Doppelverbindungen von Salzen der Sulfonsaüren mit neutralen Schwefelsäureräthern, und über die Constitution der Sulfate, sowie über den Grund ihrer Dimorphie. *Ann. Chem.*, **218**, 288–302. [33]

Ghinescu, I., Dragonic, V.-A., Saidac, S., and Crustescu, E. (1984). Interprinderea de Coloranti "Colorom". Patent RO 83 912. [379]

Ghosh, M., Banerjee, S., Kahn, M. A. S., Sikder, N., and Sikder, A. K. (2016). Understanding metastable phase transformation during crystallization of RDX, HMX and CL-20: experimental and DFT studies. *Phys. Chem. Chem. Phys.*, **18**, 23554–71. [399, 408]

Ghosh, M., Venkatesan, V., Mandave, S., Banerjee, S., Sikder, N., Sikder, A. K., and Bhattacharya, B. (2014). Probing crystal growth of epsilon- and alpha-CL-20 polymorphs via metastable phase transition using microscopy and vibrational spectroscopy. *Cryst. Growth Des.*, **14**, 5053–63. [408]

Ghosh, M., Venkatesan, V., Sikder, A. K., and Sikder, N. (2012). Preparation and characterization of ϵ-CL-20 by solvent evaporation and precipitation methods. *Def. Sci. J.*, **62**, 390–8. [408]

Giacovazzo, C. (ed.) (1992). *Fundamentals of Crystallography*. International Union of Crystallography, Oxford. [59, 159]

Gibbs, J. W. (1876). On the equilibrium of heterogeneous substances. *Trans. Connecticut Acad. Arts Sci.*, **3**, 108–248. [42]

Gibbs, T. R. and Popolato, A. (1980). *Los Alamos Scientific Laboratory Explosive Property Data*. University of California Press, Berkeley. [404t]

Gieren, A. and Hoppe, W. (1971). X-ray crystal structure analysis of bisphthalocyanatouranium(IV). *J. Chem. Soc., Chem. Commun.*, 413–14. [389]

Gigg, J., Gigg, R., Payne, S., and Conant, R. (1987). The allyl group for protection in carbohydrate chemistry. Part 21. (±)-1,2:5,6- and (±)-1,2:3,4-Di-*O*-isopropylidene-*myo*-inositol. The unusual behavior of crystals of (±)-3,4-di-*O*-acetyl-1,2,5,6-tetra-*O*-benzyl-*myo*-inositol. *J. Chem. Soc., Perkin Trans. I*, 2411–14. [319]

Gilardi, R. D. and Butcher, R. J. (1998a). A new class of flexible energetic salts. 3. The crystal structures of the 3,3-dinitroazetidinium dinitramide and 1-isopropyl-3,3-dinitroazetidinium dinitramide salts. *J. Chem. Crystallogr.*, **28**, 163–9. [416]

Gilardi, R. D. and Butcher, R. J. (1998b). A new class of flexible energetic salts. Part 5. The structures of two hexaammonium polymorphs and the ethan-1,2-diamonium salts of dinitramide. *J. Chem. Crystallogr.*, **28**, 673–81. [415]

Gilardi, R. D., Axenrod, T., Sun, J., and Flippen-Anderson, J. L. (2001). 1,3,5,7-Tetranitro-3,7-diazabicyclo(3.3.1)nonane. *Acta Crystallogr. E*, **57**, o1107. [414]

Gilardi, R., Flippen-Anderson, J., George, C., and Butcher, R. J. (1997). A new class of flexible energetic salts: the crystal structures of the ammonium, lithium, potassium, and cesium salts of dinitramide. *J. Am. Chem. Soc.*, **119**, 9411–16. [416]

Gilmore, C. J., Barr, G., and Paisley, J. (2004). High-throughput powder diffraction. I. A new approach to qualitative and quantitative powder diffraction analysis using pattern profiles. *J. Appl. Crystallogr.*, **37**, 231–42. [271]

Girlando, A., Marzola, F., Pecile, C., and Torrance, J. B. (1983). Vibrational spectroscopy of mixed stack organic semiconductors: neutral and ionic phases of tetrathiafulvalene-chloranil (TTF-CA) charge transfer complex. *J. Chem. Phys.*, **79**, 1075–85. [284]

Giron, D. (1981). Polymorphism. *Labo-Pharma Problems Tech.*, **307**, 151–60. [343]

Giron, D. (1988). Impacts of solid-state reactions on medicaments. *Mol. Cryst. Liq. Cryst.*, **161**, 77–100. [367]

Giron, D. (1990). Thermal analysis in pharmaceutical routine analysis. *Acta Pharm. Jugoslav.*, **40**, 95–157. [366]

Giron, D. (1995). Thermal analysis and calorimetric methods in the characterisation of polymorphs and solvates. *Thermochim. Acta*, **248**, 1–59. [147, 150, 151f, 154, 363, 366, 375]

Giron, D. (1997). Thermal analysis of drugs and drug products. In *Encyclopedia of Pharmaceutical Technology. Volume 15. Thermal Analysis of Drugs and Drug Products to Unit Processes in Pharmacy: Fundamentals* (ed. J. Swarbrick and J. C. Boylan), pp. 1–79. Marcel Dekker, New York. [363, 366, 370]

Giron, D. (1998). Contribution of thermal methods and related techniques to the rational development of pharmaceuticals—Part I. *Pharm. Sci. Technol. Today*, **1**, 191–9. [366]

Giron, D. (1999). Thermal analysis, microcalorimetry and combined techniques for the study of pharmaceuticals. *J. Therm. Anal. Calorim.*, **56**, 1285–304. [366]

Giron, D. and Goldbronn, C. (1997). Use of DSC and TG for identification and quantification of the dosage form. *J. Therm. Anal.*, **48**, 473–83. [366]

Giron, D., Draghi, M., Goldbronn, C., Pfeffer, S., and Peichon, P. (1997). Study of the polymorphic behavior of some local anesthetic drugs. *J. Therm. Anal.*, **49**, 913–27. [251]

Giron, D., Edel, B., and Piechon, P. (1990). X-ray quantitative determination of polymorphism in pharmaceuticals. *Mol. Cryst. Liq. Cryst.*, **187**, 557–67. [367]

Giron, D., Monnier, S., Mutz, M., Piechon, P., Buser, T., Stowasser, F., Schulze, K., and Bellus, M. (2007). Comparison of quantitative methods for analysis of polyphasic pharmaceuticals. *J. Therm. Anal. Calorim.*, **89**, 729–43. [352]

Giron, D., Piechon, P., Goldbronn, S., and Pfeffer, S. (1999). Thermal analysis, microcalorimetry and combined techniques for the study of the polymorphic behaviour of a purine derivative. *J. Therm. Anal. Calorim.*, **57**, 61–73. [366, 375]

Giron-Forest, D. (1984). Anwendung der thermischen Analyse in der Pharmazie. *Pharm. Ind.*, **46**, 851–9. [365]

Glasstone, S. (1940). *Text-book of Physical Chemistry*, pp. 465–70. MacMillan and Co., London. [42]

Glaxo, Inc. v. Boehringer-Ingelheim Corp. (1996). U.S. District Court (Docket No. 3: 95CV01342) May 14, 1996. [441]

Glaxo, Inc. v. Geneva Pharmaceuticals et al. (1994, 1996). Nos. 94–1921, 94–4589 and 96–3489, D. NJ). [441]

Glaxo, Inc. et al. v. Torpharm, Inc. et al. (1997). NO. 95 C 4686 (ND IL Eastern Div., May 18, 1997; No. Civ. AMD 96–455, Nov. 4, 1998). [441]

Glaxo, Inc. et al. v. Torpharm, Inc. et al. (1998), 153 F.3d 1366. [441]

Glowka, M. L., Olubek, Z., and Olczak, A. (1995). 3-Nitro-N-phenyl-4-(phenylamino) benzenesulfonamide. *Acta Crystallogr. C*, **51**, 1639–41. [379t]

Glusker, J. P. and Trueblood, K. N. (2010). *Crystal Structure Analysis – A Primer*, 3rd edn. Oxford University Press, Oxford. [159]

Glusker, J. P., Lewis, M., and Rossi, M. (1994). *Crystal Structure Analysis for Chemists and Biologists*. VCH Publishers, New York. [59, 159, 208, 215]

Gnutzmann, T., Thi, Y. N., Rademann, K., and Emmerling, F. (2014). Solvent triggered crystallization of polymorphs studied in situ. *Cryst. Growth Des.*, **14**, 6445–50. [352]

Goede, P., Latypov, N. V., and Ostmark, H. (2004). Fourier transform Raman spectroscopy of the four crystallographic phases of alpha, beta, gamma and epsilon 2,4,6,8,10,12-hexanitro-2,4,6,8,10,12-hexaazatetracyclo[5.5.0.0(5,9).0(3,11)]dodecane (HNIW, CL-20). *Propellants, Explos., Pyrotech.*, **29**, 205–8. [408]

Goetz, F. and Brill, T. B. (1979). Laser Raman spectra of α-, β-, χ-, and δ-octahydro-1,3,5,7-tetranitro-1,3,5,7-tetrazocine and their temperature dependence. *J. Phys. Chem.*, **83**, 340–6. [406]

Goldberg, I. and Swift, J. (2012). New insights into the metastable form of RDX. *Cryst. Growth Des.*, **12**, 1040–5. [410]

Goldschmidt, V. M. (1929). Crystal structure and chemical constitution. *Trans. Faraday Soc.*, **25**, 253–83. [37]

Golovina, N. I., Raevsky, A. V., Chukanov, N. V., Korsunsky, B. L., Atovmyan, L. O., and Aldoshin, S. M. (2004). *Russ. Chem. J.*, **48**, 41–51. [407]

Golovina, N. I., Titkov, A. N., Raevskii, A. V., and Atovmyan, L. O. (1994). Kinetics and mechanism of phase transitions in the crystals of 2,4,6-trinitrotoluene and benzotrifuroxane. *J. Solid State Chem.*, **113**, 229–38. [418, 418t, 419f, 420, 421f]

Gorelik, T. E., Czech, C., Hammer, S. M., and Schmidt, M. U. (2016). Crystal structure of disordered nanocrystalline α^{II}-quinacridone determined by electron diffraction. *CrystEngComm*, **18**, 529–35. [385]

Gorelik, T. E., Sarfraz, A., Kolb, U., Emmerling, F., and Rademan, K. (2012). Detecting crystalline nonequilibrium phases on the nanometer scale. *Cryst. Growth Des.*, **12**, 3229–42. [352]

Goto, H. and Osawa, E. (1989). Corner flapping: a simple and fast algorithm for exhaustive generation of ring conformations. *J. Am. Chem. Soc.*, **111**, 8950–1. [243]

Goto, H. and Osawa, E. (1993). An efficient algorithm for searching low-energy conformers of cyclic and acyclic molecules. *J. Chem. Soc., Perkin Trans. 2*, 187–98. [243]

Goto, H., Fujinawa, T., Asahi, H., Ogata, H., Miyajima, S., and Maruyama, Y. (1996). Crystal structures and physical properties of 1,6-diaminopyrene-p-chloranil (DAP-CHL) charge transfer complex. Two polymorphs and their unusual electrical properties. *Bull. Chem. Soc. Jpn.*, **69**, 85–93. [124, 280]

Grabar, D. G. and McCrone, W. C. (1950). Antabuse (tetraethyl thiuram disulfide). *Anal. Chem.*, **22**, 620–1. [24]

Grabar, D. G., Rauch, F. C., and Fanelli, A. J. (1969). Observation of a solid–solid polymorphic transformation in 2,4,6-trinitrotoluene. *J. Phys. Chem.*, **73**, 3514–16. [417]

Gracin, S., Uusu-Pentillä, M., and Rasmussen, Å. C. (2015). Influence of ultrasound on the nucleation of polymorphs of *p*-aminobenzoic acid. *Cryst. Growth Des.*, **15**, 1787–94. [101]

Graeber, E. J. and Morosin, B. (1974). The crystal structures of 2,2′,4,4′,6,6′-hexanitroazobenzene (HNAB) Forms I and II. *Acta Crystallogr. B*, **30**, 310–17. [222, 425]

Graham, J. A., Grim, N. M., and Fately, W. G. (1985). Fourier Transform infrared photoacoustic spectroscopy of condensed-phase samples. In *Fourier Transform Infrared Spectroscopy*, Vol. 4 (ed. J. R. Ferraro and L. J. Basil), pp. 345–92. Academic Press, New York. [174]

Grainger, C. T. and McConnell, J. F. (1969). Crystal structure of 1-*p*-nitrobenzeneazo-2-naphthol (parared) from overlapped twin-crystal data. *Acta Crystallogr. B*, **25**, 1962–70. [380t]

Grant, D. J. W. and Brittain, H. G. (1995). Solubility of pharmaceutical Solids. In *Physical Characterization of Pharmaceutical Solids* (ed. H. G. Brittain), pp. 321–86. Marcel Dekker, New York. [346]

Grant, D. J. W. and Higuchi, T. (1990). Solubility behavior of organic compounds. In *Techniques of Chemistry*, Vol. 21 (ed. W. H. Saunders Jr.). John Wiley and Sons, New York. [155, 346]

Graser, F. and Hädicke, E. (1980). Crystal structure and perylene-3,4:9,10-bis(dicarboxymid) pigments. *Justus Liebigs Ann. Chem.*, 1994–2011. [387]

Graser, F. and Hädicke, E. (1984). Crystal structure of perylene-3,4:9,10-bis(dicarboxymid) pigments. 2. *Justus Liebigs Ann. Chem.*, 483–94. [387]

Green, B. S. and Knossow, M. (1981). Lamellar twinning explains the nearly racemic composition of chiral, single crystals of hexahelicene. *Science*, **214**, 795–7. [42]

Grell, J., Bernstein, J., and Tinhofer, B. (1999). Graph-set analysis of hydrogen-bond patterns: some mathematical concepts. *Acta Crystallogr. B*, **55**, 1030–43. [69, 72]

Griesser, U. J. (2000). Private communication to the author. [161, 170, 171f]

Griesser, U. J. (2006). The importance of solvates. In *Polymorphism in the Pharmaceutical Industry* (ed. R. Hilfiker), pp. 211–34. Wiley-VCH, Weinheim. [8, 355]

Griesser, U. J. (2011). Personal communication to the author. [17t, 104f]

Griesser, U. J. and Burger, A. (1993). The polymorphic drug substances of the European Pharmacopoeia. Part 8. Thermal analytical and FTIR-microscopic investigations of etofylline crystal forms. *J. Pharm. Sci.*, **61**, 133–43. [170]

Griesser, U. J. and Burger, A. (1999). Statistical aspects of the occurrence of crystal forms among organic drug substances. In *Abstracts, XVIII Congress and General Assembly of the International Union of Crystallography*, Glasgow, Scotland. [345]

Griesser, U. J., Auer, M. E., and Burger, A. (2000). Micro-thermal analysis, FTIR- and Raman-microscopy of (R,S)-proxyphylline crystal forms. *Microchem. J.*, **65**, 283–92. [365, 368]

Griesser, U. J., Szelagiewicz, M., Hofmeier, U. C., Pitt, C., and Cianferani, S. (1999). Vapor pressure and heat of sublimation of crystal polymorphs. *J. Therm. Anal. Calorim.*, **57**, 45–60. [47]

Griffiths, C. H. and Monahan, A. R. (1976). Polymorphism and spectroscopic characterization of an azo-pigment. *Mol. Cryst. Liq. Cryst.*, **33**, 175–87. [117, 380t]

Groth, P. (1969). Crystal structure of 3,3,6,6,9,9-hexamethyl-1,2,4,5,7,8-hexaoxacyclononane (trimeric acetone peroxide). *Acta Chem. Scand.*, **23**, 1311–29. [411]

Groth, P. H. R. v. (1906a). *An Introduction to Chemical Crystallography* (trans. H. Marshall), pp. 28–31. Gurney & Jackson, London. [34]

Groth, P. H. R. v. (1906b). *Chemische Kristallographie. Erster Teil. Elemente—anorganische Verbindungen ohne Salzcharakter—einfache und komplexe Halogenide, Cyanide und Azide der Metalle, nebst den zugehörgen Alkylverbindungen.* W. Engelemann, Leipzig. [21, 29, 114, 116, 120]

Groth, P. H. R. v. (1908). *Chemische Kristallographie. Zweiter Teil. Die anorganischen Oxo- und Sulfosalze.* W. Engelemann, Leipzig. [21, 29]

Groth, P. H. R. v. (1910). *Chemische Kristallographie. Dritter Teil. Aliphatische und Hydroaromatische Kohlenstoffverbindungen.* W. Engelemann, Leipzig. [21, 36]

Groth, P. H. R. v. (1917). *Chemische Kristallographie. Vierter Teil. Aromatische Kohlenstoffverbindungen mit einem Benzolringe.* W. Engelemann, Leipzig. [21, 36, 420, 422f]

Groth, P. H. R. v. (1919). *Chemische Kristallographie. Fünfter Teil. Aromatische Kohlenstoffverbindungen mit meheren Benzolringen. Heterocyclische Verbindungen*, pp. 104–5. Engelmann, Leipzig. [21, 36, 302]

Grubb, P. W. (1986). *Patents in Chemistry and Biotechnology.* Clarendon Press, Oxford. [430]

Grubb, P. W. and Thomsen, P. R. (2010). *Patents for Chemicals, Pharmaceuticals and Biotechnology.* Oxford University Press, Oxford. [430]

Grunenberg, A., Henck, J.-O., and Siesler, H. W. (1996). Theoretical derivation and practical applications of energy/temperature diagrams as an instrument in preformulation studies of polymorphic drug substances. *Int. J. Pharm.*, **129**, 147–58. [45, 46, 47f, 48f, 52, 150, 176, 230]

Grunenberg, A., Keil, B., and Henck, J.-O. (1995). Polymorphism in binary mixtures, as exemplified by nimodipine. *Int. J. Pharm.*, **118**, 11–21. [176, 184f]

Gruno, M., Wulff, H., and Pflegel, P. (1993). Polymorphism of benzocaine. *Pharmazie*, **48**, 834–7. [9]

Grzesiak, A. L. and Matzger, A. J. (2007). New form discovery for the analgesics flurbiprofen and sulindac facilitated by polymer-induced heteronucleation. *J. Pharm. Sci.*, **96**, 2978–86. [100]

Grzesiak, A. L., Uribe, F. J., Ockwig, N. W., Yaghi, O. M., and Matzger, A. J. (2006). Polymer-induced heteronucleation for the discovery of new extended solids. *Angew. Chem. Int. Ed. Engl.*, **45**, 2553–6. [100]

Gu, W. (1993). Factor analysis of phase transitions and conformational changes in pentaerythritol tetrastearate. *Anal. Chem.*, **65**, 823–7. [170]

Gu, X. J. and Jiang, W. (1995). Characterization of polymorphic forms of fluconazole using Fourier transform Raman spectroscopy. *J. Pharm. Sci.*, **84**, 1438–41. [177]

Gu, X., Yao, J., Zhang, G., Yan, Y., Zhang, C., Peng, Q., Liao, Q., Wu, Y., Xu, Z., Zhao, Y., Fu, H., and Zhang, D. (2012). Polymorphism-dependent emission for di(p-methoxyphenyl)dibenzofulvene and analogues: optical waveguide/amplified spontaneous emission behaviors. *Adv. Mater.*, **22**, 4862–72. [340]

Gudmunsdottir, A. D., Lewis, T. J., Randall, L. H., Scheffer, J. R., Rettig, S. J., Trotter, J., and Wu, C. H. (1996). Geometric requirements for hydrogen abstractability and 1,4-biradical reactivity in the Norrish/Yang type II reaction: studies based on the solid state photochemistry and X-ray crystallography of medium-sized ring and macrocyclic diketones. *J. Am. Chem. Soc.*, **118**, 6167–84. [334]

Guillory, J. K. (1999). Generation of polymorphs, hydrates, solvates, and amorphous solids. In *Physical Characterization of Pharmaceutical Solids* (ed. H. G. Brittain), pp. 183–226. Marcel Dekker, New York. [92, 95, 105, 354, 370, 371]

Guillory, J. K. and Erb, D. M. (1985). Using solution calorimetry to quantitate binary mixtures of three crystalline forms of sulfamethoxazole. *Pharm. Manuf.*, **2**, 30–3. [154]

Guinot, S. and Leveiller, F. (1999). The use of MTDSC to assess the amorphous phase content of a micronised drug substance. *Int. J. Pharm.*, **192**, 63–75. [372]

Guo, Y. S., Byrn, S. R., and Zografi, G. (2000). Physical characteristics and chemical degradation of amorphous quinapril hydrochloride. *J. Pharm. Sci.*, **89**, 128–43. [371]

Gussenhoven, E. M., Olmstead, M. M., Fettinger, J. C., and Balch, A. L. (2008). Interplay of supramolecular organization, metallophilic interactions, phase changes and luminescence in four polymorphs of Ir^I(CO)$_2$(OC(CH$_3$)CHC(CH$_3$)N(p-tol)). *Inorg. Chem.*, **47**, 4570–8. [338]

Guzman, J. and Largo-Cabrerizo, J. (1978). Polymorphism in the 2-(4-morpholinothio) benzothiazole. *J. Heterocycl. Chem.*, **15**, 1531–3. [222]

Ha, J.-M., Wolf, J. H., Hillmyer, M. A., and Ward, M. D. (2004). Polymorph selectivity under nanoscopic confinement. *J. Am. Chem. Soc.*, **126**, 3382–3. [100]

Haaland, D. M. and Thomas, E. V. (1998). Partial least squares methods for spectral analysis. 1. Relation to other quantitative calibration methods and the extraction of qualitative information. *Anal. Chem.*, **60**, 1193–202. [175]

Hädicke, E. and Graser, F. (1986a). Structures of eleven perylene-3,4:9,10-bis(dicarboxymid) pigments. *Acta Crystallogr. C*, **42**, 189–95. [380t, 387]

Hädicke, E. and Graser, F. (1986b). Structures of three perylene-3,4:9,10-bis(dicarboxymid) pigments. *Acta Crystallogr. C*, **42**, 195–8. [387]

Hagemann, J. W. and Rothfus, J. A. (1993). Transitions of saturated monoacid triglycerides—modeling conformational change at glycerol during alpha → beta′ → beta transformation. *J. Am. Oil Chem. Soc.*, **70**, 211–17. [40]

Hagler, A. T. and Lifson, S. (1974). Energy functions for peptides and proteins. II. Amide hydrogen bond and calculation of amide crystal properties. *J. Am. Chem. Soc.*, **96**, 5327–35. [217]

Hagler, A. T., Huler, E., and Lifson, S. (1974). Energy functions for peptides and proteins. I. Deviation of a consistent force field including the hydrogen bond from amide crystals. *J. Am. Chem. Soc.*, **96**, 5319–27. [217, 227]

Hahn, T. (ed.) (1987). *International Tables for X-ray Crystallography. Vol. A. Space-Group Symmetry*, 2nd edn, revised. D. Reidel, Dordrecht. [68, 159, 161t]

Hahn, T. and Klapper, H. (1992). Point groups and crystal classes. In *International Tables for Crystallography. Vol. A. Space Group Symmetry*, 3rd edn (ed. T. Hahn), pp. 752–892. International Union of Crystallography, Kluwer Academic Publishers, Dordrecht. [59]

Hähnle, R. and Optiz, K. (1976). Hoechst AG Patent DE 2 524 187. [379t]

Hakey, P., Ouellette, W., Zubieta, J., and Korter, T. (2008). Redetermination of cyclo-trimethylene-trinitramine. *Acta Crystallogr. E*, **64**, o1428. [410]

Haleblian, J. and McCrone, W. C. (1969). Pharmaceutical applications of polymorphism. *J. Pharm. Sci.*, **58**, 911–29. [7, 8, 38, 38f, 342, 349f]

Haleblian, J. K. (1975). Characterization of habits and crystalline modification of solids and their pharmaceutical applications. *J. Pharm. Sci.*, **64**, 1269–88. [208, 343]

Hall, P. (1971). Thermal decomposition and phase transitions in solid nitramides. *Trans. Faraday Soc.*, **67**, 556–62. [410, 414]

Hall, R. C., Paul, I. C., and Curtin, D. Y. (1988). Structures and interconversion of polymorphs of 2,3-dichloroquinazirin. Use of second harmonic generation to follow the change of a centrosymmetric to a polar structure. *J. Am. Chem. Soc.*, **110**, 2848–54. [302]

Hall, S. R., Kolinsky, P. V., Jones, R., Allen, S., Gordon, P., Bothwell, B., Bloor, D., Norman, P. A., Hursthouse, M., Karaulov, A., Baldwin, J., Goodyear, M., and Bishop, D. (1986). Polymorphism and nonlinear optical activity in organic crystals. *J. Cryst. Growth*, **79**, 745–51. [297, 302]

Haller, T. M., Rheingold, A. L., and Brill, T. B. (1983). The structure of the complex between octahydro-1,3,5,7-tetranitro-1,3,5,7-tetrazocine (HMX) and *N,N*-dimethylformamide (DMF), $C_4H_8N_8O_8 \cdot C_3H_7NO$. A second polymorph. *Acta Crystallogr. C*, **39**, 1559–63. [403]

Halliwell, R. A., Bhardwaj, R. M., Brown, C. J., Briggs, N. E. B., Dunn, J., Robertson, J., Nordon, A., and Florence, A. J. (2017). Spray drying as a reliable route to produce metastable carbamazepine Form IV. *J. Pharm. Sci.*, **106**, 1874–80. [369]

Hamaed, H., Pawlowski, J. M., Jenna, M., Cooper, B. F. T., Fu, R. Q., Eichorn, S. H., and Schurko, R. W. (2008). Application of solid-state ^{35}Cl NMR to the structural characterization of hydrochloride pharmaceuticals and their polymorphs. *J. Am. Chem. Soc.*, **130**, 11056–65. [193]

Haman, C. and Wagner, H. (1971). Textures of evaporated copper phthalocyanine films. *Krist. Tech.*, **6**, 307–20. [391]

Hamilton, B. D., Ha, J.-M., Hillmyer, M. A., and Ward, M. D. (2012). Manipulating crystal growth and polymorphism by confinement in nanoscale crystallization chambers. *Acc. Chem. Res.*, **45**, 414–23. [100, 105]

Hamilton, W. C. and Ibers, J. A. (1968). *Hydrogen Bonding in Solids*, pp. 19–21. W. A. Benjamin, New York. [69]

Hammond, R. B., Ma, C., Roberts, K. J., Ghi, P. Y., and Harris, R. (2003). Application of systematic search methods to studies of the structures of urea–dihydroxybenzene co-crystals. *J. Phys. Chem.*, **107**, 11820–6. [243]

Hammond, R. B., Roberts, K. J., Docherty, R., Edmondson, M., and Gairns, R. (1996). X-form metal-free phthalocyanine: crystal structure determination using a combination of high-resolution X-ray powder diffraction and molecular modeling technique. *J. Chem. Soc. Perkin Trans. 2*, 1527–8. [383t, 393]

Hamzaoui, F., Baert, F., and Wojcik, G. (1996). Electron-density study of *m*-nitrophenol in the orthorhombic structure. *Acta Crystallogr. B*, **52**, 159–64. [121, 302]

Han, J., Gupte, S., and Suryanarayanan, R. (1998). Application of pressure differential scanning calorimetry in the study of pharmaceutical hydrates. II. Ampicillin trihydrate. *Int. J. Pharm.*, **170**, 63–72. [367]

Hancock, B. C. and Parks, M. (2000). What is the true solubility advantage for amorphous pharmaceuticals? *Pharm. Res.*, **17**, 397–404. [370]

Hancock, B. C. and Zografi, G. (1997). Characteristics and significance of the amorphous state in pharmaceutical systems. *J. Pharm. Sci.*, **86**, 1–12. [370]

Hanson, J. R., Hitchcock, P. B., and Saberi, H. (2004). Steric factors in the preparation of nitrostilbenes. *J. Chem. Res.*, 667–9. [427]

Hantzsch, A. (1907a). Concerning chromoisomers. *Angew. Chem.*, **20**, 1889–92. [303]

Hantzsch, A. (1907b). Yellow, red, green, violet and colorless salts from dinitro compounds. *Ber. Dtsch. Chem. Ges.*, **40**, 1533–55. [303]

Hantzsch, A. (1908). Concerning chromo isomers. *Z. Zentralbl. Tech Chem.*, **20**, 1889. [303]

Hao, H., Barrett, M., Hu, Y., Su, W., Ferguson, S., Wood, B., and Glennon, B. (2012). The use of in situ tools to monitor the enantiotropic transformation of *p*-aminobenzoic acid polymorphs. *Org. Process Res. Dev.*, **16**, 35–41. [199]

Hao, Z. and Iqbal, A. (1997). Some aspects of organic pigments. *Chem. Soc. Rev.*, **26**, 203–13. [377, 382t, 395]

Hao, Z., Iqbal, A., and Herren, F. (1999a). 1,4-Diketo-3,6-bis-4-chloro-phenyl-pyrrolopyrrole in β-modification—useful as pigment with more yellow-red nuance than α-modification. Prepd. by acid hydrolysis of soluble carbamate. Derivation and preparation by cooling. Ciba-Geigy Ltd. EP 690 057 B1. [394]

Hao, Z., Iqbal, A., and Herren, F. (1999b). 1,4-Diketo-3,6-bis-4-chloro-phenyl-pyrrolopyrrole in β and γ modification, prepd. by acid hydrolysis of soluble carbamate. Derivation and preparation by cooling. Ciba-Geigy Ltd. EP 690 059 B1. [394]

Hao, Z., Schloeder, I., and Iqbal, A. (1999c). 1,4-Diketo-3,6-bis-4-chloro-phenyl-pyrrolopyrrole in β-modification—useful as pigment with more yellow-red nuance than α-modification is. Prepd. by acid hydrolysis of soluble carbamate. Derivation and preparation by cooling. Ciba-Geigy Ltd. EP 690 058 A1. [394]

Hapgood, K., Litster, J. D., and Wang, C. H. (2015). Pharmaceutical particles. *Chem. Eng. Sci.*, **125**, 1–3. [244]

Harano, K., Homma, T., Niimi, Y., Koshino, M., Suenaga, K., Leibler, L., and Nakamura, E. (2012). Heterogeneous nucleation of organic crystals mediated by-single-molecule template. *Nat. Mater.*, **11**, 877–81. [80]

Hardy, G. E., Zink, J. I., Kaska, W. C., and Baldwin, J. C. (1978). Structure and triboluminescence of polymorphs of $(PH_3P)_2C$. *J. Am. Chem. Soc.*, **100**, 8001–2. [315]

Hargittai, I. and Levy, J. B. (1999). Accessible geometric changes. *Struct. Chem.*, **10**, 387–9. [216, 216t]

Harmon, P., Li, L., Marsac, P., McKelvey, C., Variankaval, N., and Xu, W. (2009). Amorphous solid dispersions: analytical challenges and opportunities. *AAPS News*, September 2009, 14–20. [370]

Harper, J. K. and Grant, D. M. (2000). Solid state C-13 chemical shift tensors in terpenes. 3. Structural characterization of polymorphous verbenol. *J. Am. Chem. Soc.*, **122**, 3708–14. [186]

Harper, M. (2012). The truly staggering cost of inventing new drugs. *Forbes*, February 10. [432]

Harris, K. D. M. and Thomas, J. M. (1991). Probing polymorphism and reactivity in the organic solid-state using ^{13}C NMR spectroscopy: studies of *p*-formyl-*trans*-cinnamic acid. *J. Solid State Chem.*, **93**, 197–205. [186]

Harris, R. K. (1985). Quantitative aspects of high-resolution solid-state nuclear magnetic resonance spectroscopy. *Analyst*, **110**, 649–55. [190]

Harris, R. K. (1993). State-of-the-art for solids. *Chem. Br.*, 601–4. [182]

Harris, R. K. (2006). NMR studies of organic polymorphs and solvates. *Analyst*, **131**, 351–73. [182]

Harris, R. K. (2007). Applications of solid-state NMR to pharmaceutical polymorphism and related matters. *J. Pharm. Pharmacol.*, **59**, 225–39. [182]

Harris, R. K., Wasylishen, R. E., and Duer, M. J. (2009). *NMR Crystallography*. Wiley, New York. [191]

Harris, R. K., Yeung, R. R., Lamont, R. B., Lancester, R. W., Lynn, S. M., and Staniforth, S. E. (1997). "Polymorphism" in a novel antiviral agent: lamivudine. *J. Chem. Soc., Perkin Trans. 2*, 2653–9. [173f, 173t]

Hartauer, K. J., Miller, E., and Guillory, J. K. (1992). Diffuse reflectance infrared Fourier transform spectroscopy for the quantitative analysis of mixtures of polymorphs. *Int. J. Pharmacol.*, **85**, 163–74. [173, 176]

Hartley, H. (1902). *Polymorphism. An Historical Account*, p. 15. Holywell Press, Oxford. [30, 32, 33]

Hartshorne, N. H. and Stuart, A. (1964). *Practical optical crystallography*, pp. 1–46. American Elsevier, New York. [61f, 137]

Hasa, D. and Jones, W. (2017). Screening for new pharmaceutical solid forms using mechanochemistry: a practical guide. *Adv. Drug Deliv. Rev.*, **117**, 147–61. [352]

Hasa, D., Miniussi, E., and Jones, W. (2016). Mechanochemical synthesis of multicomponent crystals: one liquid for one polymorph? A myth to dispel. *Cryst. Growth Des.*, **16**, 4582–8. [360]

Hasegawa, M. (1986). Topochemical photopolymerization of diolefin crystals. *Pure Appl. Chem.*, **58**, 1179–88. [333]

Haselbach, E. and Heilbronner, E. (1968). Electronic structure and physical chemical properties of azo compounds. Part XIV. The conformation of benzalaniline. *Helv. Chim. Acta*, **51**, 16–34. [323]

Hatada, M., Jancarik, J., Graves, B., and Kim, S. -H. (1985). Crystal structure of aspartame, a peptide sweetener. *J. Am. Chem. Soc.*, **107**, 4279–82. [453, 454, 455]

Hayashi, Y. and Sakaguchi, I. (1981). Dioxazine violet pigment with stable β-crystalline form—prepared by heating metastable dioxazine violet pigment with aromatic cdp which is sparingly soluble in water. Sumitomo Chem. Co. Ltd Patent DE 3 031 444 A1. [379t]

Hayashi, Y. and Sakaguchi, I. (1982). Stable dioxazine violet pigment production from metastable form—by heating aqueous suspension with aliphatic or alicyclic ketone or acetate. Sumitomo Chem. Co. Ltd Patent DE 3 211 607 A1. [379t]

Hayward, I. P., Batchelder, D. N., and Pitt, G. D. (1994). Applications of Raman spectroscopy and imaging to industrial quality control. *Analyst*, **22**, M22–8. [177]

He, G. W., Wong, A. B. H., Chow, P. S., and Tan, R. B. H. (2011). Effects of rate of supersaturation generation on polymorphic crystallization of *m*-hydroxybenzoic acid and *o*-aminobenzoic acid. *J. Cryst. Growth*, **314**, 220–6. [369]

He, X., Stowell, J. G., Morris, K. R., Pfeiffer, R. R., Li, H., Stahly, G. P., and Byrn, S. R. (2001). Stabilization of a metastable polymorph of 4-methyl-2-nitroscetanilide by isomorphic additives. *Cryst. Growth Des.*, **1**, 305–12. [369]

He, Z., Zhang, L., Mei, J., Zhang, T., Lam, J. W. Y., Shuai, Z., Dong, Y. O., and Tang, B. Z. (2015). Polymorphism-dependent and switchable emission of butterfly-like bis(diarylmethylene)dihydroanthracenes. *Chem. Mater.*, **27**, 6601–7. [339]

Healy, A. M., Worku, Z. A., Kumar, D., and Madi, A. M. (2017). Pharmaceutical solvates, hydrates and amorphous forms: a special emphasis on cocrystals. *Adv. Drug Deliv. Syst.*, **117**, 25–46. [360]

Hean, D., Gelbrich, T., Griesser, U. J., Michael, J. P., and Lemmerrer, A. (2015). Structural insights into the hexamorphic system of an isoniazid derivative. *CrystEngComm*, **17**, 5143–53. [365]

Hegedus, B. and Görög, S. (1985). The polymorphism in cimetidine. *J. Pharm. Biomed. Anal.*, **3**, 303–13. [374]

Heijna, M. C. R., van Enckevort, W. J. P., and Vlieg, E. (2008). Growth inhibition of protein crystals: a study of lysozyme polymorphs. *Cryst. Growth Des.*, **8**, 270–4. [62]

Heinz, A., Strachan, C. J., Gordon, K. C., and Rades, T. (2009). Analysis of solid-state transformations of pharmaceutical compounds using vibrational spectroscopy. *J. Pharm. Pharmacol.*, **61**, 971–88. [170, 177, 179t]

Hemminger, W. and Höhne, G. (1984). *Calorimetry: Fundamentals and Practice*. Verlag Chemie, Weinheim. [155]

Henck, J.-O. and Kuhnert-Brandstätter, M. (1999). Demonstration of the terms enantiotropy and monotropy in polymorphism research exemplified by flurbiprofen. *J. Pharm. Sci.*, **88**, 103–8. [46]

Henck, J.-O., Bernstein, J., Ellern, A., and Boese, R. (2001). Disappearing and reappearing polymorphs. The case of benzocaine:picric acid. *J. Am. Chem. Soc.*, **123**, 1834–41. [134, 145, 367]

Henck, J.-O., Finner, E., and Burger, A. (2000). Polymorphism of tedisamil dihydrochloride. *J. Pharm. Sci.*, **89**, 1151–9. [365]

Henck, J.-O., Griesser, U. J., and Burger, A. (1997). Polymorphism of drug substances. An economic challenge? *Pharm. Ind.*, **59**, 165–9. [8, 343]

Hendrickson, B. A., Preston, M. S., and York, P. (1995). Processing effects on crystallinity of cephalexin—characterization by vacuum microbalance. *Int. J. Pharm.*, **118**, 1–10. [9]

Henisch, H. K. (1996). *Crystal Growth in Gels*. Dover Publications, New York. [101, 356]

Henson, B. F., Asay, B. W., Sander, R. K., Son, S. F., Robinson, J. M., and Dickson, P. M. (1999). Dynamic measurement of the HMX beta–delta phase transition by second harmonic generation. *Phys. Rev. Lett.*, **82**, 1213–16. [405]

Herbst, W. and Hunger, K. (1997). *Industrial Organic Pigments*, 2nd edn. VCH, Weinheim. [376, 384t, 393]

Herbstein, F. H. (1971). Crystalline π-molecular compounds. Chemistry, spectroscopy and crystallography. In *Perspectives in Structural Chemistry*, Vol. 4. (ed. J. D. Dunitz and J. A. Ibers), pp. 166–395. John Wiley & Sons, New York. [275, 282]

Herbstein, F. H. (2000). How precise are measurements of unit-cell dimensions from single crystals? *Acta Crytallogr. B*, **56**, 547–57. [147]

Herbstein, F. H. (2001). Varieties of polymorphism. In *Advances in Structure Analysis* (ed. R. Kucel and J. Hasek), pp. 114–54. Czech and Slovak Crystallographic Association, Prague. [12, 41, 63, 159, 161, 208]

Herbstein, F. H. (2004). Diversity amidst similarity: a multidisciplinary approach to phase relationships, solvates and polymorphs. *Cryst. Growth Des.*, **4**, 1419–29. [45]

Herbstein, F. H. (2005). *Crystalline Molecular Complexes and Compounds* (2 volumes). Oxford University Press, Oxford. [5, 275, 359]

Herper, M. (2012). The truly staggering costs of inventing new drugs. *Forbes Pharma and Healthcare*, February 10. https://www.forbes.com/sites/matthewherper/2012/02/10/the-truly-staggering-cost-of-inventing-new-drugs/#75eabfe94a94 [344]

Herrera, M. L. and Rocha, F. J. M. (1996). Effects of sucrose ester on the kinetics of polymorphic transition in hydrogenated sunflower oil. *J. Am. Oil Chem. Soc.*, **73**, 321–6. [40]

Hertel, E. (1931). Kristallstruktur. *Z. Electrochem.*, **37**, 536–8. [332]

Hertel, E. and Römer, G. H. (1930). Der strukturelle Aufbau organischer Molekülverbindungen mit zwei- und eindimensionalen Abwechselungsprinzip. *Z. Phys. Chem. B*, **11**, 77–89. [417]

Herz, A. (1974). Dye–dye interactions of cyanines in solution and at silver bromide surfaces. *Photogr. Sci. Eng.*, **18**, 323–35. [328]

Hibbert, H. (1914). Nitroglycerin. *Z. Sprengstoffw.*, **9**, 305–7. [412]

Higashi, K., Ueda, K., and Moribe, K. (2017). Recent progress of structural study of polymorphic pharmaceutical drugs. *Adv. Drug Deliv. Rev.*, **117**, 71–85. [345]

Hildebrand, M., Hamaed, H., Namespetra, A. M., Donohue, J. M., Fu, R. Q., Hung, I., Gan, Z. H., and Schurko, R. W. (2014). ^{35}Cl solid-state NMR of HCl salts of active pharmaceutical ingredients: structural prediction, spectral fingerprinting and polymorph recognition. *CrystEngComm*, **16**, 7334–56. [193]

Hilfiker, R. (ed.) (2006). *Polymorphism in the Pharmaceutical Industry*. Wiley-VCH Verlag, Weinheim. [136, 343, 353]

Hilfiker, R., De Paul, S. M., and Szelagiewicz, M. (2006). Approaches to polymorphism screening. In *Polymorphism in the Pharmaceutical Industry* (ed. R. Hilfiker), pp. 287–308. Wiley-VCH Verlag, Weinheim. [92, 99t, 99, 108]

Hill, S. A., Jones, K. H., Seager, H., and Taskis, C. B. (1975). Dissolution and bioavailability of the anhydrate and trihydrate forms of ampicillin. *J. Pharm. Pharmacol.*, **27**, 594–8. [348]

Hiremath, R., Basile, J. A., Varney, S. W., and Swift, J. A. (2005). Controlling molecular crystal polymorphism with self-assembled monolayer templates. *J. Am. Chem. Soc.*, **127**, 18321–7. [100]

Hirsch, D. A., Rossini, A. J., Emsley, L., and Schurko, R. W. (2016). Cl-35 dynamic nuclear solid-state NMR of active pharmaceutical ingredients. *Phys. Chem. Chem. Phys.*, **18**, 25893–25904. [193]

Hirshfeld, F. L. (1977). Bond-atom fragments for describing molecular charge densities. *Theor. Chim. Acta*, **44**, 129–38. [248]

Hirshfeld, F. L. and Mirsky, K. (1979). The electrostatic term in lattice-energy calculations: acetylene, carbon dioxide and cyanogen. *Acta Crytallogr. A*, **35**, 366–70. [227]

Hisaki, I., Kometani, E., Shigemitsu, H., Saeki, A., Seki, S., Tohnai, N., and Miyata, M. (2011). Polymorphism of dehydrobenzo[14]annulene possessing two methyl ester groups in noncentrosymmetric positions. *Cryst. Growth Des.*, **11**, 5488–97. [339f]

Hiszpanski, A. M. and Loo, Y.-L. (2014). Directing the film structure of organic semiconductors via post-deposition processing for transistor and solar cell applications. *Energy Environ. Sci.*, **7**, 592–608. [281]

Hiszpanski, A. M., Baur, R. M., Kim, B., Tremblay, N. J., Nuckolls, C., Woll, A. R., and Loo, Y.-L. (2014). Tuning polymorphism and orientation in organic semiconductor thin films via post-deposition processing. *J. Am. Chem. Soc.*, **136**, 15749–56. [281]

Hiszpanski, A. M., Woll, R. M., Kim, B., Nuckolls, C., and Loo, Y.-L. (2017). Altering the polymorphic accessibility of polycyclic aromatic hydrocarbons with fluorination. *Chem. Mater.*, **29**, 4311–16. [281]

Hitchens, A. L. and Garfield, F. M. (1941). Basic lead styphnate and a process of making it. Western Cartridge U.S. Patent 2 265 230. [426]

Hoard, M. S. and Elakovich, S. D. (1996). Grinding-induced polymorphism in the aporphine alkaloid magnoflorine. *Phytochemistry*, **43**, 1129–33. [170]

Hoechst (1982). Crystalline modification of pigment red 53:1. European Patent 97 913. [380t]

Hoechst (1988). Monoazo pigment, its preparation process and its use. European Patent 320 774. [382t]

Hoffmann-La Roche (1937). Verfahren zur Darstellung von 2,4-Dioxo-3,3-diäthyl-tetrahydropyridin. Swiss Patent 187 826. [360]

Hofmann, D. W. M. and Kuleshova, L. (2005). New similarity index for crystal structure determination from X-ray powder diagrams. *J. Appl. Crystallogr.* **38**, 861–6. [271]

Höhener, A. and Smith, R. E. (1987). New crystalline form of sulphonated diazo dye—for dyeing and printing wool, polyamide, etc., more easily filtered than amorphous form. Ciba-Geigy AG Patent EP 222 697. [379t]

Holden, J. R., Du, Z., and Ammon, H. L. (1993). Prediction of possible crystal structures for C-, H-, N-, O-, and F-containing organic compounds. *J. Comput. Chem.*, **14**, 422–37. [238, 239]

Holman, K. T., Pivovar, A. M., and Ward, M. D. (2001). Engineering crystal symmetry and polar order in molecular host frameworks. *Science*, **294**, 1907–11. [296]

Holston Defense Corp. (1962). *Physical and chemical properties of RDX and HMX*. Control No. 20-P-26 Series B. Eastman Kodak, Kingsport, TN. [403, 404t, 406]

Honigmann, B. (1964). Modification and physical-particle shapes of organic pigments. *Farbe Lack*, **70**, 789–91. [391]

Honjo, S., Yokota, M., Doki, N., and Shimizu, K. (2008). Magnetic field influence on the crystal structure of 2,2′:6′,2″-terpyridine. *Kagaku Kogaku Ronbun.*, **34**, 383–7. [292]

Hooks, D. E., Fritz, T., and Ward, M. D. (2001). Epitaxy and molecular organization on solid substrates. *Adv. Mater.*, **13**, 227–41. [100]

Horiuchi, S. and Tokura, Y. (2008). Organic ferroelectrics. *Nat. Mater.*, **7**, 357–66. [285]

Horn, D. and Honigman, B. (1974). Polymorphism of copper phthalocyanine. In *XII Fatipec Kongress*, Garmisch-Parkinkirchen, Germany, May 1974, pp. 181–9. [389, 390]

Hoshino, A., Takenaka, Y., and Miyaji, H. (2003). Redetermination of the crystal structure of α-copper phthalocyanine. *Acta Crystallogr. B*, **59**, 393–403. [391]

Hostettler, M., Birkedal, H., and Schwarzenbach, D. (2001). Polymorphs and structures of mercuric iodide. *Chimia*, **55**, 541–5. [20]

Huang, H., Zhou, Z., Song, J., Liang, L., Wang, K., Cao, D., Sun, W., Dong, X., and Xue, M. (2011). Energetic salts based on dipicrylamine and its amino derivative. *Chem. Eur. J.*, **17**, 13593–602. [427]

Huang, K.-S., Britton, D., Etter, M. C., and Byrn, S. R. (1995). Polymorphic characterization and structural comparisons of the non-linear optically-active and inactive forms of two polymorphs of 1,3-bis(*m*-nitrophenyl)urea. *J. Mater. Chem.*, **5**, 379–83. [116]

Huang, K.-S., Britton, D., Etter, M. C., and Byrn, S. R. (1996). Synthesis, polymorphic characterization and structural comparisons of the non-linear optically active and inactive forms of polymorphs of 3-(nitroanilino)cycloalk-2-en-1-ones. *J. Mater. Chem.*, **6**, 123–9. [302]

Huber, P. W. (1991). *Galileo's Revenge: Junk Science in the Courtroom*. Basic Books, New York. [79, 429]

Hughes, C. E. and Harris, K. D. M. (2009). The effect of deuteration on polymorphic outcome in the crystallization of glycine from aqueous solution. *New J. Chem.*, **33**, 713–16. [6]

Hughes, C. E. and Harris, K. D. M. (2010). Direct observation of a transient polymorph during crystallization. *Chem Commun.*, **46**, 4982–4. [193]

Hughes, C. E., Williams, P. A., Peskett, T. R., and Harris, K. D. M. (2012). Exploiting in situ solid state NMR for the discovery of new polymorphs during crystallization processes. *J. Phys. Chem. Lett.*, **3**, 3176–81. [193]

Hulliger, J. (1994). Chemistry and crystal growth. *Angew. Chem. Int. Ed. Engl.*, **33**, 143–62. [375]

Hulliger, J., Wüst, T., Brahimmi, K., and Garcia, J. C. (2012). Can mono domain polar crystals exist? *Cryst. Growth Des.*, **12**, 5211–18. [297]

Hultgren, R. (1936). An X-ray study of symmetrical trinitrotoluene and cyclotrimethylnitramine. *J. Chem. Phys.*, **4**, 84. [417]

Hunger, K. (1999). The effect of crystal structure on color application properties of organic pigments. *Rev. Prog. Color. Relat. Top.*, **29**, 71–84. [378]

Hunger, K. and Schmidt, M. U. (2018). *Industrial Organic Pigments*, 4th edn. Wiley-VCH, Weinheim. [377, 378, 385, 386, 390, 392, 393, 394]

Hunter, S., Coster, P. L., Davidson, A. J., Miller, D. I. A., Parker, S. F., Marshall, W. G., Smith, R. I., Morrison, C. A., and Pulham, C. R. (2015). High-pressure experimental and DFT-D structural studies of the energetic material FOX-7. *J. Phys. Chem. C*, **119**, 2322–34. [414]

Hunter, S., Sutinen, T., Parker, S. F., Morrison, C. A., Williamson, D. M., Thompson, S., Gould, P. F., and Pulham, C. R. (2013). Experimental and DFT-D studies of the molecular organic energetic material RDX. *J. Phys. Chem.*, **117**, 8062–71. [410]

ICDD (2016). *Power Diffraction File*. International Center for Diffraction Data, Newton Square, PA. Available software for powder diffraction may be found through this website: http://www.icdd.com/resources/websites.htm [28, 28, 165]

IHS Markit (2017). *Organic Color Chemical Economics Handbook*. https://www.ihs.com/products/organic-color-chemical-economics-handbook.html [376]

Imaeda, K., Enoki, T., Mori, T., Inokuchi, H., Sasaki, M., Nakasuji, K., and Murata, I. (1989). Electronic properties of new organic conductors based on 2,7-bis(methylthio)-1,6-dithiapyrene (MTDTRY) with TCNQ and *p*-benzoquinone derivatives. *Bull. Chem. Soc. Jpn.*, **62**, 372–9. [276]

Imahori, S. and Hirako, S. (1976). β-Crystalline phase perylene pigment—prepn by condensing with xylidine derived in organic solvent. Mitsubishi Chemical Industries Co. Ltd. Patent JP 51-7025; *Chem. Abstr.*, **84**, 166251s. [381t]

Inabe, T., Goto, H., Fujinawa, T., Ahashi, H., Ogata, H., Miyajima, S., and Maruyama, Y. (1996). Unusual electrical properties of 1,6-diaminopyrene charge-transfer complex crystals. *Mol. Cryst. Liq. Cryst.*, **284**, 283–90. [278]

International Commission on Harmonization (1996). *Q2B validation of analytical procedures: methodology*. ICH Secretariat, Geneva. [168]

Iqbal, A., Jost, M., Kirchmayr, R., Pfenninger, J., Rochat, A., and Wallquist, O. (1988). The synthesis and properties of 1,4-diketo-pyrrolo[3,4-c]pyrroles. *Bull. Soc. Chim. Belg.* **97**, 615–44. [382t]

Ismailsky, W. A. (1913). Thesis, Technische Hochschule Dresden, Germany. [376]

Ito, S., Nishimura, M., Kobayashi, Y., Itai, S., and Yamamoto, K. (1997). Characterization of polymorphs and hydrates of GK-128, a serotonin(3) receptor antagonist. *Int. J. Pharm.*, **151**, 133–43. [9]

Ito, T. (1950). *X-ray Studies on Polymorphism*, pp. 111–21. Maruzen, Tokyo. [417]

IUCr (2001). Available software for powder diffraction may be found through this website: http://www.iucr.ac.uk/sincris-top/logiciel/ [165]

Iuzzolino, L., Reilly, A. M., McCabe, P., and Price, S. L. (2017). Crystal structure informatics for defining the conformational space needed for predicting crystal structures of pharmaceutical molecules. *J. Chem. Theory Comput.*, **13**, 5163–71. [359]

Ivanisevic, I., Bugay, D. E., and Bates, S. (2005). On pattern matching of X-ray powder diffraction data. *J. Phys. Chem. B*, **109**, 7781–7. [271]

Ivashevskaya, S. N., van de Streek, J., Djanhan, J. E., Bruning, J., Alig, E., Bolter, M., Schmidt, M. U., Blaschka, P., Hoffken, H. W., and Erk, P. (2009). Structure determination of seven phases and solvated of Pigment Yellow 183 and Pigment Yellow 191 from X-ray powder and single-crystal data. *Acta Crystallogr. B*, **65**, 212–22. [383t]

Iwaihara, C., Kitagawa, D., and Kobatake, S. (2015). Polymorphic crystallization and thermodynamic phase transition between the polymorphs of a photochromic diarylethene. *Cryst. Growth Des.*, **15**, 2017–23. [310]

Jacewicz, V. W. and Nayler, J. H. C. (1979). Can metastable crystal forms "disappear"? *J. Appl. Crystallogr.*, **12**, 396–7. [134]

Jacob, G., Toupet, L., Ricard, L., and Cagnon, G. (1999). Private communication to CSD. [408]

Jacobsen, C. S. and Torrance, J. B. (1983). Behavior of charge-transfer absorption upon passing through the neutral–ionic phase transition. *J. Chem. Phys.*, **78**, 112–15. [283, 284]

Jaeger, G. (1998). The Ehrenfest classification of phase transitions: introduction and evolution. *Arch. Hist. Exact Sci.*, **53**, 51–81. [210]

Jaffe, E. E. (1992). Quinacridone and some of its derivatives. *JOCCA-Surf. Coat. Int.*, **75**, 24–31. [385]

Jaffe, E. E. (1996). *Encyclopedia of Chemical Technology*, 4th edn, Vol. 19. Wiley, New York. [377]

Jahansouz, H., Thompson, K. C., Brenner, G. S., and Kaufman, M. J. (1999). Investigation of the polymorphism of the angiotensin II antagonist agent MK-996. *Pharm. Dev. Technol.*, **4**, 181–7. [95]

Jakeway, S. C., de Mello, A. J., and Russell, E. L. (2000). Miniaturized total analysis systems for biological analysis. *Fresenius J. Anal. Chem.*, **366**, 525–39. [208]

Jameson, G. B., Oswald, H. R., and Beer, H. R. (1984). Structural phase transition in dihalo(N,N'-di-*tert*-butyldiazabutadiene)nickel) complexes. Structures of bis[dibromo (N,N'-di-*tert*-butyldiazabutadiene)nickel] and dibromo(N,N'-di-*tert*-butyldiazabutadiene) nickel. *J. Am. Chem. Soc.*, **106**, 1669–75. [335]

Janczak, J. (2000). Comment on the polymorphic forms of metal-free phthalocyanine. Refinement of the crystal structure of α-H$_2$Pc at 160 K. *Pol. J. Chem.*, **74**, 157–62. [383t, 393]

Janczak, J. and Kubiak, R. (1992). Crystal and molecular structures of metal-free phthalo-cyanines, 1,2-dicyanobenzene tetramers. II. α Form. *J. Alloys Compd*, **190**, 121–4. [393]

Jang, M.-S., Nakamura, T., Takashige, M., and Kojima, S. (1980). Crystal growth and polymorphism of 2,2,6,6-tetramethyl-piperidino oxy. *Jpn. J. Appl. Phys.*, **19**, 1413–14. [291]

Jarvis, L. M. (2016). The year in new drugs. *Chem. Eng. News*, **54**(5), 12–17. [432]

Jaslovsky, G. S., Egyed, O., Holly, S., and Hegedus, B. (1995). Investigation of the morpho-logical composition of cimetidine by FT-Raman spectroscopy. *Appl. Spectrosc.*, **49**, 1142–5. [176]

Jeffrey, G. A. and Saenger, W. (1991). *Hydrogen Bonding in Biological Structures*. Springer-Verlag, Berlin. [217, 218, 269]

Jelinek, Z. K., Maly, J., and Pizl, J. (1964). Veränderungen der krystalinischen delta-modifikation con indanthren-blau RS unter dem electronenmikroskop. In *Proceedings, Third European Regional Conference on Electron Microscopy*, Czechoslovak Academy of Science, Prague, pp. 357–8. [394]

Jelley, E. E. (1936). Spectral absorption and fluorescence of dyes in the molecular state. *Nature*, **138**, 1009–10. [328]

Jenkins, R. and Snyder, R. L. (1996). Introduction to X-ray powder diffractometry. In *Chemical Analysis*, Vol. 138 (ed. J. D. Winefordner), pp. 324–35. Wiley-Interscience, New York. [28, 29, 156, 161, 208]

Jensen, W. B. (1998). Logic, history and the chemistry textbook. *J. Chem. Educ.*, **75**, 817–28. [5]

Ji, Y., Pend, Z., Tong, B., Shi, J., Zhi, J., and Ding, Y. (2017). Polymorphism-dependent aggregation-induced emission of pyrrolopyrrole-base derivative and its multi-stimuli response behaviors. *Dyes Pigm.*, **139**, 664–71. [319]

Jiang, Q. and Ward, M. D. (2014). Crystallization under nanoscale confinement. *Chem. Soc. Rev.*, **43**, 2066–79. [115]

Jin, Y. S. (2012). *Discovering new crystalline forms of atorvastatin – new strategies for screening*. Ph.D. Thesis, Martin-Luther Universität Halle-Wittenberg, Germany. [94, 264, 347, 433]

Joachim, J., Opota, D. O., Joachim, G., Reynier, J. P., Monges, P., and Maury, L. (1995). Effect of solvate formation on lyoavailability of lorazepam during wet granulation. *STP Pharma Sci.*, **5**, 486–8. [9]

Johansson, D., Bergenstaahl, B., and Lundgren, E. (1995). Wetting off at crystals by trigly-ceride oil and water. 1. The effect of additives. *J. Am. Oil Chem. Soc.*, **72**, 921–31. [40]

Johansson, K. E., van de Streek, J. Revision of the crystal structure of the first molecular polymorph in history, Cryst. Growth Des. 2016, 16, 1366–1370. [156]

Johnston, A., Bhardwaj-Miglani, R., Gurung, R., Vassileiou, A. D., Florence, A., and Johnston, B. (2017). Combined chemoinformatics approach to solvent library design using cluster Sim and multidimensional scaling. *J. Chem. Inf. Model.*, **57**, 1807–15. [354]

Jojart-Laczkovich, O. and Szabo-Revesz, P. (2011). Formulation of tablets containing an "in-process" amorphized active pharmaceutical ingredient. *Drug Dev. Ind. Pharm.*, **37**, 1272–81. [371]

Jones, A. O. F., Chattopadhyay, B., Geerts, Y. H., and Resel, R. (2016). Substrate-induced and thin-film phases: polymorphism of organic materials on surfaces. *Adv. Funct. Mater.*, **26**, 2233–55. [207, 274]

Jovanovic, O., Karlovic, D. J., and Jakovljevic, J. (1995). Chocolate precrystallization—a review. *Acta Aliment.*, **24**, 225–39. [40]

Julian, Y. and McCrone, W. C. (1971). Accurate use of hot stages. *Microscope*, **19**, 225–34. [138]

Kachi, S., Terada, M., and Hashimoto, H. (1998). Effects of amorphous and polymorphs of PF1022A, a new antinematode drug, on *Angiostrongylus costaricensis* in mice. *Jpn. J. Pharmacol.*, **77**, 235–45. [347]

Kagawa, H., Sagawa, M., Hama, T., Kakuta, A., Kaji, M., Nakayama, H., and Ishli, K. (1996). New crystal form of an organic nonlinear optical material, 8-(4'-acetylphenyl)-1,4-dioxa-8-azaspiro[4.5]decan (APDA). *Chem. Mater.*, **8**, 2622–7.

Kahn, O., Garcia, Y., Létard, J. F., and Mathoniere, C. (1999). Hysteresis and memory effect in supramolecular chemistry. In *Supramolecular Engineering of Synthetic Metallic Materials* (ed. J. Veciana, C. Rovira, and D. B. Amabilino), NATO ASI Series, Vol. C518, pp. 127–44. Kluwer Academic Publishers, Dordrecht. [285]

Kahr, B. and McBride, M. (1992). Optically anomalous crystals. *Angew. Chem. Int. Ed. Engl.*, **31**, 1–26. This reference contains a historical account of the phenomenon of optical anomalies, a field which went dormant for nearly half a century for reasons similar to those involving activity in the field of polymorphism. [20, 303]

Kalinkova, G. N. and Hristov, St. I. (1996). Infrared spectroscopic and thermal analysis of different modifications of calcium valproate. *Vibrational Spectroscopy*, **11**, 143–49. [9]

Kalinkova, G. N. and Stoeva, S. (1996). Polymorphism of azlocillin sodium. *Int. J. Pharmacol.*, **135**, 111–14. [347]

Kalipharma, Inc. v. Bristol-Myers Co. (1989). No. 88 CIV. 4640; 707 F. Supp. 741. [434, 436]

Kálmán, A., Parkanyi, L., and Argay, G. (1993a). On the isostructuralism of organic molecules in terms of Kitaigorodskii's early perceptions. *Acta Chim. Hung. Models Chem.*, **130**, 279–98. [395]

Kálmán, A., Parkanyi, L., and Argay, G. (1993b) Classification of the isostructurality of organic molecules in the crystalline state. *Acta Crystallogr. B*, **49**, 1039–49. [395]

Kampschulte, L., Lackinger, M., Maier, A. K., Kishore, R. S. K., Griessl, S., Schmittel, M., and Heeckl, W. M. (2006). Solvent induced polymorphism in supramolecular 1,3,5-benzenetribenzoic acid monolayers. *J. Phys. Chem. B*, **110**, 10829–36. [207]

Kaneko, F., Sakashita, H., Kobayashi, M., Kitagawa, Y., Matsuura, Y., and Suzuki, M. (1994a). Double-layered polytypic structure of the B form of octadecanoic acid, $C_{18}H_{36}O_2$. *Acta Crystallogr. C*, **50**, 245–7. [174]

Kaneko, F., Sakashita, H., Kobayashi, M., Kitagawa, Y., Matsuura, Y., and Suzuki, M. (1994b). Double-layered polytypic structure of the E form of octadecanoic acid, $C_{18}H_{36}O_2$. *Acta Crystallogr. C*, **50**, 247–50. [174]

Kaneko, K., Shirai, O., Miyamoto, H., Kobayashi, M., and Suzuki, M. (1994c). Oblique infrared transmission spectroscopic study on the E → C and B → C phase transitions of stearic acid: effects on polytypic structure. *J. Phys. Chem.*, **98**, 2185–91. [174]

Karfunkel, H. R., Rohde, B., Leusen, F. J. J., Gdanitz, R. J., and Rihs, G. (1993). Continuous similarity measure between non-overlapping X-ray powder diagrams of different crystal modifications. *J. Comput. Chem.*, **14**, 1125–35. [271]

Karfunkel, H., Wilts, H., Hao, Z. M., Iqbal, A., Mizuguchi, J., and Wu, Z. J. (1999). Local similarity in organic crystals and the non-uniqueness of X-ray powder patterns. *Acta Crystallogr. B*, **55**, 1075–89. [161]

Karothu, D. P., Weston, J., Desta, I. T., and Naumov, P. (2016). Shape-memory and self-healing effects in mechanosalient molecular crystals. *J. Am. Chem. Soc.*, **138**, 13298–306. [320]

Karpowicz, R. J. and Brill, T. B. (1982). The β–δ transformation of HMX: its thermal analysis and relationship to propellants. *Am. Inst. Aeronaut. Astronaut. J.*, **20**, 1586–91. [405]

Karpowicz, R. J. and Brill, T. B. (1984). Comparison of the molecular structure of hexahydro-1,3,5-trinitro-*s*-triazine in the vapor, solution and solid phases. *J. Phys. Chem.*, **88**, 348–52. [410]

Karpowicz, R. J., Sergio, S. T., and Brill, T. B. (1983). β-Polymorph of hexahydro-1,3,5-trinitro-*s*-triazine. A Fourier Transform infrared spectroscopy study of an energetic material. *Ind. Eng. Chem. Prod. Res. Dev.*, **22**, 363–5. [410]

Kashchiev, D. (2000). *Nucleation: Basic Theory with Applications*. Butterworth-Heinemann, Oxford. [80]

Katan, C. and Koenig, C. (1999). Charge-transfer variation caused by symmetry breaking in a mixed-stack organic compound: TTF-2, 5Cl$_2$BQ. *J. Phys. Condens. Matter*, **11**, 4163–77. [283]

Kato, R., Kobayashi, H., Kobayashi, A., Moriyama, S., Nishio, Y., Kajita, K., and Sasaki, W. (1987). A new ambient-pressure superconductor, κ-(BEDT-TTF)$_2$I$_3$. *Chem. Lett.*, **16**, 507–10. [119, 277]

Katrusiak, A. (1990). High pressure X-ray diffraction study on the structure and phase-transition of 1,3-cyclohexanedione crystals. *Acta Crystallogr. B*, **46**, 246–56. [336]

Katrusiak, A. (1991). High pressure X-ray diffraction studies on organic crystals. *Cryst. Res. Tech.*, **26**, 523–31. [336]

Katrusiak, A. (1995). High pressure X-ray diffraction of pentaerythritol. *Acta Crystallogr. B*, **51**, 873–9. [336]

Katrusiak, A. (2000). Conformational transformation coupled with the order-disorder phase transition in 2-methyl-1,3-cyclohexanedione crystals. *Acta Crystallogr. B*, **56**, 872–81. [335]

Katrusiak, A. (2008). High-pressure crystallography. *Acta Crystallogr. A*, **64**, 135–48.

Katrusiak, A. and Szafranski, M. (1996). Structural phase transitions in guanidinium nitrate. *J. Mol. Struct.*, **378**, 205–23. [312]

Kaufmann, L., Kennedy, S. R., Jones, C. D., and Steed, J. W. (2016). Cavity-containing supramolecular gels as a crystallization tool for hydrophobic pharmaceuticals. *Chem. Commun.*, **52**, 10113–16. [356]

Kawakami, K. (2007). Reversibility of enantiotropically related polymorphic transformations from a practical viewpoint: thermal analysis of kinetically reversible/irreversible polymorphic transformations. *J. Pharm. Sci.*, **96**, 982–9. [45]

Kawakami, K. (2010). Parallel thermal analysis technology using an infrared camera for high-throughput evaluation of active pharmaceutical ingredients: a case study of melting point determination. *AAAS Pharmscitech*, **11**, 1202–5. [352]

Kawamura, T. (1987). New diazo compound. Crystal polymorph used for yellow diazo pigment—has good transparency and tinting strength when used for printing inks. Toyo Ink Patent JP 62 153 353. [382t]

Kazantsev, A. V., Karamertzanis, P. G., Adjiman, C. S., Pantelides, C. C., Price, S. L., Galek, P. T., Day, G. M., and Cruz-Cabeza, A. J. (2011). Successful prediction of a model pharmaceutical in the fifth blind test of crystal structure prediction. *Int. J. Pharm.*, **418**, 168–78. [244]

Kazmaier, P. M. and Hoffmann, R. (1994). A theoretical study of crystallochromy. Quantum interference effects in the spectra of perylene pigments. *J. Am. Chem. Soc.*, **116**, 9684–91. [387, 388]

Kellens, M., Meeussen, W., and Reynaers, H. (1992). Study of the polymorphism and the crystallization kinetics of tripalmitin: a microscopic approach. *J. Am. Oil Chem. Soc.*, **69**, 906–11. [40]

Keller, A. and Cheng, S. Z. D. (1998). The role of metastability in polymer phase transitions. *Polymer*, **39**, 4461–87. [40]

Kelly, J. J. and Giambalvo, V. A. (1966). 2,9-Dimethylquinacridone in a "yellow" crystalline form. American Cyanamid Co. Patent US 3 264 300. [381t]

Kelly, P. F., Man, S.-M., Slawin, A. M. Z., and Waring, K. W. (1999). Preparation of a range of copper complexes of diphenylsulfimide: X-ray crystal structures of [Cu(Ph$_2$SNH)$_4$]Cl$_2$ and [Cu$_4$(μ_4-O)(μ-Cl)$_6$(Ph$_2$SNH)$_4$]. *Polyhedron*, **18**, 3173–9. [226]

Kelly, P. F., Slawin, A. M. Z., and Waring, K. W. (1997). Preparation and crystal structures of two forms of trans-[CuCl$_2${N(H)SPh$_2$}$_2$], an unusual example of square planar/pseudo-tetrahedral isomerism in a neutral copper(II) complex. *J. Chem. Soc., Dalton Trans.*, 2853–4. [115, 226]

Kelly, P. F., Slawin, A. M. Z., Waring, K. W., and Wilson, S. (2001). Further investigation into the isomerism of Cu(II) complexes of diphenylsulfimide: the preparation and X-ray crystal structure of {[Cu(Ph$_2$SNH)$_4$][CuBr$_2$(Ph$_2$SNH)$_2$]Br$_2$}, the first example of a {[CuL$_4$][CuX$_2$L$_2$]X$_2$} structure. *Inorg. Chim. Acta*, **312**, 201–4. [115]

Keng, E. Y. H. (1968). Air and helium pycnometer. *Powder Technol.*, **3**, 179–80. [208]

Kennard, O. (1993). From data to knowledge—use of the Cambridge Structural Database for studying molecular interactions. *Supramol. Chem.*, **1**, 277–95. [26]

Kern, A., Madsen, I. C., and Scarlett, N. V. Y. (2012). Quantifying amorphous phases. In *Uniting Electron Crystallography and Powder Diffraction* (ed. U. Kolb, K. Shankland, L. Meshi, A. Avilov, and W. David), pp. 219–31. Springer, Dordrecht. [164]

Kettner, F., Huter, L., Schafer, J., Roder, K., Purgahn, U., and Krautscheid, H. (2011). Selective crystallization of indigo B by a modified sublimation method and its redetermined structure. *Acta Crystallogr. E*, **67**, o2867. [383t]

Khalafallah, N., Khalil, S. A., and Moustafa, M. A. (1974). Bioavailability of determination of two crystal forms of sulfameter in humans from urinary excretion data. *J. Pharm. Sci.*, **63**, 861–4. [349]

Khalil, S. A., Moustafa, M. A., Ebian, A. R., and Motawi, M. M. (1972). GI absorption of two crystal forms of sulfameter in man. *J. Pharm. Sci.*, **61**, 1615–17. [348]

Ki, H., Oang, K. Y., Kim, J., and Ihee, H. (2017). Ultrafast X-ray crystallography and liquidography. *Annu. Rev. Phys. Chem.*, **68**, 473–97. [319]

Kiers, C. Th., de Boer, J. L., Olthof, R., and Spek, A. L. (1976). The crystal structure of a 2,2-diphenyl-1-picrylhydrazyl (DPPH) modification. *Acta Crystallogr. B*, **32**, 2297–305. [291]

Kikuchi, K., Honda, Y., Ishikawa, Y., Saito, K., Ikemoto, I., Murata, K., Hiroyuki, A., Ishiguro, T., and Kobayashi, K. (1988). Polymorphism and electrical conductivity of the organic superconductor (DMET)$_2$AuBr$_2$. *Solid State Commun.*, **66**, 405–8. [277]

Kim, H. S., Jeffrey, G. A., and Rosenstein, R. D. (1968). The crystal structures of the κ form of D-mannitol. *Acta Crystallogr. B*, **24**, 1449–55. [88]

Kim, J.-H., Park, Y.-C., Yim, Y.-J., and Han, J.-S. (1998). Crystallization behavior of hexanitrohexaazaisowurtzitane at 298 K and quantitative analysis of mixtures by FTIR. *J. Chem. Eng. Jpn.*, **31**, 478–81. [408]

Kim, S. H., Quigley, G. J., Suddath, F. L., McPherson, A., Sneden, D., Kin, J. J., Weinzirl, J., and Rich, A. (1973). X-ray crystallographic studies of polymorphic forms of yeast phenylalanine transfer RNA. *J. Mol. Biol.*, **75**, 421–8. [30]

Kim, S.-J., Lee, B.-M., Lee, B.-C., Kim, H.-S., Kim, H., and Lee, Y.-W. (2011). Recrystallization of cyclotetramethylenetetranitramine (HMX) using gas anti-solvent (GAS) process. *J. Supercrit. Fluids*, **59**, 108–16. [404]

Kimura, K., Hirayama, F., and Uekama, K. (1999). Characterization of tolbutamide polymorphs (Burger's forms II and IV) and polymorphic transition behavior. *J. Pharm. Sci.*, **88**, 385–91. [346]

Kimura, K., Hirayama, F., Arima, H., and Uekama, K. (2000). Effects of ageing on crystallization, dissolution, and absorption characteristics of amorphous tolbutamide-2-hydroxypropyl-beta-cyclodextrin complex. *Chem. Pharm. Bull.*, **48**, 646–50. [370]

King, M. V., Bello, J., Pgnatano, E. H., and Harder, D. (1962). Crystalline forms of bovine pancreatic ribonuclease. Some new modifications. *Acta Crystallogr.*, **15**, 144–7. [30]

King, M. V., Magdoff, B. S., Adelman, M. B., and Harker, D. (1956). Crystalline forms of bovine pancreatic ribonuclease: techniques of preparation, unit cells and space groups. *Acta Crystallogr.*, **9**, 460–5. [30]

Kinoshita, M. (1994). Ferromagnetism of organic radical crystals. *Jpn. J. Appl. Phys.*, **33**, 5718–33. [285, 288]

Kirk, J. H., Dann, S. E., and Blatchford, C. G. (2007). Lactose: a definitive guide to polymorphic determination. *Int. J. Pharm.* **334**, 103–14. [363]

Kishimoto, S. and Naruse, M. (1987). The bundling phenomenon in aspartame crystallisation. *Chem. Ind.*, 16 February, 127–8. [454]

Kishimoto, S. and Naruse, M. (1988). A process development for the bundling crystallization of aspartame. *J. Chem. Technol. Biotechnol.*, **43**, 71–82. [453]

Kishimoto, S., Nagashima, N., Naruse, M., and Toyokura, K. (1989). The "bundle-like" crystals in aspartame crystallization. *Process Technol. Proc.*, **6**, 511–14. [453, 455]

Kistenmacher, T. J., Emge, T. J., Bloch, A. N., and Cowan, D. O. (1982). Structure of the red, semiconductor form of 4,4′,5,5′-tetramethyl-$\Delta^{2,2'}$-bis-1,3-diselenole and 7,7,8,8-tetracyano-*p*-quinodimethane. *Acta Crystallogr. B*, **38**, 1193–9. [275]

Kitagawa, D. (2014). *Photochromic reaction behavior and solid state property changes of diarylethene crystals*. PhD Thesis, Osaka State University, Japan. [307]

Kitaigorodskii, A. I. (1961). *Organic Chemical Crystallography*. Consultants Bureau, New York. [53, 218, 237, 239, 360, 411]

Kitaigorodskii, A. I. (1970). General view on molecular packing. *Adv. Struct. Res. Diffr. Methods*, **3**, 173–247. [45, 215, 219, 227, 249]

Kitaigorodskii, A. I. (1973). *Molecular Crystals and Molecules*, pp. 2–3, 163–7, 184–90. Academic Press, New York. [227]

Kitamura, M. (1989). Polymorphism in the crystallization of L-glutamic acid. *J. Cryst. Growth*, **96**, 541–6. [87, 89]

Kitamura, S., Chang, L. C., and Guillory, J. K. (1994). Polymorphism of mefloquine hydrochloride. *Int. J. Pharm.*, **101**, 127–44. [9, 363]

Kitaoka, H. and Ohya, K. (1995). Pseudopolymorphism and phase stability of 7-piperidino-1,2,3,5-tetrahydroimidazo-[2,1-b]quinazolin-2-one (DN-9693). *J. Therm. Anal.*, **44**, 1047–56. [9]

Kitaoka, H., Wada, C., Moroi, R., and Hakusui, H. (1995). Effect of dehydration on the formation of levofloxacin pseudopolymorphs. *Chem. Pharm. Bull.*, **43**, 649–53. [9]

Klapötke, T. M. (2017). *Chemistry of High-Energy Materials*, 4th edn. De Gruyter, Berlin. [398, 399]

Klapper, H., Kutzke, H., and Haussühl, S. (2000). Physical properties of stable monoclinic and metastable trigonal 4-methylbenzophenone. *Z. Kristallogr.* **215**, 187–9. [321]

Klaproth, M. H. (1798). *Bergmannische J.* I, 294–9. [cited by Partington, J. R. (1964). *A History of Chemistry*, Vol. 4, p. 203. MacMillan & Co., London] [31]

Klebe, G., Graser, F., Hädicke, E., and Berndt, J. (1989). Crystallochromy as a solid state effect: correlation of molecular conformation, crystal packing and colour in perylene-3,4:9,10-bis(carboximide) pigments. *Acta Crystallogr. B*, **45**, 69–77. [387]

Klein, J., Lehmann, C. W., Schmidt, H. W., and Maier, W. F. (1998). Combinatorial material libraries on the microgram scale with an example of hydrothermal synthesis. *Angew. Chem. Int. Ed. Engl.*, **37**, 3369–72. [209]

Klug, H. P. and Alexander, L. E. (1974). *X-ray Diffraction Procedures for Polycrystalline and Amorphous Materials.* John Wiley & Sons, New York. [164, 370, 371]

Knepper, R., Tappan, A. S., Rodriguez, M. A., Alam, M. K., Martin, L., and Marquex, M. P. (2012). Crystallization behavior of vapor-deposited hexanitroazobenzene (HNAB) films. *AIP Conf. Proc.* **1426**, 1589–92. [426]

Knopp, M. M., Löbmann, K., Elder, D. P., Rades, T., and Holm, R. (2016). Recent advances and potential applications of modulated differential scanning calorimetry (mDSC) in drug development. *Eur. J. Pharm. Sci.*, **87**, 164–73. [154]

Knudsen, B. I. (1966). Copper phthalocyanine. Infrared absorption spectra of polymorphic modifications. *Acta Chem. Scand.*, **20**, 1344–50. [392]

Kobayashi, A., Kato, R., Kobayashi, H., Moriyama, S., Nishio, Y., Kajita, K., and Sasaki, W. (1987). Crystal and electronic structure of a new molecular superconductor, κ-(BEDT-TTF)$_2$I$_3$. *Chem. Lett.*, **16**, 459–62. [277]

Kobayashi, M., Matsumoto, Y., Ishida, A., Ute, K., and Hatada, K. (1994). Polymorphic structures and molecular vibrations of linear oligomers of polyoxymethylene studied by polarized infrared and Raman spectra measured on single crystals. *Spectrochim. Acta A*, **50**, 1605–17. [170]

Kobayashi, N. and Ando, H. (1988a). Red monoazo lake pigment. Dainippon Inc. Patent JP 63-225 666; *Chem. Abstr.*, **110**, 116666r. [381t]

Kobayashi, N. and Ando, H. (1988b). Monoazo lake for printing inks etc.—prepared by diazotizing 2-amino-5-methyl benzenesulphonic acid, coupling with 2-hydroxy-3-naphtholic acid and adding barium chloride solution. Dainippon Inc. Patent JP 63-225 667; *Chem. Abstr.*, **110**, 77504q. [381t]

Kobayashi, N. and Ando, H. (1988c). New monoazo lake dye—comprises barium salt of 1-(4-methyl-2-sulphonylphenyl azo)2-hydroxy-3-naphthoic acid. Dainippon Inc. Patent JP 63-225 668; *Chem. Abstr.*, **110**, 116665q. [381t]

Kobayashi, Y., Ito, S., Itai, S., and Yamamoto, K. (2000). Physicochemical properties and bioavailability of carbamazepine polymorphs and dihydrate. *Int. J. Pharm.*, **193**, 137–46. [351]

Koch, U., Kuhnt, R., Lück, M., and Modrow, H.-W. (1987a). C.I. Disperse Yellow 42 α- or β-modification production—by adding salt to hot aqueous alkaline solution and acidifying at specified temperatures. Martin-Luther Universität, Halle-Wittenberg, Patent DD 251 359. [379t]

Koch, U., Modrow, H.-W., and Wallascheck, G. (1987b). New metastable α-modification of C.I. Disperse Yellow 23 production—by conversion to color stable β-modification by dissolving crude product in aqueous sodium hydroxide, cooling filtered solution etc. Martin-Luther Universität, Halle-Wittenberg, Patent DD 251 359. [379t]

Kofler, A. (1941). Thermal analysis with a hot-stage microscope. *Z. Phys. Chem. Abt. A*, **187**, 201–10. [102]

Kofler, A. and Kofler, L. (1948). The estimation of the melting points of unstable modifications of organic materials. *Monatsh. Chem.*, **78**, 13–22. [142]

Kofler, L. and Kofler, A. (1954). *Thermo-mikro-methoden zur Kennzeichnung organischer Stoffe und Stoffgemiche.* Innsbruck, Wagner. A 1980 translation of this book by Walter C. McCrone is available from McCrone Associates, Inc. [12, 26, 137, 138, 142, 342]

Kofler, L. and Winkler, H. (1950a). A new method for the analysis of mixtures of pharmaceuticals. *Arch. Farm.*, **283**, 176–83. [146]

Kofler, L. and Winkler, H. (1950b). A rapid method for constructing melting diagrams. II. *Monatsh. Chem.*, **81**, 746–50. [146]

Kohno, Y., Ueda, K., and Imamura, A. (1996). Molecular dynamics simulations of initial decomposition process on the unique N—N bond in nitramines in the crystalline state. *J. Phys. Chem.*, **100**, 4701–12. [404, 406]

Koizumi, S. and Matsunaga, Y. (1972). Polymorphism and physical properties of the 1,6-diaminopyrene–*p*-chloranil and related molecular complexes. *Bull. Chem. Soc. Jpn.*, **45**, 423–8. [124]

Kolb, U., Mugnaioli, E., and Gorelik, T. E. (2011). Automated electron diffraction tomography – a new tool for nano crystal structure analysis. *Cryst. Res. Technol.*, **46**, 542–54. [195]

Kolinsky, P. V. (1992). New materials and their characterization for photonic device applications. *Opt. Eng.*, **31**, 1676–84. [296]

Kolossváry, I. and Guida, W. C. (1996). Low mode search. An efficient, automated computational method for conformational analysis: application to cyclic and acyclic alkanes and cyclic peptides. *J. Am. Chem. Soc.*, **118**, 5011–19. [243]

Komai, A., Shirane, N., Ito, Y., and Terui, S. (1977). Copper phthalocyanine in ρ crystal form—which is redder than α type and used for textile printing, resin coloring and in printing ink. Nippon Shokubai Kagaku Kogyo, Ltd. Patent DE 2 659 211. [389, 390]

Komorski, R. A. (ed.) (1986). *High Resolution NMR Spectroscopy of Synthetic Polymers in Bulk.* Methods in Stereochemical Analysis Series, Vol. 7. VCH, Deerfield Beach, FL. [182]

Konek, C. T., Mason, B. P., Hooper, J. P., Stoltz, C. A., and Wilkinson, J. (2010). Terahertz absorption spectra of 1,3,5,7-tetranitro-1,3,5,7-tetrazocane (HMX) polymorphs. *Chem. Phys. Lett.*, **489**, 48–53. [406]

Koshelev, V. I., Shelyapin, O. P., Shtanov, N. P., Moroz, V. A., Kovalenko, S. A., and Paramonova, L. N. (1987). 4,4′-Dibenzamido-1,1′-dianthraquinonyl and its polymorphic modifications. *J. Appl. Chem. USSR*, **60**, 559–62; English translation of *Zh. Prikl. Khim.*, **60**, 596–9. [381t]

Koyama, H., Scheel, H. J., and Laves, F. (1966). Kristallstrukturen organischer Pigmentfarbstoffe. I. γ-Chinacridon $C_{20}H_{12}O_2N_2$. *Naturwissenschaften*, **53**, 700–7. [380t]

Kozin (1964). *Zh. Prikl. Khim.*, **5**, 324. [427]

Kratochvil, B. (2007). Crystallization of pharmaceutical substances. *Chem. Listy*, **101**, 3–12. [353]

Krc, J., Jr. (1955). *Crystallographic properties of primary explosives.* Quarterly Progress Report No. 2, Armour Research Foundation, Chicago, IL. [404t]

Krishnan, K. and Ferraro, J. R. (1982). *Fourier Transform Infrared Spectroscopy*, Vol. 4. Academic Press, New York. [170]

Krishnan, M., Namasivayan, V., Lin, R. S., Pal, R., and Burns, M. A. (2001). Microfabricated reaction and separation systems. *Curr. Opin. Biotechnol.*, **12**, 92–8. [208]

Kristl, A., Srcic, S., Vrecer, F., Sustar, B., and Vojnovic, D. (1996). Polymorphism and pseudo-polymorphism—influencing the dissolution properties of the guanine derivative acyclovir. *Int. J. Pharm.*, **139**, 231–5. [9]

Kronick, P. L. and Labes, M. M. (1961). Organic semiconductors. V. Comparison of measurements on single-crystal and compressed microcrystalline molecular complexes. *J. Chem. Phys.*, **35**, 2016–19. [124, 278]

Kronick, P. L., Scott, H., and Labes, M. M. (1964). Composition of some conducting complexes of 1,6-diaminopyrene. *J. Chem. Phys.*, **40**, 890–4. [124, 193]

Kruse, H. and Sommer, K. (1976). Color stable modification of monoazo dye of benzene series—made by heating in aqueous or organic suspension. Hoechst AG Patent DE 2 520 577. [379t]

Kubiak, R. and Janczak, J. (1992). Crystal and molecular structures of metal-free phthalo-cyanines, 1,2-dicyanobenzene tetramers. I. β Form. *J. Alloys Compd.*, **190**, 117–20. [393]

Kuhnert-Brandstätter, M. (1962). The Kofler method in chemical microscopy. *Microchem. J. Symp. Ser.*, **2**, 221–31. [364]

Kuhnert-Brandstätter, M. (1965). The status and future of chemical microscopy. *Pure Appl. Chem.*, **10**, 133–44. [344]

Kuhnert-Brandstätter, M. (1971). *Thermomicroscopy in the Analysis of Pharmaceuticals*. Monographs in Analytical Chemistry, Vol. 45. Pergamon, Oxford. [12, 24, 38, 137, 138, 140f, 141f, 142, 143f, 146, 219, 344, 364]

Kuhnert-Brandstätter, M. (1973). Polymorphie von Arzneistoffen und ihre Bedeutung in der pharmazeutischen Technologie. *Informationsdienst APV*, **19**, 73–90. [346]

Kuhnert-Brandstätter, M. (1975). Polymorphism in pharmaceuticals. *Pharm. Unserer Zeit*, **4**, 131–7. [343]

Kuhnert-Brandstätter, M. (1982). Thermomicroscopy of organic compounds. In *Comprehensive Analytical Chemistry*, Vol. 16 (ed. G. Svehla), pp. 329–513, Elsevier, Amsterdam. [137, 138, 143f, 144f, 146]

Kuhnert-Brandstätter, M. (1996). Thermoanalytical methods and their pharmaceutical applications. *Pharmazie*, **51**, 443–57. [147, 366]

Kuhnert-Brandstätter, M. and Junger, E. (1976). IR Spektroskopische Untersuchungen polymorphen Kristallmodifikationen von Alkoholen und Phenolen. *Spectrochim. Acta A*, **23**, 1453–61. [170]

Kuhnert-Brandstätter, M. and Martinek, A. (1965). Statistics on polymorphism of drugs. *Mikrochim. Acta*, **5–6**, 909–19. [344]

Kuhnert-Brandstätter, M. and Riedmann, M. (1989). Thermal analytical and infrared spectroscopic investigations on polymorphic organic compounds. III. *Mikrochim. Acta*, **1**, 373–485. [170, 379t]

Kuhnert-Brandstätter, M. and Sollinger, H. W. (1989). Thermal analytical and infrared spectroscopic investigations on polymorphic organic compounds. 5. *Mikrochim. Acta*, **3**, 125–36. [45]

Kuleshova, L. N. and Zorky, P. M. (1980). Graphical enumeration of hydrogen-bonded structures. *Acta Crystallogr. B*, **36**, 2113–15. [69]

Kumar, R., Siril, P. F., and Soni, P. (2015). Optimized synthesis of HMX nanoparticles using antisolvent precipitation method. *J. Energ. Mater.*, **33**, 277–87. [406]

Kumar, V. S., Addlagatta, A., Nangia, A., Robinson, W. T., Broder, C. K., Mondal, R., Evans, I. R., Howard, J. A. K., and Allen, F. H. (2002). 4,4-Diphenyl-2,5-cyclohexanedinone: four polymorphs and nineteen independent molecular conformations. *Angew. Chem. Int. Ed. Engl.*, **41**, 3848–51. [254]

Kwon, O.-P., Kwon, S.-J., Jazbinsek, M., Choubey, A., Loslo, P. A., Gramlich, V., and Günter, P. (2006). Morphology and polymorphism control of organic polyene crystals by tailor-made auxiliaries. *Cryst. Growth Des.*, **6**, 2327–32. [101]

Lackinger, M., Griessl, S., Heckl, W. A., Hietschold, M., and Flynn, G. W. (2005). Self-assembly of trimesic acid at the liquid–solid interface: a study of solvent-induced polymorphism. *Langmuir*, **21**, 4984–8. [207]

Lahav, M. and Leiserowitz, L. (1993). Tailor-made auxiliaries for the control of nucleation growth and dissolution of two- and three-dimensional crystals. *J. Phys. D: Appl. Phys.*, **26**, B22–B31. [61]

Lahav, M. and Leiserowitz. L. (2006). A stereochemical approach that demonstrates the effect of solvent on the growth of polar crystals: a perspective. *Cryst. Growth Des.*, **6**, 618–24. [62]

Lahav, M., Green, B. S., and Rabinovich, D. (1979). Asymmetric synthesis via reactions in chiral crystals. *Acc. Chem. Res.*, **12**, 191–7. [333]

Laine, E., Pirttimaki, J., and Rajala, R. (1995). Thermal studies on polymorphic structures of ibopamin. *Thermochim. Acta*, **248**, 205–16. [365]

Lancaster, R. (1999). Lamuvidine—development with a sting in the tail! Presented at the 1st International Symposium on Aspects of Polymorphism and Crystallisation—Chemical Development Issues, Hinckley, UK, June 23–25, 1999. Scientific Update, Mayfield, UK. [173f, 187f]

Landers, A. G., Apple, T. M., Dybowski, C., and Brill, T. B. (1985). ^{1}H nuclear magnetic resonance of α-hexahydro-1,3,5-trinitro-s-triazine (RDX) and the α-, β-, γ-, and δ-polymorphs of octahydro-1,3,5,7-tetranitro-1,3,5,7-tetrazocine (HMX). *Magn. Reson. Chem.*, **23**, 158–60. [406]

Landers, A. G., Brill, T. B., and Marino, R. A. (1981). Electronic effects and molecular motion in β-octahydro-1,3,5,7-teranitro-1,3,5,7-tetrazocine based on ^{14}N nuclear quadrupole resonance spectroscopy. *J. Phys. Chem.*, **85**, 2618–23. [406]

Landes, B. G., Malanga, M. T., and Thill, B. P. (1990). Polymorphism in syndiotactic polystyrene. *Adv. X-ray Anal.*, **33**, 433–44. [209]

Lang, M., Kampf, J. W., and Matzger, A. J. (2002). Form IV of carbamazepine. *J. Pharm. Sci.*, **91**, 1186–90. [100]

Lang, M. D., Grzesiak, A. L., and Matzger, A. J. (2002). The use of polymer heteronuclei for crystalline polymorph selection. *J. Am. Chem. Soc.*, **124**, 14834–5. [369]

Langkilde, F. W., Sjoeboem, J., Tekenbergs-Hjelte, L., and Mark, J. (1997). Quantitative FT-Raman analysis of two crystal forms of a pharmaceutical compound. *J. Pharm. Biomed. Anal.*, **15**, 687–96. [177]

Latimer, W. M. and Rodebush, W. H. (1920). Polarity and ionization from the standpoint of the Lewis theory of valence. *J. Chem. Soc.*, **42**, 1419–33. [68]

Latypov, N. V., Bergman, J., Langlet, A., Wellmar, U., and Bemm, U. (1998). Synthesis and reactions of 1,1-diamino-2,2-dinitroethylene. *Tetrahedron*, **54**, 11525–36. [414]

Lautz, H. (1913). The relationship of instable forms with stable forms. *Z. Phys. Chem.*, **84**, 611–41. [142]

Law, K. Y. (1993). Organic photoconductive materials—recent trends and developments. *Chem. Rev.*, **93**, 449–86. [293, 294, 296, 383t]

Law, K. Y., Tarnawsky, J. I. W., and Popovic, Z. D. (1994). Azo pigments and their interme-diates—a study of the structure-sensitivity relationship of photogenerating bisazo pig-ments in bilayer xerographic devices. *J. Imag. Sci. Technol.*, **38**, 118–24. [293]

Le Cointe, M., Gallier, J., Cailleau, H., Gourdji, M., Peneau, A., and Guibe, L. (1995a). ^{35}Cl NQR study on TTF-CA crystals: symmetry lowering and hysteresis at the neutral-to-ionic transition. *Solid State Commun.*, **94**, 455–9. [284]

Le Cointe, M., Lemee-Cailleau, M. H., Cailleau, H., and Toudic, B. (1996). Structural aspects of the neutral-to-ionic transition in mixed stack charge-transfer complexes. *J. Mol. Struct.*, **374**, 147–53. [285]

Le Cointe, M., Lemee-Cailleau, M. H., Cailleau, H., Toudic, B., Toupet, L., Heger, G., Moussa, F., Schweiss, P., Kraft, K. H., and Karl, N. (1995b). Symmetry breaking and structural changes at the neutral-to-ionic transition in tetrathiafulvalene-*p*-chloranil. *Phys. Rev. B*, **51**, 3374–86. [284]

Lee, A. Y., Erdemir, D., and Myerson, A. S. (2011a). Crystal polymorphism in chemical process development. *Annu. Rev. Chem. Biomol. Eng.*, **2**, 259–80. [91, 92, 343]

Lee, B.-M., Kim, S.-J., Lee, B.-C., Kim, H.-S., Kim, H., and Lee, Y.-W. (2011b). Preparation of micronized β-HMX using supercritical carbon dioxide as antisolvent. *Ind. Eng. Chem. Res.*, **50**, 9107–15. [403]

Lee, I. S., Lee, A. Y., and Myerson, A. S. (2008). Concomitant polymorphism in confined environment. *Pharm. Res.*, **25**, 960–8. [115, 359]

Lee, J.-E., Kim, H. J., Han, M. R., Lee, S. Y., Jo, W. J., Lee, S. S., and Lee, J. S. (2009). Crystal structures of C.I. Disperse Red 65 and C.I. Disperse Red 73. *Dyes Pigm.*, **80**, 181–6. [379t]

Lee, J. H., Naumov, P., Chung, I. H. and Lee, S. C. (2011c). Solid state thermochromism and phase transition of charge transfer 1,3-diamino-4,6-dinitro dyes. *J. Phys. Chem. A*, **115**, 10087–96. [313]

Lee, K. and Gilardi, R. (1993). NTO polymorphs. *Mater. Res. Soc. Symp. Proc.*, **296**, 237–42. [413]

Legrand, J. F., Lajzerowicz, J., Lajzerowicz-Bonneteau, J., and Capiomont, A. (1982). Ferroelastic and ferroelectric phase transition in a molecular crystal: tanane. 3. From *ab initio* computation of the intermolecular forces to statistical mechanics of the transition. *J. Phys.*, **43**, 1117–25. [291]

Legros, J.-P. and Valade, L. (1988). Polymorphism of the highly conducting metal complex TTF[Pd(dmit)$_2$]$_2$. *Solid State Commun.*, **68**, 599–604. [118, 276]

Lehmann, M. S., Koetzle, T. F., and Hamilton, W. (1972). Precision neutron diffraction structure determination of protein and nucleic acid components. VIII. Crystal and molecular structure of the β-form of the amino acid, L-glutamic acid. *J. Cryst. Mol. Struct.*, **2**, 225–33. [66]

Lehmann, O. (1877a). Über die Dimorphie des Hydrochinons und des Paranitrophenols. *Z. Kristallogr.*, **1**, 43–8. [33, 138]

Lehmann, O. (1877b). Ueber physikalische Isomerie. *Z. Kristallogr.*, **1**, 97–131. [33, 138, 139f]

Lehmann, O. (1885). XX. Mikrokrystallographische Untersuchungen. *Z. Kristallogr.*, **10**, 321. [332]

Lehmann, O. (1888). *Molekularphysik*. Engelmann, Leipzig. [138, 144]

Lehmann, O. (1891). *Die Krystallanalyse oder die chemische Analyse durch Beobachtung der Krystallbildung mit Hülfe des Mikroskops*. Wilhelm Engelmann, Leipzig. [33, 36, 138]

Lehmann, O. (1910). *Das Kristallisationsmikroskop*. Vieweg, Braunschweig. [138]

Leites, L. A., Buklov, S. S., Mangette, J. E., Schmedake, T. A., and West, R. (2003). Conformational polymorphism of solid tetramesityldisilene Mes$_2$SiSiMes$_2$ (Raman, UV-vis, IR and fluorescence study). *Spectrochim. Acta A*, **59**, 1975–88. [306]

Lemmerer, A., Adsmond, D. A., Esterhuysen, C., and Bernstein, J. (2013). Polymorphic co-crystals from polymorphic co-crystal formers: competition between carboxylic acid···pyridine and phenol···pyridine hydrogen bonds. *Cryst. Growth Des.*, **13**, 3935–52. [360]

Lemmerer, A., Bernstein, J., Griesser, U. J., Kahlenberg, V., Tobbens, D. M., Lapidus, S. H., Stephens, P. W., and Esterhuysen, C. (2011). A tale of two polymorphic pharmaceuticals: pyrithyldione and propyphenazone and their 1937 co-crystal patent. *Chem. Eur. J.*, **17**, 13445–60. [360, 365]

Leon, R. A. L., Badruddoza, A. Z. M., Zhenf, L., Yeap, E. W. Q., Toldy, A. I., Wong, K. Y., Hatton, T. A., and Khan, S. A. (2015). Highly selective, kinetically driven polymorphic selection in microfluidic emulsion-based crystallization and formulation. *Cryst. Growth Des.*, **15**, 212–18. [100]

LePage, Y. and Donnay, G. (1976). Refinement of the crystal structure of low-quartz. *Acta Crystallogr.*, **32**, 2456–9. [159]

Leung, S. S., Padden, B. E., Munson, E. J., and Grant, D. J. W. (1998a). Solid-state characterization of two polymorphs of aspartame hemihydrate. *J. Pharm. Sci.*, **87**, 501–7. [453]

Leung, S. S., Padden, B. E., Munson, E. J. and Grant, D. J. W. (1998b). Hydration and dehydration behavior of aspartame hemihydrate. *J. Pharm. Sci.*, **87**, 508–13. [453]

Leusen, F. J. J. (1996). *Ab initio* prediction of polymorphs. *J. Cryst. Growth*, **166**, 900–3. [385]

Levene, P. A. (1935). Note on the preparation of crystalline D-mannose and of crystalline D-ribose. *J. Biol. Chem.*, **108**, 419–20. [86]

Levitas, V. I., Smilowitz, L. B., Henson, B. F., and Asay, B. W. (2009). HMX polymorphism: virtual melting growth mechanism, cluster-to-cluster nucleation mechanism and physically based kinetics. *Int. J. Energ. Mater. Chem. Propuls.*, **8**, 571–93. [405]

Levy, G. C., Lichter, R. L., and Nelson, G. L. (1980). *Carbon-13 Nuclear Magnetic Resonance Spectroscopy*. John Wiley & Sons, New York. [182]

Lewis, J. P., Sewell, T. D., Evans, R. B., and Voth, G. A. (2000). Electronic structure calculation of the structures and energies of the three polymorphic forms of crystalline HMX. *J. Phys. Chem. B*, **104**, 1009–13. [334]

Lewis, N. (2000). Shedding some light on crystallization issues. Lecture transcript from the first international symposium on aspects of polymorphism and crystallization—chemical development issues. *Org. Process Res. Dev.*, **4**, 407–12. [406]

Leznoff, C. C. and Lever, A. P. B. (eds.) (1996). *Phthalocyanines—Properties and Applications*, Vol. 4. VCH, New York. [Vols 1–3 published 1989, 1992, and 1993, respectively]. [388]

Leznoff, D. B., Rancurel, C., Sutter, J.-P., Rettig, S. J., Pink, M., Paulsen, C., and Kahn, O. (1999). Ferromagnetic interactions and polymorphism in radical-substituted gold phosphine complexes. *J. Chem. Soc., Dalton Trans.*, 3593–9. [289]

Li, J. and Brill, T. B. (2007). Kinetics of solid polymorphic phase transitions of CL-20. *Propellants, Explos., Pyrotech.*, **32**, 326–30. [409]

Li, R., Xiao, S., Li, Y., Lin, Q., Zhang, R., Zhao, J., Yang, C., Zou, K., Li, D., and Yi, T. (2014). Polymorphism-dependent and piezochromic luminescence based on molecular packing of a conjugated molecule. *Chem. Sci.*, **5**, 3922–8. [317]

Li, W., Ikni, A., Scouflaire, P., Shi, X., El Hassan, N., Gémeiner, P., Gillet, J.-M., and Spasojević-de Biré, A. (2016). Non-photochemical laser-induced nucleation of sulfathiazole in a water/ethanol mixture. *Cryst. Growth Des.*, **16**, 2514–26. [101]

Li, Y. H., Han, J., Zhang, G. G. Z., Grant, D. W. J., and Suryanarayanan, R. (2000). In situ dehydration of carbamazepine dihydrate: a novel technique to prepare amorphous anhydrous carbamazepine. *Pharm. Dev. Technol.*, **5**, 257–66. [371]

Li, Y.-X., Zhou, H.-B., Miao, J.-L., Sun, G.-X., Li, G.-B., Nie, Y., Chen, C.-L., and Chen, Z. (2012). Conformation twisting induced orientational disorder, polymorphism and solid-state emission properties of 1-(9-anthryl)-2-(1-napthyl)ethylene. *CrystEngComm*, **14**, 8286–91. [338]

Lieberman, H. F., Davey, R. J., and Newsham, D. M. T. (2000). Br···Br and Br···H interactions in action: polymorphism, hopping, and twinning in 1,2,4,5-tetrabromobenzene. *Chem. Mater.*, **12**, 490–4. [311, 320]

Liebermann, H. (1935). Über die Bildung von Chinakridonen aus p-Di-arylmeno-terephtalsäuren. *Annalen*, **518**, 245–59. [384]

Lima-de-Faria, J. (ed.) (1990). *Historical Atlas of Crystallography*, pp. 68–9. Kluwer Academic Publishers, Dordrecht. This reference contains a well-documented historical account of the development of the microscope for the study of crystals. [30, 31, 36]

Lin, S. Y. (1992). Isolation and solid-state characteristics of a new crystal form of indomethacin. *J. Pharm. Sci.*, **81**, 572–6. [209]

Lin-Vien, D., Colthup, N. B., Fateley, W. G., and Grasselli, J. G. (1991). *Infrared and Raman Characteristic Frequencies of Organic Molecules*. Academic Press, New York. [176]

Lindenbaum, S. and McGraw, S. E. (1985). The identification and characterization of polymorphism in drug solids by solution calorimetry. *Pharm. Manuf.*, January, 26–30. [154, 154t]

Lindenbaum, S., Rattie, E. S., Zyber, G. E., Miller, M. E., and Ravin, L. J. (1985). Polymorphism of Auranofin. *Int. J. Pharmacol.*, **26**, 123–32. [155]

Lingafelter, E. C., Simmons, G. L., Morosin, B., Scheringer, C., and Freiburg, C. (1961). The crystal structure of the α-form of bis-(N-methylsalicylaldiminato)-copper. *Acta Crystallogr.*, **14**, 1222–5. [225]

Lipinski, C. A. (2004). Lead- and drug-like compounds: the rule-of-five revolution. *Drug Disc. Today: Technol.*, **1**, 337–41. [20, 344]

Litvinov, I. A., Struchkov, Yu. T., Arbuzov, B. A., Makarova, N. A., and Mukmenev, E. T. (1985). Crystal and molecular structure and quantum-chemical calculation for the stable β-modification of 1,2,3-propanetriol trinitrate $C_3H_5(ONO_2)_3$. *Russ. Chem. Bull.*, **34**, 1425–9. Translated from *Isvestia Akad Nauk SSSR, Ser. Khim.*, 1558–63. [412]

Liu, G., Hansen, T. B., Qu, H., Yang, M., Pajander, J. P., Rantanen, J., and Christensen, L. P. (2014). Crystallization of piroxicam solid forms and the effects of additives. *Chem. Eng. Technol.*, **37**, 1297–1304. [101]

Liu, Y., Du, H., Wang, G, Gong, X., and Wang, L. (2012). Comparative theoretical studies of high pressure effect on polymorph I of 2,2′,4,4′,6,6′-hexanitroazobenzene crystal. *Struct. Chem.*, **23**, 1631–42. [426]

Liu, Y., Li, S., Wang, Z., Xu, J., Sun, J., and Huang, H. (2016). Thermally induced polymorphic transformation of hexanitrohexaazaisowurtzitane (HNIW) investigated by *in-situ* X-ray powder diffraction. *Cent. Eur. J. Energ. Mater.*, **13**, 1023–37. [409]

Llinàs, A. and Goodman, J. M. (2007) Polymorph control: past, present and future. *Drug Disc. Today*, **13**, 198–210. [92, 99, 368]

Löbbert, G. (2000). Phthalocyanines. In *Ullmann's Encyclopedia of Industrial Chemistry*, 6th edn. Wiley-VCH Verlag, Weinheim. [388, 389, 390]

Lockhart, T. P. and Manders, W. F. (1986). Solid state ^{13}C NMR probe for organotin(IV) structural polymorphism. *Inorg. Chem.*, **25**, 583–5. [189]

Loisel, C., Keller, G., Lecq, G., Bourgaux, C., and Ollivon, M. (1998). Phase transitions and polymorphism of cocoa butter. *J. Am. Oil Chem. Soc.*, **75**, 425–39. [40, 209, 335]

Lomax, S. Q. (2010). The application of X-ray powder diffraction for the analysis of synthetic organic pigments. Part 1. Dry pigments. *J. Coat. Technol. Res.*, 7, 331–46. [378]

Lommerse, J. P. M., Motherwell, W. D. S., Ammon, H. L., Dunitz, J. D., Gavezzotti, A., Hofmann, D. W. M., Leusen, F. J. J., Mooij, W. T. M., Price, S. L., Schweizer, B., Schmidt, M. U., van Eijck, B. P., Verwer, P., and Williams, D. E. (2000). A test of crystal structure prediction of small organic molecules. *Acta Crystallogr. B*, **56**, 697–714. [210, 228, 238, 241, 387]

Long, S., Parkin, S., Siegler, M. A., Cammers, A., and Li, T. (2008). Polymorphism and phase behaviors of 2-(phenylamino)nicotinic acid. *Cryst. Growth Des.*, **8**, 4006–13. [306]

Lopez-Mejias, V., Kampf, J. W., and Matzger, A. (2012). Nanomorphism in flufenamic acid and a new record for a polymorphic compound with solved structures. *J. Am. Chem. Soc.*, **134**, 9872–5. [237, 264, 374]

Lord, R. C. and Yu, N. Y. (1970a). Laser-excited Raman spectroscopy of biomolecules. I. Native lysozyme and its constituent amino acids. *J. Mol. Biol.*, **50**, 509–24. [322]

Lord, R. C. and Yu, N. Y. (1970b). Laser-excited Raman spectroscopy of biomolecules. II. Native ribonuclease and α-chymotrypsin. *J. Mol. Biol.*, **5**, 203–13. [322]

Loschen, C. and Klamt, A. (2016). Computational screening of drug solvates. *Pharm. Res.*, **33**, 2794–804. [355]

Lotz, B. (2000). What can polymer crystal structure tell about polymer crystallization processes? *Eur. Phys. J. E*, **3**, 185–94. [40]

Loutfy, R. O. (1981). Bulk optical properties of phthalocyanine pigment particles. *Can. J. Chem.*, **59**, 549–54. [393]

Loutfy, R. O., Hor, A.-M., Hsiao, C. K., Baranyi, G., and Kazmaier, P. (1988). Organic photoconductive materials. *Pure Appl. Chem.*, **60**, 1047–54. [393]

Love, J. C., Estroff, L. A., Kriebel, J. K, Nuzzo, R. G., and Whitesides, G. M. (2005). Self-assembled monolayers of thiolates on metals as a form of nanotechnology. *Chem. Rev.*, **105**, 1103–70. [100]

Lovinger, A. J., Forrest, S. R., Kaplan, M. L., Schmidt, P. H., and Venkatesan, T. (1984). Structural and morphological investigation of the development of electrical conductivity in ion irradiated thin films of an organic material. *J. Appl. Phys.*, **55**, 476–82. [386]

Lowe-Ma, C. (2000). Private communication to the author. [416, 417, 427]

Lowe-Ma, C. K., Fischer, J. W., and Willer, R. L. (1990). Structures of four related 4,5,6,7-tetrahydro-1,2,5-oxadiazolo[3,4-b]pyrazines. *Acta Crystallogr. C*, **46**, 1853–9. [426]

Lowe-Ma, C. K., Nissan, R. A., and Wilson, W. S. (1989). *The synthesis and properties of picryldinitrobenzimidazoles and the "trigger linkage" in picryldinitrobenzotriazoles.* NWC Technical Publication 7008, Naval Weapons Center, China Lake, CA. [426]

Lu, J. and Rohani, S. (2009a). Polymorphic crystallization and transformation of the antiviral/HIV drug stavudine. *Org. Process. Res. Dev.*, **13**, 1262–8. [369]

Lu, J. and Rohani, S. (2009b). Polymorphism and crystallization of active pharmaceutical ingredients (APIs). *Curr. Med. Chem.*, **16**, 884–905. [100, 369]

Lu, J., Litster, J., and Nagy, Z. K. (2015). Nucleation studies of active pharmaceutical ingredients in an air-segmented microfluidic drop-based crystallizer. *Cryst. Growth Des.*, **15**, 3645–51. [84]

Lu, Z., Xue, X., Meng, L., Zeng, Q., Chi, Y., Fan, G., Li, H., Zhang, Z., Nie, F., and Zhang, C. (2017). Heat-induced solid–solid phase transformation of TKX-50. *J. Phys. Chem. C*, **121**, 8262–71. [415]

Ludlam-Brown, I. and York, P. (1993). The crystalline modification of succinic acid by variations in crystallization conditions. *J. Phys. D: Appl. Phys.*, **26**, B60–5. [134]

Luque de Castro, M. D. and Priego-Capote, F. (2007). Ultrasound-assisted crystallization (sonocrystallization). *Ultras. Sonochem.*, **14**, 717–24. [101]

Łużny, W. and Czarnecki, W. (2014). Application of genetic algorithms to model the structure of molecular crystals. *Polimery*, **59**, 542–8. [243]

Ma, Y., Zhang, A., Zhang, C., Jiang, D., Zhu, Y., and Zhang, C. (2014). Crystal packing of low-sensitivity and high-energy explosives. *Cryst. Growth Des.*, **14**, 4703–13. [399]

Macicek, J. and Yordanov, A. (1992). BLAF—a robust program for tracking out admittable Bravais lattices from the experimental unit-cell data. *J. Appl. Crystallogr.*, **25**, 73–80. [159]

MacLean, E. J., Tremayne, M., Kariuki, B. M., Cameron, J. R. A., Roberts, M. A., and Harris, K. D. M. (2009). Lessons on the discovery and assignment of polymorphs, highlighted by the case of latent pigment DPP-Boc. *Cryst Growth Des.*, **9**, 853–7. [384t, 394]

Macrae, C. F., Bruno, I. J., Chisolm, J. A., Edginton, P. R., McCabe, P., Pidcock, E., Rodriguez-Monge, L., Taylor, R., van de Streek, J., and Wood, P. (2008). New features for the visualization and investigation of crystal structures. *J. Appl. Crystallogr.*, **41**, 466–70. [200]

Maddox, J. (1988). Crystals from first principles. *Nature*, **335**, 201. [450]

Madsen, I. C., Scarlett, N. V. Y., and Kern, A. (2011). Description and survey of methodologies for the determination of amorphous content via X-ray powder diffraction. *Z. Kristallogr.*, **226**, 944–55. [164]

Madsen, I. C., Scarlett, N. V. Y., Riley, D. P., and Raven, M. D. (2013). Quantitative analysis using the Rietveld method. In *Modern Diffraction Methods* (ed. E. J. Mittemeijer and U. Welzel), pp. 285–320. Wiley-VCH Verlag, Weinheim. [156, 168]

Maeda, K. (1991). Photochromism in organized media. *J. Synth. Chem. Jpn.*, **49**, 554–65. [307]

Mahieu, A., Willart, J. F., Dudognon, E., Eddleston, M. D., Jones, W., Danede, F., and Descamps, M. (2013). On the polymorphism of griseofulvin: identification of two additional polymorphs. *J. Pharm. Sci.*, **102**, 462–8. [357]

Main, P., Cobbledick, R. E., and Small, R. W. H. (1985). Structure of the fourth form of 1,3,5,7-tetranitro-1,3,5,7-tetraazacyclooctane (δ-HMX) $2C_4H_8N_8O \cdot O \cdot 5H_2O$. *Acta Crystallogr. C*, **41**, 1351–4. [404t]

Malamatari, M., Ross, S. A., Douroumis, D., and Velaga, S. P. (2017). Experimental cocrystal screening and solution based scale-up cocrystallization methods. *Adv. Drug. Deliv. Rev.*, **117**, 162–77. [353, 360]

Mallard, F. E. (1876). Explication des phenomes optiques anomaux. *Annales Mines*, **10**, 60–196. [32]

Mallard, F. E. (1879). *Traité de cristallographie géometrie et physique*. Dunod, Paris. [32]

Malwitz, M. A., Lim, S. H., White-Morris, R. L., Pham, D. M., Olmstead, M. M., and Balch, A. L. (2012). Crystallization and interconversions of vapor-sensitive, luminescent polymorphs of $[(C_6H_{11}NC)_2Au^I](AsF_6)$ and $[(C_6H_{11}NC)_2Au^I](PF_6)$. *J. Am. Chem. Soc.*, **134**, 10885–93. [340]

Manger, C. W. and Struve, W. S. (1958). Organic pigments. E. I. du Pont de Nemours and Company U.S. Patent 2 844 581. [385]

Mangin, D., Puel, F., and Veesler, S. (2009). Polymorphism in processes of crystallization in solution: a practical review. *Org. Process Res. Rev.*, **13**, 1241–53. [100]

Marchetti, A. P., Salzberg, C. D., and Walker, E. I. P. (1976). The optical properties of crystalline 1,1′-diethyl-2,2′-cyanine iodide. *J. Chem. Phys.*, **64**, 4693–7. [294]

Marcus, Y. (1998). *The Properties of Solvents*, p. 95. Wiley, New York. [95]

Marom, N., DiStasio, R. A., Jr., Atalla, V., Levchenko, S., Reilly, A. M., Chelikowsky, J. R., Leiserowitz, L., and Tkatchenko, A. (2013). Many-body dispersion interactions in molecular crystal polymorphism. *Angew. Chem. Int. Ed.*, **52**, 6629–32. [245]

Marsh, R. E. (1984). Concerning a second polymorph of the HMX-DMF complex. *Acta Crystallogr. C*, **40**, 1632–3. [403]

Martin, D. W. and Waters, T. N. (1973). Conformational influences in copper coordination compounds. IV. Crystal structure of a fourth crystalline isomer of bis(2-hydroxy-*N*-methyl-1-naphthyl methyleneiminato)copper(II). *J. Chem. Soc., Dalton Trans.*, 2440–3. [225]

Martins, I. C. B., Gomes, J. R. B., Duarte, M. T., and Mafra, L. (2017). Understanding polymorphic control of pharmaceuticals using imidazolium-based ionic liquid mixtures as crystallization directing agents. *Cryst. Growth Des.*, **17**, 428–32. [357]

Maruyama, S., Ooshima, H., and Kato, J. (1999). Crystal structures and solvent-mediated transformation of taltireline polymorphs. *Chem. Eng. J.*, **75**, 193–200. [367]

Masamura, M. (2000). Error of atomic charges derived from electrostatic potential. *Struct. Chem.*, **11**, 41–5. [227]

Masciocchi, N., Ardizzoia, G. A., La Monica, G., Moret, M., and Sironi, A. (1997). Polymorphism in coordination chemistry. Selective synthesis and *ab-initio* X-ray powder diffraction characterization of two new crystalline phases of solid [Pd(dmpz)$_2$(Hdmpz)$_2$]$_2$ (Hdmpz = 3,5-dimethylpyrazole). *Inorg. Chem.*, **36**, 449–54. [126]

Mataka, S., Moriyama, H., Sawada, T., Takahashi, K., Sakashita, H., and Tashiro, M. (1996). Conformational polymorphism of mechanochromic 5,6-di(*p*-chlorobenzoyl)-1,3,4,7-tetraphenylbenzo[*c*]thiophene. *Chem. Lett.*, **25**, 363–4. [315]

Materazzi, S. (1997). Thermogravimetry infrared spectroscopy (TG-FTIR) coupled analysis. *Appl. Spectrosc. Rev.*, **32**, 385–404. [209, 367]

Materazzi, S. (1998). Mass spectrometry coupled to thermogravimetry (TG-MS) for evolved gas characterization: a review. *Appl. Spectrosc. Rev.*, **33**, 189–218. [367]

Materials Studio, v 7.0.100, ©2013, Accelrys Software Inc. All rights reserved. [200]

Mathieu, J. P. (1973). Vibration spectra and polymorphism of chiral compounds. *J. Raman Spectrosc.*, **1**, 47–51. [321]

Matsumoto, S., Matsuhama, K., and Mizuguchi, J. (1999). β Metal-free phthalocyanine. *Acta Crystallogr. C*, **55**, 131–3. [393]

Matsumoto, S., Uchida, Y., and Yanagita, M. (2006). A series of polymorphs with different colors in fluorescent 2,5-diamino-3,6-dicyanopyrazine dyes. *Chem. Lett.*, **35**, 654–5. [338]

Matsunaga, Y. (1965). Electrical resistivities and the type of bonding in some quinone complexes. *Nature*, **205**, 72–3. [124]

Matsunaga, Y. (1966). Polymorphic forms of the diaminopyrene-*p*-chloranil and related complexes. *Nature*, **205**, 183–4. [82 to 124]

Matsushima, R., Tatemura, M., and Okamoto, N. (1992). Second harmonic generation from 2-arylideneindan-1,3-diones studied by the powder method. *J. Mater. Chem.*, **2**, 507–10. [302]

Matsushita, M. M., Izuoka, A., Sugawara, T., Kobayashi, T., Wada, N., Takeda, N., and Ishikawa, M. (1997). Hydrogen bonded organic ferromagnet. *J. Am. Chem. Soc.*, **119**, 4369–79. [289]

Matthews, J. H., Paul, I. C., and Curtin, D. Y. (1991). Configurational isomerism in crystalline forms of benzophenone anils. *J. Chem. Soc., Perkin Trans. 2*, 113–18. [9, 125]

Mayersohn, M. and Endrenyi, L. (1973). Relative bioavailability of commercial ampicillin formulations. *Can. Med. Assoc. J.*, **109**, 989–93. [348]

Maynard, J. Y. and Peters, H. M. (1991). *Understanding Chemical Patents*, 2nd edn. American Chemical Society, Washington, DC. [430]

McCabe, J. F. (2010). Application of design of experiment (DOE) to polymorph screening and subsequent data analysis. *CrystEngComm*, **12**, 1110–19. [359]

McCauley, J. A., Varsolona, R. J., and Levorse, D. A. (1993). The effect of polymorphism and metastability on the characterization and isolation of two pharmaceutical compounds. *J. Phys. Sect. D*, **26**, B85–9. [89]

McClure, R. J. and Craven, B. M. (1974). X-ray data for four crystalline forms of serum albumin. *J. Mol. Biol.*, **83**, 551–5. [30]

McCrone, W. C. (1949). Crystallographic data 25. 2,4,6-trinitrotoluene (TNT). *Anal. Chem.*, **21**, 1583–4. [420, 422f]

McCrone, W. C. (1950a). Crystallographic data 36. Cyclotetramethylene tetranitramine (HMX). *Anal. Chem.*, **22**, 1225–6. [403, 404t, 406]

McCrone, W. C. (1950b). Crystallographic data 32. RDX cyclotrimethylenetrinitramine. *Anal. Chem.*, **22**, 954–5. [410]

McCrone, W. C. (1951). Crystallographic data 47. 1,8-Dinitronaphthalene I. *Anal. Chem.*, **23**, 1188–9. [427]

McCrone, W. C. (1952). Crystallographic data 54. 2,4,6-2′,4′,6′-Hexanitrodiphenylamine (HND). *Anal. Chem.*, **24**, 592–3. [427]

McCrone, W. C. (1954). Crystallographic data 89. 2,6-Dinitrotoluene. *Anal. Chem.*, **26**, 1997–8. [427]

McCrone, W. C. (1957). *Fusion Methods in Chemical Microscopy*. Interscience Publishers, New York. [11, 26, 37, 38, 137, 138, 140, 144, 146, 210, 230, 232, 403, 410, 417, 427]

McCrone, W. C. (1965). Polymorphism. In *Physics and Chemistry of the Organic Solid State*, Vol. 2 (ed. D. Fox, M. M. Labes, and A. Weissberger), pp. 725–67. Wiley Interscience, New York. [1, 3, 4, 5, 6, 7, 8, 9, 10, 11, 12, 15, 38, 38f, 48, 49f, 88, 114, 124, 126, 210, 342, 378, 403]

McCrone, W. C. (1967). *Crystallographic Study of SC-101*. Project 883, Chicago, IL. [425]

McCrone, W. C. (1983). Particle characterization by PLM. 1. Single polar. *Microscope* **31**, 187–206. [96f, 96]

McCrone, W. C. (1983) Particle Characterization by PLM. III. Crossed Polars. *Microscope*, 31:2, 187–206 Courtesy of the McCrone Research Institute. [96]

McCrone, W. C. and Adams, W. C. (1955). Crystallographic data 101. Lead styphnate (normal). *Anal. Chem.*, **27**, 2014–15. [426]

McCrone, W. C., McCrone, L. B., and Delly, J. G. (1978). *Polarized Light Microscopy*. Ann Arbor Science Publishers, Inc., Ann Arbor, MI. [137, 138]

McGeorge, G., Harris, R. K., Chippendale, A. M., and Bullock, J. F. (1996). Conformational analysis by magic-angle spinning NMR spectroscopy for a series of polymorphs of a disperse azobenzene dyestuff. *J. Chem. Soc., Perkin Trans.*, 1733–8. [222]

McHale, J. M., Navrotsky, A., and Perrotta, A. J. (1997). Effects of increased surface area and chemisorbed H_2O on the relative stability of nanocrystalline gamma Al_2O_3 and alpha-Al_2O_3. *J. Phys. Chem. B*, **101**, 603–13. [77]

McIntosh, D. (1919). Crystallization of supersaturated solutions and supercooled liquids. *Trans. Roy. Soc. Can. Sect. 3*, **13**, 265–72. [80]

McKay, R. B., Iqbal, A., and Medinger, B. (1994). Organic pigments. In *Technological Applications of Dispersions* (ed. R. B. McKay), pp. 143–76. Marcel Dekker, New York. [377]

McKerrow, A. J., Buncel, E., and Kazmaier, P. M. (1993). Molecular modeling of photoactive pigments in the solid state: investigations of polymorphism. *Can. J. Chem.*, **71**, 390–8. [387]

McKinnon, J. J., Fabbiani, F. P. A., and Spackman, M. A. (2007a). Comparison of polymorphic molecular crystal structure through Hirshfeld surface analysis. *Cryst. Growth Des.*, 7, 755–69. [248, 255]

McKinnon, J. J., Jayatilaka, D., and Spackman, M. A. (2007b). Towards quantitative analysis of intermolecular interactions with Hirshfeld surfaces. *Chem. Commun.*, 3814–16. [248, 255]

McKinnon, J. J., Spackman, M. A., and Mitchell, A. S. (2004). Novel tools for visualizing and exploring intermolecular interactions in molecular crystals. *Acta Crystallogr. B*, **60**, 627–68. [248]

McLafferty, F. W. (1990). Analytical chemistry: historic and modern. *Acc. Chem. Res.*, **23**, 63–4. [37, 137]

McNaughton, J. L. and Mortimer, C. T. (1975). Differential scanning calorimetry. (Reprinted from IRS; Phys. Chem. Series 2. Vol. 10). Perkin-Elmer, Norwalk, CT. [147]

McPherson, A. (1982). *Preparation and Analysis of Protein Crystals*, pp. 127–59. Wiley, New York. [16, 29, 30]

McPherson, A. (1989). *Preparation and Analysis of Protein Crystals*. Krieger Publishing Company, Florida. [29]

McPherson, A. (1999). *Crystallization of Biological Macromolecules*. Cold Spring Harbor Laboratory Press, Cold Spring Harbor, NY. [29]

McPherson, A. (2004). Introduction to protein crystallization. *Methods*, **34**, 254–65. [102]

McPherson, A. (2016). *Introduction to Macromolecular Crystallography*, 2nd edn, Chapter 2. Wiley-Blackwell, Hoboken, NJ. [102]

McPherson, A. and Gavira, J. A. (2014). Introduction to protein crystallization. *Acta Crystallogr. F*, **70**, 2–20. [16, 29]

Medek, A. (2006). Solid-state nuclear magnetic resonance spectroscopy. In *Spectroscopy of Pharmaceutical Solids* (ed. H. G. Brittain), pp. 413–557. Taylor and Francis, New York. [182]

Meents, A., Dittrich, B., Johnas, S. K. J., Thome, V., and Weckert, E. F. (2008). Charge-density studies of energetic materials: CL-20 and FOX-7. *Acta Crystallogr. B*, **64**, 42–9. [407, 414]

Meguro, T., Kashiwagi, T., and Satow, Y. (2000). Crystal structure of the low-humidity form of aspartame sweetener. *J. Peptide Res.*, **56**, 97–104. [454]

Mehuns, S. M., Kale, U. J., and Qu, X. (2005). Statistical analysis in the Raman spectra of polymorphs. *J. Pharm. Sci.*, **94**, 1354–67. [177]

Menéndez-Taboada, L., Roces, L., Presa Soto, A., García-Granda, S. The use of polystyrene organic-based gels for the diffraction quality single-crystal growth of small molecules, Acta Crystallog. A, 2012, 68, s261-s261. [101]

Merck Index (2016). Available at https://www.rsc.org/merck-index

The Merriam-Webster.com Dictionary, s.v. "pharmacognosy (n.)," accessed January 15, 2020, https://www.merriam-webster.com/dictionary/pharmacognosy [431, 432]

Merritt, V. Y. (1978). Organic photovoltaic materials: squarylium and cyanine-TCNQ dyes. *IBM J. Res. Dev.*, **22**, 353–71. [293]

Merton, R. K. and Barber, E. (2004). *The Travels and Adventures of Serendipity. A Study in Sociological Semantic and the Sociology of Science*. Princeton University Press, Princeton, NJ. [438]

Mesley, R. J., Clements, R. L., Flaherty, B., and Goodhead, K. (1968). The polymorphism of phenobarbitone. *J. Pharm. Pharmacol.*, **20**, 329–40. [374]

Meyer, J. C., Girit, C. O., Crommie, M. F., and Zetti, A. (2008). Imaging and dynamics of light atoms and molecules on graphene. *Nature*, **454**, 319–22. [195]

Meyer, R. (1987). *Explosives*. VCH, Weinheim. [399, 404t, 410, 426, 427]

Meyers, M. A. (2007). *Happy Accidents. Serendipity in Major Medical Breakthroughs in the 20th Century*. Arcade Publishing, New York. [438]

Miao, J., Nie, Y., Hu, C., Zhang, Z., Li, G., Xu, M., and Sun, G. (2012). Polymorphism in 3,4-bis(2-benimidazolyl)pyridine: controlled synthesis, crystal structures and photophysical properties. *J. Mol. Struct.*, **1014**, 97–101. [306]

Miclaus, M., Grosu, I. G., Filip, X., Tripon, C., and Filip, C. (2014). Optimizing structure determination from powders of crystalline organic solids with high molecular flexibility: the case of lisinopril dehydrate. *CrystEngComm*, **16**, 299–303. [191]

Mighell, A. D. (1976). The reduced cell: its use in identification of crystalline materials. *J. Appl. Crystallogr.*, **9**, 491. [159]

Millar, D. I. A., Maynard-Casely, H. E., Kleppe, A. K., Marshall, W. G., Pulham, C. R., and Cumming, A. S. (2010a). Putting the squeeze on energetic materials—structural characterisation of a high-pressure phase of CL-20. *CrystEngComm*, **12**, 2524–7. [409]

Millar, D. I. A., Oswald, I. D. H., Barry, C., Francis, D. J., Marshall, W. G., Pullman, C. R., and Cumming, A. S. (2010b). Pressure-cooking of explosives—the crystal structure of ϵ-RDX as determined by X-ray and neutron diffraction. *Chem. Commun.*, **46**, 5662–4. [410]

Millar, D. I. A., Oswald, I. D. H., Francis, D. J., Marshall, W. G., Pulham, C. R., and Cumming, A. S. (2009). The crystal structure of β-RDX—an elusive form of an explosive revealed. *Chem. Commun.*, **45**, 562–4. [410]

Miller, G. R. and Garroway, A. N. (2001). *A review of the crystal structures of common explosives part I: RDX, HMX, TNT, PETN, and Tetryl.* NRL/MR/6120-01-8585, United States Naval Research Laboratory, Washington, DC. [398]

Miller, J. M., Collman, B. M., Greene, L. R., Grant, D. W. J., and Blackburn, A. C. (2005). Identifying the stable polymorph early in the drug discovery-development process. *Pharm. Dev. Technol.*, **10**, 291–7. [95]

Miller, J. S. (1998). Polymorphic molecular materials: the importance of tertiary structures. *Adv. Mater.*, **10**, 1553–7. [285, 286, 288]

Miller, J. S. (2011). Magnetically ordered molecule-based materials. *Chem. Soc. Rev.*, **40**, 3266–96. [292]

Miller, J. S. and Epstein, A. S. (1994). Organic and organometallic molecular magnetic materials—designer magnets. *Angew. Chem. Int. Ed. Engl.*, **33**, 385–415. [285]

Miller, J. S. and Epstein, A. S. (1996). Magnets based on the molecular solid state. *Mol. Cryst. Liq. Cryst. Sci. Technol. A*, **278**, 145–54. [285]

Miller, J. S., Zhang, J. H., Reiff, W. M., Doxon, D. A., Preston, L. D., Reis, A. H., Gebert, E., Extine, M., Troup, J., Epstein, A. J., and Ward, M. D. (1987). Characterization of the charge-transfer reaction between decamethylferrocene and 7,7,8,8-tetracyano-*p*-quinodimethane (1:1). The ^{57}Fe Mössbauer spectra and structures of the paramagnetic and metamagnetic one-dimensional salts of the molecular and electronic structures of [TCNQ]^n ($n = 0, -1, -2$). *J. Phys. Chem.*, **91**, 4344–60. [286]

Miller, P. J., Block, S., and Piermarini, G. J. (1991). Effects of pressure on the thermal-decomposition kinetics, chemical-reactivity and phase behavior of RDX. *Combust. Flame*, **83**, 174–84. [410]

Miller, R. S. (1995). *Decomposition, combustion and detonation chem. of energetic materials.* In *Materials Research Society Symposium Proceedings*, Vol. 418 (ed. T. B. Brill, T. P. Russell, W. C. Tao, and R. B. Wardle), p. 3. Materials Research Society, Pittsburgh. [403]

Milton, J. (1667). *Paradise Lost*, 6th Book, Verses 499–501. The Harvard Classics. Harvard University Press, Cambridge, MA. [433]

Mirmehrabi, M., Rohani, S., Murthy, K. S. K., and Radatus, B. (2006). Polymorphic behavior and crystal habit of an anti-viral/HIV drug: stavudine. *Cryst. Growth Des.*, **6**, 141–9. [62]

Mishra, M. K., Ramamurty, U., and Desiraju, G. R. (2015). Solid solution hardening of molecular crystals: tautomeric polymorphs of omeprazole. *J. Am. Chem. Soc.*, **137**, 1794–7. [125]

Mislow, K. (1966). *Introduction to Stereochemistry*, pp. 33–6. W. A. Benjamin, New York. [216]

Mitra, A. K., Ghosh, L. K., and Gupta, B. K. (1993). Development of methods for the preparation and evaluation of chloramphenicol palmitate ester and its biopharmaceutically effective metastable polymorph. *Drug Dev. Ind. Pharm.*, **19**, 971–80. [349]

Mitscherlich, E. (1820). Sur la relations qui existe entre la forme cristalline et les proportion chimiques. *Ann. Chim. Phys.*, **14**, 172–90. [395]

Mitscherlich, E. (1822). Sur la relation qui existe entre la forme cristalline et les proportions chimiques, I. Mémóire sure les arseniates et les phosphates. *Ann. Chim. Phys.*, **19**, 350–419. [30]

Mitscherlich, E. (1823). Über die Körper, welche in zwei verschiedenen krystallisieren Formen. *Abhl. Akad. Berlin*, 43–8. [3, 31]

Mitsubishi Chemical Industries (1976). Beta-crystalline phase perylene pigment preparation. Japanese Patent JA 51–7025. [386]

Mittapalli, S., Perumalla, D. S., and Nangia, A. (2017). Mechanical synthesis of N-salicylidene-aniline: thermosalient effect of polymorphic crystals. *IUCr J.*, **4**, 243–50. [320]

Miyahara, T., Hasegawa, H., Takahashi, Y., and Inabe, T. (2013). Electrochemical crystallization of organic molecular conductors: electrode surface conditions for crystal growth. *Cryst. Growth Des.*, **13**, 1955–60. [101]

Mizuguchi, I., Grubenmann, A., and Rihs, G. (1993). Structures of 3,6-bis(3-chlorophenyl)-pyrrolo[3,4-c]pyrrole-1,4-dione and 3,6-bis(4-chlorophenyl)pyrrolo[3,4-c]pyrrole-1,4-dione. *Acta Crystallogr. B*, **49**, 1056–60. [382t]

Mizuguchi, J. (1997). Structural and optical properties of 5,15-diaza-6,16-dihydroxy-tetrabenzo[b,e,k,n]perylene. *Dyes Pigm.*, **35**, 347–60. [387]

Mizuguchi, J. (1998a). Electronic characterization of N,N′-bis(2-phenylethyl)perylene-3,4:9,10-bis(dicarboximide) and its application to optical disks. *J. Appl. Phys.*, **84**, 4479–85. [387, 388f]

Mizuguchi, J. (1998b). N,N′-bis(2-phenethyl)perylene-3,4:9,10-bis-(dicarboximide). *Acta Crystallogr. C*, **54**, 1479–81. [380, 388]

Mizuguchi, J. (2003). Refinement of the crystal structure of α-1,4-dioxo-3,6-diphenylpyrrolo-[3,4-c]pyrrole-2,5(1H,4H)-dicarboxylic acid bis(1,1-dimethylethyl) ester, $C_{28}H_{28}N_2O_6$. *Z. Kristallogr.*, **218**, 134–6. [384t]

Mizuguchi, J. and Tojo, K. (2001). Crystal structure of N,N′-bis(3,5-xylyl)perylene-3,4:9,10-bis(dicarboximide), $C_{40}H_{26}N_2O_4$. *Z. Kristallogr. New Cryst. Struct.*, **3**, 375–6. [381t, 387]

Mizuguchi, J. and Tojo, K. (2002). Electronic structure of perylene pigments as viewed from the crystal structure and excitonic interactions. *J. Phys. Chem. B*, **106**, 767–72. [386, 387]

Mizuguchi, J., Grubenmann, A., and Rihs, G. (1993). Structures of 3,6-bis(3-chlorophe-nyl)pyrrolo[3,4-c]pyrrole-1,4-dione and 3,6-bis(4-chlorophenyl)pyrrolo[3,4-c]pyrrole-1,4-dione. *Acta Crystallogr. B*, **49**, 1056–60. [394]

Mizuguchi, J., Grubenmann, A., Wooden, G., and Rihs, G. (1992). Structures of 3,6-diphenylpyrrolo[3,4-c]pyrrole-1,4-dione and 2,5-dimethyl-3,6-diphenylpyrrolo[3,4-c]pyrrole-1,4-dione. *Acta Crystallogr. B*, **48**, 696–700. [382t]

Mizuguchi, J., Rochat, A. C., and Rohs, G. (1994). Electronic spectra of 7,14-dithioketo-5,7,12,14-tetrahydroquinolino-[2,3-b]-acridine in solution and in the solid state. *Ber. Bunsenges. Phys. Chem.*, **98**, 19–28. [330]

Mizuguchi, J., Sasaki, T., and Tojo, K. (2002a). Refinement of the crystal structure of 5,7,12,14-tetrahydro[2,3-*b*]-quinolinoacridine (γ-form), $C_{20}H_{12}N_2O_2$, at 223 K. *Z. Kristallogr. New Cryst. Struct.*, **217**, 249–50. [380t, 381t, 385]

Mnyukh, Yu. V., Panfilova, N. A., Petropavlov, N. N., and Ukhvatova, N. S. (1975). Polymorphic transitions in molecular crystals. III Transitions exhibiting unusual behavior. *J. Phys. Chem. Solids*, **36**, 127–44. [161]

Möbus, M., Karl, N., and Kobayashi, T. (1992). Structure of perylenetetracarboxylic dianhydride thin films on alkali halide crystal substances. *J. Cryst. Growth*, **116**, 495–504. [386]

Moers, O., Wijaya, K., Lamge, I., Blaschette, A., and Jones, P. G. (2000). Polysulfonamines, CXXVI—hydrogen bonding in crystalline onium dimesylamines: a robust eight-membered ring synthon in coexistence with a third hydrogen binding motif—cyclodimers, supramolecular linkage isomers, and a quintuply interwoven three-dimensional C-H···O network. *Z. Naturforsch. B*, **55**, 738–52. [69]

Mohapatra, H. and Eckhardt, C. J. (2008). Elastic constants and related mechanical properties of the monoclinic polymorph of the carbamazepine molecular crystal. *J. Phys. Chem. B*, **112**, 2293–8. [320]

Molt, R. W., Watson, T., Bazante, A. P., and Bartlett, R. J. (2013). The great diversity of HMX conformers: probing the potential energy surface using CCSD(T). *J. Phys. Chem. A*, **117**, 3467–74. [405]

Momoi, Y., Yamane, M., Yamaguchi, I., and Matsushita, H. (1976). Yellow pigments. Japanese Patent 510 88516 (Application 75-13362). [382t]

Montgomery, L. K., Geiser, U., Wang, H. H., Beno, M. A., Schultz, A. J., Kini, A. M., Carlson, K. D., Williams, J. M., Whitworth, J. R., Gates, B. D., Cariss, C. S., Pipan, C. M., Donega, K. M., Wenz, C., Kwok, W. K., and Crabtree, G. W. (1988). How well do we understand the synthesis of $(ET)_2I_3$ by electrocrystallization? ESR and X-ray identification of $(ET)_2I_3$ crystals which are mixtures of phases and observation of high-T_c states of $(ET)_2I_3$, ranging from 2.5–6.9 K. *Synth. Met.*, **27**, A195–A207. [119]

Mooij, W. T. M. (2000). *Ab initio prediction of crystal structures*. PhD Thesis, University of Utrecht, The Netherlands. [238]

Mooij, W. T. M., van Eijck, B. P., Price, S. L., Verwer, P., and Kroon, J. (1998). Crystal structure prediction for acetic acids. *J. Comput. Chem.*, **19**, 459–74. [238]

Moorthy, J. N. and Venkatesan, K. (1994). Photobehavior of crystalline 4-styrylcoumarin dimorphs: structure–reactivity correlations. *Bull. Chem. Soc. Jpn.*, **67**, 1–6. [333, 334f]

Morel, D. L., Stogryn, E. L., Ghosh, A. K., Feng, T., Purin, P. E., Shaw, R. F., Fishman, C., Bird, G. R., and Piechowsky, A. P. (1984). Organic photovoltaic cells. Correlations between cell performance and molecular structure. *J. Phys. Chem.*, **88**, 923–33. [293]

Moreno, A. and Mendoza, M. E. (2015). Crystallization in gels. In *Handbook of Crystal Growth*, 2nd edn (ed. P. Rudolph), pp. 1277–315. Elsevier, Amsterdam. [101]

Morimoto, M., Kobatake, S., and Irie, M. (2003). Polymorphism of 1,2-bis(2-methyl-5-*p*-methoxyphenyl-3-thienyl)perfluorocyclopentene and photochromic reactivity of the single crystals. *Chem. Eur. J.*, **9**, 621–7. [310]

Morissette, S. L., Almarsson, Ö., Peterson, M. L., Remenar, J. F., Read, M. J., Lemmo, A. V., Ellis, S., Cima, M. J., and Gardner, C. R. (2004). High-throughput crystallization: polymorphs, salts, co-crystals and solvates of pharmaceutical solids, *Adv. Drug. Deliv. Rev.*, **56**, 275–300. [14t, 105, 106t, 108, 444]

Morissette, S. L., Soukasene, S., Levinson, D., Cima, M. J., and Almarsson, Ö. (2003). Elucidation of crystal form diversity of the HIV protease inhibitor ritonavir by high-throughput crystallization. *Proc. Natl. Acad. Sci. U.S.A.*, **100**, 2180–4. [105]

Morris, K. R. (1999). Structural aspects of hydrates and solvates, in *Polymorphism in Pharmaceutical Solids* (ed. H. G. Brittain), Marcel Dekker, New York, pp. 125–82. [8, 355]

Morris, K. R., Nail, S. L., Peck, G. E., Byrn, S. R., Griesser, U. J., Stowell, J. G., Hwang, S. J., and Park, K. (1998). Advances in pharmaceutical material and processing. *Pharm. Sci. Technol. Today*, **1**, 235–45. [364]

Morrison, H., Quan, B. P., Walker, S. D., Hansen, K. B., Nagapudi, K., and Cui, S. (2015). Appearance of a new hydrated form during development: a case study in process and solid-state optimization. *Org. Process Res. Dev.*, **19**, 1842–8. [438]

Morrison, P., Morrison, P., and the Office of Charles and Ray Eames (1982). *Powers of ten*, Scientific American, New York. [133]

Moser, F. H. and Thomas, A. L. (1963). *Phthalocyanine Compounds*. Reinhold Publishing Co., New York. [388]

Moser, F. H. and Thomas, A. L. (1983). *The Phthalocyanines*, Vols. I and II, CRC Press, Boca Raton, FL. [388, 389]

Mossinghoff, G. J. (1999). Overview of the Hatch–Waxman Act and its impact on the drug development process. *Food Drug Law J.*, **54**, 187–94. [434]

Motherwell, W. D. S. (1997). Distribution of molecular centers in crystallographic unit cells. *Acta Crystallogr. B*, **53**, 726–36. [239]

Motherwell, W. D. S. (2001). Crystal structure prediction and the Cambridge Structural Database. *Mol. Cryst. Liq. Cryst.*, **356**, 559–67. [387]

Motherwell, W. D. S., Ammon, H. L., Dunitz, J. D., Dzyabchenko, A., Erk, P., Gavezzotti, A., Hofmann, D. W. M., Leusen, F. J. J., Lommerse, J. P. M., Mooij, W. T. M., and Price, S. L. (2002). Crystal structure prediction of small organic molecules: a second blind test. *Acta Crystallogr. B*, **58**, 647–61. [241]

Motherwell, W. D. S., Shields, G. P., and Allen, F. H. (1999). Visualization and characterization of non-covalent networks in molecular crystals: automated assignment of graph-set descriptors for asymmetric molecules. *Acta Crystallogr. B*, **55**, 1044–56. [72]

Moulton, B. and Zaworotko, M. J. (2001). From molecules to crystal engineering: supramolecular isomerism and polymorphism in network solids. *Chem. Rev.*, **101**, 1629–58. [444]

Moustafa, M. A., Ebian, A. R., Khalil, S. A., and Motawi, M. M. (1971). Sulfamethoxydiazine. *J. Pharm. Pharmacol.*, **23**, 868–74. [348]

Moynihan, H. A. and Claudon, G. (2010). Direction of copper phthalocyanine crystallization using in situ generated tethered phthalocyanines. *CrystEngComm*, **12**, 2695–6. [391]

Mulliken, R. S. and Person, W. B. (1962). Donor–acceptor complexes. *Annu. Rev. Phys. Chem.*, **13**, 107–26. [359]

Mullin, J. W. (2001). *Crystallization*, 4th edn. Butterworth-Heinemann Ltd., Oxford. [79, 80, 95, 133]

Munn, R. W. and Ironside, C. N. (eds.) (1993). *Principles and Applications of Nonlinear Optical Materials*. Blackie A & P, Chapman and Hall, Glasgow. [296]

Murrell, J. N. and Tanaka, J. (1964). The theory of the electronic spectra of aromatic hydrocarbon dimers. *Mol. Phys.*, **7**, 363–80. [330]

Nagai, Y., Goto, N., and Nishi, H. (1965). Recent survey of quinacridones and related compounds. *Yuki Gosei Kagaku Kyokai Shi*, **23**, 318; *Chem. Abstr.*, **62**, 16221. [385]

Nagashima, N., Sano, C., Kishimoto, S., and Iitaka, Y. (1987). The characterization of various crystalline forms of aspartame (a dipeptide sweetener). *Acta Crystallogr. A*, **43**, C54. [453]

Nair, U. R., Sivabalan, R., Gore, G. M., Geetha, M., Asthana, S. N., and Singh, H. (2005). Hexanitrohexaazaisowurtzitane (CL-20) and CL-20-based formulations (Review) *Combust. Explos. Shock Waves*, **41**, 121–32. [409]

Nakasuji, K., Sasaki, M., Kotani, T., Murata, I., Enoki, T., Imaeda, K., Inokuchi, H., Kaawamoto, A., and Tanaka, J. (1987). Methylthio- and ethanediyldithio-substituted 1,6-dithiapyrene and their charge-transfer complexes: new organic molecular metals. *J. Am. Chem. Soc.*, **109**, 6970–5. [276]

Nakatsu, K., Yoshie, N., Yoshioka, H., Nogami, T., Shirota, Y., Shimizu, Y., Uemiya, T., and Yasuda, N. (1990). Polymorphism and the molecular and crystal structures of a 2nd-order nonlinear optical-compound containing a 1,3-dithiole ring. *Mol. Cryst. Liq. Cryst.*, **182**, 59–69. [302]

Nakazawa, Y., Tamura, M., Shirakawa, N., Shiomi, D., Yakahishi, M., Kinoshita, M., and Ishikawa, M. (1992). Low-temperature magnetic properties of the ferromagnetic organic radical, *p*-nitrophenyl nitronyl nitroxide. *Phys. Rev. B*, **46**, 8906–14. [287]

Nangia, A. (2006). Pseudopolymorph: retain this widely accepted term. *Cryst. Growth Des.*, **6**, 2–4. [8]

Nangia, A. (2008). Conformational polymorphism in organic crystals. *Acc. Chem. Res.*, **41**, 595–604. [6, 19, 254]

Nanubolu, J. B. and Burley, J. C. (2015). In situ Raman mapping for identifying transient solid forms. *CrystEngComm*, **17**, 5280–7. [352]

Nanubolu, J. B., Sridhar, B., Babu, V. S. P., Jagadeesh, B., and Ravikumar, K. (2012). Sixth polymorph of aripiprazole – an antipsychotic drug. *CrystEngComm*, **14**, 4677–85. [13t]

Nash, C. P., Olmstead, M. M., Weiss-Lopez, B., Musker, W. K., Ramasubbu, N., and Parthasarathy, R. (1985). Structures and Raman spectra of two crystalline modifications of dithioglycolic acid. *J. Am. Chem. Soc.*, **107**, 7194–5. [322]

Nass, K. K. (1991). Process implication of polymorphism in organic compounds. *AIChE Symp. Ser.*, **284**, 72–81. [39]

Nassau, K. (1983). *The Physics and Chemistry of Color. Fifteen Causes of Color*. Wiley-Interscience, New York. [303, 307, 328, 377]

Näther, C., Bock, H., and Claridge, R. F. C. (1996). Solvent-shared radical ion pairs [pyrene⁻Na⁺ O(C₂H₅)₂]⁻: ESR evidence for two different aggregates in solution, room temperature crystallization, and structural proof of another polymorphic modification. *Helv. Chim. Acta*, **79**, 84–91. [82]

Nauha, E. and Bernstein, J. (2015). "Predicting" polymorphs of pharmaceuticals using hydrogen bond propensities: probenecid and its two single-crystal-to-single crystal phase transitions. *J. Pharm. Sci.*, **104**, 2056–61. [140]

Nauha, E., Naumov, P., and Lusi, M. (2016). Fine-tuning of a thermosalient phase transition by solid solutions. *CrystEngComm*, **18**, 4699–703. [320]

Nauha, E., Ojala, A., Nissinen, M., and Saxell, H. (2011). Comparison of the polymorphs and solvates of two analogous fungicides—a case study of the applicability of a supramolecular synthon approach in crystal engineering. *CrystEngComm*, **13**, 4956–64. [361]

Naumov, P., Chizik, S., Panda, M. K., Nath, N. K., and Boldyreva, E. (2015). Mechanically responsive molecular crystals. *Chem. Rev.*, **115**, 12440–90. [311, 320, 358]

Naumov, P., Lee, S. C., Ishizawa, N., Jeong, Y. G., Chung, I. H., and Fukuzumi, S. (2009). New type of dual solid-state thermochromism: modulation of intramolecular charge transfer by intermolecular π–π interactions, kinetic trapping of the aci-nitro group, and reversible molecular locking. *J. Phys. Chem. A*, **113**, 11354–66. [313]

Navon, O., Bernstein, J., and Khodorkovsky, V. (1997). Chains, ladders and two-dimensional sheets with halogen–halogen and halogen–hydrogen interactions. *Angew. Chem. Int. Ed. Engl.*, **36**, 601–3. [69]

Navon, O., Bernstein, J., MacDonald, J. C., and Reutzel, S. (2001). The polymorphism of a barbiturate: 5,5'-diethylbarbituric acid. Unpublished work. [190f]

Navrotsky, A., Mazeina, L., and Majzlan, J. (2008). Size-driven structural and thermodynamic complexity in iron oxides. *Science*, **319**, 1635–8. [77]

Nedelko, V. V., Chukanov, N. V., Raevskii, A. V., Korsounskii, B. L., Larikova, T. S., and Kolesova, O. I. (2000). Comparative investigation of thermal decomposition of various modifications of hexanitrohexaazaisowurtzitane (CL-20). *Propellants, Explos., Pyrotech.*, **25**, 255–9. [407, 409]

Neumann, M. A., Leusen, F. J. J., and Kendrick, J. (2008). A major advance in crystal structure prediction. *Angew. Chem. Int. Ed. Engl.*, **47**, 2417–30. [244]

Neumann, M. A., van de Streek, J., Fabbiani, P. A., Hidber, P., and Grassman, O. (2015). Combined crystal structure prediction and high-pressure crystallization in rational pharmaceutical screening. *Nat. Commun.*, **6**, 7793. [336]

Neville, G. A., Beckstead, H. D., and Shurvell, H. F. (1992). Utility of Fourier transform Raman and Fourier transform infrared diffuse reflectance spectroscopy for differentiation of polymorphic spirolactone samples. *J. Pharm. Sci.*, **81**, 1141–6. [173, 177]

Newman, A. (2010). Basics of amorphous and amorphous solid dispersions. Presented at PPXRD-9, February 22–25, 2010. Available at http://www.seventhstreetdev.com [105]

Newman, A. (2011). An overview of solid form screening drug development. Presented at PPXRD-10, May 18, 2011. Available at http://www.seventhstreetdev.com [103]

Newman, A. (2013). Specialized solid form screening techniques. *Org. Process Res. Dev.*, **17**, 457–71. [92, 93f, 95, 104f, 110, 111f, 354]

Newman A. and Wenslow, R. (2016). Solid form changes during drug development: good, bad and ugly test cases. *AAPS Open*, **2**, 2. [103]

Newman, A. W., Stephens, P. W., Morrison, H. G., Andres, M. C., Shatly, G. P., and Thomas, A. S. (1999). Quantitation of two polymorphic forms using Rietveld analysis, synchrotron X P R D and traditional X P R D. In *Pharmaceutical Powder X-ray Diffraction Symposium*, September 27–30, 1999, pp. 85–96. International Centre for Diffraction Data, Newtown Square, PA. [166, 166f, 167f, 168f]

Ng, W. L. (1990). A study of the kinetics of nucleation in a palm oil melt. *J. Am. Oil Chem. Soc.*, **67**, 879–82. [40]

Nguyen, N. A. T., Ghosh, S., Gatlin, L. A., and Grant, D. J. W. (1994). Physicochemical characterization of the various solid forms of carbovir, an antiviral nucleoside. *J. Pharm. Sci.*, **83**, 1116–23. [9]

Nichols, G. (1998). Optical properties of polymorphic forms I and II of paracetamol. *Microscope*, **46**, 117–22. [143f, 194f]

Nichols, G. (1999). Thermodynamic and kinetic control of conformational polymorphism: a case study. In *1st International Symposium on Aspects of Polymorphism and Crystallization—Chemical Development Issues*, Hinckley, UK, June 23–25, 1999, pp. 199–209. Scientific Update, Mayfield, UK. [143f, 144f]

Nichols, G. (2006). Light microscopy. In *Polymorphism in the Pharmaceutical Industry* (ed. R. Hilfiker), pp. 167–209. Wiley-VCH, Weinheim. [137, 138, 364]

Nichols, G. and Frampton, C. S. (1998). Physicochemical characterization of the orthorhombic polymorph of paracetamol crystallized from solution. *J. Pharm. Sci.*, **87**, 684–93. [157f, 194f]

Nie, J.-J., Xu, D.-J., Li, Z.-Y., and Chiang, M. Y. (2001). 2,6-Dinitrotoluene. *Acta Crystallogr. E*, **57**, o827–8. [427]

Nielsen, A. T., Chafin, A. P., Christian, S. L., Moore, D. W., Nadler, M. P., Nissan, R. A., Vanderah, D. J., Gilardi, R. D., George, C. F., and Flippen-Anderson, J. L. (1998). Synthesis of polyazapolycyclic caged polynitramines. *Tetrahedron*, **54**, 11793–812. [407, 408]

Niementowski, S. (1896). Über das Chinakridin. *Chem. Ber.*, **29**, 76–83. [384]

Niementowski, S. (1906). Oxy-chinakridin und phlorchinyl. *Chem. Ber.*, **39**, 385–92. [384]

Niggli, P. (1924). *Lehrbuch der Mineralogie*, Zweite Auflage. Verlag von Gebrüder Borntraeger, Berlin. [37]

Ninomiya, M., Komai, A., Shirane, N., Ito, Y., and Terui, S. (1979). Nippon Shokubai Kagaku Kogyo Co. Patent JP 53–118427 [*Chem. Abstr.*, **90**, 56431s]; Patent JP 53–36036 [*Chem. Abstr.*, **90**, 105642x]; Patent JP 54–10331 [*Chem. Abstr.*, **90**, 170162e]; Patent JP 54–11135 [*Chem. Abstr.*, **91**, 92990c]; Patent JP 54–11136 [*Chem. Abstr.*, **90**, 205787s]. [390]

Nishimura, N., Senju, T., and Mizuguchi, J. (2006). 5,7,12,14-Tetrahydro[2,3-b]quinolinoacridine (β form). *Acta Crystallogr. E*, **62**, o4683–5. [380, 385, 386]

NIST (2001). *NIST Crystal Data*. U.S. National Institute of Standards and Technology. Gaithersburg, MD (http://www.nist.gov/srd/nistz.htm). [26]

Nordquist, K. A., Schaab, K. M., Sha, J., and Bond, A. H. (2017). Crystal nucleation using surface-energy-modified glass substrates. *Cryst. Growth Des.*, **17**, 4049–55. [359]

Norris, T., Aldridge, P. K., and Sekulic, S. S. (1997). Near-infrared spectroscopy. *Analyst*, **122**, 549–52. [170]

Nyvlt, J., Söhnel, O., Matuchová, M., and Broul, M. (1985). *The Kinetics of Industrial Crystallization*. Academia, Prague. [81]

O'Connor, R. T. (1960). X-ray diffraction and polymorphism. In *Fatty Acids*, Part I, 2nd edn (ed. K. S. Marklay), pp. 285–378. Interscience, New York. [40]

O'Mahoney, M. A., Croker, D. M., Rasmuson, A. C., Veesler, S., and Hodnett, B. K. (2013). Measuring the solubility of a quickly transforming metastable polymorph. *Org. Process Res. Dev.*, **17**, 512–18. [358]

Offerdahl, T. J. (2006). Solid-state nuclear magnetic resonance spectroscopy for analyzing polymorphic drug forms and formulations. *Pharm. Tech.*, **30**, S25–S42. [182]

Ogawa, K. and Scheel, H.-J. (1969). The crystal structure of 4,4′-diamino-1,1′-bianthroquinone, $C_{28}H_{16}N_2O_4$. *Z. Kristallogr., Kristallgeom., Kristallphys., Kristallchem.*, **130**, 405–19. [381t]

Ogawa, T., Kuwamoto, K., Isoda, S., Kobayashi, T., and Karl, N. (1999). 3,4:9,10-Perylen etetracarboxylic acid dianhydride (PTCDA) by electron crystallography. *Acta Crystallogr. B*, **55**, 123–30. [382t, 386]

Ogawa, Y. and Tasumi, M. (1978). Raman spectroscopic studies on conformational polymorphism of 1-bromopentane. *Chem. Lett.*, **7**, 947–50. [222]

Ogawa, Y. and Tasumi, M. (1979). Raman and infrared spectroscopic studies on conformational polymorphism of *n*-propyl acetate. *Chem. Lett.*, **8**, 1411–16. [221]

Ohashi, Y. (2014). *Crystalline State Photoreaction – Direct Observations of Reaction Processes and Metastable Intermediates*, pp.159–65. Springer, Tokyo. [310]

Ojala, C. R., Ojala, W. H., Pennamon, S. Y., and Gleason, W. B. (1998). Conformational polymorphs of acetone tosylhydrazone. *Acta Crystallogr. C*, **54**, 57–60. [124]

Oliver, S. N., Pantelis, P., and Dunn, P. L. (1990). Polymorphism and crystal–crystal transformations of the highly optically nonlinear organic compound α-[(4′-methoxyphenyl) methylene]-4-nitro-benzene-acetonitrile. *Appl. Phys. Lett.*, **56**, 307–9. [302]

Ono, T., ter Horst, J. H., and Jansens, P. J. (2004). Quantitative measurement of the polymorphic transformation of L-glutamic acid using in-situ Raman spectroscopy. *Cryst. Growth Des.*, **4**, 465–9. [177]

Orndorff, W. R. (1893). *Laboratory Manual*. D.C. Heath & Co., Boston. [20]

Orola, L., Veidis, M. V., Sarcevica, I., Actins, A., Belyakov, S., and Platonenko, A., (2012). The effect of pH on polymorph formation of the pharmaceutically active compound tianeptine. *Int. J. Pharm.*, **432**, 50–6. [352]

Orpen, A. G., Brammer, L., Allen, F. H., Kennard, O., Watson, D. G., and Taylor, R. (1989). Tables of bond lengths determined by X-ray and neutron diffraction. Part 2. Organometallic compounds and coordination complexes of the d- and f-block metals. *J. Chem. Soc., Dalton Trans.*, S1–S83. [215]

Ostmark, H., Langlet, A., Bergman, H., Wingborg, N., Wellmar, U., and Bemm, U. (1998). FOX-7 – a new explosive with low sensitivity and high performance. *Proc. Eleventh Int. Det. Symp. Preprints*, 18–21. [414]

Ostwald, W. (1902). *Lehrbuch der Allgemeinen Chemie*, Vol. II, pp. 383ff, 710ff, 773ff. Engelmann, Leipzig. [80]

Ostwald, W. F. (1897). Studien über die Bildung und Umwandlung fester Körper. *Z. Phys. Chem.*, **22**, 289–330. [12, 34, 57]

Oyumi, Y. and Brill, T. B. (1985). Thermal decomposition of energetic materials. 6. Solid-phase transitions and the decomposition of 1,2,3-triaminoguanidinium nitrate. *J. Phys. Chem.*, **89**, 4325–9. [412]

Oyumi, Y., Brill, T. B., and Rheingold, A. L. (1986a). Thermal decomposition of energetic materials. 9. Polymorphism, crystal structures, and thermal decomposition of poly-nitroazabicyclo[3.3.1]nonanes. *J. Phys. Chem.*, **90**, 2526–33. [413, 414]

Oyumi, Y., Brill, T. B., and Rheingold, A. L. (1987a) Thermal decomposition of energetic materials. A comparison of energetic materials and thermal reactivity of an acyclic and cyclic tetramethylenetetranitramine pair. *Thermochim. Acta*, **114**, 209–25. [412]

Oyumi, Y., Rheingold, A. L., and Brill, T. B. (1986b). Thermal decomposition of energetic materials. 16. Solid-phase structural analysis and the thermolysis of 1,4-dinitrofurzano[3,4-b]piperazine. *J. Phys. Chem.*, **90**, 4686–90. [426]

Oyumi, Y., Rheingold, A. L., and Brill, T. B. (1987b). Thermal decomposition of energetic materials. 19. Unusual condensed-phase and thermolysis properties of a mixed azido-methyl nitramine: 1,7-diazido-2,4,6-trinitro-2,4,6-triazaheptane. *J. Phys. Chem.*, **91**, 920–5. [412]

Ozawa, Y., Pressprich, M. R., and Coppens, P. (1998). On the analysis of reversible light-induced changes in molecular crystals. *J. Appl. Crystallogr.*, **31**, 128–35. [332]

Pace, M. D. (1991). EPR spectra of photochemical NO_2 formation in monocyclic nit-ramines and hexanitrohexaazaisowurtzitane. *J. Phys. Chem.*, **95**, 5858–64. [410]

Pace, M. D. (1992). Free radical mechanisms in high-density nitrocompounds—hexa-nitrohexaazaisowurtzitane, a new high energy nitramine. *Mol. Cryst. Liq. Cryst.*, **219**, 139–48. [410]

Padden, B. E., Zell, M. T., Dong, Z., Schroeder, S. A., Grant, D. J. W., and Munson, E. J. (1999). Comparison of solid-state ^{13}C NMR spectroscopy and powder X-ray diffrac-tion for analyzing mixtures of polymorphs of neotame. *Anal. Chem.*, **71**, 3325–31. [189]

Padrela, L., Rodrigues, M. A., Tiago, J., Velaga, S. P., Matos, H. A., and Gomes de Azevedo, E. (2015). Insight into the mechanisms of cocrystallization of pharmaceutical in super-critical solvents. *Cryst. Growth Des.*, **15**, 3175–81. [101]

Palacio, F., Antorenna, G., Casatro, M., Burriel, R., Rawson, J., Smith, J. N. B., Bricklebank, N., Novoa, J. J., and Ritter, C. (1997a). High temperature magnetic ordering in a new organic magnet. *Phys. Rev. Lett.*, **79**, 2336–9. [287]

Palacio, F., Antorenna, G., Casatro, M., Burriel, R., Rawson, J., Smith, J. N. B., Bricklebank, N., Novoa, J. J., and Ritter, C. (1997b). Spontaneous magnetization at 36 K in a sulfur–nitrogen radical. *Mol. Cryst. Liq. Cryst., Sci. Technol. A*, **306**, 293–300. [287]

Palmore, G. T. R., Luo, T. J., Martin, T. L., McBride-Wieser, M. T., and Voong, N. T. (1998). Using the atomic force microscope to study the assembly of molecular solids. *Trans. Am. Crystallogr. Assoc.*, **33**, 45–57. [207]

Panagiotopoulos, N. C., Jeffrey, G. A., LaPlaca, S. J., and Hamilton, W. C. (1974). The crystal structures of the A and B forms of potassium D-gluconate monohydrate. *Acta Crystallogr. B*, **30**, 1421–30. [4, 219, 221, 257]

Panagopoulou-Kaplani, A. and Malamataris, S. (2000). Preparation and characterisation of a new insoluble polymorphic form of glibenclamide. *Int. J. Pharm.*, **195**, 239–46. [351]

Panda, M. K., Etter, M., Dinnebier, R. E., and Naumov, P. (2017). Acoustic emission from organic martensites. *Angew. Chem. Int. Ed. Engl.*, **56**, 8104–9. [320]

Pandarese, F., Ungaretti, L., and Coda, A. (1975). The crystal structure of a monoclinic phase of *m*-nitrophenol. *Acta Crystallogr. B*, **31**, 2671–5. [121, 302]

Pando, C., Cabanas, A., and Cuadra, I. A. (2016). Preparation of pharmaceutical co-crystals through sustainable processes using supercritical carbon dioxide: a review. *RSC Adv.*, **6**, 71134–50. [101]

Panina, N., Leusen, F. J. J., Janssen, F. F. B. J., Verwer, P., Mekes, H., Vlieg, E., and Deroover, G. (2007). Crystal structure prediction of organic pigments: quinacridone as an example. *J. Appl. Crystallogr.*, **40**, 105–14. [380t, 385]

Park, A., Chyall, L., Dunlap, J., Scherz, C., Jonaitis, D., Stahly, B. C., Bates, S., Shipplett, R., and Childs, C. (2007). New solid-state chemistry technologies to bring better drugs to market: knowledge-based decision making. *Expert Opin. Drug. Discov.*, **2**, 145–54. [271]

Park, S. A., Lee, S., and Kim, W. S. (2015). The polymorphic crystallization of sulfamerazine in Taylor vortex flow: polymorphic nucleation and phase transformation. *Cryst. Growth Des.*, **15**, 3617–27. [358]

Parkin, A., Barr, G., Dong, W., Gilmore, C. J., Jayatilaka, D., McKinnon, J. J., Spackman, M. A., and Wilson, C. C. (2007). Comparing entire crystal structures: structural genetic fingerprinting. *CrystEngComm*, **9**, 648–52. [255]

Parrish, D. A., Deschamps, J. R., Gilardi, R. D., and Butcher, R. J. (2008). Polymorphs of picryl bromide. *Cryst. Growth Des.*, **8**, 57–62. [428]

Parsons, S. and Karaliota, A. (2015). Private communication. Deposition CCDC #1417464. [411]

Partington, J. R. (1952). *An Advanced Treaty on Physical Chemistry. Volume 3. The Properties of Solids*, pp. 512–13. Longmans, Green and Co., London. [31]

Pasquali, I., Bettini, R., and Giordano, F. (2008). Supercritical fluid technologies: an innovative approach for manipulating the solid-state of pharmaceuticals. *Adv. Drug Deliv. Rev.*, **60**, 399–410. [101]

Patel, A. D., Luner, P. E., and Kemper, M. S. (2001). Low-level determination of poly-morph composition in physical mixtures by near-infrared reflectance spectroscopy. *J. Pharm. Sci.*, **90**, 360–70. [174]

Patel, J., Jagia, M., Bansal, A. K., and Patel, S. (2015). Characterization and thermodynamic relationship of three polymorphs of xanthine oxidase inhibitor, Febusoxostat. *J. Pharm. Sci.*, **104**, 3722–30. [237]

Patil, D. G. and Brill, T. B. (1993). Thermal-decomposition of energetic materials. 59. Characterization of the residue of hexanitrohexaazaisowurtzitane. *Combust. Flame*, **92**, 456–8. [410]

Patyk, E., Skumiel, J., Podsiadlo, M., and Katrusiak, A. (2012), High-pressure (+)-sucrose polymorph *Angew. Chem. Int. Ed. Eng.*, **51**, 2146–50. [16, 85]

Paul, I. C. and Curtin, D. Y. (1973). Thermally induced organic reactions in the solid state. *Acc. Chem. Res.*, **6**, 217–25. [3, 310, 319. 335]

Paul, I. C. and Curtin, D. Y. (1975). Reactions of organic crystals with gases. *Science*, **187**, 19–26. [3, 336]

Paul, I. C. and Curtin, D. Y. (1987). Gas–solid reactions and polar crystals. In *Organic Solid State Chemistry* (ed. G. Desiraju), pp. 331–70. Studies in Organic Chemistry, Vol. 32, Elsevier, Amsterdam. [336]

Paul, I. C. and Go, K. T. (1969). The crystal and molecular structure of 5-methyl-1-thia-5-azacyclooctane 1-oxide perchlorate. *J. Chem. Soc. B*, 33–42. [67, 68f]

Pauling, L. (1960). *The Nature of the Chemical Bond*, 3rd edn. Cornell University Press, Ithaca, NY. [237, 249]

Paulus, E. F., Dietz, E., Kroh, A., Prokschy, F., and Lincke, G. (1989). Crystal structure of quinacridones. In *Collected abstracts of twelfth European crystallographic meeting, Moscow, Z. Kristallogr.*, **2** (Suppl 2). [385, 395]

Paulus, E. F., Leusen, F. J. J., and Schmidt, M. U. (2007). Crystal structures of quinacridones. *CrystEngComm*, **9**, 131–43. [380, 385]

Pavia, D. L., Lampman, G. L., and Kriz, G. S. (1988). *Introduction to Organic Laboratory Techniques*, 3rd edn. Saunder College Publishing, Philadelphia, PA. [133]

Pecharsky, V. K. and Zavalij, P. Y. (2009). *Fundamentals of Powder Diffraction and Structural Characterization of Materials*, 2nd edn. Springer, New York. [156]

Pella, P. A. (1976). Generator for producing trace vapor concentrations of 2,4,6-trinitro-toluene, 2,4-dinitrotoluene, and ethylene glycol dinitrate for calibrating explosives vapor detectors. *Anal. Chem.*, **48**, 1634–7. [420]

Pella, P. A. (1977). Measurements of the vapor pressures of TNT, 2,4-DNT, 2,6-DNT and EGDN. *J. Chem. Thermodyn.*, **9**, 301–5. [420]

Penfold, B. R. and White, J. C. B. (1959). The crystal and molecular structure of benza-mide. *Acta Crystallogr.*, **12**, 130–5.

Pepe, G., Perbost, R., Courcambeck, J., and Jouanna, P. (2009). Prediction of molecular crystal structures using a genetic algorithm: validation by GenMolTM on energetic compounds. *J. Cryst. Growth*, **311**, 3498–510. [410]

Percino, M. J., Cerón, M., Ceballos, P., Soriano-Moro, G., Castro, M. E., Chapela, V. M., Bonilla-Cruz, M., and López-Sandoval, R. (2014). Important role of molecular packing and intermolecular interactions in two polymorphs of (Z)-2-phenyl-3-(4-(pyridine-2-yl)phenyl)acrylonitrile. Preparation, structures, and optical properties. *J. Mol. Struct.*, **1078**, 74–82. [337]

Peresypkina, E. V., Bushuev, M. B., Virovets, A. V., Krivopalov, V. P., Lavrenova, L. G., and Larionov, S. V. (2005). Three differently coloured concomitant polymorphs: synthesis, structure and packing analysis of (4-(3′,5′-dimethyl-1*H*-pyrazol-1′-yl)-6-methyl-2-phenylpyrimidine)dichlorocopper(II). *Acta Crystallogr. B*, **61**, 164–73. [305]

Perlstein, J. (1992). Molecular self-assemblies: Monte Carlo predictions for the structure of the one-dimensional translation aggregate. *J. Am. Chem. Soc.*, **114**, 1955–63. [239]

Perlstein, J. (1994a). Molecular self-assemblies. 2. A computational method for the predic-tion of the structure of one-dimensional screw, glide and inversion molecular aggregates and implications for the packing of molecules in monolayers and crystals. *J. Am. Chem. Soc.*, **116**, 455–70. [239]

Perlstein, J. (1994b). Molecular self-assemblies. 4. Using Kitaigorodskii's aufbau principle for quantitatively predicting the packing geometry of semiflexible organic molecules in translation monolayer aggregates. *J. Am. Chem. Soc.*, **116**, 11420–32. [239]

Perlstein, J. (1996). Molecular self-assemblies. 5. Analysis of the vector properties of hydrogen bonding in crystal engineering. *J. Am. Chem. Soc.*, **118**, 8433–43. [239]

Perlstein, J. (1999). Introduction to packing patterns and packing energetics for crystalline self-assembled structures. In *From Molecules and Crystals to Materials* (ed. D. Braga, F. Grepioni, and A. G. Orpen), NATO ASI Series, Vol. C538, pp. 23–42. Kluwer Academic Publishers, Dordrecht. [239]

Perrenot, B. and Widmann, G. (1994). Polymorphism by differential scanning calorimetry. *Thermochim. Acta*, **234**, 31–9. [9, 146, 150]

Perrin, R. and Lamartine, R. (1990). Organic reactions in the solid state. In *Structure and properties of molecular crystals—studies in physical and theoretical chemistry*, Vol. 69, pp. 107–59. Elsevier, Amsterdam. [336]

Pertsin, A. J. and Kitaigorodskii, A. I. (1987). *The Atom–Atom Potential Method. Applications to Organic Molecular Solids.* Springer-Verlag, Berlin. [217, 227]

Peterson, G. R., Bassett, W. P., Weeks, B. L., and Hope-Weeks, L. J. (2013). Phase pure triacetone triperoxide: the influence of ionic strength, oxidant source, and acid catalyst. *Cryst. Growth Des.*, **13**, 2307–11. [412]

Peterson, M. L., Morissette, S. L., McNulty, C., Goldsweig, A., Shaw, P., LeQuesne, M., Monagle, J., Encina, N., Marchionna, J., Johnson, A., Gonzalez-Zugasti, J., Lemmo, A. V., Ellis, S. J., Cima, M. J., and Almarsson, Ö. (2002). Iterative high-throughput polymorphism studies on acetaminophen and an experimentally derived structure for form III. *J. Am. Chem. Soc.*, **124**, 10958–9. [108, 445]

Peterson, P. D., Lee, K.-Y., Moore, D. S., Scharff, R. J., and Avilucea, G. R. (2007). The evolution of sensitivity in HMX-based explosives during the reversion from delta to beta phase. *AIP Conf. Proc.*, **955**, 987. [407]

Petit, S. and Coquerel, G. (2006). The amorphous state. In *Polymorphism in the Pharmaceutical Industry* (ed. R. Hilfiker), pp. 259–85. Wiley-VCH, Weinheim. [370]

Petit, S., Coquerel, G., and Perez, G. (1994). Influence of water molecules on the nucleation rate of polymorphic complexes with different conformations in solution. In *Hydrogen Bond Networks* (ed. M.-C. Bellissent-Funel and J. C. Dore), pp. 255–9. Kluwer Academic Publishers, Dordrecht. [82]

Petty, C. J., Bugay, D. E., Findlay, W. P., and Rodriguez, C. (1996). Applications of FT-Raman spectroscopy in the pharmaceutical industry. *Spectroscopy*, **11**, 41–5. [177]

Pfeiffer, F. L. (1962). Pigmentary copper phthalocyanine in the "R" form and its preparation. American Cyanamid Patent US 3 051 721. [389]

Pfeiffer, P. (1922). *Organische Molekülverbindungen.* Verlag von Ferdinand Enke, Stuttgart. [5, 26]

Pfeiffer, P. (1927). *Organische Molekülverbindungen*, 2nd edn. Verlag von Ferdinand Enke, Stuttgart. [359]

Phillips, J. W. C. and Mumford, A. (1934). Dimorphism of certain aliphatic compounds. V. Primary alcohols and their acetates. *J. Chem. Soc.*, 1657–65. [142]

Pidcock, E., van de Streek, J., and Schmidt, M. (2009). The crystal structure of Pigment Yellow 181 from laboratory powder diffraction data. *Z. Kristallogr.*, **222**, 713–17. [382t]

Piecha, A., Bialonska, A., Jakubas, R., and Medycki, W. (2008). Structural polymorphism in new organic-inorganic hybrid: pyrazolium bromoantimonates(III) $[C_3N_2H_5]_6Sb_4Br_{18}\cdot2H_2O$

(tetragonal and triclinic forms). Thermal, dielectric and proton magnetic resonance (¹H NMR) studies on the tetragonal form. *Solid State Sci.*, **10**, 1469–79. [6]

Pierce-Butler, M. A. (1982). Structures of the barium salt of 2,4,6-trinitro-1,3-benzenediol monohydrate and the isomorphous lead salt (β-polymorph). *Acta Crystallogr. B*, **38**, 3100–4. [426]

Pierce-Butler, M. A. (1984). The structure of the lead salt of 2,4,6-trinitro-1,3-benzenediol monohydrate (α-polymorph), $\alpha - Pb^{2-+}C_6HN_3O_8^{2-} \cdot H_2O$. *Acta Crystallogr. C*, **40**, 63–5. [426]

Piermarini, G. J., Mighell, A. D., Weir, C. E., and Block, S. (1969). Crystal structure of benzene II at 25 kilobars. *Science*, **165**, 1250–5. [336]

Pillardy, J., Wawak, R. J., Arnautova, Y. A., Czaplewski, C., and Scheraga, H. A. (2000). Crystal structure prediction by global optimization as a tool for evaluating potentials: role of the dipole moment correction term in successful predictions. *J. Am. Chem. Soc.*, **122**, 907–21. [238]

Pimentel, G. C. and McClellan, A. L. (1960). *The Hydrogen Bond.* Freeman, San Francisco, CA. [68]

Pindelska, E., Szeleszczuk, L., Pisklak, D. M., Mazurek, A., and Kolodziejski, W. (2015). Solid-state NMR as an effective method of polymorphic analysis: solid dosage forms of clopidogrel hydrogensulfate. *J. Pharm. Sci.*, **104**, 106–13. [352]

Pinon, A. C., Rossini, A. J., Widdifield, C. M., Gajan, D., and Emsley, L. (2015). Polymorphs of theophylline characterized by DNP enhanced solid-state NMR. *Mol. Pharm.*, **12**, 4146–53. [193]

Pogoda, D., Janczak, J., and Videnova-Adrabinska, V. (2016). New polymorphs of an old drug: conformational and synthon polymorphism of 5-nitrofurazone. *Acta Crystallogr. B*, **72**, 263–73. [345]

Poole, J. W. and Bahal, C. K. (1968). Dissolution behavior and solubility of anhydrous and trihydrate forms of ampicillin. *J. Pharm. Sci.*, **57**, 1945–8. [348]

Poole, J. W., Owen, G., Silverio, J., Freyhof, J. N., and Rosenman, S. B. (1968). Physicochemical factors influencing the absorption of the anhydrous and trihydrate forms of ampicillin. *Curr. Ther. Res. Clin. Exp.*, **10**, 292–303. [348]

Poornachary, S. K., Chow, P. S., and Tan, R. B. H. (2008). Impurity effects on the growth of molecular crystals: experiments and modeling. *Adv. Powder Technol.* **19**, 459–73. [62]

Poornachary, S. K., Parambil, J. V., Chow, P. S., Tam, R. B. H., and Heng, J. Y. Y. (2013). Nucleation of elusive crystal polymorphs at the solution–substrate contact line. *Cryst. Growth Des.*, **13**, 1180–6. [101, 352, 369]

Popelier, P., Lenstra, A. T. H., van Alsenoy, C., and Geise, H. J. (1989). An *ab initio* study of crystal field effects: solid state and gas phase geometry of acetamide. *J. Am. Chem. Soc.*, **111**, 5658–60. [218]

Popov, E. V., Shvets, V. I., Shalimova, G. V., and Shtanov, N. P. (1981). Effects of crystal structure on coloristic properties of vat and disperse dyes. III. Morphological features of such vat dyes as violanthrone derivatives. *J. Appl. Chem.*, USSR, **54**, 2090–4; English translation of *Zh. Prikl. Khim.* (Leningrad), **54**, 2362–6. [383t]

Popovitz-Biro, R., Addadi, L., Leiserowitz, L., and Lahav, M. (1991). Strategies for the design of solids with polar arrangement. *ACS Symp. Ser.*, **455**, 472–83. [298]

Porter, M. W. and Codd, L. W. (1963). Crystals of the anorthic system. In *The Barker Index of Crystals. A Method for the Identification of Crystalline Substances*, Vol. 3. W. Heffer & Sons, Cambridge. [24, 25]

Porter, M. W. and Spiller, R. C. (1951). Crystals of the tetragonal, hexagonal, trigonal and orthorhombic systems. In *The Barker Index of Crystals. A Method for the Identification of Crystalline Substances*, Vol. 1. W. Heffer & Sons, Cambridge. [24, 25, 138]

Porter, M. W. and Spiller, R. C. (1956). Crystals of the monoclinic system. In *The Barker Index of Crystals. A Method for the Identification of Crystalline Substances*, Vol. 2. W. Heffer & Sons, Cambridge. [24, 25]

Porter, W. W., Elie, S. C., and Matzger, A. J. (2008). Polymorphism in carbamazepine co-crystals. *Cryst. Growth Des.*, **8**, 14–16. [100]

Potter, B. S., Palmer, R. A., Withnall, R., Chowdhry, B. Z., and Price, S. L. (1999). Aza analogues of nucleic acid bases: experimental determination and computational prediction of the crystal structure of anhydrous 5-azauracil. *J. Mol. Struct.*, **486**, 349–61. [239]

Potticary, J., Terry, L. R., Bell, C., Papanikolopoulos, A. N., Christianen, P. C. M., Engelkamp, H. E., Collins, A. M., Fontanesi, C., Kociok-Köhn, G., Crampin, S., Da Como, E., and Hall, S. R. (2016). An unforeseen polymorph of coronene by the application of magnetic fields during crystal growth. *Nat. Commun.*, **7**, 11555. [292]

Potts, G. D., Jones, W., Bullock, J. F., Andrews, S. J., and Maginn, S. J. (1994). The crystal structure of quinacridone: an archetypal pigment. *J. Chem. Soc., Chem. Commun.*, 2565–6. [161, 380, 385]

Power, L. F., Turner, K. E., and Moore, F. H. (1976). The crystal and molecular structure of α-glycine by neutron diffraction—a comparison. *Acta Crystallogr. B*, **32**, 11–16. [72]

Powers, H. E. C. (1971). Nucleation and the sugar industry. *Z. Zukerind.*, **21**, 272–7. [82, 85]

Pravica, M., Galley, M., Park, C., Ruiz, H., and Wojno, J. (2010). A high pressure, high temperature study of 1,1-diamino-2,2,-dinitro ethylene. *High Press. Res.*, **31**, 80–5. [414]

Pravica, M. G., Shen, Y. R., and Nicol, M. F. (2004). High pressure Raman spectroscopic study of structural polymorphism in cyclohexane. *Appl. Phys. Lett.*, **84**, 5452–4. [6]

Price, B. J., Clitherow, J. W., and Bradshaw, J. (1978). Aminoalkyl furan derivatives. Allen & Hanburys Ltd U.S. Patent 4 128 658. [438]

Price, C. P., Grzesiak, A. L., and Matzger, A. J. (2005). Crystalline polymorph selection and discovery with polymer heteronuclei. *J. Am. Chem. Soc.*, **127**, 5512–17. [100, 384t]

Price, S. L. (2000). Toward more accurate model of intermolecular potentials for organic molecules. *Rev. Comput. Chem.*, **14**, 225–89. [153, 227]

Price, S. L. (2008a). Computed crystal energy landscapes for understanding and predicting organic crystal structures and polymorphism. *Acc. Chem. Res.*, **41**, 595–604. [6]

Price, S. L. (2008b). Computational prediction of organic crystal structures and polymorphs. *Int. Rev. Phys. Chem.*, **27**, 541–68. [240]

Price, S. L. (2013). Why don't we find more polymorphs? *Acta Crystallogr. B*, **69**, 313–28. [359, 370]

Price, S. L. (2014). Predicting crystal structures of organic compounds. *Chem. Soc. Rev.*, **43**, 2098–111. [240, 359]

Price, S. L. and Reutzel-Edens, S. M. (2016). The potential of computed crystal energy landscapes to aid solid form development. *Drug Disc. Today*, **21**, 912–23. [114, 344, 354, 362, 362f]

Price, S. L. and Stone, A. J. (1992). Electrostatic models for polypeptides: can we assume transferability? *J. Chem. Soc., Faraday Trans.*, **88**, 1755–63. [227]

Price, S. L. and Wiley, K. S. (1997). Predictions of crystal packings for uracil, 6-azauracil and allopurinol: the interplay between hydrogen bonding and close packing. *J. Phys. Chem. A*, **101**, 2198–206. [239]

Price, S. L., Braun, D. E., and Reutzel-Edens, S. M. (2016). Can computed crystal energy landscapes help understand pharmaceutical solids? *Chem. Commun.*, **52**, 7065–77. [114, 354]

Price, S. L., Leslie, M., Welch, G. W., Habgood, M., Price, L. S., Karamertzanis, P. G., and Day, G. M. (2010). Modelling organic crystal structures using distributed multipole and polarizability-based model intermolecular potentials. *Phys. Chem. Chem. Phys.*, **12**, 8478–90. [244]

Prokop'eva, I. A. and Davidov, A. A. (1975). Some features of the structure of solid solutions of α,α-diphenyl-β-picrylhydrazyl in α,α-diphenyl-β-picrylhydrazine. *Akad. Nauk SSSR, Ural Nauk Is.* 34. [292]

Pu, L., Xu, J.-J., Liu, X.-F., and Sun, J. (2016). Investigation on the thermal expansion of four polymorphs of crystalline CL-20. *J. Energ. Mater.*, **34**, 205–15. [409]

Purdum, G. E., Yao, N., Woll, A., Gessner, T., Weitz, R. T., and Loo, Y. L. (2016). Understanding polymorph transformations in core-chlorinated naphthalene diimides and their impact on thin-film transistor performance. *Adv. Funct. Mater.*, **26**, 2357–64. [280]

Pushkina, L. L. and Shelyapin, O. P. (1988). Study of preparation of a pigment form of *cis*-naphthoylenebisbenzene. *J. Appl. Chem. USSR*, **61**, 2296–301; English translation of *Zh. Prikl. Khim.* (Leningrad), **61**, 2515–19. [381t]

Putz, H., Schon, J. C., and Jansen, M. (1999). Combined method for ab initio structure solution from powder diffraction data. *J. Appl. Crystallogr.*, **32**, 864–70. [155]

Pyszczynski, S. J. and Munson, E. J. (2013). Generation and characterization of a new solid form of trehalose. *Mol. Pharm.*, **10**, 3323–32. [363]

Qi, Q., Zhang, J., Xu, B., Li, B., Zhang, S. X.-A., and Tian, W. (2013). Mechanochromism and polymorphism-dependent emission of tetrakis(4-(dimethylamino)phenyl)ethylene. *J. Phys. Chem. C*, **117**, 24997–5003. [317]

Qiu, H., Stepanov, V., Chou, T., Surapaneni, A., and Di Stasio, A. R. (2012). Single-step production and formulation of HMX nanocrystals. *Powder Technol.*, **226**, 235–8. [405]

Radhakrishnan, T. P. (2008). Molecular structure, symmetry, and shape as design elements in the fabrication of molecular crystals for second harmonic generation and the role of molecules-in-materials. *Acc. Chem. Res.*, **41**, 367–76. [296]

Rafilovich, M. and Bernstein, J. (2006). Serendipity and the polymorphs of benzidene. *J. Am. Chem. Soc.*, **128**, 12185–91. [255, 258f, 354]

Rafilovich, M., Bernstein, J., Harris, R. K., Apperley, D. C., Karamertzanis, P. G., and Price, S. L. (2005). Groth's original concomitant polymorphs revisited. *Cryst. Growth Des.*, **5**, 2197–209. [96, 116, 117]

Ramakrishnan, S., Yarraguntia, S. R., Pedireddy, S. R., Kanniah, S. L., Mudapaka, V. K., Shekhawat, L. K., Mahapatra, S., Mohammad, A. B., Vishweshwar, P., and Stephens, P. W. (2017). Development of pharmaceutically acceptable crystalline forms of drug substances via solid-state solvent exchange. *Org. Process Res. Dev.*, **21**, 1478–87. [358]

Ramamurthy, V. and Venkatesan, K. (1987). Photochemical reactions of organic crystals. *Chem. Rev.*, **87**, 433–81. [333]

Raman, C. V. and Krishnan, K. S. (1928). A new type of secondary radiation. *Nature*, **121**, 501–2. [176]

Ramdas, S., Thomas, J. M., Jordan, M. E., and Eckhardt, C. J. (1981). Enantiomeric intergrowths in hexahelicenes. *J. Phys. Chem.*, **85**, 2421–5. [42]

Rao, C. N. R. (1984). Phase transitions and the chemistry of solids. *Acc. Chem. Res.*, **17**, 83–9. [274]

Rao, C. N. R. (1987). Phase transitions in organic solids. In *Organic Solid State Chemistry* (ed. G. Desiraju), pp. 371–432. Studies in Organic Chemistry, Vol. 32, Elsevier, Amsterdam. [274]

Rao, C. N. R. and Gopalakrishnan, J. (1997). *New Directions in Solid State Chemistry*, 2nd edn, pp. 168–228. Cambridge University Press, Cambridge. [274]

Rao, C. N. R. and Rao, K. J. (1978). *Phase Transitions in Solids*. McGraw-Hill, New York. [319]

Rastogi, S. and Kurelec, L. (2000). Polymorphism in polymers; its implications for polymer crystallization. *J. Mater. Sci.*, **35**, 5121–38. [40]

Readings, M. (1993). Modulated differential scanning calorimetry—a new way forward in material characterization. *Trends Polym. Sci.*, **1**, 248–53. [154]

Reany, O., Kapon, M., Botoshansky, M., and Keinan, E. (2009). Rich polymorphism in triacetone-triperoxide. *Cryst. Growth Des.*, **9**, 3661–70. [411]

Reed, S. M., Weakley, T. J. R., and Hutchison, J. E. (2000). Polymorphism in a conformationally flexible substituted anthraquinone: a crystallographic, thermodynamic, and molecular modeling study. *Cryst. Eng.*, **3**, 85–99. [114, 222, 304, 305]

Reetz, M. T., Höger, S., and Harms, K. (1994). Proton-transfer-dependent reversible phase changes in the 4,4′-bipyridinium salt of squaric acid. *Angew. Chem. Int. Ed. Engl.*, **33**, 181–3. [304, 310]

Reichelt, H., Faunce, C. A., and Paradies, H. H. (2007). Elusive forms and structures of *N*-hydroxyphthalimide: the colorless and yellow forms of *N*-hydroxyphthalimide. *J. Phys. Chem. A*, **111**, 2587–2601. [306]

Reilly, A. M. and Tkatchenko, A. (2015). van der Waals dispersion interactions in molecular materials: beyond pairwise additivity. *Chem. Sci.*, **6**, 3289–301. [217, 245]

Reilly, A. M., Cooper, R. I., Adjiman, C. S., Bhattacharya, S., Boese, A. D., Brandenburg, J. G., Bygrave, P. J., Bylsma, R., Campbell, J. E., Car, R., and Case, D. H. (2016). Report on the sixth blind test of organic crystal structure prediction methods. *Acta Crystallogr. B*, **72**, 439–59. [241, 244, 245]

Reilly, J. and Rae, W. N. (1954). *Physicochemical Methods*, 5th edn, Vol. 1, pp. 507–608. van Nostrand, New York. [54]

Reinke, H., Dehne, H., and Hans, M. (1993). A discussion of the term "polymorphism." *J. Chem. Educ.*, **70**, 101. [5, 29]

Remanar, J. F., Morissette, S. L., Peterson, M. L., Moulton, B., MacPhee, J. M., Guzman, H. R., and Almarsson, Ö. (2003). Crystal engineering of novel cocrystals of a triazole drug with 1,4-dicarboxylic acids. *J. Am. Chem. Soc.*, **125**, 8456–7. [360]

Reutzel-Edens, S. M. (2006). Achieving polymorph selectivity in the crystallization of pharmaceutical solids: basic considerations and recent advances. *Curr. Opin. Drug Deliv. Dev.*, **9**, 806–15. [375]

Reznichenko, V. V., Chesnovskaya, E. S., Podrezova, T. N., and Ezhkova, Z. I. (1984). Effect of the method of synthesis on allotropic properties of [2,2]-bi(naphtho[2,1-b] thiophenylidene]-1,1′-dione. *J. Appl. Chem. USSR*, **57**, 194–6; English translation of *Zh. Prakt. Khim.*, **57**, 205–7. [390]

Ribka, J. (1961). German Patent 1 208 435. [381t]

Ribka, J. (1969). German Patent 1 287 731. [381t]

Ribka, J. (1970). German Patent 2 043 482. [381t]

Richards, F. M. and Lindley, P. F. (1999). Determination of the density of solids. In *International Tables for Crystallography* (ed. A. J. C. Wilson and E. Prince), Vol. C, 2nd edn, pp. 156–9. Kluwer Academic Publishers, Dordrecht. [54, 207]

Richardson, M. R., Yang, Q., Novotny-Bregger, E., and Dunitz, D. J. (1990). Conformational polymorphism of dimethyl 3,6-dichloro-2,5-dihydroxyterphthalate. II. Structural, thermodynamic, kinetic and mechanical aspects of phase transformations among the three crystal forms. *Acta Crystallogr. B*, **46**, 653–60. [305]

Rieper, W. and Baier, E. (1992). Neue Kristallmodifikationen von C.I. Pigment Yellow 16. European Patent EP 054072. [382t]

Riley, P. E. and Davis, R. E. (1976). Crystal and molecular structures at 35°C of two crystal forms of bis(2,6-dimethylpyridine)chromium, a bis heterocyclic sandwich complex. *J. Inorg. Chem.*, **15**, 2735–40. [224]

Ripmeester, J. A. (1980). Application of ^{13}C NMR to the study of polymorphs, clathrates and complexes. *Chem. Phys. Lett.*, **74**, 536–8. [184]

Roberts, R. J. and Rowe, R. C. (1996). Influence of polymorphism on the Young's modulus and yield stress of carbamazepine, sulfathiazole and sulfanilamide. *Int. J. Pharm.* **129**, 79–94. [320]

Roberts, R. M. (1989). *Serendipity: Accidental Discoveries in Science.* Wiley, New York. [438]

Robertson, J. M. (1935). An X-ray study of the structure of the phthalocyanines. I. The metal-free, nickel, copper and platinum compounds. *J. Chem. Soc.*, 615–21. [389]

Robertson, J. M. (1936). An X-ray study of the phthalocyanines. II. Structure determination of the metal-free compound. *J. Chem. Soc.*, 1195–209. [389, 393]

Robertson, J. M. (1953). *Organic Crystals and Molecules.* Cornell University Press, Ithaca, NY. [393]

Robertson, J. M. and Woodward, I. (1937). An X-ray study of the phthalocyanines. Part III. Quantitative structure determination of nickel phthalocyanines. *J. Chem. Soc.*, 219–30. [389]

Robles, L., Mondieig, D., Haget, Y., and Cuevas-Diarte, M. A. (1998). Review on the energetic and crystallographic behaviour of *n*-alkanes. II. Series from $C_{22}H_{46}$ to $C_{27}H_{56}$. *J. Chim. Phys. Phys.-Chim. Biol.*, **95**, 92–111. [40]

Rockley, N. L., Woodard, M. K., and Rockley, M. G. (1984). The effect of particle size on FT-IR-PAS spectra. *Appl. Spectrosc.*, **38**, 329–34. [174]

Rodriguez-Hornedo, N. and Murphy, D. (1999). Significance of controlling crystallization mechanisms and kinetics in pharmaceutical systems. *J. Pharm. Sci.*, **88**, 651–60. [198f, 275]

Rodriguez, M. A., Campana, C. F., Rae, A. D., Graeber, E., and Morosin, B. (2005). Form II of 2,2′,4,4′,6,6′-hexanitroazobenzene (HNAB-III). *Acta Crystallogr. C*, **61**, o127–30. [425]

Rodriguez, M. A., Tiago, J., Duart, A., Geraldes, V., Matos, H. A., and Azevedo, E. G. (2016). Polymorphism in pharmaceutical drugs by supercritical processing: clarifying the role of the antisolvent effect and atomization enhancement. *Cryst. Growth Des.*, **16**, 6222–9. [101]

Romé de l'Isle, J. B. L. (1783). *Cristallographie ou description des formes propres a tous les corps de Règne Minéral dans l'Etat dede Combinasion Saline, Pierreuse ou Metallique* (4 vols). Imprimerie de Monsieur, Paris. [30]

Roozeboom, H. W. B. (1911). *Die Heterogenen Gleichgewichte von Standpunkte der Phasen lehre.* Friedrich Vieweg & Son, Braunschweig. [42]

Rosenstein, S. and Lamy, P. P. (1969). Polymorphism. *Am. J. Hos. Pharm.*, **26**, 598–601. [5]

Ross, S. A., Lamprou, D. A., and Douroumis, D. (2016). Engineering and manufacturing of pharmaceutical co-crystals: a review of solvent-free manufacturing technologies. *Chem. Commun.*, **52**, 8772–86. [360]

Roston, D. A., Walters, M. C., Rhineberger, R. R., and Ferro, L. J. (1993). Characterization of polymorphs of a new antiinflammatory drug. *J. Pharm. Biomed. Anal.*, **11**, 293–300. [174]

Roth, K. (2005). Chocolate – the noblest polymorphism II. *Chem. Unserer Zeit*, **39**, 416–28. [40]

Rouvray, D. H. (1995). That fuzzy feeling in chemistry. *Chem. Brit.*, July, 544–6. [7, 271]

Rouvray, D. H. (ed.) (1997). *Fuzzy Logic in Chemistry.* Academic Press, San Diego. [7, 271]

Rovira, C. and Novoa, J. J. (2001). A first principles computation of the low-energy polymorphic form of the acetic acid crystal. A test of the atom–atom force field predictions. *J. Phys. Chem. B*, **105**, 1710–19. [240]

Row, T. N. G. (1999). Hydrogen and fluorine in crystal engineering: systematics from crystallographic studies of hydrogen bonded tartrate-amine complexes and fluoro-substituted coumarins, styrylcoumarins and butadienes. *Coord. Chem. Rev.*, **183**, 81–100. [333]

Roy, S. and Matzger, A. J. (2009). Unmasking a third polymorph of a benchmark crystal-structure prediction compound. *Angew. Chem. Int. Ed. Engl.*, **48**, 8505–8. [373]

Roy, S., Goud, N. R., and Matzger, A. J. (2016). Polymorphism in phenobarbital: discovery of a new polymorph and crystal structure of elusive form V. *Chem. Commun.*, **52**, 4389–92. [373]

Royall, P. G. and Gaisford, S. (2016). Application of solution calorimetry in pharmaceutical and biopharmaceutical research. *Curr. Pharm. Tech.*, **6**, 215–22. [154]

Ruch, T. and Wallquist, O. (1997). Crystal modification of a diketopyrrolopyrrole pigment. European Patent 0825234. [382t]

Rudel, P., Odiot, S., Mutin, J., C. and Peyrard, M. (1990). Crystal-structure and explosive power of molecular crystals. *J. Chim. Phys. Phys.-Chim. Biol.*, **87**, 1307–44. [398]

Russell, T. P., Miller, P. J., Piermarini, G. J., and Block, S. (1992). High pressure phase transition in γ-hexanitrohexaazaisowurtzitane. *J. Phys. Chem.*, **96**, 5509–12. [407]

Russell, T. P., Miller, P. J., Piermarini, G. J., and Block, S. (1993). Pressure/temperature phase diagram of hexanitrohexaazaisowurtzitane. *J. Phys. Chem.*, **97**, 1993–7. [407, 409]

Russell, T. P., Piermarini, G. J., and Miller, P. J. (1997). Pressure/temperature and reaction phase diagram for dinitro azetidinium dinitramide. *J. Phys. Chem. B*, **101**, 3566–70. [415, 416]

Russell, T. P., Piermarini, G. J., Block, S., and Miller, P. J. (1996). Pressure, temperature reaction phase diagram for ammonium dinitramide. *J. Phys. Chem.*, **100**, 3248–51. [416]

Rustichelli, C., Gamberini, G., Ferioli, V., Gamberini, M. C., Ficarra, R., and Tommasini, S. (2000). Solid state study of polymorphic drugs: carbamazepine. *J. Pharm. Biomed. Anal.*, **23**, 41–54. [209, 365]

Rykounov, A. A. (2015). Structural and thermodynamic properties of two polymorphic modifications of the insensitive high explosive 5-nitro-2,4-dihydro-1,2,4-triazol-3-one (NTO) under finite pressures and temperatures from *ab initio* calculations. *CrystEngComm*, **17**, 7653–62. [413]

Ryzhkov, L. R. and McBride, J. M. (1996). Low-temperature reactions in single crystals of two polymorphs of the polycyclic nitramine N-15-HNIW. *J. Phys. Chem.*, **100**, 163–9. [410]

Safin, D. A., Rogeyns, K., Babashkina, M. G., Filinchuk, Y., Rotaru, A., Jurescg, C., Mitoraj, M. P., Hooper, J., Brela, M., and Garcia, Y. (2016). Polymorphism driven optical properties of an anil. *CrystEngComm*, **18**, 7249–59. [310]

Saha, B. K., Nangia, A., and Nicoud, J.-F. (2006). Using halogen···halogen interactions to direct noncentrosymmetric crystal packing in dipolar organic molecules. *Cryst. Growth Des.*, **6**, 1278–81. [297]

Saindon, P. J., Cauchon, N. S., Sutton, P. A., Chang, C. J., Peck, G. E., and Byrn, S. R. (1993). Solid-state nuclear-magnetic-resonance (NMR) spectra of pharmaceutical dosage forms. *Pharm. Res.*, **10**, 197–203. [363]

Saito, G. (1997). Organic superconducting solids. In *Organic Molecular Solids. Properties and Applications* (ed. W. Jones), pp. 309–40. CRC Press, Boca Raton, FL. [277]

Sakaguchi, I. and Hayashi, Y. (1981). Sumitomo Chemical Co. Ltd. Patent DE 3 031 444 A1. [380t]

Sakaguchi, I. and Hayashi, Y. (1982). Sumitomo Chemical Co. Ltd. Patent DE 3 211 607 A1. [380t]

Saklatvala, R., Royall, P. G., and Craig, D. Q. M. (1999). The detection of amorphous material in a nominally crystalline drug using modulated temperature DSC—a case study. *Int. J. Pharm.*, **192**, 55–62. [372]

Sala, S., Elizondo, E., Moreno, E., Calvet, T., Cuevas-Diarte, M. A., Ventosa, N., and Jaume Veciana, J. (2010). Kinetically driven crystallization of a pure polymorphic phase of stearic acid from CO_2-expanded solutions. *Cryst. Growth Des.*, **10**, 1226–32. [87]

Salammal, S. T., Balandier, J.-Y., Arlin, J.-B., Olivier, Y., Lemaur, V., Wang, L., Beljonne, D., Cornil, J., Kennedy, A. R., Geerts, Y. H., and Chattopadhyay, B. (2014). Polymorphism in bulk and thin films: The curious case of dithiophene-DPP(Boc)-dithiophene. *J. Phys. Chem. C*, **118**, 657–69. [384t, 395]

Salammal, S. T., Zhang, Z., Chen, J., Chattopadhyay, B., Wu, J., Fu, L., Fan, C., and Chen, H. (2016). Polymorphic phase-dependent optical and electrical properties of a diketopyrrolopyrrole-based small molecule. *Appl. Mater. Interfaces*, **8**, 20916–27. [306]

Salje, E. K. H. (1990). *Phase Transitions in Ferroelastic and Co-Elastic Crystals*. Cambridge University Press, Cambridge. [319]

Samsonov, G. V. (ed.) (1976). *Physicheskie Svoistva Elementov*. Spravochnik, Moscow. [29]

Sanchis, M. J., Marthe, S., Díaz Calleja, R., Sánchez Martínez, E., Epple, M., and Klar, G. (1997). The thermally induced phase transition in 2,3,7,8-tetramethoxythianthrene. *Ber. Bunsen-Ges. Phys. Chem.*, **101**, 1889–95. [335]

Sandefur, C. W. and Thomas, T. J. (1984). Modified acid dyestuff. Mobay Chemical Corp. Patent US 4 474 577. [379t]

Sanders, G. H. W., Roberts, C. J., Danesh, A., Murray, A. J., Price, D. M., Davies, D. C., Tendler, S. J. B., and Wilkins, M. J. (2000). Discrimination of polymorphic forms of a drug product by localized thermal analysis. *J. Microsc.*, **198**, 77–81. [154]

Sano, N. and Tanaka, J. (1986). Electronic spectra of two polymorphs of (5-dimethylamino-2,4-pentadienylidene)dimethylammonium perchlorate. *Bull. Chem. Soc. Jpn.*, **59**, 843–51. [330]

Sappok, R. (1978). Recent progress in the physical chemistry of organic pigments with special reference to phthalocyanines. *J. Oil Colour Chem. Assoc.*, **61**, 299–308. [295f]

Saraswatula, V. and Saha, B. K. (2015). Modulation of thermal expansion by guest and polymorphism in a hydrogen bond host. *Cryst. Growth Des.*, **15**, 593–601. [321]

Sarma, B., Chen, J., His, H. Y., and Myerson, A. S. (2011). Solid forms of pharmaceuticals. Polymorphs, salts and cocrystals. *Kor. J. Chem. Eng.*, **28**, 315–22. [343]

Sarma, J. and Desiraju, G. R. (1999). Polymorphism and pseudopolymorphism in organic crystals. In *Crystal Engineering: The Design and Applications of Functional Solids* (ed. K. R. Seddon and M. J. Zaworotko), pp. 325–56. Kluwer Academic Publishers, Dordrecht. [19]

Sarma, J. A. R. P. and Desiraju, G. R. (1985). The chloro-substituent as a steering group: a comparative study of non-bonded interactions and hydrogen bonding in crystalline chloroaromatics. *Chem. Phys. Lett.*, **117**, 160–4. [227]

Sasaki, T., Ida, Y., Hisaki, I., Tsuzuki, S., Tohnal, N., Coquerel, G., Sat, H., and Miyata, M. (2016). Construction of chiral polar crystals from achiral molecules by stacking control of hydrogen-bonded layers using type II halogen bonds. *Cryst. Growth Des.*, **16**, 1626–35. [297]

Sato, H., Suzuki, K., Okada, M., and Garti, N. (1985). Solvent effects on kinetics of solution-mediated transition of stearic acid polymorphs. *J. Cryst. Growth*, **72**, 699–704. [40]

Sato, K. (1999). Solidification and phase transformation behavior of food fats—a review. *Fett/Lipid*, **101**, 467–74. [40, 155]

Sato, K. and Boistelle, R. (1984). Stability and occurrence of polymorphic modifications of stearic acid in polar and nonpolar solutions. *J. Cryst. Growth*, **66**, 441–50. [79]

Sawada, K., Hashizume, D., Sekine, A., Uekusa, H., Kato, K., Ohashi, Y., Kakinuma, K., and Ohgo, Y. (1996). Four polymorphs of a cobaloxime complex with different solid-state photoisomerization rates. *Acta Crystallogr. B*, **52**, 303–13. [335]

Schaefer, J. and Stejskal, E. O. (1976). Carbon-13 nuclear magnetic resonance of polymers spinning at the magic angle. *J. Am. Chem. Soc.*, **98**, 1031–2. [183]

Scheidt, W. R., Geiger, D. K., Hayes, R. G., and Lang, G. (1983). Control of spin state in (porphinato)iron(III) complexes. Axial ligand orientation effect leading to an intermediate-spin complex. Molecular structure and physical characterization of the monoclinic form of bis(3-chloropyridine)(octaethylporphinato)iron(III) perchlorate. *J. Am. Chem. Soc.*, **105**, 2625–32. [285]

Schinzer, W. C., Bergren, M. S., Aldrich, D. S., Chao, R. S., Dunn, M. J., Jeganathan, A., and Madden, L. M. (1997). Characterization and interconversion of polymorphs of premafloxacin, a new quinoline antibiotic. *J. Pharm. Sci.*, **86**, 1426–31. [365]

Schlueter, J. A., Carlson, K. D., Williams, J. M., Geiser, U., Wang, H. H., Welp, U., Kowk, W.-K., Fendrich, J. A., Dudek, J. D., Achenbach, C. A., Keane, P. M., Komosa, A. S., Naumann, D., Roy, T., Schirber, J. E., and Bayless, W. R. (1994). A new 9K superconducting organic salt composed of the bis(ethylenedithio)tetrathiafulvalene (ET) electron-donor molecule and the tetrakis(trifluoromethyl)cuprate(III) anion, $[Cu(CF_3)_4]^-$. *Physica C*, **230**, 378–84. [277]

Schmid, S., Muller-Goymann, C. C., and Schmidt, P. C. (2000). Interactions during aqueous film coating of ibuprofen with Aquacoat ECD. *Int. J. Pharm.*, **197**, 35–9. [363]

Schmidt, A., Kababya, S., Appel, M., Khatib, S., Botoshansky, M., and Eichen, Y. (1999). Measuring the temperature width of a first-order single crystal to single crystal phase transition using solid-state NMR: application to the polymorphism of 2-(2,4-dinitrobenzyl)-3-methylpyridine. *J. Am. Chem. Soc.*, **121**, 11291–9. [189]

Schmidt, G. M. J. (1964). Topochemistry. Part III. The crystal chemistry of some *trans*-cinnamic acids. *J. Chem. Soc.*, 2014–21. [332]

Schmidt, G. M. J. (1971). Photodimerization in the solid state. *Pure Appl. Chem.*, **27**, 647–78. [3, 333]

Schmidt, M. U. (1999). Process for preparing new crystalline modifications of C.I. Pigment Red 53:2. European Patent EP 1010732 A1. [380t]

Schmidt, M. U. (2000). Process for preparing new crystalline modifications of C. I. Pigment Red 53:2. Clariant GMBH Patent DE 19858853 A1. [380t]

Schmidt, M. U. and Metz, H. J. (1998a). Novel crystal modification of C.I. Pigment Red 53:2 (gamma-phase). German Patent DE 19827272.3. [380t]

Schmidt, M. U. and Metz, H. J. (1998b). Novel crystal modification of C.I. Pigment Red 53:2 (delta-phase). German Patent DE 19827273.1. [380t]

Schmidt, M. U. and Metz, H. J. (1999a). Novel crystal modification of C.I. Pigment Red 53:2 (gamma-phase). European Patent EP 965617 A1. [380t]

Schmidt, M. U. and Metz, H. J. (1999b). Novel crystal modification of C.I. Pigment Red 53:2 (delta-phase). European Patent EP 965616 A1. [380t]

Schmidt, M.U., Brühne, S., Wolf, A.K., Rech, A., Bruning, J., Alig, E., Fink, L., Buchsbaum, C., Glinnemann, J., van de Streek, J., Gozzo, F., Brünelli, M., Stowasser, F., Gorelik, T., Mugnaioli, E., and Kolb, U. (2009). Electron diffraction, X-ray powder diffraction and pair-distribution-function analyses to determine the crystal structure of Pigment Yellow 213 $C_{23}H_{21}N_5O_9$. *Acta Crystallogr. B*, **65**, 189–99. [382t, 383t]

Schmidt, M. U., Buchsbaum, C., Schnorr, J. M., Hofmann, D. W. M., and Emrich, M. (2007a). Pigment Orange 5: crystal structure determination from a non-indexed X-ray powder diagram. *Z. Kristallogr.*, **222**, 30–3. [380t]

Schmidt, M. U., Dinnebier, R. E., Kalkhof, H. Crystal engineering on industrial diaryl pigments using lattice energy minimizations and X-ray powder diffraction, J. Phys. Chem. B, 2007, 111, 9722–9732. [380t]

Schmidt, M. U., Hofmann, D. W. M., Buchsbaum, C., and Metz, H. J. (2006). Crystal structures of Pigment Red 170 and derivatives, as determined by X-ray powder diffraction. *Angew. Chem. Int. Ed. Engl.*, **45**, 1313–17. [381t]

Schmidt, M. U., Paulus, E. F., Rademacher, N., and Day, G. M. (2010). Experimental and predicted crystal structures of Pigment Red 168 and other dihalogenated anthanthrones. *Acta Crystallogr. B*, **66**, 515–26. [381t]

Schmitt, P. D., DeWalt, E. L., Dow, X. Y., and Simpson, G. J. (2016). Rapid discrimination of polymorphic crystal forms by nonlinear optical Stokes ellipsometric microscopy. *Anal. Chem.*, **88**, 5760–8. [359]

Schöll, J., Vicum, L., Muller, M., and Mazzotti, M. (2006). Precipitation of L-glutamic acid: determination of nucleation kinetics. *Chem. Eng. Technol.*, **29**, 257–64. [87]

Schollmeyer, D. and Ravindran, B. (2015). Private communication. Depositions CCDC #1438037 and #1438039. [411]

Schorlemmer, C. (1874). *The Carbon Compounds or Organic Chemistry*. MacMillan & Co., London. [20]

Schui, F., Deubel, R., and Wester, N. (1988). β-Modified crystalline red pigment. Hoechst AG Patent DE 3 2223 888; US 4 719 292. [380t]

Schwarz, E. and de Buhr, J. (1998). Collected applications. Thermal analysis. Pharmaceuticals, Mettler-Toledo GmbH, Schwerzenbach. [148f, 149f, 153f]

Schwarz, E. and Pfeffer, S. (1997). Use of subambient DSC for liquid and semi solid dosage forms—Pharmaceutical product development and quality control. *J. Therm. Anal.*, **48**, 557–67. [366]

Scottish Dyes (1929a). Patent GB 322 169. [389]

Scottish Dyes (1929b). Patent DE 586 906. [389]

Seddon, K. R. (2004). Pseudopolymorphs: a polemic. *Cryst. Growth Des.*, **4**, 1087. [8]

Seki, T., Sakurada, K., Muromoto, M., and Ito, H. (2015). Photoinduced single-crystal-to-single-crystal transition and photosalient effect of a gold(I) isocyanide complex with shortening of intermolecular aurophilic bonds. *Chem. Sci.*, **6**, 1491–7. [320]

Selby, M. D., de Konig, P. D., and Roberts, D. F. (2011). A perspective on synthesis and solid-form enablement of inhalation candidates. *Fut. Med. Chem.*, **3**, 1679–701. [373]

Selig, W. (1982). New adducts of octahydro-1,3,5,7-tetranitro-1,3,5,7-tetrazocine (HMX). *Propellants, Explos., Pyrotech.*, **7**, 70–7. [403]

Senechal, M. (1990). Brief history of geometrical crystallography. In *Historical Atlas of Crystallography* (ed. J. Lima-de-Faria), pp. 43–59. Kluwer Academic Publishers, Dordrecht. [20]

Senju, T., Hoki, T., and Mizuguchi, J. (2005a). 2,9-Dichloro-5-12-dihydroquino[2,3-*b*] acridine-7,14-dione. *Acta Crystallogr. E*, **61**, o1061–3. [385]

Senju, T., Hoki, T., and Mizuguchi, J. (2006). 3,10-Dichloro-5,12-dihydroquino[2,3-*b*] acridine-7,14-dione. *Acta Crystallogr. E*, **62**, o261–3. [382t]

Senju, T., Nishimura, N., Hoki, T., and Mizuguchi, J. (2005b). 2,9-Dichloro-5-12-dihydroquino[2,3-*b*]acridine-7,14-dione (red phase). *Acta Crystallogr. E*, **61**, o2596–8. [381t, 385]

Senju, T., Sakai, M., and Mizuguchi, J. (2007). Cohesion of γ-quinacridone and 2,9-dimethylquinacridone in the solid state. *Dyes Pigm.*, **75**, 449–53. [380t]

Seo, J., Jo, W. J., Choi, G., Park, K.-M., Lee, S. S., and Lee, J. S. (2007). X-ray crystal structure of C.I. Disperse Brown 1. *Dyes Pigm.*, **72**, 327–30. [379]

Serajuddin, A. T. M., Thakur, A. B., Ghoshal, R. N., Fakes, M. G., Ranadive, S. A., Morris, K. R., and Varia, S. A. (1999). Selection of solid dosage form composition through drug–excipient compatibility testing. *J. Pharm. Sci.*, **88**, 696–704. [363]

Serbutoviez, C., Nicoud, J.-F., Fischer, J., Ledoux, I., and Zyss, J. (1994). Crystalline zwitterionic stilbazolium derivatives with large quadratic optical nonlinearities. *Chem. Mater.*, **6**, 1358–68. [303]

Sergio, S. T. (1978). *Studies of the polymorphs of RDXI*. M.Sc. Thesis, University of Delaware, Newark, DE. [410]

Seyer, J. J., Luner, P. E., and Kemper, M. S. (2000). Application of diffuse reflectance near infrared spectroscopy for determination of crystallinity. *J. Pharm. Sci.*, **89**, 1305–16. [371]

Shah, J. C., Chen, J. R., and Chow, D. (1999). Metastable polymorph of etoposide with higher dissolution rate. *Drug Dev. Ind. Pharm.*, **25**, 63–7. [346, 367]

Shahar, C., Dutta, S., Weissman, H., Shimon, L., Ott, H.,. and Rybtchinski, B. (2016). Precrystalline aggregates enable control over organic crystallization in solution. *Angew. Chem. Int. Ed. Engl.*, **55**, 179–82. [84]

Shaibat, M. A., Casablanca, L. B., Siberio-Perez, D. Y., and Matzger, A. J. (2010). Distinguishing polymorphs of the semiconducting pigment copper phthalocyanine by solid-state NMR and Raman spectroscopy. *J. Phys. Chem.*, **114**, 4400–6. [193]

Shaik, S. S. (1982). On the stability and properties of organic metals and their isomeric charge-transfer complexes. *J. Am. Chem. Soc.*, **104**, 5328–34. [275]

Shalaev, E. Y. and Zografi, G. (1996). Interrelationships between phase transformation and organic chemical reactivity in the solid state. *J. Phys. Org. Chem.*, **9**, 729–38. [319]

Shamshina, J. L., Barber, P. S., and Rogers, R. (2013). Ionic liquids in drug delivery. *Expert Opin. Drug. Deliv.*, **10**, 1367–80. [356]

Shankland, K., David, W. I. F., and Sivia, D. S. (1997). Routine *ab initio* structure determination of chlorothiazole by X-ray powder diffraction using optimized data collection and analysis strategies. *J. Mater. Chem.*, **7**, 569–72. [155]

Sharma, B. D. (1987). Allotropes and polymorphs. *J. Chem. Educ.*, **64**, 404–7. [5, 29]

Sharma, S. and Radhakrishnan, T. P. (2000). Modeling polymorphism-solvated supramolecular clusters reveal the solvent selection of SHG active and inactive dimorphs. *J. Phys. Chem. B*, **104**, 10191–5. [303]

Sharp, J. H., Miller, R. L., and Lardon, M.A. (1972). Xerox Corporation Patent US 3 862 127; DE-OS 1 944 021 (1970); GB 1 268 422 (1972); FR 2 016 641 (1972). [389]

Sharpe, S. A., Celik, M., Newman, A. W., and Brittain, H. G. (1997). Physical characterization of the polymorphic variations of magnesium stearate and magnesium palmitate hydrate species. *Struct. Chem.*, **8**, 73–84. [364]

Shekunov, B. Y. and York, P. (2000). Crystallization processes in pharmaceutical technology and drug delivery design. *J. Cryst. Growth*, **211**, 122–36. [371]

Shelyapin, O. P., Pushkina, L. L., Kaugina, T. I., Tonchilova, V. F., Yaroshevich, G., and Peskova, V. I. (1987). *cis*-Naphthoylenebisbenzimidazole in a β-modification. Patent SU 1 310 415; *Chem. Abstr.*, **107**, 219146p. [381t]

Shenmin, Z., Dongzhi, L., and Shengwu, R. (1992). A study of the synergism and crystal form of some dichlorobenzidine disazo yellow pigments. *Dyes Pigm.*, **18**, 137–49. [382t, 395, 396f, 397f]

Sheokand, S., Modi, S. R., and Bansal, A. K. (2014). Dynamic vapor sorption as a tool for characterization and quantification of amorphous content in predominantly crystalline materials. *J. Pharm. Sci.*, **103**, 3364–76. [371]

Sherwood, J. N. and Gallagher, H. G. (1984). *The influence of lattice imperfections on the chemical reactivity of solids*. Final Technical Report, European Research Office of the U.S. Army, Contract Number DAJA-37-81-M-0395. Avail. NTVS Gov. Rep. Announce. Index (NS) 1984, **84**, 83. [416]

Sheskey, P. J., Cook, W. G., and Cable, C. B. (eds.) (2017). *Handbook of Pharmaceutical Ingredients*, 8th edn. Pharmaceutical Press, London. [363, 364]

Shevchenko, A., Belle, D. D., Tittanen, S., Karjalainen, A., Tolvanen, A., Tanninen, V. P., Haarala, J., Makela, M., Vikrussi, J., and Miroshnyk, I. (2011). Coupling polymorphism/solvatomorphism and physical stability evaluation with early salt synthesis optimization of an investigational drug. *Org. Process Res. Dev.*, **15**, 666–72. [356]

Shibaeva, R. P., Yagubskii, E. B., Laukhina, E. E., and Laukhin, V. N. (1990). Organic conductors and superconductors based on (BEDT-TTF)-polyiodides. In *The Physics and Chemistry of Organic Superconductors* (ed. G. Saito and S. Kagoshima), pp. 342–8. Springer-Verlag, Berlin. [118]

Shibuya, Y., Itoh, Y., and Aida, T. (2017). Jumping crystals of pyrene tweezers: crystal-to-crystal transition involving π/π-to-CH/π assembly mode switching. *Chem. Asian J.*, **12**, 811–15. [320]

Shiftan, D., Ravenelle, F., Mateescu, M. A., and Marchessault, R. H. (2000). Change in the V/B polymorph ratio and T-1 relaxation of epichlorohydrin crosslinked high amylose starch excipient. *Starch-Starke*, **52**, 186–95. [363]

Shigorin, V. D. and Shipulo, G. P. (1972). Use of the effect of second harmonic generation during a determination of crystal symmetry. *Kvantovaya Electron. (Moscow)*, **4**, 116–18. [302]

Shimura, K., Tada, K., and Imai, H. (1988). Crystal transformation of benzanthrone compound—shown by X-ray diffraction pattern, giving good dyeing properties. Nippon Kayaku, Co. Ltd. Patent JP 63010672; *Chem. Abstr.*, **109**, 8033c. [383t]

Shishkin, O. V., Dyakonenko, V. V., and Malev, A. V. (2012a). Supramolecular architecture of crystals of fused hydrocarbons based on topology of intermolecular interactions. *CrystEngComm*, **12**, 1795–804. [269]

Shishkin, O. V., Medvedev, V. V., Zubatyuk, R. L., Shyshkina, O. O., Kovalenko, N. V., and Volovenko, J. M. (2012b). Role of different molecular fragments in formation of the supramolecular architecture of the crystal of 1,1-dioxo-tetrahydro-1λ^6–thyopyran-3-one. *CrystEngComm*, **14**, 8698–707. [269]

Shriner, R. L., Hermann, C. K. F., Morrill, T. C., Curtin, D. Y., and Fuson, R. C. (1997). *The Systematic Identification of Organic Compounds*, 7th edn. John Wiley & Sons, New York. [133]

Shtanov, N. P., Moroz, V. A., Tikhonov V. I. and Rogovik, V. I. (1981). Relation of structure to the coloring properties of the pigment perylene scarlet. Khim. Tekhnol. (Kiev) 1981, 19–20; Chem. Abstr., 94, 193745v. [261t]

Shtanov, N. P., Moroz, V. A., and Tikhonov, V. I. (1980). Identification of polymorphic forms of the pigment Bordeaux anthraquinone. *Khim. Tekhnol. (Kiev)*, **1980**, 23–5; *Chem. Abstr.*, **93**, 96873f. [381t]

Silvestri, H. H. and Johnson, D. A. v. Grant, N. H. and Alburn, H. (1974). U.S. Court of Customs and Patent Appeals, Patent Appeal No. 8978; 496 F.2d 593. [438]

Simon, F., Clevers, S., Dupray, V., and Coquerel, G. (2015). Relevance of the second harmonic generation to characterize crystalline samples. *Chem. Eng. Technol.*, **38**, 971–83. [359]

Simone. E., Cenzato, M. V., and Nagy, Z. K. (2016). A study on the effect of the polymeric additive hydroxypropyl methylcellulose HPMC on morphology and polymorphism of ortho-aminobenzoic acid crystals. *J. Cryst. Growth*, **446**, 50–9. [199, 199f, 200f, 201f, 202f]

Simons, L. J., Caprather, B. W., Callahan, M., Graham, J. M., Kimura, T., Lai, Y., LeVine, H., Lipinski, W., Sakkab, A. T., and Tasaki, Y. (2009). The synthesis and structure–activity relationship of substituted *N*-phenyl anthranilic acid analogs as amyloid aggregation inhibitors. *Biorg. Med. Chem. Lett.*, **19**, 654–7. [245]

Singh, N. B., Singh, R. J., and Singh, N. P. (1994). Organic solid-state reactivity. *Tetrahedron*, **50**, 6441–93. [333]

Singhal, D. and Curatolo, W. (2004). Drug polymorphism and dosage form design: a practical perspective. *Adv. Drug Deliv. Rev.*, **56**, 335–47. [6]

Sirota, N. N. (1982). Certain problems of polymorphism (I). *Cryst. Res. Technol.*, **17**, 661–91. [16, 29]

Sitzmann, M. E., Gilardi, R., Butcher, R. J., Koppes, K. M., Stern, A. G., Thrasher, J. S., Trivedi, N. J., and Yang, Z. Y. (2000). Pentafluorosulfanylnitramide salts. *Inorg. Chem.*, **39**, 843–50. [415]

Skoko, Ž., Zamir, S., Naumov, P., and Bernstein, J. (2010). The thermosalient phenomenon. "Jumping crystals" and crystal chemistry of the anticholinergic agent oxitropium bromide. *J. Am. Chem. Soc.*, **132**, 14191–202. [320]

Smith, D. A. (1993). *Being an Effective Expert Witness*. Thames Publishing, London. [429]

Smith, D. L. (1974). Structure of dyes and aggregates. Evidence from crystal structure analysis. *Photogr. Sci. Eng.*, **18**, 309–22. [328]

Smith, F. J., Armstrong, A. T., and McGlynn, S. P. (1966). Energy of excimer luminescence. V. Excimer fluorescence of naphthalene and methylnaphthalenes. *J. Chem. Phys.*, **44**, 442–8. [330]

Smith, G. D. and Bharadwaj, R. K. (1999). Quantum chemistry based force field for simulations of HMX. *J. Phys. Chem. B*, **103**, 3570–5. [406]

Smith, G. P. S., McGoverin, C. M., Fraser, S. J., and Gordon, K. C. (2015). Raman imaging of drug delivery systems. *Adv. Drug Deliv. Syst.*, **89**, 21–41. [209]

Smith, J., MacNamara, E., Raftery, D., Borchardt, T., and Byrn, S. (1998). Application of two-dimensional ^{13}C solid-state NMR to the study of conformational polymorphism. *J. Am. Chem. Soc.*, **120**, 11710–13. [222]

Smith, K. W., Bhaggan, K., Talbot, G., and Van Malssen, K. F. (2011). Crystallization of fats: influence of minor components and additives. *J. Am. Oil Chem. Soc.*, **88**, 1085–101. [62]

Smithells, C. J. (ed.) (1976). *Metals Reference Book*, 5th edn. Butterworths, London. [29, 62]

Sobol', L. (1967). On the distribution of points in a cube and the approximate value of integrals. *USSR Comput. Math. Math. Phys.*, **7**, 86–112. [243]

Sodikoff, B. J., Masar, M. S., and Garrett, C. D. (2015). What form is it? Lessons from 13 polymorph pharmaceutical cases. *Pharm. Law Ind. Rep.*, **13**, April 17. [430, 434]

Sommer, K. and Kruse, H. (1979). Modification of monoazo dispersion dye of benzene series-with improved dyeing stability. Hoechst AG Patent DE 2 835 544. [379t]

Sommer, R., Schulze, J., and Wolfrum, G. (1974). Exhaust dyeing of (semi) synthetic fibers—uniform yellow shades with a modified diazo dye from organic solvents. Bayer AG Patent DE 2 313 356. [379t]

Soni, P., Sarkar, C., Tewari, R., and Sharma, T. D. (2011). HMX polymorphs: gamma to beta phase transformation. *J. Energ. Mater.*, **29**, 261–79. [407]

Sorescu, D. C. and Rice, B. M. (2016). RDX compression, alpha → gamma phase transition, and shock Hugoniot calculations from density-functional-theory-based molecular dynamics simulation. *J. Phys. Chem.*, **120**, 19547–57. [399]

Sorescu, D. C., Rice, B. M., and Thompson, D. L. (1997). Intermolecular potential for the hexahydro-1,3,5,7-trinitro-1,3,5-s-triazine crystal: a crystal packing, Monte Carlo, and molecular dynamics study. *J. Phys. Chem. B*, **101**, 798–808. [409]

Sorescu, D. C., Rice, B. M., and Thompson, D. L. (1998a). Molecular packing and NPT-molecular dynamics investigation of the transferability of the RDX intermolecular potential to 2,4,6,8,10,12-hexanitrohexaazaisowurtzitane. *J. Phys. Chem. B*, **102**, 948–52. [403, 407, 409]

Sorescu, D. C., Rice, B. M., and Thompson, D. L. (1998b). Isothermal–isobaric molecular dynamics simulations of 1,3,5,7-tetranitro-1,3,5,7-tetraazacyclooctane (HMX) crystals. *J. Phys. Chem. B*, **102**, 6692–5. [406, 409]

Sorescu, D. C., Rice, B. M., and Thompson, D. L. (1999). Theoretical studies of the hydrostatic compression of RDX, HMX, HNIW, and PETN crystals. *J. Phys. Chem. B*, **103**, 6783–90. [411]

Spackman, M. A. (2004). Personal communication to Joel Bernstein. [249f, 250f, 251f, 252t, 252f, 253f, 254f]

Spackman, M. A. and Byrom, P. G. (1997). A novel definition of a molecule in a crystal. *Chem. Phys. Lett.*, **267**, 215–20. [248]

Spackman, M. A. and Jayatilaka, D. (2009). Hirshfeld surface analysis. *CrystEngComm*, **11**, 19–32. [248]

Spackman, M. A. and McKinnon, J. J. (2002). Fingerprinting intermolecular interactions in molecular crystals. *CrystEngComm*, **4**, 378–92. [255]

Spackman, M. A., McKinnon, J. J., and Jayatilka, D. (2008). Electrostatic potentials mapped on Hirshfeld surfaces provide direct insight into intermolecular interactions in crystals. *CrystEngComm*, **10**, 377–88. [255]

Spek, A. L. (1990). Platon, an integrated tool for the analysis of the results of a single crystal structure determination. *Acta Crystallogr. A*, **46**, C34. [239]

Spietschta, E. and Tröster, H. (1988a). Mix-crystal pigments based on perylenetetracarbamides, process for preparing and their use. Hoechst AG Patent DE 3 436 206; US 4 769 460. [381t]

Spietschta, E. and Tröster, H. (1988b). Mix-crystal pigments based on perylenetetracarbamides, process for their preparation, and their use. Hoechst AG Patent DE 3 436 209; US 4 742 170. [381t]

Srinivasan, K. and Dhanasekaran, P. (2011). Separation and nucleation control of alpha and beta polymorphs of L-glutamic acid by swift cooling crystallization process. *Amino Acids*, **40**, 1257–60. [87]

Stahly, G. P. (2007). Diversity in single- and multiple-component crystals. The search for and prevalence of polymorphs and co-crystals. *Cryst. Growth Des.*, **7**, 1007–26. [6, 8, 17, 17t, 92, 94, 95f, 96, 97, 99, 369]

Stam, C. H. (1972). Crystal structure of a monoclinic modification and the refinement of a triclinic modification of vitamin A acid (retinoic acid), $C_{20}H_{28}O_2$. *Acta Crystallogr. B*, **28**, 2936–45. [223]

Stam, C. H. and MacGillavry, C. H. (1963). The crystal structure of the triclinic modification of vitamin A acid. *Acta Crystallogr.*, **16**, 62–8. [223]

Stanley-Wood, N. G. and Riley, G. S. (1972). The effect of temperature on the physical nature of phenobarbitone produced by acid–base precipitation. *Pharm. Acta Helv.*, **47**, 58–64. [374]

Starbuck, J., Docherty, R., Charlton, M. H., and Buttar, D. (1999). A theoretical investigation of conformational aspects of polymorphism. Part 2. Diarylamines. *J. Chem. Soc., Perkin Trans. 2*, 677–91. [219]

Starr, T. L. and Williams, D. E. (1977a). Comparison of models for molecular hydrogen–molecular hydrogen and molecular hydrogen–helium anisotropic intermolecular repulsions. *J. Chem. Phys.*, **66**, 2054–7. [217]

Starr, T. L. and Williams, D. E. (1977b). Coulombic nonbonded interatomic potential functions derived from crystal-lattice vibrational frequencies in hydrocarbons. *Acta Crystallogr. A*, **33**, 771–6. [217]

Stawasz, M. E., Sampson, D. L., and Parkinson, B. A. (2000). Scanning tunneling microscopy investigation of the ordered structures of dialkylamino hydroxylated squaraines absorbed on highly oriented pyrolytic graphite. *Langmuir*, **16**, 2326–42. [293]

Steed, J. W. (2013). The role of co-crystals in pharmaceutical design. *Trends Pharm. Sci.*, **34**, 185–93. [360]

Steed, K. M. and Steed, J. W. (2015). Packing problems: high Z' crystal structures and their relationship to cocrystals, inclusion compounds and polymorphism. *Chem Rev.*, **115**, 2895–933. [218]

Steiner, T., Hinrichs, W., Saenger, W., and Gigg, R. (1993). 'Jumping crystals': X-ray structures of the three crystalline phases of (+/−)-3,4-di-*O*-acetyl-1,2,5,6-tetra-*O*-benzyl-*myo*-inositol. *Acta Crystallogr. C*, **49**, 708–19. [311, 319]

Steinmann, S. S., Piemontesi, A., Delachat, C., and Corminboeuf, C. (2012). Why are interaction energies of charge-transfer complexes challenging for DFT? *J. Chem. Theor. Comput.*, **8**, 1629–40. [218]

Steinmetz, H. (1915). Crystallographic study of some nitro derivatives of benzene. *Z. Kristallogr. Cryst. Mater.*, **54**, 467–97. [302]

Stevens, L. L., Velisavijevic, N., Hooks, D. E., and Dattelbaum, D. M. (2008). The high-pressure behavior and compressibility of 2,4,6-trinitrotoluene. *Appl. Phys. Lett.*, **93**, 081812. [424]

Stezowski, J. J., Biedermann, P. U., Hildenbrand, T., Dorsch, J. A., Eckhardt, C. J., and Agranat, I. (1993). Overcrowded enes of the tricycloindane-1,3-dione series: Interplay of twisting, folding and pyramidalization. *J. Chem. Soc., Chem. Commun.*, 213–15. [312]

Stobbe, H. and Lehfeldt, A. (1925). Polymerization and depolymerization by light of different wavelengths. II. α and β-*trans* cinnamic acids, allo-cinnamic acids and their dimers. *Chem. Ber.*, **58B**, 2415–27. [332]

Stobbe, H. and Steinberger, F. K. (1922). Light reactions of the *cis*- and *trans*-cinnamic acids. *Chem. Ber.*, **55B**, 2225–45. [332]

Stockton, G. W., Godfrey, R., Hitchcock, P., Mendelsohn, R., Mowery, P. C., Rajan, S., and Walker, A. F. (1998). Crystal polymorphism in pendimathalin herbicide is driven by electronic delocalization in intramolecular hydrogen bonding. A crystallographic, spectroscopic and computational study. *J. Chem. Soc., Perkin Trans. 2*, 2061–71. [190, 219]

Stoltz, M., Oliver, D. W., Wessels, P. L., and Chalmers, A. A. (1991). High resolution solid state ^{13}C NMR spectra of mofebutazone, phenylbutazone and oxyphenbutazone in relation to X-ray crystallographic data. *J. Pharm. Sci.*, **80**, 357–62. [189]

Stone, A. J. (2005). Distributed multipole analysis: stability for large basis sets. *J. Chem. Theory Comput.*, **1**, 1128–32. [244]

Stone, A. J. (2013). *The Theory of Intermolecular Forces*, 2nd edn. Oxford University Press, Oxford. [217]

Störmer, H. and Laage, E. (1921). Truxillic acids. III. Natural and artificial truxillic and truxinic acids. *Chem. Ber.*, **54B**, 77–84. [332]

Störmer, R. and Förster, G. (1919). The truxillic acids and truxones. *Chem. Ber.*, **52B**, 1255–72. [322]

Stott, P. E., McCausland, C. W., and Parish, W. W. (1979). Polymorphic behavior and melting points of certain benzo crown ether compounds. *J. Heterocycl. Chem.*, **16**, 453–5. [222]

Stout, G. and Jensen, L. H. (1989). *X-ray Structure Determination. A Practical Guide*, p. 91. John Wiley & Sons, New York. [54, 59]

Stranski, I. N. and Totomanov, D. (1933). Rate of formation of (crystal) nuclei and the Ostwald step rule. *Z. Phys. Chem.*, **163**, 399–408. [57]

Streng, W. H. (1997). Physical chemical characterization of drug substances. *Drug Disc. Today*, **2**, 415–26. [137]

Strickland-Constable, R. F. (1968). *Kinetics and Mechanism of Crystallization*. Academic Press, London. [81]

Strohmeier, M., Orendt, A. M., Alderman, D. W., and Grant, D. M. (2001). Investigation of the polymorphs of dimethyl-3,6-dichloro-2,5-dihydroxyterephthalate by ^{13}C solid state NMR. *J. Am. Chem. Soc.*, **123**, 1713–22. [182]

Struve, W. S. (1959). U.S. Patent 2,844,485. [385]

Sudha, C. and Srinivasin, K. (2013). Supersaturation dependent nucleation control and separation of mono, ortho and unstable polymorphs of paracetamol by swift cooling crystallization techniques. *CrystEngComm*, **15**, 1914–21. [358]

Sudo, S., Sato, K., and Harano, Y. (1991). Growth and solvent-mediated phase transition of cimetidine polymorphic forms A and B. *J. Chem. Eng. Jpn.*, **24**, 628–32. [374]

Sugawara, T., Matsushita, M. M., Izuoka, A., Wada, N., Takeda, N., and Ishikawa, M. (1994). An organic ferromagnet: α-phase crystals of 2-(2′,5′-dihydroxyphenyl)-4,4,5,5-tetra-methyl-4,5-dihydro-1*H*-imidazolyl-1-oxy-3-oxide (α-HQNN). *J. Chem. Soc., Chem. Commun.*, 1723–4. [289]

Sugeta, H. (1975). Normal vibrations and molecular conformations of dialkyl disulfides. *Spectrochim. Acta A*, **31**, 1729–37. [322]

Sugeta, H., Go, A., and Miyazawa, T. (1972). S–S and C–S stretching vibrations and molecular conformations of dialkyl disulfides and cystine. *Chem. Lett.*, **1**, 83–6. [322]

Sugeta, H., Go, A., and Miyazawa, T. (1973). Vibrational spectra and molecular conformations of dialkyl disulfides. *Bull. Chem. Soc. Jpn.*, **46**, 3407–11. [322]

Sugita, Y.-H. (1988). Polymorphism of L-glutamic acid crystals and inhibitory substance for β-transition in beet molasses. *Agric. Biol. Chem.*, **52**, 3081–5. [40]

Suihko, E., Poso, A., Korhonen, O., Gynther, J., Ketolainen, J., and Paronen, P. (2000). Deformation behaviors of tolbutamide, hydroxypropyl-beta-cyclodextrin, and their dispersions. *Pharm. Res.*, **17**, 942–8. [375]

Suito, E. and Uyeda, N. (1963). Transformation and growth of a copper phthalocyanine crystal in organic suspension. *Kolloid Z. Z. Polym.*, **193**, 97–111. [391]

Suito, E. and Uyeda, N. (1974). *Kolloid Z. Z. Polym.*, **19**, 77. [391]

Sullivan, R. A. and Davey, R. J. (2015). Concerning the crystal morphologies of the α and β polymorphs of *p*-aminobenzoic acid. *CrystEngComm*, **17**, 1015–23. [200, 203f, 204f, 205f, 206f]

Summers, M. P., Enever, R. P., and Carless, J. E. (1977). Influence of crystal form on tensile strength of compacts of pharmaceutical materials. *J. Pharm. Sci.*, **66**, 1172–5. [375]

Sun, E. (1998). Excerpts from news conference, October 15, 1998. IAPAC, International Association of Physicians in AIDS Care, Chicago, IL. [342]

Suponitskii, K. Yu., Gusev, D. V., Kuleshova, L. N., and Antipin, M. Yu. (2002). X-ray structure investigation of two polymorphic modifications of 1-acetyl-3-(4-nitrophenyl)-5-(2′-furyl)pyrazoline. *Crystallogr. Rep.*, **47**, 610–15. [302]

Surana, R., Pyne, A., and Suryanarayanan, R. (2004). Effect of aging on the physical properties of amorphous trehalose. *Pharm. Res.*, **21**, 867–74. [371]

Suryanarayanan, R. and Wiedmann, T. S. (1990). Quantitation of the relative amounts of anhydrous carbamazepine ($C_{15}H_{12}N_2O$) and carbamazepine dihydrate ($C_{15}H_{12}N_2O\cdot2H_2O$) in a mixture by solid-state nuclear magnetic resonance (NMR). *Pharm. Res.*, **7**, 184–7. [190]

Susich, G. (1950). Identification of organic dyestuffs by X-ray powder diffraction. *Anal. Chem.*, **22**, 425–30. [377, 383t, 389, 393]

Süsse, P. and Wolf, A. (1980). A new crystalline phase of indigo. *Naturwissenschaften*, **67**, 453. [383t]

Süsse, P., Steins, M., and Kupcik, V. (1988). Indigo: crystal structure refinement based on synchrotron data. *Z. Kristallogr. Cryst. Mater.*, **184**, 269–73. [383t]

Sussich, F. and Cesaro, A. (2000). Transitions and phenomenology of α,α-trehalose polymorphs interconversion. *J. Therm. Anal. Calorim.*, **62**, 757–68. [363]

Suzuki, E., Shirotani, K.-I., Tsuda, Y., and Sekiguchi, K. (1985). Water content and dehydration behavior of crystalline caffeine hydrate. *Chem. Pharm. Bull.*, **33**, 5028–35. [161, 170]

Suzuki, M. and Ogaki, T. (1986). Crystallization and transformation mechanisms of α-, β- and γ-polymorphs of ultra-pure oleic acid. *J. Am. Oil Chem. Soc.*, **62**, 1602–4. [40]

Suzuki, Y., Fujita, T., Hayashi, Y., and Okayasu, H. (1989a). New crystal type copper phthalocyanine—is useful in ink, paint and coloring material for resins as high temperature use organic pigment. Sumitomo Chemical Co. Ltd. Patent JP 1 153 758; *Chem. Abstr.*, **111**, 235058z. [390]

Suzuki, Y., Fujita, T., Hayashi, Y., and Okayasu, H. (1989b). Novel crystal type copper phthalocyanine blue pigment—obtained from phthalic anhydride, urea, cuprous chloride, titanium tetrachloride and sulphophthalic acid. Sumitomo Chemical Co. Ltd. Patent JP 1 153 756. *Chem. Abstr.*, **111**, 235057y. [390]

Sweeting, L., Rheingold, A. L., Gingerich, J. M., Rutter, A. W., Spence, R. A., Cox, C. D., and Kim, T. J. (1997). Crystal structure and triboluminescence. 2,9-anthracenecarboxylic acid and its esters. *Chem. Mater.*, **9**, 1103–15. [315]

Szafranski, M. and Katrusiak, A. (2000). Phase transitions in the layered structure of diguanidinium tetraiodoplumbate. *Phys. Rev. B*, **61**, 1026–35. [336]

Szafranski, M., Czarnecki, P., Katrusiak, A., and Habrylo, S. (1992). DTA investigation of phase-transitions in 1,3-cyclohexanedione under high-pressures. *Solid State Commun.*, **82**, 277–81. [336]

Szelagiewicz, M., Marcoli, C., Cianferani, S., Hard, A. P., Vit, A., Burkhard, A., von Raumer, M., Hofmeier, U. C., Zilian, A., Francotte, E., and Schenker, R. (1999). In situ characterization of polymorphic forms: the potential of Raman techniques. *J. Therm. Anal. Calorim.*, **57**, 23–43. [365]

Szeverenyi, N. M., Sullivan, M. J., and Maciel, G. E. (1982). Observation of spin exchange by two-dimensional Fourier transform carbon-13 cross-polarization—magic angle spinning. *J. Magn. Reson.*, **47**, 462–75. [189]

Tabei, H., Kurihara, T., and Kaino, T. (1987). Recrystallization solvent effects on second-order nonlinear optical organic materials. *Appl. Phys. Lett.*, **50**, 1855–7. [303]

Talaczynska, A., Dzitko, J., and Cielecka-Piontek, J. (2016). Benefits and limitations of polymorphic and amorphous forms of active pharmaceutical ingredients. *Curr. Pharm. Des.*, **22**, 4975–80. [370]

Tammann, G. (1903). *Kristallisieren und Schmelzen*. Verlag von Johann Ambrosius Barth, Leipzig. [36]

Tammann, G. (1926). *The States of Aggregation* (trans. F. F. Mehl), pp. 116–57. Constable and Company, Ltd., London. [36, 37, 42, 52]

Tammann, G., Elsner, H., and von Gronow, H. E. (1931). Über die spontane Kristallisation unterkühlter schmelzen und übersättigter Lösungen. *Z. Anorg. Allg. Chem.*, **200**, 57–73. [80]

Tamura, M., Nakazawa, Y., Shiomi, D., Nozawa, K., Hosokoshi, Y., Ishikawa, M., Takahashi, M., and Kinoshita, M. (1991). Bulk ferromagnetism in the β-phase crystal of the *p*-nitrophenyl nitronyl nitroxide radical. *Chem. Phys. Lett.*, **186**, 401–4. [288]

Tamura, M., Shiomi, D., Hosokoshi, Y., Iwasawa, N., Nozawa, K., Kinoshita, M., Sawa, H., and Kato, R. (1993). Magnetic properties of phenyl nitronyl nitroxide. *Mol. Cryst. Liq. Cryst. Sci. Technol. A*, **232**, 45–52. [288]

Taneda, M., Amimoto, K., Koyama, H., and Kawato, T. (2004). Photochromism of polymorphic 4,4′-methylenebis(*N*-salicylidene-2,6-diisopropylaniline) crystals. *Org. Biol. Chem.*, **2**, 499–504. [310]

Tanninem, V. P. and Yliruusi, J. (1992). X-ray powder diffraction profile fitting in quantitative determination of two polymorphs from their powder mixture. *Int. J. Pharmacol.*, **81**, 169–77. [169]

Tarling, S. E. (1992). Declaration to European Patent Office regarding action of the Opposition Division dated 23rd April 1992 in relation to EP 0 091 787. [454]

Tarling, S. E. (2001). Private communication to the author; see also http://www.vino. demon.co.uk/ppxrd/cefad.html [436, 455f]

Tausen, H. (1935). Process of making basic lead trinitroresorcinol. U.S. Patent 2 020 665. [426]

Taylor, J. W. and Crookes, R. J. (1976). Vapor pressure and enthalpy of sublimation of 1,3,5,7-tetranitro-1,3,5,7-tetraazacyclooctane. *J. Chem. Soc., Faraday Trans. 1*, 723–30. [404]

Taylor, L. S. and Shamblin, S. L. (2009). Amorphous solids. In *Polymorphism in Pharmaceutical Solids* (ed. H. G. Brittain), pp. 587–629. Informa Healthcare, New York. [370]

Taylor, L. S. and Zografi, G. (1998). The quantitative analysis of crystallinity using FT-Raman spectroscopy. *Pharm. Res.*, **15**, 755–61. [371]

Teetsov, A. S. and McCrone, W. C. (1965). The microscopical study of polymorph stability diagrams. *Microsc. Cryst. Front.*, **15**, 13–29. [403, 404t]

Terao, H., Itoh, Y., Ohno, K., Isogai, M., Kakuta, A., and Mukoh, A. (1990). Second order nonlinear optical properties and polymorphism of benzophenone derivatives. *Opt. Commun.*, **74**, 451–3. [302]

Terol, A., Cassanas, G., Nurit, J., Pauvet, B., Bouassab, A., Rambaud, J., and Chevallet, P. (1994). Infrared, Raman and ¹³C NMR spectra of two crystalline forms of (1R,3S)-3-(*p*-thioanisoyl)-1,2,2-trimethylcyclopentanecarboxylic acid. *J. Pharm. Sci.*, **83**, 1437–42. [177]

Terpstra, P. and Codd, L. W. (1961). *Crystallometry*. Academic Press, New York. [138]

Teteruk, J. L., Glinnemann, J., Heyse, W., Johansson, K. E., van de Streek, J., and Schmidt, M. U. (2016). Local structure in the disordered solid solution of *cis*- and *trans*-perinones. *Acta Crystallogr. B*, **72**, 416–33. [381t]

Thai HVAC (2001). HVAC system – Thai Takasago Co., Ltd. Available at http://www. thaitakasago.co.th [134]

Thakar, A. L., Hirsch, C. A., and Page, J. G. (1977). Solid dispersion approach for overcoming bioavailability problems due to polymorphism of nabilone, a cannabinoid derivative. *J. Pharmacol. Pharm.*, **29**, 783–4. [347]

Thakur, T. S., Dubey, R., and Desiraju, G. R. (2015). Crystal structure and prediction. *Annu. Rev. Phys. Chem.*, **66**, 21–42. [131, 434, 441]

Thakuria, R., Eddleston, M. D., Chow, E. H. H., Lloyd, G. O., Aldous, B. J., Krzyzaniak, J. F., Bond, A. D., and Jones, W. (2013). Use of in situ atomic force microscopy to follow phase changes at crystal surfaces in real time. *Angew. Chem. Int. Ed. Engl.*, **52**, 519–32. [352]

Thenard, L. J. and Biot, J. B. (1809). Mémoire sur l'analyse comparée de l'arragonite, et du carbonate de chaux rhomboidal. *Chem. Phys. II, Soc. D'Arcueil*, **2**, 176–206. [31]

Theocharis, C. R. and Jones, W. (1984). The thermally induced phase transition of crystalline 9-cyanoanthracene dimer: a single crystal study. *J. Chem. Soc., Chem. Commun.*, 369–84. [330, 237]

Theocharis, C. R., Jones, W., and Rao, C. N. R. (1984). An unusual photo-induced conformational polymorphism: a crystallographic study of bis(p-methoxy)-*trans*-stilbene. *J. Chem. Soc., Chem. Commun.*, 1291–3. [234]

Thomas, J. M. (1979). Organic reactions in the solid state: accident and design. *Pure Appl. Chem.*, **51**, 1065–82. [333]

Thomas, J. M., Morsi, S. E., and Desvergne, J. P. (1977). Topochemical phenomena in organic solid-state chemistry. *Adv. Phys. Org. Chem.*, **15**, 63–151. [333]

Thompson, K. C. (2000). Pharmaceutical applications of calorimetric measurements in the new millennium. *Thermochim. Acta*, **355**, 83–7. [366]

Threlfall, T. (1995). Analysis of organic polymorphs. A review. *The Analyst*, **120**, 2435–60. [7, 8, 10, 12, 126, 137, 147, 150, 170, 174, 184, 343]

Threlfall, T. and Hursthouse, M. (2000). Private communication to author. [120]

Threlfall, T. L. (1999). Developments in analysis of polymorphs and solvates. Presented at the 1st International Symposium on Aspects of Polymorphism and Crystallization—Chemical Development Issues, Hinckley, UK, June 23–25, 1999. Scientific Update, Mayfield, UK. [163f, 170, 189f, 374]

Threlfall, T. L. (2000). Crystallization of polymorphs: thermodynamic insight into the role of the solvent. *Org. Process Res. Dev.*, **4**, 384–90. [3, 368, 374]

Thun, J., Seyfarth, L., Butterhof, C., Senker, J., Dinnebier, R. E., and Breu, J. (2009). Wöhler and Liebig revisited: 176 years of polymorphism in benzamide – and the story still continues! *Cryst. Growth Des.*, **9**, 2435–41. [156]

Tian, F. and Rantanen, J. (2011). Perspective of water of crystallization affecting the functionality of pharmaceuticals. *Food Biophys.*, **6**, 250–8. [355]

Tian, F., Qu, H. Y., Louhi-Kultanen, M., and Rantanen, J. (2010a). Insight into crystalline mechanisms of polymorphic hydrate systems. *Chem. Eng. Technol.*, **33**, 833–8. [355, 373]

Tian, F., Qu, H. Y., Zimmerman, A., Munk, T., Jorgenson, A. C., and Rantanen, J. (2010b). Factors affecting crystallization of hydrates. *J. Pharm. Pharmacol.*, **62**, 1535–46. [355, 373]

Tidey, J. P., O'Connor, A. E., Markevich, A., Bichoutskaia, E., Cavan, J. J. P., Lawrance, G. A., Wong, H. L. S., McMaster, J., Schroeder, M., and Blake, A. J. (2015). The epitaxial retrieval of a disappearing polymorph. *Cryst. Growth Des.*, **15**, 115–23. [134, 357]

Tilley, R. (1999). *Colour and Optical Properties of Materials*. John Wiley & Sons, Chichester. [377]

Timken, M. D., Chen, J. K., and Brill, T. B. (1990). Thermal decomposition of energetic materials. 37. SMATCH FT-IR (simultaneous mass and temperature-change FT-IR) spectroscopy. *Appl. Spectrosc.*, **44**, 701–6. [209]

Timofeeva, T. V., Chernikova, N. Yu., and Zorkii, P. M. (1980). Theoretical calculation of the spatial distribution of molecules in crystals. *Russ. Chem. Rev.*, **49**, 509–25. [237]

Timofeeva, T. V., Nesterov, V. N., Dolgushin, F. M., Zubavichus, Y. V., Goldshtein, J. T., Sammeth, D. M., Clark, R. D., Penn, B., and Antipin, M. Yu. (2000). One-pot polymorphism of nonlinear optical materials. First example of organic polytypes. *Cryst. Eng.*, **3**, 263–88. [303]

Tipson, R. S. (1956). Crystallization and recrystallization. In *Techniques of Organic Chemistry* (ed. A. Weissberger) Vol. III, 2nd edn, pp. 395–561. Interscience, New York. [61, 79]

Tishmack, P. A. (2009). Solid-state nuclear magnetic resonance spectroscopy. In *Polymorphism in Pharmaceutical Solids,* 2nd edn (ed. H. G. Brittain), pp. 381–435. Informa Healthcare, New York. [182]

Tishmack, P. A., Bugay, D. E., and Byrn, S. R. (2003). Solid-state nuclear magnetic resonance spectroscopy – pharmaceutical applications. *J. Pharm. Sci.*, **92**, 441–74. [182]

Tojo, K. and Mizuguchi, J. (2002a). Refinement of the crystal structure of 3,4:9,10-perylene-bis(dicarboximide), $C_{24}H_{10}N_2O_4$, at 263 K. *Z. Kristallogr. New Cryst. Struct.*, **217**, 45–6. [382t, 386]

Tojo, K. and Mizuguchi, J. (2002b). Refinement of the crystal structure of a 3,4:9,10-pery lenetetracarboxylic dianhydride $C_{24}H_8O_6$ at 223 K. *Z. Kristallogr. New Cryst. Struct.*, **217**, 253–4. [386]

Tojo, K. and Mizuguchi, J. (2002c). Refinement of the crystal structure of β-3,4:9,10-perylenetetracarboxylic dianhydride, $C_{24}H_8O_6$, at 223 K. *Z. Kristallogr. New Cryst. Struct.*, **217**, 255–6. [386]

Tokura, Y., Kaneko, Y., Okamoto, H., Tanuma, S., Koda, T., Mitani, T., and Saito, G. (1985). Spectroscopic study of the neutral-to-ionic phase transition in TTF-chloranil. *Mol. Cryst. Liq. Cryst.*, **125**, 71–80. [284]

Toma, P. H., Kelley, M. P., Borchardt, T. B., Byrn, S. R., and Kahr, B. (1994). Chloroisomers and polymorphs of 9-phenylacridinium hydrogen sulfate. *Chem. Mater.*, **6**, 1317–24. [303]

Tomita, Y., Ando, T., and Ueno, K. (1965). Three crystalline forms of iminodiacetic acid. *Bull. Chem. Soc. Jpn.*, **38**, 138–9. [222]

Torbeev, V. Yu., Shavit, E., Weissbuch, I., Leiserowitz, L., and Lahav, M. (2005). Control of crystal polymorphism by tuning the structure of auxiliary molecules as nucleation inhibitors. The β-polymorph of glycine grown in aqueous solutions. *Cryst. Growth Des.*, **5**, 2190–6. [101]

Torrance, J. B., Vasquez, J. E., Mayerle, J. J., and Lee, V. Y. (1981). Discovery of a neutral-to-ionic phase transition in organic materials. *Phys. Rev. Lett.*, **46**, 253–7. [283]

Toscani, S. (1998). An up-to-date approach to drug polymorphism. *Thermochim. Acta*, **321**, 73–9. [367]

Tremayne, M., Kariuki, B., and Harris, K. D. M. (1997). Structure determination of a complex organic solid from X-ray powder diffraction data by a generalized Monte Carlo method: the crystal structure of red fluorescein. *Angew. Chem. Int. Ed. Engl.*, **36**, 770–2. [155]

Trevor, J. E. (1902). The nomenclature of variance. *J. Phys. Chem.*, **6**, 136–7. [43]

Tristani-Kendra, M. and Eckhardt, C. J. (1984). Influence of crystal fields on the quasimetallic reflection spectra of crystals: optical spectra of polymorphs of a squarylium dye. *J. Chem. Phys.*, **90**, 1160–73. [294, 329f, 330]

Tristani-Kendra, M., Eckhardt, C. J., Bernstein, J., and Goldstein, E. (1983). Strong coupling in the optical spectra of a squarylium dye. *Chem. Phys. Lett.*, **98**, 57–61. [328, 328f]

Trotter, J. (1963). Bond lengths and angles in pentaerythritol tetranitrate. *Acta Crystallogr.*, **16**, 698–9. [411]

Tsuboi, M., Ueda, T., and Ikeda, T. (1991). Use of a Raman microscope in conformational analysis of a peptide with polymorphism. In *Proceedings of the Fourth European Conference on the Spectroscopy of Biological Molecules* (ed. R. E. Hester and R. B. Girling), pp. 29–30. Special Publication #94, Royal Society of Chemistry, London. [453]

Tuck, B., Stirling, J. A., Farnoochi, C. J., and McKay, R. B. (1997). Polymorph of a yellow diarylide pigment. European Patent EP 0790282. [381t]

Tudor, A. M., Church, S. J., Hendra, P. J., Davies, M. C., and Melia, C. D. (1993). The qualitative and quantitative analysis of chlorpropanide polymorphic mixtures by near-infrared Fourier transform Raman spectroscopy. *Pharm. Res.*, **10**, 1772–6. [176, 177]

Tudor, A. M., Davies, M. C., Melia, C. D., Lee, D. C., Mitchell, R. C., Hendra, P. J., and Church, S. J. (1991). The applications of near-infrared Fourier transform Raman spectroscopy to the analysis of polymorphic forms of cimetidine. *Spectrochim. Acta A*, **47**, 1389–93. [177]

Tung, M. and Gallagher, D. T. (2009). The Biomolecular Crystallization Database version 4: expanded content and new features. *Acta Crystallogr. D*, **65**, 18–23. [30]

Turek, P., Nozawa, K., Shiomi, D., Awaga, K., Inabe, T., Maruyama, M., and Kinoshita, M. (1991). Ferromagnetic coupling in a new phase of the *p*-nitrophenyl nitronyl nitroxide radical. *Chem. Phys. Lett.*, **180**, 327–31. [287]

Turner, M. J., Grabowsky, S., Jayatilaka, D., and Spackman, M. A. (2014). Accurate and efficient model energies for exploring intermolecular interactions in molecular crystals. *J. Phys. Chem. Lett.*, **5**, 4249–55. [269, 271f]

Turner, M. J., McKinnon, J. J., Jayatilaka, D., and Spackman, M. A. (2011). Visualization and characterization of voids in crystalline materials. *CrystEngComm*, **13**, 1804–13. [255]

Turner, M. J., Thomas, S. P., Shi, M. W., Jayatilaka, D., and Spackman, M. A. (2015). Energy frameworks: insights into interaction anisotropy and the mechanical properties of molecular crystals. *Chem. Commun.*, **51**, 3725–38. [270f]

Tutton, A. E. H. (1911a). The work of Eilhardt Mitscherlich and his discovery of isomorphism. In *Crystals*, pp. 70–97. Kegan Paul, Trench, Trubner & Co. Ltd., London. [30]

Tutton, A. E. H. (1911b). *Crystals*, p. 140, Kegan Paul, Trench, Trubner & Co. Ltd., London. [33, 36]

Tutton, A. E. H. (1922). *Crystallography and Practical Crystal Measurement*, Vol. 1, pp. 625–39. MacMillan, London. [54, 209]

Ullman, F. and Maag, R. (1906). Ueber Chinacridon. *Chem. Ber.*, **39**, 1693–6. [384]

Upadhyay, P., Dantuluri, A. K., Kumar, L., and Bansal, A. (2012). Estimating relative stability of polymorphs by generation of configurational free energy phase diagrams. *J. Pharm. Sci.*, **101**, 1843–51. [45]

Upadhyay, P., Khorriane, K. S., Kumar, L., and Bansal, A. K. (2013). Relationship between crystal structure and mechanical properties of ranitidine hydrochloride polymorphs. *CrystEngComm*, **15**, 3959–64. [320]

Urbelis, J. H. and Swift, J. A. (2014). Solvent effects on the growth morphology and phase purity of CL-20. *Cryst. Growth Des.*, **14**, 1642–9. [409]

Užarević, K., Rubčić, M., Đilović, I., Kokan, Z., Matković-Čalogović, D., and Cindrić, M. (2009). Concomitant conformational polymorphism: mechanochemical reactivity and phase relationships in the (methanol)*cis*-dioxo(*N*-salicylidene-2-amino-3-hydroxypyridine) molybdenum(VI) trimorph. *Cryst. Growth Des.*, **9**, 5327–33. [337]

Vachon, M. G. and Grant, D. J. W. (1987). Enthalpy–entropy compensation in pharmaceutical solids. *Int. J. Pharm.*, **40**, 1–14. [346]

van de Streek, J. and Motherwell, S. (2005). Searching the Cambridge Structural Database for polymorphs. *Acta Crystallogr. B*, **61**, 504–10. [27, 272]

van de Streek, J., Bruning, J., Ivashevskaya, S. N., Ermrich, M., Paulus, E. F., Bolte, M., and Schmidt, M. U. (2009). Structures of six industrial benzimidazolone pigments from laboratory powder diffraction data. *Acta Crystallogr.*, **65**, 200–11. [380t, 382t]

van den Ende, J. A., Ensing, B., and Cuppen, H. M. (2016). Energy barriers and mechanisms in solid–solid polymorphic transitions exhibiting cooperative motion. *CrystEngComm*, **18**, 4420–30. [320]

van der Sluis, P. and Kroon, J. (1989). Solvents and X-ray crystallography. *J. Cryst. Growth*, **97**, 645–56. [8]

van Eijck, B. and Kroon, J. (1999). UPACK program package for crystal structure prediction: force field and crystal structure generation for small carbohydrate molecules. *J. Comput. Chem.*, **20**, 799–812. [238, 240]

van Eijck, B., Mooij, W. T. M., and Kroon, J. (1995). Attempted prediction of the crystal structures of six mono-saccharides. *Acta Crystallogr. B*, **51**, 99–103. [238]

van Eijck, B., Mooij, W. T. M., and Kroon, J. (2001). Ab initio crystal structure predictions for flexible hydrogen-bonded molecules. Part II. Accurate energy minimization. *J. Comput. Chem.*, **22**, 805–15. [245]

van Enckevort, W. J. P. and Los, J. H. (2008). "Tailor-made" inhibitors in crystal growth: a Monte Carlo simulation study. *J. Phys. Chem.*, **112**, 6380–9. [63]

van Hook, A. (1961). *Crystallization in theory and practice*. Reinhold, New York, and references therein. [79, 87]

Van Wart, H. E. and Scheraga, H. A. (1976a). Raman spectra of cystine-related disulfides. Effect of rotational isomerism about carbon–sulfur bonds on sulfur–sulfur stretching frequencies. *J. Phys. Chem.*, **80**, 1812–22. [322]

Van Wart, H. E. and Scheraga, H. A. (1976b). Raman spectra of strained disulfides. Effect of rotation about sulfur–sulfur bonds on sulfur–sulfur stretching frequencies. *J. Phys. Chem.*, **80**, 1823–32. [322]

Van Wart, H. E. and Scheraga, H. A. (1977). Stable conformations of aliphatic disulfides: influence of 1,4 interactions involving sulfur atoms. *Proc. Natl. Acad. Sci. U.S.A.*, **74**, 13–17. [322]

Van Wart, H. E. and Scheraga, H. A. (1986). Agreement with the disulfide stretching frequency-conformation correlation of Sugeta, Go, and Miyazawa. *Proc. Natl. Acad. Sci. U.S.A.*, **83**, 3064–7. [322]

Vanhoorne, V., Bekaert, B., Peeters, E., De Beer, T., Remon, J.-P., and Vervaet, C. (2016). Improved tabletability after a polymorphic transition of delta-mannitol during twin-screw granulation. *Int. J. Pharm.*, **506**, 13–24. [364]

Variankaval, N., Cote, A. S., and Doherty, M. F. (2008). From form to function: crystallization of active pharmaceutical ingredients. *AIChE J.*, **54**, 1682–8. [375]

Varughese, S. (2014). Non-covalent routes to tune optical properties of molecular materials. *J. Mater. Chem. C*, **2**, 3499–516. [115, 337]

Vasseur, K., Rand, B. P., Cheyns, D., Temst, K., Froyen, L., and Heremans, P. (2012). Correlating the polymorphism of titanyl phthalocyanine thin films with solar cell performance. *J. Phys. Chem. Lett.*, **3**, 2395–400. [294]

Veciana, J., Cirujeda, J., Rovira, C., Molins, E., and Novoa, J. J. (1996). Organic ferromagnets. Hydrogen bonded supramolecular magnetic organizations derived from hydroxylated phenyl α-nitronyl nitroxide radicals. *J. Phys. I*, **6**, 1967–86. [289]

Veinberg, S. L., Johnston, K. E., Jaroszewicz, M. J., Kispal, B. M., Mireault, C. R., Kobayashi, T., Pruski, M., and Schurko, R. W. (2016). Natural abundance ^{14}N and ^{15}N solid-state NMR of pharmaceuticals and their polymorphs. *Phys. Chem. Chem. Phys.*, **18**, 17713–30. [194]

Vekilov, P. G. (2010). Nucleation. *Cryst. Growth Des.*, **10**, 5007–19. [82, 83]

Venkatesan, K. and Ramamurthy, V. (1991). Bimolecular photoreactions in crystals. In *Photochemistry in Organized and Constrained Media* (ed. V. Ramamurthy), pp. 133–84. VCH Publishers, Weinheim. [333]

Veregin, R. P., Fyfe, C. A., Marchessault, R. H., and Taylor, M. G. (1986). Characterization of the crystalline A and N starch polymorphs and investigation of starch crystallization by high-resolution ^{13}C CP/MAS NMR. *Macromolecules*, **19**, 1030–4. [189]

Vergnat, C., Landais, V., Legrand, J. F., and Brinkman, M. (2011). Orienting semiconducting nanocrystals on nanostructured polycarbonate. *Macromolecules*, **44**, 3817–27. [392]

Verma, A. R. and Krishna, P. (1966). *Polytypism and Polymorphism in Crystals*. Wiley, New York. [3, 30, 273, 303]

Verwer, P. and Leusen, F. J. J. (1998). Computer simulations to predict possible crystal polymorphs. In *Reviews in Computational Chemistry* (ed. K. B. Lipkowitz and D. B. Boyd), Vol. 12, pp. 327–65. Wiley-VCH, New York. [238]

Vidrine, D. W. (1982). Photoacoustic Fourier transform infrared spectroscopy of solids and liquids. In *Fourier Transform Infrared Spectroscopy*, Vol. 3 (ed. J. R. Ferraro and L. J. Basile), pp. 125–48. Academic Press, New York. [174]

Vioglio, P. C., Chierotti, M. R., and Gobetto, R. (2017). Pharmaceutical aspects of salt and cocrystal forms of APIs and characterization challenges. *Adv. Drug. Deliv. Rev.*, **117**, 86–110. [353]

Vishnimurthy, K., Row, T. N. G., and Venkatesan, K. (1996). Studies in crystal engineering: effect of fluorine substitution in crystal packing and topological photodimerization of styryl coumarins in the solid state. *J. Chem. Soc., Perkin Trans. 2*, 1475–8. [333]

Vishweshwar, P., McMahon, J. A., Peterson, M. L., Hickey, M. B., Shattock, T. R., and Zaworotko, M. J. (2005). Crystal engineering of pharmaceutical co-crystals from polymorphic active pharmaceutical ingredients. *Chem. Commun.*, 4601–3. [360]

Vogt, F. G., Clawson, J. S., Strohmeier, M., Pham, T. N., Watson, S. A., and Edwards, A. J. (2010). New approaches to the characterization of drug candidates by solid-state NMR. In *Pharmaceutical Sciences Encyclopedia: Drug Discovery, Development, and Manufacturing* (ed. S. C. Gad), pp. 1–77. John Wiley & Sons, New York. [182]

Vogt, F. G., Vena, J. A., Chavda, M., Clawson, J. S., and Stroheimer, M. (2011). Solid state nuclear magnetic resonance of polymorphic materials. In *Encyclopedia of Analytical Chemistry* (ed. R. A. Meyers). John Wiley & Sons, New York. [182]

Vogt, F. G., Vena, J. A., Chavda, M., Clawson, J. S., Stroheimer, M., and Barnett, M. E. (2009). Structural analysis of 5-fluorouracil and thymine solid solutions. *J. Mol. Struct.*, **932**, 16–30. [42]

Voigt-Martin, I. G., Yan, D. H., Yakimansky, A., Schollmeyer, D., Gilmore, C. J., and Bricogne, G. (1995). Structure determination by electron crystallography using both maximum-entropy and simulation approaches. *Acta Crystallogr.*, **51**, 849–68. [156]

Volmer, M. (1939). *Kinetik der Phasenbildung*. Steinkopf, Leipzig. [56, 80]

von Eller, H. (1955). Structure de colorants indigoïdes. III. Structure cristalline de l'indigo. *Bull. Soc. Chim. Fr.*, **106**, 1433–8. [383t]

Von Rambach, L., Daubach, E., and Honigman, B. (1974). Crystalline modification of monoazo dye—from diazotised 2-cyan-4-nitro-aniline coupled with *N*-ethyl *N*-cyanoethyl aniline, with improved stability. BASF AG Patent DE 2 249 739. [379t]

von Stackelber, M. (1947). X-ray investigations of inner-complex copper salts. *Z. Anorg. Allg. Chem.*, **253**, 136–60. [225]

Vrcelj, R. M., Gallagher, H. G., and Sherwood, J. N. (2001). Polymorphism in 2,4,6-tri-nitrotoluene crystallized from solution. *J. Am. Chem. Soc.*, **123**, 2291–5. [424]

Vrcelj, R. M., Sherwood, J. N., Kennedy, A. R., Gallagher, H. G., and Gelbrich, T. (2003). Polymorphism in 2,4,6-trinitrotoluene. *Cryst. Growth Des.*, **3**, 1027–32. [420, 423f, 424f, 425f]

Wagener, A. P. and Meisters, G. J. (1970a). 4,11-Dichloroquinacridone pigments. Sherwin-Williams Co. Patent US 3 524 856. [381t]

Wagener, A. P. and Meisters, G. J. (1970b). 4,11-Dichloroquinacridone pigments. Sherwin-Williams Co. Patent US 3 547 927–8. [381t]

Wagner, T. and Englert, U. (1997). Packing effects in organometallic compounds: a study of five mixed crystals. *Struct. Chem.*, **8**, 357–65. [218]

Wahlstrom, E. E. (1969). *Optical Crystallography*, 4th edn. John Wiley & Sons, New York. [138]

Waki, S. J. (1970). *Lipid Metabolism*. Academic Press, New York. [350]

Wall, M., Hodkiewicz, J., and Henson, P. (2007). *Polymorphic characterization using Raman spectroscopy in high throughput crystallization studies*. Application Note 50873, Thermo Scientific, Madison, WI. [177]

Wang, C., Xu, B., Li, M., Chi, Z., Xie, Y., Li, Q., and Li, Z. (2016). A stable tetraphenyl-ethene derivative: aggregation-induced emission, different crystalline polymorphs, and totally different mechanoluminescence properties. *Mater. Horiz.* **3**, 220–5. [317]

Wang, F., Wachter, J. A., Antosz, F. J., and Berglund, K. A. (2000). An investigation of solvent-mediated polymorphic transformation of progesterone using in situ Raman spectroscopy. *Org. Process Res. Dev.*, **4**, 391–5. [177]

Wang, H., Barton, R. J., Robertson, B. E., Weil, J. A., Brown, K. C., Crystal and molecular structures of two polymorphs of 2,2-di(p-nitrophenyl)-1-picrylhydrazine dichloromethane, $C_{18}H_{11}N_7O_{10}\cdot CH_2Cl_2$. Can. J. Chem., 1991, 69, 1306–1314. [292]

Wang, H. H., Ferraro, J. K., Carlson, K. D., Montgomery, L. K., Geiser, U., Williams, J. M., Whitworth, J. R., Schlueter, J. A., Hill, S., Whangbo, M.-H., Evain, M., and Novoa, J. J. (1989). Electron spin resonance, infrared spectroscopic, and molecular packing studies of the thermally induced conversion of semiconducting α- to semiconducting α_t-(BEDT-TTF)$_2$I$_3$. *Inorg. Chem.*, **28**, 2267–71. [101, 118, 277]

Wang, K., Zhang, H., Chen, S., Yang, G., Zhang, J., Tian, W., Su, Z., and Wang, Y. (2014). Organic polymorphs: pre-compound-based crystals with molecular conformation and packing dependent luminescent properties. *Adv. Mater.*, **17**, 6168–73. [340]

Wang, Y., Tang, G.-M., Li, T.-D., Yu, J.-C., Wei, Y.-Q., Ling, J.-B., and Long, X.-F. (2009). Novel acentric organic solids with non-linear optical and ferroelectric properties. *Aust. J. Chem.*, **63**, 336–42. [297]

Ward, M. D. (1997). Organic crystal surfaces: structure, properties and reactivity. *Curr. Opin. Colloid Interf. Sci.*, **2**, 51–64. [87, 207]

Ward, M. D. (2005). Directing the assembly of molecular crystals. *MRS Bull.*, **30**, 705–12. [297]

Wardle, R. B., Hinshaw, J. C., and Hajik, R. M. (1990). Gas generating compositions containing nitrotriazolone. U.S. Patent 4 931 112. [413]

Webb, J. and Anderson, B. (1978). Problems with crystals. *J. Chem. Educ.*, **55**, 644. [31, 134]

Weber, C., Rustemeyer, F., and Durr, H. (1998). A light-driven switch based on photochromic dihydroindolizines. *Adv. Mater.*, **10**, 1348–51. [222, 307]

Webster, S., Smith, D. A., and Batchelder, D. N. (1998). Raman microscopy using a scanning near-field optical probe. *Vib. Spectrosc.*, **18**, 51–9. [177]

Weers, J. G. and Miller, D. P. (2015). Formulation of dry powders for inhalation. *J. Pharm. Sci.*, **104**, 3259–88. [373]

Wei, X., Xu, J., Li, H., Long, X., and Zhang, C. (2016). Comparative study of experiments and calculations on the polymorphisms of 2,4,6,8,10,12-hexanitro-2,4,6,8,10,12-hexaazaisowurtzitane (CL-20) precipitated by solvent/antisolvent method *J. Phys. Chem. C*, **120**, 5042–51. [408]

Weil, J. A. and Anderson, J. K. (1965). The determination and reaction of 2,2-diphenyl-1-picryl-hydrazyl with thiosalicylic acid. *J. Chem. Soc.*, 5567–70. [291]

Weissberger, A. (1956). *Technique of Organic Chemistry. Volume 3, Part 1*. Interscience Publishers, New York. [40]

Weissbuch, I., Lahav, M., and Leiserowitz, L. (2003). Toward stereochemical control, monitoring and understanding of crystal nucleation. *Cryst. Growth Des.*, **3**, 125–50. [81]

Weissbuch, I., Popovitz-Biro, R., Lahav, M., and Leiserowitz, L. (1995). Understanding and control of nucleation, growth, habit, dissolution and structure of two- and three-dimensional crystals using "tailor-made" auxiliaries. *Acta Crystallogr. B*, **51**, 115–48. [42, 101, 368]

Weiswasser, E. S. and Danzis, S. D. (2003). The Hatch–Waxman Act: history, structure, and legacy. *Antitrust Law J.*, **71**, 585–608. [439]

Wells, A. F. (1946). Crystal habit and internal structure. I. *Philos. Mag.*, **37**, 184–99. [61f]

Wells, A. F. (1984). *Structural Inorganic Chemistry*, 5th edn., pp. 294–315. Oxford University Press, Oxford. [2, 29, 69]

Weng, Z. F., Motherwell, W. D. S., and Cole, J. M. (2008). Tormat: a program for the automated structural alignment of molecular conformations. *J. Appl. Crystallogr.*, **41**, 955–7. [259]

Westrum, E. F. Jr. and McCullough, J. P. (1963). Thermodynamics of crystals. In *Physics and Chemistry of the Organic Solid State* (ed. D. Fox, M. M. Labes, and A. Weissberger), Vol. I, pp. 1–178. Interscience, New York. [41, 50, 51f]

Wheeler, I. (1979). Copper phthalocyanine pigment production—by reacting compound forming phthalocyanine ring system, benzophenone-tetracarboxylic acid, copper compound, catalyst and nitrogen source. Ciba-Geigy Patent GB 1 544 171. [390]

Whitaker, A. (1977a). Fresh X-ray data for two forms of linear *trans*-quinacridone, C. I. Pigment Violet 19. *J. Soc. Dyers Colour.*, **93**, 15–17. [385]

Whitaker, A. (1977b). X-ray powder diffraction of synthetic organic colorants. In *The Analytical Chemistry of Synthetic Dyes* (ed. K. Venkataraman), pp. 269–98. Wiley-Interscience, New York. [390, 393]

Whitaker, A. (1979). Crystal data for a second polymorph (β) of C. I. Pigment Red 1,1-[4-nitrophenyl)azo]-2-naphthol. *J. Appl. Crystallogr.*, **12**, 626–7. [380t]

Whitaker, A. (1980a). The crystal structure of a second polymorph (β) of C.I. Pigment Red 1,1-[(4-nitrophenyl)azo]-2-naphthol. *Z. Kristallogr.*, **152**, 227–37. [380t]

Whitaker, A. (1980b). Crystal data for a third polymorph (γ) of C.I. Pigment Red 1,1-[(4-nitrophenyl)azo]-2-naphthol. *J. Appl. Crystallogr.*, **13**, 458–9. [380t]

Whitaker, A. (1981). The crystal structure of a third polymorph (γ) of C.I. Pigment Red 1,1-[(4-nitrophenyl)azo]-2-naphthol. *Z. Kristallogr.*, **156**, 125–36. [380t]

Whitaker, A. (1982). The polymorphism of C.I. Pigment Red 1. *J. Soc. Dyers Colour.*, **98**, 436–9. [380t]

Whitaker, A. (1985a). The polymorphism of C.I. Pigment Yellow 5. *J. Soc. Dyers Colour.*, **101**, 21–4. [382t]

Whitaker, A. (1985b). The crystal structure of aceto-acetanilide azo pigments VII. A polymorph (α) of C.I. Pigment Yellow 5, α-(1-hydroxyethylidene)acetanilide-α-azo-(2′-nitrobenzene). *Z. Kristallogr.*, **171**, 17–22. [382t]

Whitaker, A. (1988). C.I. Pigment Yellow 10, 4-[2′5′-(dichlorophenyl)hydrazono]-5-methyl-2-phenyl-2-3*H*-pyrazol-3-one. *Acta Crystallogr. C*, **44**, 1767–70. [382t]

Whitaker, A. (1990). Crystal data for Solvent Yellow 18. *J. Soc. Dyers Colour.*, **106**, 108–9. [383t]

Whitaker, A. (1992). The polymorphism of C.I. Solvent Yellow 18. *J. Soc. Dyers Colour.*, **108**, 282–4. [383t]

Whitaker, A. (1995). X-ray powder diffraction of synthetic organic colorants. In *Analytical Chemistry of Synthetic Colorants* (ed. A. T. Peters and H. S. Freeman), pp. 1–48. Blackie, Glasgow. [12, 378, 383t, 384t, 389, 393]

Whitesell, J. K., Davis, R. E., Saunders, L. L., Wilson, R. J., and Feagins, J. P. (1991). Influence of molecular dipole interactions on solid state organization. *J. Am. Chem. Soc.*, **113**, 3267–70. [235]

Wiberg, K. B. (1960). *Laboratory Techniques in Organic Chemistry*, p. 104. McGraw-Hill, New York. [133]

Willart, J.-F., Carpentier, L., Danede, F., and Descamps, M. (2012). Solid-state vitrification of crystalline griseofulvin by mechanical milling. *J. Pharm. Sci.*, **101**, 1570–7. [373]

Willcock, J. D., Price, S. L., Leslie, M., and Catlow, C. R. A. (1995). The relaxation of molecular crystal structures using a distributed multipole electrostatic model. *J. Comput. Chem.*, **16**, 628. [239]

Williams, D. E. (1965). Crystallographic data for 2,2-diphenyl-l-picrylhydrazyl. *J. Chem. Soc.*, 7535–6. [291]

Williams, D. E. (1967). Crystal structure of 2,2-diphenyl-1-picrylhydrazyl free radical. *J. Am. Chem. Soc.*, **89**, 4280–7. [291]

Williams, D. E. (1974). Coulombic interactions in crystalline hydrocarbons. *Acta Crystallogr. A*, **30**, 71–7. [217]

Williams, D. E. (1983). *PCK83*. QCPE Program 548. Quantum Chemistry Program Exchange, Chemistry Department, Indiana University, Bloomington, IN. [238]

Williams, D. E. and Hsu, L.-Y. (1985). Transferability of nonbonded Cl···Cl potential energy function to crystalline chlorine. *Acta Crystallogr. A*, **41**, 296–301. [227]

Williams, G. K. and Brill, T. B. (1995). Thermal-decomposition of energetic materials. 68. Decomposition and sublimation kinetics of NTO and evaluation of prior kinetic data. *J. Phys. Chem.*, **99**, 12536–9. [413]

Williams, J. M., Beno, M. A., Wang, H. H., Leung, P. C. W., Enge, Y. J., Geiser, U., and Carlson, K. D. (1985). Organic superconductors: structural aspects and design of new materials. *Acc. Chem. Res.*, **18**, 261–7. [275]

Williams, J. M., Schultz, A. J., Geiser, U., Carlson, K. D., Kini, A. M., Wang, H. H., Kwok, W.-K., Whangbo, M.-H., and Schirber, J. E. R. (1991). Organic super-conductors—new benchmarks. *Science*, **252**, 1501–8. [118, 119, 276, 277]

Williams, K. P. J., Pitt, G. D., Batchelder, D. N., and Kip, B. J. (1994a). Confocal Raman microspectrometry using a stigmatic spectrograph and CCD detector. *Appl. Spectrosc.*, **48**, 232–5. [177]

Williams, K. P. J., Pitt, G. D., Smith, B. J. E., Whitley, A., Batchelder, D. N., and Hayward, I. P. (1994b). Use of rapid scanning stigmatic Raman imaging spectrograph in the industrial environment. *J. Raman Spectrosc.*, **25**, 131–8. [177]

Willis, M. R. (1986). Talk at British Crystallographic Association annual meeting, Herriot-Watt University, Edinburgh, Scotland. [273]

Winchell, A. N. (1943). *The Optical Properties of Organic Compounds*. University of Wisconsin Press, Madison. [24, 138]

Winchell, A. N. (1987). *The Optical Properties of Organic Compounds*. McCrone Research Institute, Chicago. [25]

Winter, G. (1999). Polymorphs and solvates of molecular solids in the pharmaceutical industry. In *Reactivity of Molecular Solids, the Molecular Solid State* (ed. E. Boldyreva and V. Boldyrev), Vol. 3, pp. 241–70. John Wiley & Sons, Chichester. [343]

Wirth, D. D. and Stephenson, G. A. (1997). Purification of dirithromycin. Impurity reduction and polymorph manipulation. *Org. Process Res. Dev.*, **1**, 55–60. [375]

Woehrle, D. (1989). Polymeric phthalocyanines. In *Phthalocyanines: Properties and Applications* (ed. C. C. Leznoff and A. B. P. Lever), pp. 55–132. VCH Verlagsgesellschaft, Weinheim. [388]

Wöhler, F. (1844). Untersuchungen uber das Chinon. *Annalen*, **51**, 153. [359]

Wöhler, F. and Liebig, J. (1832). Unter suchengen über das Radikal der Benzosäure. *Ann. Pharm.*, **3**, 249–82. [114, 155]

Wojcik, G. and Marqueton, Y. (1989). The phase transition of *m*-nitrophenol. *Mol. Cryst. Liq. Cryst.*, **168**, 247–54. [121, 302]

Wojtyk, J., McKerrow, A., Kazmaier, P., and Buncel, E. (1999). Quantitative investigations of the aggregation behavior of hydrophobic anilino squaraine dyes through UV/vis spectroscopy and dynamic light scattering. *Can. J. Chem.*, **77**, 903–12. [131, 293]

Wolf, F., Koch, U., and Modrow, H.-W. (1986b). Verfahren zur Herstellung einer neuen Modifikation Gamma des Dispersionsfarbstoffes C.I. Disperse Red 73. Martin-Luther Universität, Halle-Wittenberg Patent DD 236 544. [379t]

Wolf, F., Koch, U., Heitrich, W., and Modrow, H.-W. (1986a). Verfahren zur Herstellung einer färbestabilen Modifikation von 4,4′-Dihydroxy-Bis-Azotriphenylen (C.I. Disperse Yellow 68). Martin-Luther Universität Halle-Wittenberg Patent DD 236 543. [379]

Wolff, J. J. (1996). Crystal packing and molecular geometry. *Angew. Chem. Int. Ed. Engl.*, **35**, 2195–7. [218]

Wolff, S. K., Grimwood, D. J., McKinnon, J. J., Turner, M. J., Jayatilaka, M. A., and Spackman, M. A. (2012). *CrystalExplorer (Verson 3.1)*. University of Western Australia. Available at http://www.hirshfeldsurface.net/wiki/index.php/CrystalExplorer_3.1_Release_Notes [254]

Wolffenstein, R. (1895). Über die Einwirkung von Wasserstoffsuperoxyd auf Aceton und Mesityloxyd. *Ber. Dtsch. Chem. Ges.*, **28**, 2265–9. [411]

Wolfrom, M. L. and Kohn, E. J. (1942). Crystalline xylitol. *J. Am. Chem. Soc.*, **64**, 1739. [86, 103]

Wollmann, H. and Braun, V. (1983). Die Anwendung der Differenzthermoanalyse in der Pharmazie. *Pharmazie*, **38**, 5–20. [365]

Wood, W. M. L. (1997). Crystal science techniques in the manufacture of chiral compounds. In *Chirality in Industry II. Developments in the Commercial Manufacture and Applications of Optically Active Compounds* (ed. A. N. Collins, G. N. Sheldrake, and J. Crosby), pp. 119–56. John Wiley & Sons, New York. [5, 61]

Woodard, G. D. (1970). Calibration of the Mettler FP2 hot stage. *Microscope*, **18**, 105–8. [138]

Woodard, G. D. and McCrone, W. C. (1975). Unusual crystallization behavior. *J. Appl. Crystallogr.*, **8**, 342. [134]

Woollam, G. R., Neumann, M. A., Wagner, T., and Davey, R. J. (2018). The importance of configurational disorder in crystal structure prediction: the case of loratadine. *Faraday Discuss.*, **211**, 209–34. [370]

Wozniak, K., Mallinson, P. R., Wilson, C. C., Hovestreydt, E., and Grech, E. (2002). Charge density studies of weak interactions in dipicrylamine. *J. Phys. Chem. A*, **106**, 6897–903. [427]

Wudl, F. (1984). From organic metals to superconductors: managing conducting electrons in organic solids. *Acc. Chem. Res.*, **17**, 227–32. [275]

Würthner, F., Kaiser, T. E., and Saha-Möller, C. R. (2011). J-aggregates: from serendipitous discovery to supramolecular engineering of functional dye materials. *Angew. Chem. Int. Ed. Engl.*, **50**, 3376–410. [328]

Würthner, F., Saha-Möller, C. R., Fimmel, B., Ogi, S., Leowanawat, P., and Schmidt, D. (2016). Perylene bisimide dye assemblies and archetype functional supramolecular materials. *Chem. Rev.*, **116**, 962–1052. [293]

Wyrouboff, M. G. (1890). Recherches sur le polymorphisme et la pseudosymétrie. *Bull. Soc. Miner. Fr.*, **13**, 277–319. [33]

Xia, D., Yu, S., Shen, R., Ma, C., Cheng, C., Ji, D., Fan, Z., Wang, X., and Du, G. (2008). A novel method for the direct synthesis of crystals of copper phthalocyanine. *Dyes Pigm.*, **78**, 84–8. [383t]

Xu, J., Tian, Y., Liu, Y., Zhang, H., Shu, Y., and Sun, J. (2012). Polymorphism in hexanitrohexaazaisowurtzitane crystallized from solution. *J. Cryst. Growth*, **354**, 13–19. [408]

Xu, J.-X., Zhu, W.-H., and Xiao, H. M. (2007). DFT Studies on the four polymorphs of crystalline CL-20 and the influences of hydrostatic pressure on ϵ-CL-20 crystal. *J. Phys. Chem. B*, **111**, 2090–7. [409]

Xu, M. and Harris, K. D. M. (2008). Triple quantum ^{23}Na MAS NMR spectroscopy as a technique for probing polymorphism in sodium salts. *Cryst. Growth Des.*, **8**, 6–10. [193]

Yagi, S. and Nakazumi, H. (2008). Squarylium dyes and related compounds. *Top. Heterocycl. Chem.*, **14**, 133–81. [330]

Yagishita, T., Ikegami, K., Narusawa, T., and Okuyama, H. (1984). Photoconduction of copper phthalocyanine-binder photoconductor sensitized with poly-*N*-vinylcarbazole. *IEEE Trans. Ind. Appl.*, **IA-20**, 1642–6. [296]

Yakobson, B. I., Boldyreva, E. V., and Sidelnikov, A. A. (1989). Bending of needle crystals caused by a photochemical-reaction—quantitative description. *Izv. Sibirsk. Potdel Akad. Nauk SSSR Ser. Khim. Nauk*, 6–10. [413]

Yamada, H., Masuda, K., Ishige, T., Fujii, K., Uekusa, H., Miura, K., Yonemochi, E., and Terada, K. (2011). Potential of synchrotron X-ray powder diffractometry for detection and quantification of small amounts of crystalline drug substances in pharmaceutical tablets. *J. Pharm. Biomed. Anal.*, **56**, 448–53. [169]

Yamamoto, H., Katogi, S., Watanabe, T., Sato, H., Miyata, S., and Hosomi, T. (1992). New molecular design approach for non-centrosymmetric crystal structures: Lambda ($\wedge$)-shaped molecules for frequency doubling. *Appl. Phys. Lett.*, **60**, 935–7. [302]

Yan, Q.-L., Zeman, S., Elbeih, A., Song, Z.-W., and Malek, J. (2013). The effect of crystal structure on the thermal reactivity of CL-20 and its C4 bonded explosives (I): thermodynamic properties and decomposition kinetics. *J. Therm. Anal. Calorim.*, **112**, 823–36. [409]

Yang, J., Hu, C.T., Zhu, X., Ward, M. D., and Kahr, B. (2017a). DDT polymorphism and the lethality of crystal forms. *Angew. Chem. Int. Ed.*, **56**, 10165–9. [351]

Yang, J., Ward, M. D., and Kahr, B. (2017b). Abuse of Rachel Carson and misuse of DDT science in the service of environmental deregulation. *Angew. Chem. Int. Ed. Engl.*, **56**, 10026–32. [351]

Yang, Q.-C., Richardson, M. F., and Dunitz, J. D. (1989). Conformational polymorphism of dimethyl 3,6-dichloro-2,5-dihydroxyterephthalate. I. Structures and atomic displacement parameters between 100 and 350 K for three crystal forms. *Acta Crystallogr. B*, **15**, 312–23. [305]

Yannoni, C. S. (1982). High resolution NMR in solids: the CPMAS experiment. *Acc. Chem. Res.*, **15**, 201–8. [182]

Yano, J., Ueno, S., Sato, K., Arishimia, T., Sagi, N., Kaneko, F., and Kobayashi, M. (1993). Polymorphic transformations in SOS, POP and POS. *J. Phys. Chem.*, **97**, 12967–73. [174]

Yano, K., Takamatsu, N., Yamazaki, S., Sako, K., Nagura, S., Tomizawa, S., Shimaya, J., and Yamamoto, K. (1996). Crystal forms, improvements of dissolution and absorption of poorly water-soluble (R)-1-[2,3-dihydro-1-(2′-methylphenacyl)-2-oxo-5-phenyl-1H-1,4-benzodiazepin-3-yl]-3-(3-methylphenyl)urea (YM022). *Yakugaku Zasshi*, **116**, 639–46. [347]

Yazawa, H. and Momonaga, M. (1994). Reaction crystallisation with additive agents. *Pharm. Manuf. Int.*, 107–10. [375]

Yokota, M., Mochizuki, M., Saito, K., Sato, A., and Kubota, N. (1999). Simple batch operation for selective crystallization of metastable crystalline phase. *Chem. Eng. Commun.*, **174**, 243–56. [368]

Yokoyama, T., Umeda, T., Kuroda, K., Kuroda, T., and Asada, S. (1980). Studies on drug nonequivalence. X. Bioavailability of 6-mercaptopurine polymorphs. *Chem. Pharm. Bull.*, **29**, 194–9. [347]

Yokoyama, T., Umeda, T., Kuroda, K., Nagafuku, T., Yamamoto, T., and Asada, S. (1979). Studies on drug nonequivalence. IX. Relationship between polymorphism and rectal absorption of indomethacin. *Yakugaku Zasshi*, **99**, 837–42. [347]

Yoon, S.-J. and Park, S.-Y. (2011). Polymorphic and mechanochromic luminescence modulation in the highly emissive dicyanodistyrylbenzene crystal: secondary bonding interaction in molecular stacking assembly. *J. Mater. Chem.*, **21**, 8338–46. [318]

Yordanov, N. D. and Christova, A. G. (1997). Quantitative spectrophotometric and EPR-determination of 1,1-diphenyl-2-picryl-hydrazyl (DPPH). *Fresenius Z. Anal. Chem.*, **358**, 610–13. [291]

Yoshimoto, N., Takayuki, S., Gamachi, H., and Yoshizawa, M. (1999). Polymorphism and crystal growth of organic conductor $(BEDT\text{-}TTF)_2I_3$. *Mol. Cryst. Liq. Cryst., Sci. Technol. A*, **327**, 233–6. [277]

Youming, C., Shengqing, L., and Renyuan, Q. (1986). Polymorphism and electronic properties of 3-ethyl-5-[2-(3-ethyl-2-benzthiazolinylidene)-ethylidene]-rhodamine. *Phys. Status Solidi A: Appl. Res.*, **98**, 37–42. [293]

Young, R. A. (ed.) (1993). *The Rietveld Method.* International Union of Crystallography, Oxford Science Publications, Oxford. [155, 168]

Young, R. H., Marchetti, A. P., and Newhouse, E. I. P. (1989). Optical spectra of a planar aggregate and of a lamellar crystal containing it—how are they related? *J. Chem. Phys.*, **91**, 5743–55. [294]

Yu, L. (1995). Inferring thermodynamic stability relationship of polymorphs from melting data. *J. Pharm. Sci.*, **84**, 966–74. [150, 171, 229]

Yu, L. (2001). Amorphous pharmaceutical solids: preparation, characterization and stabilization. *Adv. Drug Rev.*, **48**, 27–42. [370]

Yu, L. (2010). Polymorphism in molecular solids: an extraordinary system of red, orange and yellow crystals. *Acc. Chem. Res.*, **43**, 1257–66. [105, 115, 229, 230f, 374]

Yu, L., Reutzel, S. M., and Stephenson, G. A. (1998). Physical characterization of polymorphic drugs: an integrated characterization strategy. *Pharm. Sci. Technol. Today*, **1**, 118–27. [343]

Yu, L., Stephenson, G. A., Mitchell, C. A., Bunnell, C. A., Snorek, S. V., Bowyer, J. J., Borchardt, T. B., Stowell, J. G., and Byrn, S. R. (2000a). Thermochemistry and conformational polymorphism of a hexamorphic crystal system. *J. Am. Chem. Soc.*, **122**, 585–91. [142, 229, 231f, 232f, 232, 233f, 234t, 235f, 236f, 403]

Yu, R. C., Yakimansky, A. V., Kothe, H., Voigt-Martin, I. G., Schollmeyer, D., Jansen, J., Zandbergen, H., and Tenkovtsev, A. V. (2000b). Strategies for structure solution and refinement of small organic molecules from electron diffraction data and limitations of the simulation approach. *Acta Crystallogr. A*, **56**, 436–50 [156].

Zaccaro, J., Matic, J., Myerson, A. S., and Garetz, B. A. (2000). Nonphotochemical, laser-induced nucleation of supersaturated aqueous glycine produces unexpected γ-polymorph. *Cryst. Growth Des.*, **1**, 5–8. [101]

Zakharov, V. V., Chukanov, N. V., Chervonnyi, A. D., Vozchikova, S. A., and Korsounskii, B. L. (2015). Kinetics of reversible polymorphic transition in energetic materials. Phase transitions $\alpha \rightarrow \beta$ and $\beta \rightarrow \alpha$ in 1,1-diamino-2,2-dinitroethylene. *Russ. J. Phys. Chem. B*, **8**, 822–8. [414]

Zakharov, V. V., Chukanov, N. V., Dremova, N. N., Chervonnyi, A. D., Shilov, G. V., Korsounskii, B. L., Kazakov, A. I., and Volkova, N. N. (2016). High-temperature structural transformations of 1,1-diamino-2,2-dinitroethene (FOX-7). *Propellants, Explos., Pyrotech.*, **41**, 1006–12. [414]

Zamir, S., Bernstein, J., and Greenwood, D. J. (1994). A single crystal to single crystal reversible phase transition which exhibits the "hopping effect." *Mol. Cryst. Liq. Cryst.*, **242**, 193–200. [311, 319, 320]

Zbačnik, M., Nogalo, I., Cinčić, D., and Kaitner, B. (2015). Polymorphism control in the mechanochemical and solution-based synthesis of a thermochromic Schiff base. *CrystEngComm*, **17**, 7870–7. [313]

Zeidan, T. A., Trotta, J. T., Chiarella, R. A., Oliveira, M. A., Hickey, M. B., Almarsson, Ö., and Ramenar, J. F. (2013). Active metabolite of the antipsychotic drug aripiprazole (Abilify). *Cryst. Growth Des.*, **13**, 2036–46. [237]

Zeidan, T. A., Trotta, J. T., Tilak, P. A., Oliveira, M. A., Chiarella, R. A., Foxman, B. M., Almarsson, Ö., and Hickey, M. B. (2016). An unprecedented case of dodecamorphism: the twelfth polymorph of aripiprazole formed by seeding with its active metabolite. *CrystEngComm*, **18**, 1486–8. [13t, 374]

Zell, M. T., Padden, B. E., Grant, D. J. W., Chapeau, M.-C., Prakash, I., and Munson, E. J. (1999). Two-dimensional high-speed CP/MAS NMR spectroscopy of polymorphs. 1. Uniformly ^{13}C-labeled aspartame. *J. Am. Chem. Soc.*, **121**, 1372–8. [189, 453]

Zeman, S. (1980). The relationship between differential thermal analysis data and the detonation characteristics of polynitroaromatic compounds. *Thermochim. Acta*, **41**, 199–212. [398]

Zeman, S. and Jungova, M. (2016). Sensitivity and performance of energetic materials. *Propellants, Explos., Pyrotech.*, **41**, 426–51. [399]

Zencirci, N., Gelbrich, T., Kahlenberg, V., and Griesser, U. J. (2009). Crystallization of metastable polymorphs of phenobarbital by isomorphic seeding. *Cryst. Growth Des.*, **9**, 3444–56. [373]

Zencirci, N., Griesser, U. J., Gelbrich, T., Kahlenber, V., Jetti, R. K. R., Apperley, D. C., and Harris, R. K. (2014). New solvates of an old drug compound (phenobarbital) and stability. *J. Phys. Chem. B*, **118**, 3267–80. [356]

Zeng, Q., Mukherjee, A., Müller, P., Rogers, R., and Myerson, A. S. (2018). Exploring the role of ionic liquids to tune the polymorphic outcome of organic compounds. *Chem. Sci.*, **9**, 1510–20. [356]

Zenith Laboratories, Inc. v. Bristol-Meyers Squibb Co. (1992). Civ. A. No. 91-3423; 1991 WL 267892 (D. N. J.) [436]

Zenith Laboratories, Inc. v. Bristol-Meyers Squibb Co. (1994). No. 92-1527; 19 F.3d 1418. [175, 429]

Zepharovich, V. V. (1888). Die Kristallformen des Mannit. *Z. Kristallogr.*, **13**, 145–9. [120]

Zerkowski, J. A., MacDonald, J. C., and Whitesides, G. M. (1997). Polymorphic packing arrangements in a class of engineered organic crystals. *Chem. Mater.*, **9**, 1933–41. [74, 75f]

Zevin, L. S. and Kimmel, G. (1995). *Quantitative X-ray Diffractometry*. Springer, New York. [164, 371]

Zhang, G. Q., Lu, J. W., Sabat, M., and Fraser, C. L. (2010). Polymorphism and reversible mechanochromic luminescence for solid-state difluoroboron avobenzone. *J. Am. Chem. Soc.*, **132**, 2160–2. [317]

Zhang, L., Jiang, S.-L., Yu, Y., Long, Y., Zhao, H.-Y., Peng, L.-J., and Chen, J. (2016). Phase transition in octahydro-1,3,5,7-tetranitro-1,3,5,7-tetrazocine (HMX) under static compression: an application of the first-principles method specialized for CHNO solid explosives. *J. Phys. Chem. B*, **120**, 11510–22. [340, 407]

Zhang, P., Xu, J., Guo, X.-Y., Jiao, Q.-J., and Zhang, J. (2014). Effect of addictives (*sic*) on polymorphic transition of ϵ-CL-20 in castable systems. *J. Therm. Anal. Calorim.*, **117**, 1001–8. [408]

Zhang, Y., Wu, G., Wenner, B. R., Bright, F. V., and Coppens, P. (1999). Engineering crystals for excited-state studies: two polymorphs of 4,4′-dihydroxybenzophenone/4, 13-diaza-18-crown-6. Abstracts of the American Crystallographic Association Meeting, Abstract 03.02.09, p. 48. [332]

Zheng, C., Penmetcha, A. R., Cona, B., Spencer, S. D., Zhu, B., Heaphy, P., Cody, J. A., and Collson, C. J. (2015). Contribution of aggregate states and energetic disorder to a squaraine system targeted for organic photovoltaic devices. *Langmuir*, **31**, 7717–26. [293, 330]

Zhitomirskaya, N. G., Eremenko, L. T., Golovina, N. I., and Atovmayan, L. O. (1987). Structural and electronic parameters of certain cyclic nitramines. *Izv. Akad. Nauk SSSR, Ser. Khim.*, 576–80. [404t]

Zhou, J., Kye, Y.-S., and Harbisan, G. S. (2004). Isotopomeric polymorphism. *J. Am. Chem. Soc.*, **126**, 8392–3. [6]

Zhou, T.-T. and Huang, F.-L. (2012). Thermal expansion behaviors and phase transitions of HMX polymorphs via ReaxFF molecular dynamics simulations. *Acta Phys. Sinica*, **61**, 246501. [405]

Zhu, M. and Yu, L. (2017). Polyamorphism of D-mannitol. *J. Chem. Phys.*, **146**, 244503. [372]

Zhuang, Z. Y., Shen, P., Ding, S., Luo, W., He, B., Nie, H., Wang, B., Huang, T., Hu, R., Qin, A., and Zhao, Z., (2016). Synthesis, aggregation-induced emission, polymorphism and piezochromism of TPE-cored foldamers with through-space conjugation. *Chem. Commun.*, **52**, 10842–5. [317]

Zhurov, V. V., Zhurova, E. A., Stash, A. I., and Pinkerton, A. A. (2011). Importance of the consideration of anharmonic motion in charge-density studies: a comparison of variable-temperature studies on two explosives, RDX and HMX. *Acta Crystallogr. A*, **67**, 160–73. [410]

Zhurova, E. A. and Pinkerton, A. A. (2001). Chemical bonding in energetic materials: beta-NTO. *Acta Crystallogr. B*, **57**, 359–65. [413]

Zhurova, E. A., Stash, A. I., Tsirelson, V. G., Zhurov, V. V., Bartashevich, E. V., Potemkin, V. A., and Pinkerton, A. A. (2006). Atoms-in-molecules study of intra- and intermolecular bonding in pentaerythritol tetranitrate crystal. *J. Am. Chem. Soc.*, **128**, 14728–34. [411]

Zilka, M., Dudenko, D. V., Hughes, C. E., Willams, P. A., Sturniolo, S., Franks, W. T., Pickard, C. J., Yates, J. R., Harris, K. D. M., and Brown, S. P. (2017). Ab initio random structure searching of organic molecular solids: assessment and validation against experimental data. *Phys. Chem. Chem. Phys.*, **19**, 25949–60. [354]

Zink, J. I. (1978). Triboluminescence. *Acc. Chem. Res.*, **11**, 289–95. [315]

Zollinger, H. (2004). *Color Chemistry: Synthesis, Properties and Applications of Organic Dyes and Pigments*, 3nd edn. Wiley-VCH, Weinheim. [377]

Zorky, P. M. and Kuleshova, L. N. (1980). Comparison of the hydrogen bonds in polymorphic modifications of organic substances. *Zh. Strukt. Khim.*, **22**, 153–6. [69]

Zugenmaier, P., Bluhm, T. L., Deslandes, Y., Orts, W. J., and Hamer, G. K. (1997). Diffraction studies on metal free phthalocyanines (β-H$_2$Pc and X-H$_2$Pc). *Mater. Sci.*, **32**, 5561–8. [383t]

Zupan, J. and Gasteiger, J. (1991). Neural networks: a new method for solving chemical problems or just a passing phase? *Anal. Chim. Acta*, **248**, 1–30. [169]

Zvoníček, V., Skořepová, E.,Dušek, M., Babor, M., Žvátora, P., and Šoóš, M. (2017). First crystal structures of pharmaceutical ibrutinib: systematic solvate screening and characterization. *Cryst. Growth Des.*, **17**, 3116–27. [355]

Zykova-Timan, T., Raiteri, P., and Parrinello, M. (2008). Investigating the polymorphism in PR179: a combined crystal structure prediction and metadynamics study. *J. Phys. Chem. B*, **112**, 13231–7. [382t]

Zyss, J. (ed.) (1994). *Molecular Nonlinear Optics: Materials, Physics and Devices*. Academic Press, Boston. [296]

Index